D1826899

Cardiac and Vascular Biology

Volume 7

Editor-in-Chief

Markus Hecker
Inst. of Physiology & Pathophysiology, Heidelberg University, Heidelberg,
Baden-Württemberg, Germany

Series Editors

Johannes Backs
Department of Molecular Cardiology and Epigenetics, Heidelberg University,
Heidelberg, Baden-Württemberg, Germany

Marc Freichel
Institute of Pharmacology, Heidelberg University, Heidelberg, Germany

Thomas Korff
Inst. of Physiology & Pathophysiology, Heidelberg University, Heidelberg,
Germany

Dierk Thomas
Department of Cardiology and HCR, University Hospital Heidelberg, Heidelberg,
Germany

The book series gives an overview on all aspects of state-of-the-art research on the cardiovascular system in health and disease. Basic research aspects of medically relevant topics are covered and the latest advances and methods covering diverse disciplines as epigenetics, genetics, mechanobiology, platelet research or stem cell biology are featured. The book series is intended for researchers, experts and graduates, both basic and clinically oriented, that look for a carefully selected collection of high quality review articles on their respective field of expertise.

More information about this series at http://www.springer.com/series/13128

Jeanette Erdmann • Alessandra Moretti
Editors

Genetic Causes of Cardiac Disease

 Springer

Editors
Jeanette Erdmann
IIEG
University of Lubeck
Lübeck, Germany

Alessandra Moretti
I Medical Dept (Cardiology)
Klinikum Rechts der Isar of the Technical
University Munich
München, Bayern, Germany

ISSN 2509-7830   ISSN 2509-7849   (electronic)
Cardiac and Vascular Biology
ISBN 978-3-030-27373-6   ISBN 978-3-030-27371-2   (eBook)
https://doi.org/10.1007/978-3-030-27371-2

Contents

Genetics of Adult and Fetal Forms of Long QT Syndrome

1

Lia Crotti, Alice Ghidoni, and Federica Dagradi

Contents

Abstract

Long QT syndrome (LQTS) is an inherited cardiac disease characterized by prolongation of QT interval at surface ECG, T-wave abnormalities, and high risk of life-threatening arrhythmias in otherwise healthy young individuals. Currently the LQTS diagnosis is genetically confirmed in nearly 75–85% of LQTS patients, revealing a good knowledge of the genetic bases of the disease.

L. Crotti (✉)
Istituto Auxologico Italiano, IRCCS, Laboratory of Cardiovascular Genetics, Center for Cardiac Arrhythmias of Genetic Origin, Milan, Italy

Department of Medicine and Surgery, University of Milano-Bicocca, Milan, Italy

Istituto Auxologico Italiano, IRCCS, Department of Cardiovascular, Neural and Metabolic Sciences, San Luca Hospital, Milan, Italy

A. Ghidoni · F. Dagradi
Istituto Auxologico Italiano, IRCCS, Laboratory of Cardiovascular Genetics, Center for Cardiac Arrhythmias of Genetic Origin, Milan, Italy

© Springer Nature Switzerland AG 2019
J. Erdmann, A. Moretti (eds.), *Genetic Causes of Cardiac Disease*, Cardiac and Vascular Biology 7, https://doi.org/10.1007/978-3-030-27371-2_1

1

The main LQTS genes are *KCNQ1*, *KCNH2*, and *SCN5A* encoding potassium and sodium cardiac ion channels responsible of the cardiac action potential duration. Minor contributors of LQTS genetic background include genes encoding other cardiac ion channels, ancillary subunits, and protein components forming channels' macromolecular complexes.

Fetal and neonatal forms of LQTS are the most aggressive form of the disease, frequently associated with typical ECG features as very prolonged QTc, 2:1 functional atrioventricular block, T-wave alternans, and life-threatening arrhythmias. The genetic basis of these early-onset cases is peculiar. Indeed, while potassium channel mutations are the most commonly observed causes of adult LQTS, fetal and neonatal forms of the disease are mainly due to aggressive sodium channel mutations or to mutations affecting calcium channel activity, as in Timothy syndrome, triadin knockout syndrome, and calmodulin-LQTS. Aggressive forms of LQTS can also cause sudden infant death syndrome (SIDS) or intrauterine fetal death.

1.1   Introduction

Long QT syndrome (LQTS) is an inherited cardiac disease characterized by prolongation of QT interval at surface ECG, T-wave abnormalities (biphasic or notched T waves), and high risk of life-threatening arrhythmias. The typical ventricular tachyarrhythmia that underlies cardiac events in LQTS is the torsades de pointes (TdP). This type of ventricular tachycardia can produce transient syncope, when it is self-limited, or can degenerate into ventricular fibrillation and cardiac arrest, mainly precipitated by emotional or physical stress. A sign of major electrical instability in LQTS patients is represented by T-wave alternans, a beat-to-beat alteration in polarity, and amplitude of the T wave [1]. Long QT syndrome is considered one of the leading causes of sudden death in young (<35 years) [2]. Unfortunately, the disease can remain clinically silent for a long time, and sudden cardiac death (SCD) may be the first manifestation in some cases.

The congenital form of the disease has been largely studied over the years and includes two main hereditary variants. The Romano-Ward (RW) variant, described for the first time in 1964 [3], represents the autosomal dominant form of the disease, and it is relatively common, with a prevalence of 1:2000 live births [4]. The Jervell and Lange-Nielsen (JLN) syndrome is an extremely severe form of the disease, associated with congenital deafness and higher mortality [5, 6]. The JLN has an autosomal recessive mode of inheritance, more frequently associated with homozygous and rarely compound heterozygous mutations. This syndrome is very rare and affects around 2–3 out of 1000 individuals with congenital deafness [6].

Since the main feature of LQTS is the prolongation of QT interval, it is not surprising that cardiac ion channels responsible for action potential (AP) duration are the main molecular players of the syndrome.

In particular, three genes (*KCNQ1*, *KCNH2*, *SCN5A*), encoding cardiac sodium and potassium channels, are the major genetic contributors underlying LQTS.

Table 1.1 Long QT syndrome variant types so far described and relative LQTS-associated genes

LQTS variant type	Syndrome	Gene	OMIM ID	Locus	Protein	Functional effect
LQT1	RWS, JLNS	KCNQ1	*607542	11p15.5-p15.4	$K_V7.1$	$\downarrow I_{Ks}$
LQT2	RWS	KCNH2	*152427	7q36.1	$K_V11.1$	$\downarrow I_{Kr}$
LQT3	RWS	SCN5A	*600163	3p22.2	$Na_V1.5$	$\uparrow I_{Na}$
LQT4	RWS, ANKB syndrome	ANKB	*106410	4q25-q26	Ankyrin B	$\uparrow [Ca^{2+}]_i$
LQT5	RWS, JLNS	KCNE1	*176261	21q22.12	MinK	$\downarrow I_{Ks}$
LQT6	RWS	KCNE2	*603796	21q22.11	MiRP1	$\downarrow I_{Kr}$
LQT7	ATS	KCNJ2	*600681	17q24.3	Kir2.1	$\downarrow I_{K1}$
LQT8	TS	CACNA1C	*114205	12p13.33	$Ca_V1.2$	$\uparrow I_{CaL}$
LQT9	RWS	CAV3	*601253	3p25.3	Caveolin-3	$\uparrow I_{Na}$
LQT10	RWS	SCN4B	*608256	11q23.3	Sodium channel β4-subunit	$\uparrow I_{Na}$
LQT11	RWS	AKAP9	*604001	7q21.2	Yotiao	$\downarrow I_{Ks}$
LQT12	RWS	SNTA1	*601017	20q11.21	α1-Syntrophin	$\uparrow I_{Na}$
LQT13	RWS	KCNJ5	*600734	11q24.3	Kir3.4	$\downarrow I_{KACh}$
LQT14	Calmodulinopathy	CALM1	*114180	14q32.11	CaM	$\uparrow I_{CaL}$
LQT15	Calmodulinopathy	CALM2	*114182	2p21	CaM	$\uparrow I_{CaL}$
LQT16	Calmodulinopathy	CALM3	*114183	19q13.32	CaM	$\uparrow I_{CaL}$
LQT17	TRDN knockout syndrome	TRDN	*603283	6q22.31	Trisk32	$\uparrow I_{CaL}$

RWS Romano-Ward syndrome, *JLNS* Jervell and Lange-Nielsen syndrome, *ATS* Andersen-Tawil syndrome, *TS* Timothy syndrome. For each gene, the Online Mendelian Inheritance in Man (OMIM) gene ID, the locus, and the encoded protein are reported. Functional effect and impact of gene mutations on cardiac ionic current derived from in vitro cellular studies are reported as $\downarrow$ loss of function or $\uparrow$ gain of function

However, many other genes, detailed in Table 1.1, have been so far associated with the disease and will be described in Sects. 1.2 and 1.3.

Besides congenital LQTS, an acquired form of the disease (aLQTS) has been described as well [7] and refers to patients in which QT prolongation is secondary to hypokalemia or QT-prolonging drugs (www.azcert.org). A genetic basis of aLQTS is recognized as well. Indeed, a third of these patients carries rare variants in the three main congenital LQTS-associated genes, with *KCNH2* being the gene most frequently involved [8]. Furthermore, a sum of common polymorphisms, known to modulate QT interval in the general population [9], has been shown to predict the degree of drug-induced QT prolongation in acquired LQTS patients [10].

The therapy of choice in congenital LQTS is represented by beta-blockers (BBs), which are effective in preventing life-threatening arrhythmias in the vast majority of patients, with the highest efficacy obtained with propranolol and nadolol [11]. Whenever a failure of BB therapy is observed, left cardiac sympathetic denervation (LCSD) offers additional protection with a 91% reduction in cardiac events [12]. ICD therapy is rarely indicated in LQTS, as available therapies are highly effective [13]. A subgroup of LQTS patients that represent an exception at what previously stated are those patients with cardiac events in the first year of life. These patients represent a small subgroup of LQTS cohorts, 2% in the LQTS International Registry [14], but they are at very high risk to have a subsequent cardiac arrest/ sudden cardiac death in the following 10 years of life and are poor responders to beta-blocker therapy [14]. The genetic basis of these most severe forms of the disease will be treated in details in Sect. 1.3.

1.2 Part I: Genetics of Adult Forms of Long QT Syndrome

1.2.1 Major LQTS Genes

The three main genes responsible for LQTS (*KCNQ1*, *KCNH2*, *SCN5A*) were identified between 1995 and 1996 [15–17]. They encode the $Na_v1.5$ sodium channel (SCN5A) and the two alpha subunits of the delayed-rectifier potassium channels (KCNQ1, KCNH2), respectively, involved in the depolarization ($Na_v1.5$) and repolarization (K^+ channels) phases of AP. They represent the major genes responsible for LQTS as they account for approximately 90% of all genotype-positive cases [18].

The *KCNQ1* gene, located on chromosome 11, encodes the α-subunit of the slow delayed-rectifier potassium channel ($K_v7.1$) responsible for the depolarizing I_{Ks} current, which is essential for QT adaptation when heart rate increases [15]. Four alpha subunits encoded by *KCNQ1* co-assemble with two beta subunits to form the functional K^+ channel. The typical effect of *KCNQ1* mutations is a decrease of the outward potassium current (loss of function), leading to ventricular repolarization delay and QT prolongation. Since I_{Ks} current is the major determinant of QT adaptation during heart rate increase, when I_{Ks} is diminished or dysfunctional, the QTc fails to adequately shorten during sympathetic activation, and this creates a potential arrhythmogenic substrate. Heterozygous *KCNQ1* mutations cause the dominant Romano-Ward LQT1 syndrome, while *KCNQ1* homozygous or compound heterozygous mutations cause the recessive JLN variant, characterized also by deafness due to the reduced I_{Ks} in the inner ear.

The gene responsible for LQTS type 2 (LQT2) is **KCNH2** [16], encoding the α-subunit of the rapid delayed-rectifier potassium channel ($K_v11.1$, hERG), which conduces I_{Kr} current. Similar to the slow rectifier potassium channel, four alpha subunits, each encoded by *KCNH2* gene, co-assemble to form a functional channel. Mutations in *KCNH2* gene mainly cause a rapid closure of potassium channels and I_{Kr} decrease (loss of function), resulting in delayed ventricular repolarization and QT prolongation.

There are different mechanisms through which mutations in *KCNQ1* and *KCNH2* can cause reduction or complete loss of the I_K current, the major determinant of the phase 3 of the cardiac AP. The first two mechanisms described, haploinsufficiency and dominant negative effect, are relevant to both *KCNQ1* and *KCNH2*. Haploinsufficiency is a mechanism causing a ~50% reduction of current density due to an overall decreased production of functional channels into the cell membrane, whereas dominant negative effect is elicited by the negative interaction of mutated subunits with the wild-type ones and can cause more than 50% reduction of current density [19]. More recently, mutations in *KCNH2* have been classified into four types on the basis of the channel biophysical property that was impaired. Specifically, class 1 mutations disrupt the synthesis or the translation of $K_v11.1$ α-subunits, class 2 mutations reduce the intracellular transport or trafficking of $K_v11.1$ proteins to the cell membrane, and class 3 and 4 mutations affect $K_v11.1$ channel gating and permeation [20].

The third major LQTS gene is **SCN5A** [17], encoding the α-subunit of the cardiac sodium channel ($Na_v1.5$) involved in the genesis of depolarizing sodium inward current (I_{Na}) and responsible for the phase 0 of AP. In vitro expression studies showed that *SCN5A* mutations lead to LQTS phenotype (LQT3 variant type) through a gain-of-function mechanism, by increasing the delayed Na^+ inward current, resulting in the prolongation of AP duration and QT interval.

Alterations in the sodium channel are also associated with other genetic disorders like Brugada syndrome, atrial fibrillation, sick sinus node syndrome, and the Lev-Lenègre disease. As a further complexity, some *SCN5A* mutations can show a pleiotropic behavior, i.e., the same mutation may associate with more than one phenotype, leading to the so-called overlap syndromes [21, 22].

Overall, the yield of genetic testing for the three main genes in clinically definite LQTS patients is approximately 75% [23], while the prevalence of LQTS variant types among genotype-positive patients is estimated to be 43% for LQT1 (*KCNQ1*), 32% for LQT2 (*KCNH2*), and 13% for LQT3 (*SCN5A*) [18].

These three major LQTS variant types have been associated with specific arrhythmic triggers [24]. LQT1 patients are at higher risk during physical or emotional stress, with swimming being particularly dangerous and specific [24]. Indeed, the majority of patients (99%) that experienced cardiac events while swimming were LQT1. By contrast, LQT2 and LQT3 patients, who have a normal level of I_{Ks}, are at low risk during physical exercise and sport activity. LQT2 patients are more sensitive to sudden noises, such as alarm clocks or telephone ringing, especially during sleep, whereas LQT3 patients tend to have their events at rest or while asleep, when the heart rate decreases [24].

The clinical manifestations of LQTS may also vary according to the different genetic background. The first large study suggesting interactions between genotype, QTc, and gender reported that the risk of cardiac events was higher for LQT2 females and LQT3 males and further increases in the presence of marked QT prolongation (QTc > 500 ms) [25]. LQT1 patients experienced less frequently cardiac events, probably because a very high percentage of them has a QTc < 440 ms [25]. These findings were confirmed some years later, in another study that showed that female gender, QTc interval > 500 ms, and syncopal events were associated with significantly increased risk of life-threatening cardiac events in adulthood [26]. However, the severity of the disease and the relative risk of cardiac events are also influenced by the type of mutation, the location of the mutation in the protein, and the effect produced on cellular function [19, 27, 28].

1.2.2   Minor LQTS Genes

After the identification of the three main LQTS genes, several others have been associated with the disease. They collectively account for a small portion of LQTS (nearly 5%); thereby they are considered as minor genes [23].

Some of the minor LQTS genes concern auxiliary beta subunits that co-assemble with alpha channel subunits encoded by *KCNQ1*, *KCNH2*, and *SCN5A*, to recapitulate sodium and potassium currents. These genes are *KCNE1*, *KCNE2*, and *SCN4B*.

KCNE1 encoding MinK is the single-transmembrane β-subunit of KCNQ1 potassium channel, which contributes as well to generate I_{Ks} current [29]. Mutations in *KCNE1* gene may cause either the dominant RW syndrome (LQT5) when present in heterozygosity or the recessive JLN syndrome if present in homozygosity or compound heterozygosity [30].

KCNE2 gene encodes MiRP1 (MinK-related peptide 1), a small peptide that co-assembles with hERG alpha subunits to form I_{Kr} channel. Mutations in this gene are responsible for the LQT6 variant type and have been associated both with congenital [31] and acquired LQTS [32].

The *SCN4B* gene, underlying LQT10 variant type, encodes the beta auxiliary subunit of $Na_v1.5$ channel and contributes to modulate I_{Na} current. The first mutation identified in this gene (*SCN4B*-p.Leu179Phe) segregated in a family whose proband presented with intermittent 2:1 atrioventricular (AV) block and a corrected QT interval of 712 ms, while other two members died for SCD [33]. The mutation showed in vitro to increase the I_{Na} current, resembling LQT3 phenotype [33].

Other minor genes associated with LQTS encode some components of the sodium channel macromolecular complex, such as *CAV3* and *SNTA1*, and represent LQT9 and LQT12 variant types. Mutations in these genes almost mimic the LQT3 phenotype.

The gene *CAV3* encodes caveolin-3, a small protein that localizes on caveolae, small microdomains of the plasmalemma involved in vesicular trafficking and in the regulation of signal transduction pathways. Mutations in this gene were first described in adult patients, and it was hypothesized that caveolin proteins associated with sodium channel may influence the I_{Na} depolarizing current [34].

The α1-syntrophin, belonging to dystrophin-associated protein family, is part of the sodium channel macromolecular complex, together with neuronal nitric oxide synthase (nNOS) and the nNOS inhibitor Ca^{2+} ATPase PMCA4b. The gene *SNTA1* was firstly implicated in the disease in 2008, with the identification of the p. Ala390Val mutation in a LQTS subject symptomatic for cardiac events, with a QTc of 529 ms [35]. This mutation localizes in the PMCA4b binding domain, resulting in $Na_v1.5$ channel function impairment and I_{Na} current increasing [35].

Additional genes associated with LQTS cases were *AKAP9* (LQT11), *KCNJ5* (LQT13), *KCNJ2* (LQT7), and *ANKB* (LQT4).

The A-kinase anchor protein 9, also known as yotiao, is involved in the phosphorylation of KCNQ1 via PKA and is responsible for the LQT11 variant type [36]. The first *AKAP9* mutation (p.Ser1570Leu) identified in a LQTS patient was predicted to weaken the interaction between PKA and KCNQ1, making the channel not responsive to AMPc, lastly causing QT prolongation [36].

More recently another potassium channel, Kir3.4, encoded by *KCNJ5* gene, was implicated in LQTS type 13. The Kir3.4 is a G protein-coupled inwardly rectifying potassium channel, with a greater tendency to allow potassium to flow into the cell rather than out of the cell. The gene was identified through a genome-wide linkage analysis performed in a family with autosomal dominant LQTS [37]. Heterologous expression studies of Kir3.4-p.Gly387Arg mutation revealed a loss-of-function phenotype resulting from reduced plasma membrane expression [37].

Long QT syndrome types 4 and 7 refer to *ANKB* and *KCNJ2* genes. Mutations in these genes were associated with complex disorders, in which the QT interval prolongation is a minor feature of the heterogeneous patients' phenotype; therefore they are atypical forms of LQTS. The *ANKB* gene encodes a membrane adapter, anchoring different proteins and ion channels to plasmatic membrane. The *ANKB*-p. Glu1425Gly mutation was identified in a large family with modest QT prolongation associated with severe sinus bradycardia and episodes of atrial fibrillation [38].

KCNJ2 gene, encoding Kir2.1 channel, is referred like LQT7. However, mutations in this gene result in Andersen-Tawil syndrome, a multisystem disease that includes modest QT interval prolongation secondary to reduction of the potassium repolarization currents (I_{K1}), and polymorphic tachycardia [39]. This current contributes both to the repolarization phase 3 of AP and to the maintenance of resting membrane potential; therefore, channel dysfunctions may lead to a reduction of I_{K1} with consequent QT prolongation.

Finally, the role of different proteins involved in Ca^{2+} transport, signalling, and homeostasis is currently emerging. Most of the Ca^{2+}-related forms are characterized by extremely severe phenotypes, manifesting in perinatal period or during infancy. Therefore, they will be presented in details in Sect. 1.3.2. The following paragraphs describe in brief the gene function and the first studies that demonstrated an association with LQTS disease.

The main gene regulating Ca^{2+} cellular load is *CACNA1C*, coding the L-type voltage-dependent Ca^{2+} calcium channel $Ca_V1.2$. This gene refers specifically to a malignant form of LQTS known as Timothy syndrome (TS) (LQT8), described for the first time in 1992 as a novel arrhythmia syndrome associated with syndactyly (webbing of fingers and toes) [40, 41]. The molecular basis of the syndrome was described in 2004 by Splawski's group, who identified mutations in *CACNA1C* affecting a single amino acid (p.Gly406Arg), co-segregating with TS phenotype in several families [42]. They also provided exhaustive clinical characterization of the syndrome, including long QT syndrome, life-threatening arrhythmias, congenital cardiac defects, syndactyly, variable penetrance of autism features, craniofacial abnormalities, and hypoglycemia [42]. The spectrum of mutations associated with TS has been enlarged during the following years [43]; however, for some of them, functional evidences supporting their causative role are less clear.

Calmodulin (CaM) is a multifunctional Ca^{2+} binding protein (Fig. 1.1, panel a) essential for intracellular signalling processes in eukaryotic cells [50] that has been recently identified as an additional causative factor for LQTS. It is a ubiquitous protein which transduces Ca^{2+} signals in excitable tissues such heart and brain and therefore influences the activity of ion channels, kinases, and other target proteins [51]. Human calmodulin is highly conserved among vertebrates and is encoded by three separate genes (*CALM1*, *CALM2*, and *CALM3*), producing proteins with identical amino acid sequence [52]. *CALM* genes, when mutated, can cause LQTS (LQT14–16, Table 1.1), cathecolaminergic polymorphic ventricular tachycardia (CPVT) [53], idiopathic ventricular fibrillation, and sudden cardiac death. CALM-LQTS [44] is characterized by very severe forms of the disease with early-onset presentation and recurrent life-threatening arrhythmias [44, 49]. Calmodulinopathy will be discussed in details in Sect. 1.3.2.2.

TRDN is another gene encoding a protein (triadin) implicated in Ca^{2+} channel regulation that was associated to both CPVT [54] and LQTS [55]. Triadin syndrome will be discussed in Sect. 1.3.2.3.

1.2.3 Genetic Modifiers of Long QT Syndrome

Long QT syndrome is a Mendelian disorder, in which the phenotype is primarily explained by a single mutation in one of the main cardiac ion channel genes. However, the disease is also characterized by high clinical heterogeneity within families and among carriers of the same disease-causing mutation. This phenomenon, usually attributed to incomplete penetrance and variable expressivity [56], could be partially due to genetic modifiers. Genetic modifiers are genes or loci, distinct from the primary disease-causing mutation, associated with arrhythmia susceptibility. They act as fine regulators of the arrhythmic risk modulating the effect of the primary disease-causing mutation in a protective or detrimental way. The role of genetic modifiers in LQTS has been largely studied in the last years, and

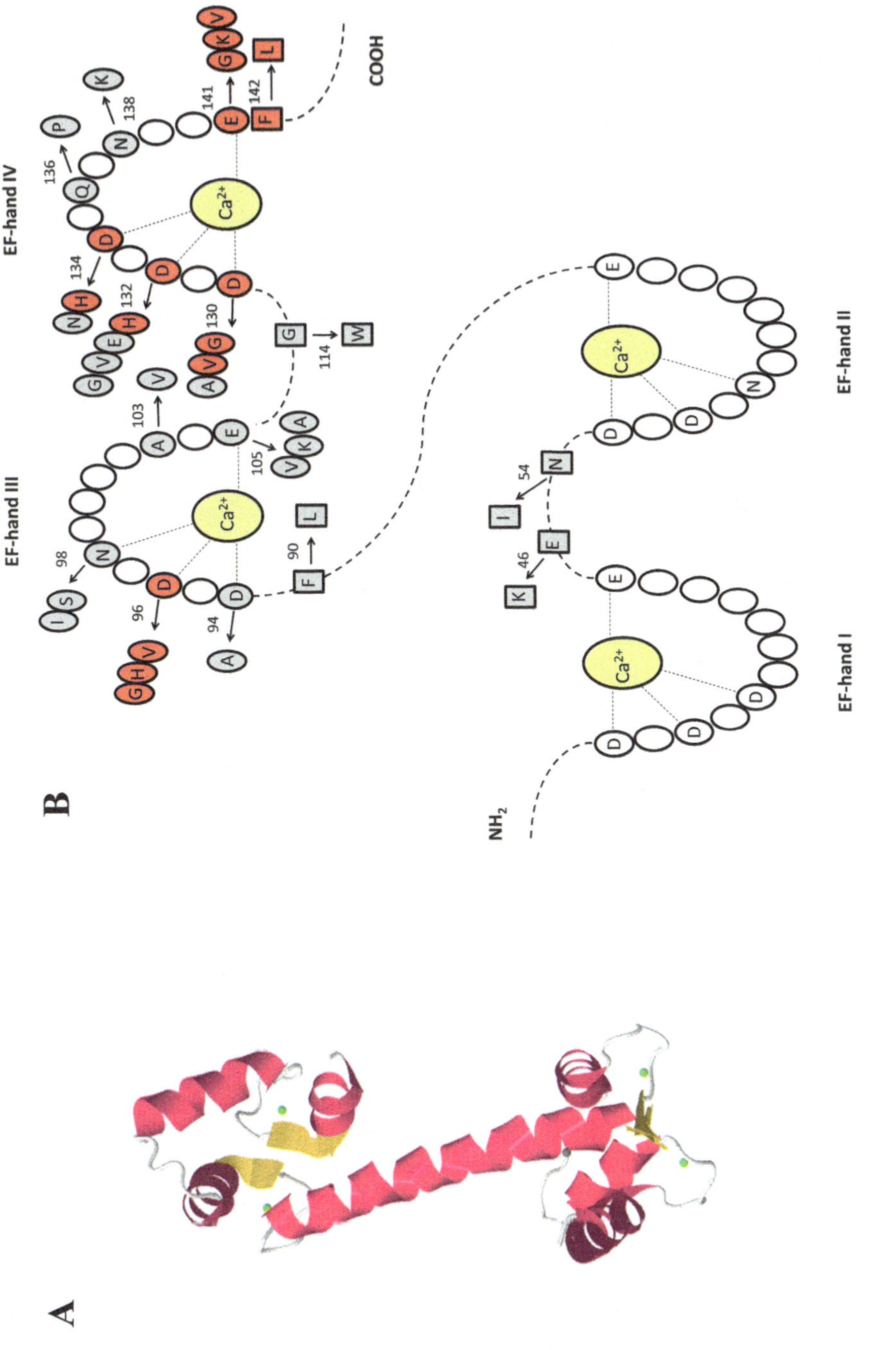
A
B
EF-hand III
EF-hand IV
EF-hand I
EF-hand II
Ca2+
NH2
COOH

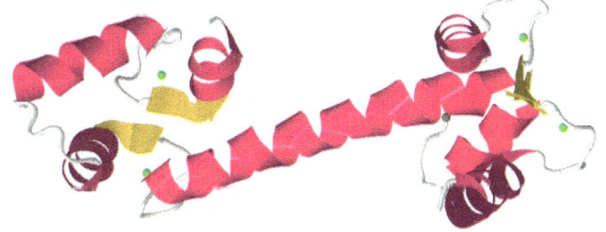

for few of them, application in clinical practice for risk stratification should be carefully evaluated.

The first genetic modifiers of LQTS identified are single nucleotide polymorphisms (SNPs) located in genes already associated with LQTS and generally detected in a specific genetic context. For example, the role of the common variant *KCNH2*-p.Lys897Thr was initially assessed in a LQT2 family which segregates the *KCNH2*-p.Ala1116Val mutation. In vitro functional studies demonstrated that the reduction of potassium current (loss of function) caused by the *KCNH2*-Ala1116Val mutation was further exaggerated by the co-expression of the mutation with the Lys897Thr common variant [57]. This could explain why the proband, carrying both variants, had a severe phenotype, while the rest of Ala1116Val carriers were asymptomatic with borderline/normal QTc. Interestingly, the deleterious effect conferred by the *KCNH2*-Lys897Thr variant in combination with a *KCNH2* mutation was subsequently confirmed by another group in a single family [58]. Furthermore, the presence of *KCNH2*-Lys897Thr was associated with a longer QT interval during maximal exercise in a group of Finnish LQTS patients carrying the *KCNQ1*-G589D mutation [59]. The aforementioned observations suggest that *KCNH2*-Lys897Thr may impair repolarization reserve in the setting of LQT1 or LQT2 mutations. However, the impact of this common variant in the general population may be more complex. Indeed, data from genome-wide association studies conducted in the general population showed that Thr897 is associated with a decreased QTc interval [60–62], and a functional study performed by Bezzina et al. reported that Thr897 channel displayed a shift in voltage dependence of activation and increased rates of current activation and deactivation, expected to increase I_{Kr} [63]. Similar studies have identified functional interactions between a common *SCN5A* variant (p.His558Arg) and LQTS-associated mutations in the same gene [64, 65]. A novel finding concerns the *KCNQ1*-p.Leu353Leu synonymous variant, able to impact splicing efficiency resulting in the generation of alternatively spliced transcripts (25% of the total) in which exon 8 is missing, ultimately decreasing functional tetramer formation [66]. Although the variant by itself is unlikely to

---

Fig. 1.1 (Panel **a**) Calmodulin secondary structure refined at 1.7 A resolution, stoichiometry A. Image from the RCSB PDB (www.rcsb.org; Berman HM, Westbrook J, Feng Z, Gilliland G, Bhat TN, Weissig H, Shindyalov IN, Bourne PE. The Protein Data Bank. Nucleic Acids Res. 2000 Jan 1;28(1):235–42) of PDB ID 1CLL (Chattopadhyaya R, Meador WE, Means AR, Quiocho FA. Calmodulin structure refined at 1.7 A resolution. J Mol Biol. 1992;228(4):1177–92). The four binding domains and pockets correspond to coiled segments, with small marbles representing ligands: 4 Ca^{2+} at structure extremes, 1 EtOH in the middle. The C-terminal region of the protein corresponds to the upper part. (Panel **b**) Schematic model of calmodulin protein showing Ca^{2+} binding loops (EF hand I–IV). Amino acids directly involved in the binding of Ca^{2+} ions are identified with a dotted line. Dark substituted amino acids represent mutations so far identified in LQTS patients with perinatal manifestation of the disease [44–49], whereas light substituted amino acids represent the rest of published mutations associated with different arrhythmic phenotypes and/or identified in patients more than 1 year age [49]. Amino acid substitutions in Ca^{2+} binding loops are identified with circles and those in linker/N-terminal/C-terminal regions with squares

result in a LQTS phenotype, it may act as a modifier and have a synergistic effect on the QTc when inherited with other variants, such as the *KCNQ1*-p.Val205Met [66].

Even though observations within a single family may occasionally provide important clues, as it was for the modifiers previously described, a structured approach more likely to yield interesting returns is that of searching modifier genes in patients carrying the same disease-causing mutation, to avoid the confounding effect of different mutations (each of which with its own impact on arrhythmic risk). Therefore, founder populations, that are group of patients derived from a common ancestor and sharing the same disease-causing mutation, are ideal population to search for genetic modifiers. In the Finnish founder population segregating the *KCNQ1*-p.Gly589Asp, the variant *KCNE1*-p.Asp85Asn, previously associated both with acquired and congenital LQTS [67, 68], was described as a gender-specific genetic modifier of the QT interval with an increased risk for male carriers [69]. The same variant was later associated with longer QT interval and higher risk of cardiac events in LQT2 patients [70]. Several other meaningful findings have been obtained through the study of the South African founder population, segregating the *KCNQ1*-p.Ala341Val mutation [71]. Indeed, this was the first population in which the role of *NOS1AP* as a genetic modifier of LQTS was demonstrated [72]. *NOS1AP* is a gene originally identified through GWAS as affecting the QT interval in the general population [73]. This observation was confirmed in all GWAS performed to search for common SNPs influencing QT duration in the general population, including the most recent one on more than 70,000 individuals [9]. *NOS1AP* is probably the strongest modifier so far identified. Specifically, two common variants in *NOS1AP* (rs4657139, rs16847548) were associated with an increased QT interval duration and a greater probability of cardiac arrest or SCD in LQTS patients carrying the *KCNQ1*-p.Ala341Val mutation [72]. The results of our original study were subsequently validated in two independent LQTS populations [70, 74]. From a mechanistic point of view, *NOS1AP*, encoding a nitric oxide synthase adaptor protein, appears as a potential interesting new player in the regulation of cardiac contractility [75]. In addition, nitric oxide signalling may be an important effector of cardiac repolarization, with a role in balancing nitric oxide and superoxide production [76]. Although the specific molecular mechanism through which *NOS1AP* may confer arrhythmic risk has not yet elucidated, all these findings indicate that *NOS1AP* can be regarded as a strong modifier of clinical severity in LQTS.

The South African founder population helped in the identification of other four common variants with a modifying effect on LQT1 phenotype in *AKAP9* gene encoding yotiao, a protein involved in the regulation of KCNQ1-KCNE1 channel [77]; two variants (rs11772585, rs7808587) were associated with an increased risk of cardiac events, one variant (rs2961024) was associated with QT interval prolongation, and one variant (rs2282972) seemed to have a protective effect and was associated with a decreased risk of cardiac events [77].

Finally, it has been recently postulated that some variants in the 3′ UTR of *KCNQ1* may have a modifier role and may be able to influence QT duration and the risk of cardiac events when located in trans to the LQTS-causing allele

[78]. Unfortunately, this finding was not replicated in any of three LQT1 founder populations [79].

Another approach used to identify genetic modifiers was a matched case-control study including 112 patient duos with LQTS, tested for polymorphisms previously associated with QTc duration in healthy populations or potentially involved in the modulation of adrenergic responses [80]. The results obtained in duos, including one symptomatic and one asymptomatic patient sharing the same mutation either in *KCNQ1* or *KCNH2* gene, were then validated in two independent LQT1 founder population cohorts (South African *KCNQ1*-p.Ala341Val and Finnish *KCNQ1*-p.Gly589Asp). The study provided evidence that the *KCNQ1* rs2074238 polymorphism was an independent risk modifier of LQTS, with the minor T-allele conferring protection against cardiac events [80].

Recently, a novel strategy combining the most up-to-date genetic and cellular technologies has proven to be successful for the identification of modifier genes in a LQT2 large family with variable disease expression among mutation carriers [81]. Electrophysiological studies in patient-derived iPSC-CMs highlighted the presence of a larger L-type calcium current (I_{CaL}) probably contributing to AP prolongation in subjects with more severe phenotypes despite all cells had lower levels of rapid delayed-rectifier current (I_{Kr}) consistent with a loss-of-function mutation in KCNH2. In parallel, whole exome sequencing (WES) allowed to identify two variants with opposite modifier effect in *KCNK17*, a two-pore domain potassium channel gene, and in a GTP-binding protein encoded by *REM2*, a suspected physiological modulator of I_{CaL}. The protective variant in *KCNK17* was absent in all severely affected individuals, carrying indeed the *REM2* aggravating variant [81].

In conclusion, all the aforementioned genetic studies, despite carried on through different approaches, gave strong support to the role of modifiers as potential interplayers of the phenotypic variability in patients with long QT syndrome [82]. Genetic modifiers of long QT syndrome so far discovered are listed in Table 1.2, with details about genes, modifier alleles, and related effects. While most of the variants so far identified as modifiers are able to directly interact with the main disease-causing mutation increasing or reducing its effect [57, 64, 66, 79, 81], other variants, as *NOS1AP*, are acting through a mechanism not yet fully elucidated.

1.3   Part II: Genetics of Perinatal Forms of Long QT Syndrome

The diagnosis of long QT syndrome can be easily performed even very early in life, through a basal ECG, and the European Society of Cardiology has published guidelines to support the interpretation of the neonatal electrocardiogram [83]. Our group has always supported the role of an ECG screening program performed in the first month of life in order to identify early neonates affected by LQTS and establish preventive therapeutic strategies [4, 84].

By contrast, the diagnosis of LQTS during fetal life is more complicated, mainly because of the rarity of symptoms, the poor availability of instruments for detection,

Table 1.2 List of SNPs considered as genetic modifiers of long QT syndrome

SNP	Gene	Localization	Nucleotide substitution	Modifier allele	Effect of the modifier allele	Literature study
rs4657139	NOS1AP	Intronic	A > T	A	↑ QT ↑ ACA, SD	[72]
rs16847548	NOS1AP	Intronic	T > C	C	↑ QT ↑ CA, SCD ↑ risk of cardiac events	[72]
rs2074238	KCNQ1	Intronic	C > T	T	↓ QT ↓ symptoms	[80]
NA	KCNQ1	Exonic (c.1059G > A:p. Leu353Leu)	G > A	A	↓ functional tetramer formation ↑QT in LQT1	[66]
rs2519184	KCNQ1	3′ UTR	G > A	A	Contrasting results on QT duration and risk of cardiac event	[78, 79]
rs8234	KCNQ1	3′ UTR	A > G	G	Contrasting results on QT duration and risk of cardiac event	[78, 79]
rs10798	KCNQ1	3′ UTR	A > G	G	Contrasting results on QT duration and risk of cardiac event	[78, 79]
rs1805123	KCNH2	Exonic (c.269A > C:p. Lys897Thr)	A > C	C	↑ symptoms in LQT2 ↑ I$_{Kr}$ current reduction	[57, 58]
rs1805124	SCN5A	Exonic (c.1673A > G:p. His558Arg)	A > G	G	Rescue normal phenotype in LQT3	[64, 65]
rs1805128	KCNE1	Exonic (c.253G > A:p. Asp85Asn)	G > A	A	↑ QT in LQT1 males ↑ risk of cardiac events in LQT2	[69, 70]
rs11772585	AKAP9	Intronic	C > T	T	↑ risk of cardiac events	[77]
rs7808587	AKAP9	Intronic	A > G	G	↑ risk of cardiac events	[77]
rs2282972	AKAP9	Intronic	C > T	T	↓ risk of cardiac events	[77]
rs2961024	AKAP9	Intronic	T > G	G	↑ QT	[77]

(continued)

Table 1.2 (continued)

SNP	Gene	Localization	Nucleotide substitution	Modifier allele	Effect of the modifier allele	Literature study
rs10947804	*KCNK17*	Exonic (c.61A > G:p. Ser21Gly)	A > G	G	↓ QT ↓ symptoms	[81]
rs8014119	*REM2*	Exonic (c.287G > C:p. Gly96Ala)	G > C	C	↑ QT ↑ risk of cardiac events	[81]

The corresponding gene, the alternative alleles (minor allele in the general population according to NCBI dbSNP is reported as second), the modifier allele, and the corresponding associated effect are reported for each SNP. References in the last column refer to literature studies describing the modifier effect of the SNPs. *ACA* aborted cardiac arrest, *SD* sudden death, *CA* cardiac arrest, *SCD* sudden cardiac death

and the incomplete knowledge in the field. Prenatal signs of LQTS could be sinus or intermittent bradycardia <110 beats/min, fetal heart rate < third percentile for gestational age, atrioventricular (AV) block, tachyarrhythmias, or pleural effusion and exceptionally fetal hydrops [85, 86]. Currently, LQTS was estimated to account for 15–17% of fetal bradycardias among fetuses with a normally structured hearts [85]. However, in the last few years, many progresses have been made in the field of prenatal LQTS diagnosis as well as in management of pregnancy in the context of LQTS. Several methods that can be used to detect fetal arrhythmias have been described, as fetal ultrasound, fetal echocardiography, cardiotocography, fetal electrocardiography, and fetal magnetocardiography, albeit not all are widely available for physicians [87]. In particular, the advent of fetal magnetocardiography, currently a technique with full regulatory approval [88], has demonstrated to allow an accurate diagnosis of LQTS in utero [89]; however, only in few centers it is available. To study and better understand the fetal presentation of LQTS, an International Registry has been recently established, and our research group is involved as a partner. Through this International Registry, we also aim to create a communication network among scientists to improve the management of these vulnerable little patients and to create a platform to allow families to gain more information about the disease and its management.

1.3.1 Sodium and Potassium Channels

Few studies have been published so far with a specific focus on LQTS diagnosed during fetal, perinatal, and neonatal periods [90, 91]. A Japanese study collected 58 LQTS cases in which the diagnosis was made in fetal ($n = 18$) and neonatal ($n = 31$) periods or within the first year of life ($n = 9$) [90]. The genetic analysis of the three main LQTS genes (*KCNQ1*, *KCNH2*, *SCN5A*) plus *CACNA1C* gene identified mutations in 29/41 cases, with a majority of variants in potassium channel genes. The clinical phenotype observed in LQT2 and LQT3 patients was more severe than in LQT1 patients. The first group is composed of patients presenting aggressive clinical phenotypes, with ventricular tachycardia/TdP and AV block, while the main clues of the latter group were sinus bradycardia or family history for LQTS [90]. The group presenting life-threatening arrhythmias required the use of more aggressive therapies to effectively reduce the frequency of arrhythmic events. A second study aimed to describe fetal manifestation of LQTS and took advantage of in utero recording of arrhythmias characteristic of LQTS [91]. The genetic analysis performed in 43 subjects just after birth was positive in 95% of cases, with a majority of mutations in *KCNQ1* gene; *SCN5A* mutations were identified much more frequently in fetuses with TdP and second AV block, while *KCNQ1* mutations were identified more often in fetuses presenting with persistent sinus bradycardia [91]. Although the distribution of genotypes among the three main LQTS genes was slightly different in the two cohorts, few observations were common:

1. Sinus bradycardia was more frequently observed in LQT1, while life-threatening arrhythmias in LQT2/LQT3 fetuses/neonates.
2. LQT3 patients carrying mutations in *SCN5A* have the most severe phenotype and are at high risk of life-threatening events.

In addition to these two studies, reports on single or few cases manifesting LQTS across the perinatal period or within the first year of life have been published and are summarized in Table 1.3. Interestingly, while adult LQTS patients most frequently carry genetic variants in *KCNQ1* and *KCNH2* [18], in perinatal/neonatal LQTS cases, mutations were almost equally distributed among the three main genes: *KCNQ1* ($n = 38$ mutation carriers (MCs), *SCN5A* ($n = 26$ MCs), and *KCNH2* ($n = 23$ MCs).

Among the cases so far described and reported in Table 1.3, major arrhythmic events (VT, TdP, VF, CA, SCD) are more frequently observed in LQT3 and LQT2 (85% of *SCN5A* and 83% of *KCNH2* MCs) compared to LQT1 patients (2.6% of *KCNQ1* MCs). Furthermore, while in adult LQT3 patients response to beta-blocker therapy is reasonably good [94], in LQT3 patients with events in the first year of life, response to beta-blocker therapy is poor, and there is a very high incidence of cardiac arrest/sudden cardiac death despite beta-blocker therapy [115].

LQT3 patients with early and severe presentation more frequently carry variants located in the transmembrane, linker, and C-terminal regions of SCN5A (81% of *SCN5A* MCs), where variants are less frequently observed in controls [116]; similarly, at variance with controls, in LQT1 and LQT2 patients with an early-onset presentation, variants are more frequently observed in the transmembrane, linker, pore-forming, and C-terminal regions of KCNQ1 and KCNH2 potassium channels (97% of *KCNQ1* and 91% of *KCNH2* MCs) [116].

Among the different mutations identified, two appear to be more frequently associated with malignant perinatal LQTS, i.e., the *SCN5A*-p.Arg1623Gln mutation [90, 91, 100, 101] and the *KCNH2*-Thr613Met [90, 110–112], both located in key functional domain of cardiac channels (SCN5A transmembrane S4 voltage sensor of repeat IV and KCNH2 intramembrane pore-forming region). The SCN5A-p.Arg1623Gln mutation showed a particularly severe channel dysfunction through in vitro electrophysiological study [117] that could explain the severe phenotype consistently observed in all mutation carriers described [90, 91, 100, 101]. Another mechanism that could explain the early occurrence of severe clinical manifestation in some *SCN5A* mutations was provided by Murphy et al. [93], that studied the mutation *SCN5A*-p.Leu409Pro, identified in a fetus with life-threatening arrhythmias, who was also carrying the *SCN5A*-p.Arg558 common variant [93]. The sodium channel dysfunction, observed in vitro with the co-expression of the two variants, was further potentiated when the variants were expressed in a fetal SCN5A splice isoform, providing a plausible mechanism for the severe fetal presentation of the disease [93].

Table 1.3 Genetic variants so far identified in the five main LQTS-associated genes (*KCNQ1, KCNH2, SCN5A, KCNE1, KCNE2*) in LQTS patients diagnosed in the perinatal period or within the first year of life

Gene	Nucleotide change	Amino acid change	Localization	Functional effect	Number of carriers	Major arrhythmic events	Primary literature report
SCN5A	c.123G > A	p.Arg43Gln	Cytoplasmic N-terminal region	Lidocaine-induced arrhythmogenic effect	1[#]	VT	[92]
SCN5A	c.523T > G	p.Ile176Met	Transmembrane S2 of repeat I		1	TdP	[90]
SCN5A	c.1218C > G	p.Asn406Lys	Transmembrane S6 of repeat I	$Na_V1.5$ gain of function	1	TdP	[90]
SCN5A	c.1226T > C	p.Leu409Pro, p.Arg558	Transmembrane S6 of repeat I	$Na_V1.5$ and fetal $Na_V1.5$ gain of function	1	TdP	[93]
SCN5A	c.1237G > A	p.Ala413Thr	Cytoplasmic linker loop 1	NA	1	None	[94]
SCN5A	c.2821_2822delinsAA	p.Ser941Asn	Cytoplasmic linker loop 2	$Na_V1.5$ gain of function	1	TdP, VF, ACA, SCD	[95]
SCN5A	c.3556G > A	p.Ala1186Thr	Cytoplasmic linker loop 2	NA	1	None	[90]
SCN5A	c.3911C > T	p.Thr1304Met	Transmembrane S4 (voltage sensor) of repeat III	$Na_V1.5$ gain of function	1	None	[91]
SCN5A	c.3988G > C	p.Ala1330Pro	Cytoplasmic linker between S4/S5	$Na_V1.5$ gain of function	1	VT/TdP, SCD	[96]
SCN5A	c.3989C > A	p.Ala1330Asp	Cytoplasmic linker between S4/S5	NA	1	CA, SCD	[94]
SCN5A	c.3995C > T	p.Pro1332Leu	Cytoplasmic linker between S4/S5	$Na_V1.5$ gain of function	1	VT/TdP	[97]
SCN5A	c.4418 T > G	p.Phe1473Cys	Cytoplasmic linker loop 3	$Na_V1.5$ gain of function	1	TdP	[98]

(continued)

Table 1.3 (continued)

Gene	Nucleotide change	Amino acid change	Localization	Functional effect	Number of carriers	Major arrhythmic events	Primary literature report
SCN5A	c.4456_4458delTTC	p.Phe1486del	Cytoplasmic linker loop 3	Nav1.5 gain of function	1	VT, SCD	[99]
SCN5A	c.4868G > A	p.Arg1623Gln	Transmembrane S4 (voltage sensor) of repeat IV	Nav1.5 gain of function	7	TdP, SCD TdP, CA TdP, CA TdP, CA VT/TdP VT/TdP None	[91] (first–fourth case) [100] (fifth case) [101] (sixth case) [90] (seventh case)
SCN5A	c.4892G > A	p.Gly1631Asp	Transmembrane S4 (voltage sensor) of repeat IV	Nav1.5 gain of function	2	VT, VF, CA TdP, CA	[102]
SCN5A	c.5287G > A	p.Val1763Met	Transmembrane S6 of repeat IV	Nav1.5 gain of function	2	TdP CA, SCD	[103] [94]
SCN5A	c.5314C > G	p.Leu1772Val	Cytoplasmic C-terminal region	NA	2	TdP CA	[104] [90]
SCN5A	c.5320A > G	p.Asn1774Asp	Cytoplasmic C-terminal region	Nav1.5 gain of function	1	VT/TdP	[90]
KCNQ1	c.310_329dup	p. Tyr111Profs*133 (V110 fs + 132X)	Cytoplasmic N-terminal region	NA	1	None	[91]
KCNQ1	c.477 + 5G > A	Predicted splicing defect	Intronic (closed to region encoding transmembrane S2)	NA	2	None	[91]

KCNQ1	c.502G > C	p.Gly168Arg	Transmembrane S2	Kv7.1 loss of function	2	None	[91]
KCNQ1	c.568C > T	p.Arg190Trp	Cytoplasmic linker between S2 and S3	NA	1	None	[91]
KCNQ1	c.677C > T	p.Ala226Val	Transmembrane S4 (voltage sensor)	NA	2	None	[91]
KCNQ1	c.760G > A	p.Val254Met	Cytoplasmic linker between S4 and S5	Kv7.1 loss of function	2	None	[91]
KCNQ1	c.775C > T	p.Arg259Cys	Cytoplasmic linker between S4 and S5	NA	2	None	[91]
KCNQ1	c.805G > A	p.Gly269Ser	Transmembrane S5	Kv7.1 loss of function; insensitivity to PKA stimulation	1	None	[90]
KCNQ1	c.825_828delTCC	p.Ser277del (276delSer)	Transmembrane S5	NA	1	None	[90]
KCNQ1	c.914G > A	p.Trp305*	Intramembrane pore-forming region	NA	3	None	[91]
KCNQ1	NA	p.Ile313Lys	Intramembrane pore-forming region	NA	2	None	[90]
KCNQ1	c.941G > A	p.Gly314Asp	Intramembrane pore-forming region	NA	2	None	[91]
KCNQ1	c.1016 T > C	p.Phe339Ser	Transmembrane S6	Kv7.1 loss of function	1	None	[105]
KCNQ1	c.1022C > T	p.Ala341Val	Transmembrane S6	Kv7.1 loss of function	3	None	[90] (2 cases), [91]
KCNQ1	c.1022C > A	p.Ala341Glu	Transmembrane S6	Kv7.1 loss of function	1	None	[91]
KCNQ1	NA	p.Thr458Met	Cytoplasmic C-terminal domain	NA	1	None	[90]
KCNQ1	c.1552C > T	p.Arg518*	Cytoplasmic C-terminal domain	Kv7.1 loss of function	1	None	[91]

(continued)

Table 1.3 (continued)

Gene	Nucleotide change	Amino acid change	Localization	Functional effect	Number of carriers	Major arrhythmic events	Primary literature report
KCNQ1	c.1615C > T	p.Arg539Trp	Cytoplasmic C-terminal domain (interaction with KCNE1)	Gating defect; reduced affinity for PIP_2	2	None	[91, 106]
KCNQ1	c.1760C > T	p.Thr587Met	Cytoplasmic C-terminal domain	Trafficking defect	2	None	[90, 107]
KCNQ1	c.1772G > A	p.Arg591His	Cytoplasmic C-terminal domain (interaction with AKAP9)	Kv7.1 loss of function	2	None	[91]
KCNQ1	c.1781G > A	p.Arg594Gln	Cytoplasmic C-terminal domain (interaction with AKAP9)	Kv7.1 loss of function; partial trafficking defect	1	None	[91]
KCNQ1	c.1831G > T	p.Asp611Tyr	Cytoplasmic C-terminal domain (interaction with AKAP9)	NA	2	None	[90]
KCNQ1	c.1927G > A	p.Gly643Ser	Cytoplasmic C-terminal domain	Kv7.1 loss of function	1	CA	[90]
KCNH2	c.558_600dupCGG	p.Thr201Argfs*145 homozygous state	Cytoplasmic N-terminal region	NA	1	TdP	[108]
KCNH2	c.730_742del	p.Ala244Serfs*112 (Ser243 + 112X)	Cytoplasmic N-terminal region	NA	1	None	[90]
KCNH2	c.1682C > T	p.Ala561Val	Transmembrane S5	Kv11.1 loss of function; trafficking defect	2	VT VT/TdP	[109] [90]
KCNH2	c.1714G > A	p.Gly572Ser	Extracellular linker	Trafficking defect	1	TdP	[90]

KCNH2	c.1838C > T	p.Thr613Met	Intramembrane pore-forming region	Trafficking defect	5	VT VT/TdP VT/TdP VT VT, SCD	[110] (first case) [111] (second case) [90] (third and fourth case) [112] (fifth case)
KCNH2	c.1838C > A	p.Thr613Lys	Intramembrane pore-forming region	NA	1	TdP	[91]
KCNH2	c.1841C > T	p.Ala614Val	Intramembrane pore-forming region	$K_V11.1$ loss of function; trafficking defect	1	VT	[90]
KCNH2	c.1879 T > C	p.Phe627Leu	Intramembrane pore-forming region	$K_V11.1$ loss of function	1[#]	VT	[92]
KCNH2	c.1876G > A	p.Gly628Ser	Intramembrane pore-forming region	Trafficking defect	2	VT/TdP TdP	[90]
KCNH2	c.1883G > C	p.Gly628Ala	Intramembrane pore-forming region	NA	1	VT/TdP	[90]
KCNH2	c.1898A > G	p.Asn633Ser	Extracellular linker	$K_V11.1$ loss of function; trafficking defect	1	VT/TdP	[90]
KCNH2	NA	p.Ala715Ala predicted splicing defect	Cytoplasmic C-terminal region	NA	1	None	[91]
KCNH2	c.2255G > A	p.Arg752Gln homozygous state	Cytoplasmic C-terminal region	$K_V11.1$ loss of function	1	TdP	[113]
KCNH2		p. Pro926AlafsX14, p.Lys897Thr	Cytoplasmic C-terminal region	$K_V11.1$ loss of function	1	VT	[58]

(continued)

Table 1.3 (continued)

Gene	Nucleotide change	Amino acid change	Localization	Functional effect	Number of carriers	Major arrhythmic events	Primary literature report
KCNH2	NA	p.Trp1001*	Cytoplasmic C-terminal region	Decrease in mutant mRNA levels by NMD	1	None	[91]
KCNH2	NA	p.Thr1945 + 6C	Intronic (closed to region encoding C-terminus S2)	NA	1	None	[91]
KCNH2	del(7)(q32qter)	NA	Cytoplasmic C-terminal region	NA	1	TdP	[90]
KCNE1	c.226G > A	p.Asp76Asn	Cytoplasmic C-terminal region	I_{Ks} reduction	2	None	[91]
KCNE2	c.80G > A, c.59 T > A	p.Ile20Asn, p. Arg27His	Cytoplasmic N-terminal region	NA	1	VF, CA, SCD	[114]

Mutation nomenclature is conforming to Human Genome Variation Society guidelines (www.hgvs.org) and has been reported according to the longer transcripts; for some mutations, the old nomenclature version used in the original study was reported as well in brackets. For each genetic variant, the number of carriers and the presence of major cardiac events (*VT* ventricular tachycardia, *TdP* torsades de pointes, *VF* ventricular fibrillation, *CA* cardiac arrest, *SCD* sudden cardiac death) was reported. Reference number corresponds to the original study in which the perinatal LQTS case was presented. #Symbol identifies a compound heterozygous patient carrying in addition to the primary disease-causing mutation (KCNH2-p.Phe627Leu) a secondary variant in SCN5A (p. Arg43Gln). For the analysis in Sect. 1.3.1, the patient was considered KCNH2 mutation carrier. *PIP*₂ phosphatidylinositol-4,5-bisphosphate, *NMD* nonsense-mediated mRNA decay, *NA* not available

1.3.2 Calcium Channel Complex

1.3.2.1 Timothy Syndrome

Timothy syndrome (TS) is a multisystem disorder, caused by mutations in *CACNA1C*, the gene encoding the alpha subunit of the calcium channel $Ca_v1.2$. Patients are characterized by extremely prolonged QT, 2:1 AV blocks, T-wave alternans, and life-threatening arrhythmias occurring very early in life. Congenital heart defects are present in 70% of the patients, and also extra-cardiac disorders are typically observed. Hand and foot syndactyly is present in almost 90% of cases; developmental delays, autism spectrum disorder, seizures, and intermittent hypoglycemia have been reported as well as frequent infections secondary to altered immune responses [42]. In most of the cases identified within the first year of life, the disease-causing mutation arises de novo or was inherited from an asymptomatic parent with mosaicism, and interestingly, in all studies [90, 118–127] but four [119, 124, 125, 127], the identified mutation conferring the TS phenotype was always the same, p. Gly406Arg in the alternatively spliced exon 8A of *CACNA1C* gene. This finding shows that the p.Gly406Arg mutation has a strong and reproducible effect across a number of unrelated patients detected early in life, as well as in children belonging to different families [42]. This observation further suggests the importance of this amino acid residue in maintaining the correct biophysical properties of the L-type calcium channel and supports the causality of the mutation for the syndromic presentation. The p.Gly406Arg mutation is located at the end of the IS6 transmembrane segment, just upstream the calcium channel beta subunit binding site. This region forms a rigid alpha-helix that helps properly orient the auxiliary beta subunit. The hypothesized underlying mechanism led by the p.Gly406Arg mutation is the perturbation of the voltage-dependent inactivation (VDI) of the channel, due to the reorientation of the beta subunit secondary to the disruption of the alpha-helix segment [128]. A subsequent study showed that VDI is not the only L-type Ca^{2+} channel inactivation regulatory system impaired by the mutation that may impact also on Ca^{2+}/CaM-dependent inactivation (CDI) [129]. Therefore, p.Gly406Arg gain-of-function mutation is able to cause the loss of both channel inactivation mechanisms: as a consequence, calcium channels fail to close during the plateau phase of AP, leading to an increase of the depolarizing inward calcium current and to a marked delay of repolarization [42, 129]. In addition, TS mouse model showed that low levels of $Ca_v1.2$-mutated channels are sufficient to create a Ca^{2+} overload in the cytosol, favoring spontaneous sarcoplasmic Ca^{2+} release, able to increase excitation-contraction (EC) coupling and arrhytmogenesis [130]. Figure 1.2 provides a schematic model of Ca^{2+} macromolecular complexes that activate sequentially and contribute to different phases of EC coupling in cardiomyocytes.

Four different mutations were so far identified in TS patients diagnosed within the first year of life: the *CACNA1C*-p.Ala1473Gly [119], located at the end of the IVS6 of $Ca_v1.2$ channel, the p.Gly406Arg identified in exon 8 of a different splicing isoform of the protein [124, 125], the p.Gly402Ser and the p.Lys1211Glu, reported in a recent clinical study on TS [127]. Mutations in exon 8, mutually exclusive to exon 8A but encoding the same region, have been initially associated with a subtype

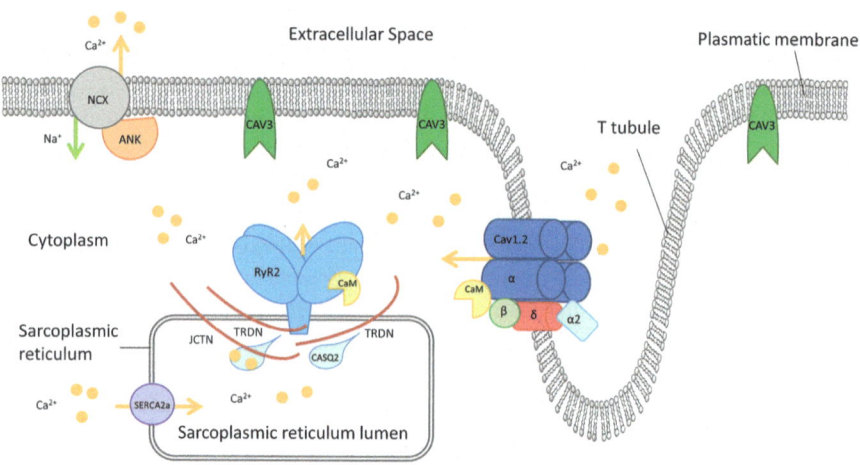

Fig. 1.2 Schematic model of calcium handling in cardiomyocytes and excitation-contraction coupling protein players. The calcium flux across three main Ca^{2+} macromolecular complexes is displayed. Calcium ions enter into the cell through $Cav_{1.2}$ channel, located in transverse T tubules of plasmalemma invaginations and acting in the plateau phase of cardiac action potential. The $[Ca^{2+}]i$ increase triggers the activation of RyR2 channel responsible of a much larger Ca^{2+} release from sarcoplasmic reticulum to cytosol. Once $[Ca^{2+}]i$ reaches the highest concentration level, the excitation phase finishes and starts the myofilament contraction mediated by Ca^{2+} binding to troponin C. Later on, NCX and SERCA2a contribute to $[Ca^{2+}]i$ decrease through extrusion of Ca^{2+} excess from cytoplasm to extracellular space (NCX) and Ca^{2+} sequestration into sarcoplasmic reticulum (SERCA2a), reporting $[Ca^{2+}]i$ in the cytoplasm to basic level. *NCX* Na^+/Ca^{2+} exchanger, *ANK* ankyrin B, *CAV3* caveolin-3, *RyR2* ryanodine receptor type 2, *CaM* calmodulin, *TRDN* triadin, *JCTN* junction, *CASQ2* calsequestrin 2, *SERCA2a* Ca^{2+}-ATPase type 2a, *Cav1.2* L-type Ca^{2+} channel (LTCC) with pore-forming alpha subunit and accessory alpha 2, beta, delta subunits

of the disease characterized by extremely severe cardiac features and a milder extra-cardiac phenotype, named Timothy syndrome type 2 [124]. The clinical study by Dufendach et al. provided data from a novel large cohort of TS cases showing heterogeneous combination of cardiac and extra-cardiac features [127]. Interestingly, 70% of them (12/17) have been diagnosed within the first year of life, and the p. Gly406Arg mutation remained the most frequent, present in 10/12 TS patients. The two cases carrying different variants showed, respectively, a complex phenotype (p. Gly402Ser) or solely QT prolongation and 2:1 AV block in the case of p. Lys1211Glu carrier [127].

1.3.2.2 Calmodulinopathies

Calmodulinopathy is a term conferred to the novel clinical entity of arrhythmia diseases due to calmodulin genetic defects. Calmodulin is able to interact and modulate the activity of different cardiac ion channels, such as the L-type calcium channel, the ryanodine receptor, and the sodium and potassium channels [131]. Therefore, it is not surprising that CaM protein alterations can explain a wide range of cardiac arrhythmia phenotypes as LQTS [44, 132], CPVT [53, 132], and IVF [133]. Most of the calmodulin variants so far described were associated with LQTS and the majority of patients manifest the disease very early in life [45, 49]. This observation supports the emerging concept that calmodulinopathies are mainly diseases of the perinatal/pediatric period with less impact on adulthood.

The first description of perinatal forms of LQTS due to CaM mutations was published by our group in 2013 [44]. Two unrelated neonates with recurrent cardiac arrest and dramatically prolonged QT interval (>600 ms), born by healthy parents, underwent a WES parent-child analysis to find a potential genetic cause. Two missense mutations altering amino acid residues in the C-terminal protein domain, important for calcium binding (*CALM2*-p.Asp96Val, *CALM1*-p.Asp130Gly), were identified in the two probands. Biochemical studies demonstrated several-fold reduction of Ca^{2+} binding affinity for the CaM-mutated proteins with respect to wild type; therefore, it was hypothesized that mutated CaM might disrupt Ca^{2+} signalling in heart cells [44]. During the following years, other *CALM*-linked LQTS cases have been described, with severe phenotypes and early arrhythmia onset detected during fetal life or just after birth [45–49]. Table 1.4 and Fig. 1.1 (panel b) show CaM mutations so far described in LQTS cases with a perinatal diagnosis. It is interesting to note that all mutations but one (p.Phe142Leu) are located in the two C-terminal Ca^{2+} binding domains of the protein and involve amino acid residues directly implicated in Ca^{2+} binding. The mutation p.Phe142Leu, however, resides in the first amino acid just downstream the fourth C-terminal Ca^{2+} binding loop and is one of the most recurrent among perinatal LQTS subjects together with the p.Asp130Gly.

A recent study conducted through WES on 38 genetically elusive LQTS patients of different ages identified CaM mutations in five cases: four with a previously described mutation (p.Asp130Gly, p.Asp130Val, p.Phe142Leu in two patients), one with a novel mutation (p.Glu141Gly) [45]. Interestingly, all the five CaM-positive patients had a more severe phenotype than other patients in the cohort, with an average age of onset of 10 months, an average corrected QT interval of 676 ms, and a high prevalence of cardiac arrest [45]. This finding suggests again that CaM mutations cause particularly severe forms of LQTS, characterized by very early onset and extreme phenotype severity, with life-threatening arrhythmias, cardiac arrest, or sudden cardiac death occurring within the first year of life.

Given the high number of CaM molecular interactions, in vitro studies tried to elucidate molecular mechanisms leading to arrhythmogenesis in patients with CaM mutations. The first studies, performed in heterologous expression systems or in mammalian ventricular myocytes, showed that Ca^{2+}/CaM-dependent inactivation (CDI) was the main mechanism impaired in LQTS patients with CaM mutations [45, 134, 135]. The slow of L-type Ca^{2+} channel inactivation and the consequent boost of Ca^{2+} influx during the plateau phase of AP would likely result in a repolarization delay and arrhythmogenesis (Fig. 1.2). Comparable results were obtained using the cellular model of human cardiomyocytes differentiated from induced pluripotent stem cells (iPS-derived CMCs) [47, 136, 137]. This type of in vitro model, developed in the last years, allows to functionally characterize a mutation in the patient-specific cellular environment and appears particularly suit-

Table 1.4 Calmodulin mutations identified so far in LQTS patients with a first manifestation in perinatal period or within the first year of life

Protein sequence variant	Gene	Number of carriers	Age at first symptoms	Major arrhythmic events/ outcome	Functional effect on LQTS-associated ion channels	Reference study
p. Asp96Val	CALM2	2	Prenatal	CA	Cav1.2 gain of function; no effect on Nav1.5	[44, 91, 134, 135]
	CALM2		Prenatal	Death due to heart failure and infection (6 months)		[49]
p. Asp96His	CALM3	2	Prenatal	None	NA	[48, 49]
p. Asp96Gly	CALM2	1	Prenatal	None	NA	[49]
p. Asp130Gly	CALM1	7	6 months	CA	Cav1.2 gain of function; fetal Nav1.5 gain of function	[44, 134, 135]
	CALM1		1 month	CA		[44, 134, 135]
	CALM2		Birth	ICD shock for VF		[45, 136]
	CALM3		Birth	None		[46]
	CALM3		Birth	Electrical storms; SCD at 4.5 years		[49]
	CALM3		Neonatal	Neonatal death due to ICD complication		[49]
	CALM3		Birth	SCD at 45 dd		[49]
p. Asp130Val	CALM2	1	Prenatal	TdP, ICD shocks for VF	NA	[45]
p. Asp132His	CALM2	1	Prenatal	None	Cav1.2 gain of function	[47]
p. Asp134His	CALM2	1	Prenatal	CA, ICD shocks		[49]
p. Glu141Val	CALM1	1	Neonatal	VT/VF, ICD shock		[49]
p. Glu141Gly	CALM2	1	Prenatal	SCD at 20 months		[49]

(continued)

Table 1.4 (continued)

Protein sequence variant	Gene	Number of carriers	Age at first symptoms	Major arrhythmic events/ outcome	Functional effect on LQTS-associated ion channels	Reference study
p. Glu141Lys	CALM3	1	Prenatale	CA, ICD shocks for VF		[49]
p. Phe142Leu	CALM1	4	Birth	VF	Cav1.2 gain of function; no effect on Nav1.5 and Kv7.1	[44, 134, 135, 137]
	CALM1		Birth	CA, SCD at 2 years		[45]
	CALM1		Birth	SCD at 1 year		[45]
	CALM3		Prenatal	None		[48]

The number of carriers as well as the age at manifestation and major events are reported for each patient. References concern the first study in which the mutation was identified and associated with LQTS and/or the in vitro functional studies of the variant. $Ca_v1.2$ L-type voltage-dependent Ca^{2+} calcium channel, $Na_v1.5$ α-subunit of the cardiac sodium channel, *CA* cardiac arrest, *TdP* torsades de pointes, *VF* ventricular fibrillation, *SCD* sudden cardiac death

able to study molecular mechanisms underlying arrhythmogenesis [136–138]. Regarding calmodulin, it is important to point out that this model allows the study of CaM mutations in the context of native allelic balance. This is quite important for calmodulin as there are three genes encoding the same calmodulin product; therefore, there is only one mutated allele with five wild-type ones, and obviously overexpression studies are not ideal in such an allelic ratio. Since it is known that CaM modulates also sodium and potassium channels [139–142], both implicated in the genesis of LQTS, it has been attempted also to determine the potential impact of CaM mutations on I_{Ks} and I_{Na} currents. However, the first three CaM mutations tested (p.Asp96Val, p.Asp130Gly, p.Phe142Leu) did not affect late sodium and potassium currents [135, 137]. Further studies are needed to better understand CaM influence on different type of cardiac ion channels and the specific effects carried on by different mutations.

Given the need to better define this novel CaM-mediated arrhythmogenic disease, our group has established an International Calmodulinopathy Registry [143], with the aim to collect an adequate number of cases, define the spectrum of the clinical features associated with CaM mutations, identify risk factors for life-threatening events, promote among physicians discussion on patients' management, and potentially come up with therapeutic guidelines. The first results of this large consortium have been recently published, with the collection of clinical and genetic data on 74 subjects carrying a variant in CALM1 ($n = 36$), CALM2 ($n = 23$), or CALM3

($n = 15$) genes [49]. Among CALM mutation carriers, LQTS is the most prevalent phenotype, accounting for 49% of the cohort. CALM-LQTS patients exhibited extremely prolonged QTc intervals (594 ± 73 ms) and a high prevalence (78%) of life-threatening arrhythmias occurring very early in life; 58% of patients have a perinatal presentation, consisting in a variable combination of striking QT prolongation (QTc 628 ± 62 ms), sinus bradycardia, 2:1 atrioventricular block, T wave alternans, and/or MAEs [49]. Additionally, the early onset is almost exclusive of the LQTS phenotype (91%) with respect to the rest of the cohort. The LQTS patients have also poor response to the available therapies, therefore a combination therapy with drugs, sympathectomy, and devices should be considered.

1.3.2.3 Triadin Knockout Syndromes

Triadin (TRDN) is an important component of the cell macromolecular calcium releasing complex, with the specific function to anchor calsequestrin 2 (CASQ2) to ryanodine receptor (RyR2) at the terminal cisternae of the sarcoplasmic reticulum (SR). The protein isoform mainly expressed in cardiac muscle is Trisk32. The first translational study focusing on *TRDN* highlighted the role of the gene in CPVT [54]. In particular, *TRDN* gene is considered responsible for a recessive form of the disease, since all the genotype-positive patients described in this first study were homozygous or compound heterozygous TRDN mutation carriers. In vitro and in vivo studies showed that all the mutations tested lead to protein instability and subsequent degradation [54]. The CPVT subjects are effectively triadin knockout cases, characterized by the absence of TRDN protein Ca^{2+} release units.

Since the calcium releasing complex is primarily responsible for the EC coupling via L-type calcium and RyR2 channels (Fig. 1.2), it is quite obvious that *TRDN* mutations impairing the Ca^{2+} release units resemble in part the CPVT phenotype [144]. This concept has been clearly proven with the aid of mouse models, showing *TRDN* mutations contribute both to create a cellular Ca^{2+} overload and a destabilization of CASQ2 anchoring [145]. The mechanism by which *TRDN* mutations lead to LQTS phenotype is less clear. Indeed, triadin knockout syndrome has also been associated more recently with autosomal recessive LQTS phenotype [55]. So far, only two isolated descriptions of *TRDN*-mutated subjects with early disease manifestation are present in literature.

The first patient is a girl of 10 years who experienced syncope at 1 and 2 years of age combined with a QTc of 500 ms, extensive T-wave inversions in precordial leads V1 through V4, and severe disease expression of exercise-induced cardiac arrest in early childhood [55]. A child-parent trio WES analysis was performed, and the *TRDN*-p.D18fs*13 frameshift mutation was identified in the proband in homozygous state [55]. Both unaffected parents were heterozygous for the mutation. Surprisingly, the mutation identified in this LQTS proband was identical to the one detected in the first CPVT cohort [54], and the two patients effectively share some common phenotypic features. This specific mutation was further identified in two siblings who had an out-of-hospital cardiac arrest at 2 years of age and subsequent recurrent episodes of ventricular fibrillation with documented QT prolongation [146]. The two siblings are both compound heterozygous carriers of the previously reported *TRDN-*

p.D18fs*13 nonsense mutation and the novel p.Glu168* mutation. The authors concluded that the combination of the two mutations is the likely explanation of the particularly malignant phenotype, with cardiac arrest occurring at a very young age and recurrent episodes of VF despite beta-blocker therapy [146]. None of the parents, carrying either one or the other mutation in isolation, was symptomatic.

These first reports showed that *TRDN* mutations may be associated both with LQTS and CPVT phenotypes. A possible explanation could be found in the hypothesized arrhythmic mechanism induced by TRDN loss, which seems to involve L-type calcium channel inactivation regulation rather than SR Ca^{2+} leak and triggered beats typical of CPVT mutations in *RYR2* and *CASQ2* [144]. According to this hypothesis, the main consequence of TRDN loss would be the reduction of the negative feedback on the L-type calcium channel, resulting in Ca^{2+} overload and increasing SR Ca^{2+} release (Fig. 1.2).

1.3.3 Sudden Infant Death Syndrome (SIDS) and Intrauterine Fetal Death (IUFD): Role of Long QT Syndrome

Sudden unexpected death of an infant occurring within the first year of life, and without identification of a possible cause despite extensive postmortem examination, is classified as SIDS (sudden infant death syndrome), while intrauterine fetal death (IUFD) refers to fetal losses occurred after the 14th week of gestation and includes late miscarriages and stillbirths. The two phenomena are very complex and have been extensively studied over the years with the aim to identify the most common causes of death, the main underlying diseases, and the maternal or environmental risk factors [147, 148].

Sudden infant death syndrome is actually considered the leading cause of mortality in the first year of life, with a prevalence of 0.5 per 1000 live births [149]. The hypothesis that a fraction of SIDS could be due to cardiac causes was advanced in 1976 by Schwartz, who suggested that LQTS could be responsible for some of these cases [150]. The proof of concept came some years later with the description of an infant resuscitated from ventricular fibrillation that would have been a typical SIDS case without prompt intervention. Molecular screening identified a mutation in *SCN5A*, one of the main genes associated with LQTS [95]. It is now well known that cardiac channelopathies and LQTS in particular may account for 10–15% of SIDS cases [151–154]. Most of the functional variants identified in population-based SIDS cohorts are related to sodium channel macromolecular complex [155]. However, *SCN5A* genetic variants identified in SIDS cases have different features (i.e., topological distribution in the Na^+ channel and prevalence in the general population) compared to *SCN5A* variants observed in LQT3 patients with documented perinatal arrhythmias. Specifically, almost all LQT3 variants were absent in the general population and were located in the most important functional regions of the channel (transmembrane and linker regions), while nearly 50% of SIDS variants were also present in the general population and spanned across N-terminal and linker regions of the protein where variants present in controls are more frequently observed

[155]. This difference points out the possibility that LQT3 variants identified in perinatal LQTS cases may be able per se to cause a severe LQTS phenotype manifesting early in life (in line with data presented in Sect. 1.3.1), while *SCN5A* variants identified in SIDS may more likely be considered as favoring factors contributing to SCD according to the triple risk hypothesis [156].

More recently, other population-based studies aimed to determine the prevalence of LQTS mutations in SIDS have been published [157, 158]. The first collected a large multi-ethnic population of sudden unexplained deaths, part of which are SIDS ($n = 141$). The screening of the main channelopathies genes brought results that are in line with previous studies: nearly 13.5% of SIDS were genotype-positive, with a majority of favoring variants in *SCN5A* gene [157]. The second study reported the same percentage of genotype-positives (around 10%) only for a subgroup of SIDS cases previously selected for genetic testing by an expert cardiac team [158]. The LQTS molecular autopsy has not proven to be useful in the rest of the cohort (unselected cases), where more common risk factors have been reported as potential causes of SIDS [158].

A third population-based study attempted to highlight the role of different cardiovascular disease in SIDS [159]:155 European SIDS cases were analyzed through WES with a focus on 192 genes associated with cardiovascular and metabolic diseases. Among SIDS infants with likely causative variants, 9% carried variants in channelopathies genes, 7% in cardiomyopathies genes, and 2.5% in genes associated with mitral valve prolapse, aortic valve disease, Marfan syndrome, or Ehlers-Danlos syndrome [159]. This comprehensive study confirmed the role of channelopathies as an underlying cause of SIDS and pointed out the potential contribution of cardiomyopathies to the phenomenon.

Finally, the role of genetic heart diseases in SIDS pathogenesis was explored through WES in the largest SIDS cohort published to date ($n = 419$) [160]; 12.6% of SIDS cases hosted at least one potentially informative variant in genetic heart disease-associated genes ($n = 90$). The percentage reduced to 4.3% of cases (18/419) considering only carriers of clinically actionable variants, i.e., pathogenic or likely pathogenic variants according to the American College of Medical Genetics and Genomics guidelines. This stringent approach allowed to recognize SIDS cases that may be safely attributed to genetic heart disease as monogenic cause of SIDS and to discriminate clinically actionable variants from those with questionable pathogenicity [161, 162]. However, the study included also an additional analysis showing the overrepresentation of ultra-rare channelopathies-associated variants in European SIDS cases versus European control subjects [160]. Although the percentage of SIDS that may now be attributed to channelopathies may be lower than previously assumed, the ultra-rare variant burden observed in SIDS population suggests that variants with milder effect may act synergistically with other factors to contribute to SIDS and still represent the underlying pathological basis for some of SIDS cases [163]. Overall, genetic factors able to modulate arrhythmia susceptibility and cardiovascular impairment should be regarded as a possible explanation in SIDS cases that remain elusive despite extensive postmortem investigations.

Sudden cardiac death during fetal life is more difficult to be investigated; therefore genetic studies supporting the cardiac theory are very few, and all focused on stillbirths. Stillbirth, defined as a fetal death occurring in the late gestational period after 20 weeks of gestation, has an incidence of 5.96 per 1000 live births [164]. A possible cause of the demise can be currently identified in 70% of cases [148]. The genetic studies aimed to find a possible contribution of arrhythmogenic diseases in stillbirth focusing on the 30% of IUFDs that remains without an explanation.

The first descriptions available in literature concerned anecdotal stillbirth cases, in which a mutation in one of the three main LQTS genes was identified [101, 108, 165]. The first stillbirth case, described by Hoorntje et al., carried a homozygous nonsense mutation in *KCNH2* gene, also identified in the sister born premature in distress due to ventricular arrhythmia in the presence of severe QT prolongation [108]. The second report concerned three siblings, two died before birth and one presenting a malignant perinatal LQTS form due to *SCN5A*-p.Arg1623Gln mutation, also carried by the mother who did not manifest the disease because of mosaicism [101]. The third description presented a family with recurrent fetal losses due to a homozygous nonsense mutation (p.Gln1070*) in *KCNH2* gene [165]. More recently, another isolate stillbirth case has been described: a woman was referred for genetic testing after the death of her 2-day-old infant and spontaneous abortion of a second baby in the first trimester [58]. The newborn infant had incessant ventricular tachycardia while in utero and a prolonged QTc (560 ms), whereas the mother was asymptomatic but displayed a prolonged QTc. The genetic analysis identified a nonsense heterozygous *KCNH2* mutation in the mother and a common *KCNH2* polymorphism with a demonstrated modifier role in the father, both inherited by the newborn and the fetus. The stronger impairment of the channel (loss of function) shown in vitro with the co-expression of both variants was retained the most likely explanation for the severe phenotype observed in children versus the mother [58].

The major study so far designed to evaluate the prevalence of LQTS mutations in a population-based IUFD cohort was published by our group in collaboration with Mayo Clinic in 2013 [166]. The study demonstrated that nearly 9% of IUFD cases (8/91) carried functional variants in the main 3 LQTS genes (*KCNQ1*, *KCNH2*, *SCN5A*) [166]. Three variants in *KCNQ1* and *KCNH2* have been classified as putative LQTS susceptibility mutations, on the basis of in vitro studies showing a loss-of-function effect consistent with LQT1 and LQT2 phenotypes, and a heterozygous frequency in publicly available databases of general population less than LQTS disease prevalence (0.05%); additionally, other five variants in *SCN5A* have been considered as predisposing factors [166]. Interestingly, the fact that two out of three putative LQTS susceptibility mutations with a demonstrated functional effect are still absent (*KCNQ1*-p.Ala283Thr) or identified in a few over 100,000 subjects (*KCNH2*[1b]-p.Arg25Trp) in online exome and whole-genome database including thousands of individuals (gnomAD browser, Broad Institute, November 2018 accessed) further supports their pathogenic role. Moreover, the impact of *KCNH2* [1b]-p.Arg25Trp mutation on cardiac I_{Kr} current and AP behavior has been further explored by another group [167]: they first confirmed our previous electrophysiological findings in HEK293 cells, and then they demonstrated for the first time that the dominant-negative effect of p.Arg25Trp was reproduced in iPS-CMCs, reflecting cellular manifestations of pro-arrhythmia [167].

1.4   Conclusions

The long QT syndrome is a channelopathy caused by different mutations in ion channel genes. Clinical manifestation can be very different ranging from no symptoms to sudden cardiac death. Such a clinical variability can be observed even among family members carrying the same disease-causing mutation, and genetic modifiers may play a role, influencing the arrhythmic risk. However, there are some subtypes of LQTS, mainly due mutations in genes encoding the calcium channel (i.e., Timothy syndrome) or protein modulating calcium channel activity (i.e., CaM and TRDN) in which the phenotype is consistently severe, with life-threatening arrhythmias occurring very early in life and a high incidence of SCD. In these cases, mutations are frequently de novo, and the role of genetic modifiers is probably less relevant.

References

1. Schwartz PJ, Malliani A. Electrical alternation of the T wave. Clinical and experimental evidence of its relationship with the sympathetic nervous system and with the long QT syndrome. Am Heart J. 1975;89:45–50.
2. Tester DJ, Ackerman MJ. Postmortem long QT syndrome genetic testing for sudden unexplained death in the young. J Am Coll Cardiol. 2007;49:240–6.
3. Ward OC. A new familial cardiac syndrome in children. J Ir Med Assoc. 1964;54:103–6.
4. Schwartz PJ, Stramba-Badiale M, Crotti L, Pedrazzini M, Besana A, Bosi G, Gabbarini F, Goulene K, Insolia R, Mannarino S, Mosca F, Nespoli L, Rimini A, Rosati E, Salice P, Spazzolini C. Prevalence of the congenital long-QT syndrome. Circulation. 2009;120:1761–7.
5. Jervell A, Lange-Nielsen F. Congenital deaf-mutism, functional heart disease with prolongation of the Q-T interval, and sudden death. Am Heart J. 1957;54:59–68.
6. Schwartz PJ, Spazzolini C, Crotti L, Bathen J, Amlie JP, Timothy K, Shkolnikova M, Berul CI, Bitner-Glindzicz M, Toivonen L, Horie M, Schulze-Bahr E, Denjoy I. The Jervell and Lange-Nielsen syndrome. Natural history, molecular basis, and clinical outcome. Circulation. 2006;113:783–90.
7. Lipka LJ, Dizon JM, Reiffel JA. Desired mechanisms of drugs for ventricular arrhythmia; class III antiarrhythmic agents. Am Heart J. 1995;130:632–40.
8. Itoh H, Crotti L, Aiba T, Spazzolini C, Denjoy I, Fressart V, Hayashi K, Nakajima T, Ohno S, Makiyama T, Wu J, Hasegawa K, Mastantuono E, Dagradi F, Pedrazzini M, Yamagishi M, Berthet M, Murakami Y, Shimizu W, Guicheney P, Schwartz PJ, Horie M. The genetics underlying acquired long QT syndrome: impact for genetic screening. Eur Heart J. 2016;37 (18):1456–64.
9. Arking DE, Pulit SL, Crotti L, van der Harst P, Munroe PB, Koopmann TT, Sotoodehnia N, Rossin EJ, Morley M, Wang X, Johnson AD, Lundby A, Gudbjartsson DF, Noseworthy PA, Eijgelsheim M, Bradford Y, Tarasov KV, Dörr M, Müller-Nurasyid M, Lahtinen AM, Nolte IM, Smith AV, Bis JC, Isaacs A, Newhouse SJ, Evans DS, Post WS, Waggott D, Lyytikäinen LP, Hicks AA, Eisele L, Ellinghaus D, Hayward C, Navarro P, Ulivi S, Tanaka T, Tester DJ, Chatel S, Gustafsson S, Kumari M, Morris RW, Naluai ÅT, Padmanabhan S, Kluttig A,

Strohmer B, Panayiotou AG, Torres M, Knoflach M, Hubacek JA, Slowikowski K, Raychaudhuri S, Kumar RD, Harris TB, Launer LJ, Shuldiner AR, Alonso A, Bader JS, Ehret G, Huang H, Kao WH, Strait JB, Macfarlane PW, Brown M, Caulfield MJ, Samani NJ, Kronenberg F, Willeit J, CARe Consortium, COGENT Consortium, Smith JG, Greiser KH, Meyer Zu Schwabedissen H, Werdan K, Carella M, Zelante L, Heckbert SR, Psaty BM, Rotter JI, Kolcic I, Polašek O, Wright AF, Griffin M, Daly MJ, DCCT/EDIC, Arnar DO, Hólm H, Thorsteinsdottir U, eMERGE Consortium, Denny JC, Roden DM, Zuvich RL, Emilsson V, Plump AS, Larson MG, O'Donnell CJ, Yin X, Bobbo M, D'Adamo AP, Iorio A, Sinagra G, Carracedo A, Cummings SR, Nalls MA, Jula A, Kontula KK, Marjamaa A, Oikarinen L, Perola M, Porthan K, Erbel R, Hoffmann P, Jöckel KH, Kälsch H, Nöthen MM, HRGEN Consortium, den Hoed M, Loos RJ, Thelle DS, Gieger C, Meitinger T, Perz S, Peters A, Prucha H, Sinner MF, Waldenberger M, de Boer RA, Franke L, van der Vleuten PA, Beckmann BM, Martens E, Bardai A, Hofman N, Wilde AA, Behr ER, Dalageorgou C, Giudicessi JR, Medeiros-Domingo A, Barc J, Kyndt F, Probst V, Ghidoni A, Insolia R, Hamilton RM, Scherer SW, Brandimarto J, Margulies K, Moravec CE, del Greco MF, Fuchsberger C, O'Connell JR, Lee WK, Watt GC, Campbell H, Wild SH, El Mokhtari NE, Frey N, Asselbergs FW, Mateo Leach I, Navis G, van den Berg MP, van Veldhuisen DJ, Kellis M, Krijthe BP, Franco OH, Hofman A, Kors JA, Uitterlinden AG, Witteman JC, Kedenko L, Lamina C, Oostra BA, Abecasis GR, Lakatta EG, Mulas A, Orrú M, Schlessinger D, Uda M, Markus MR, Völker U, Snieder H, Spector TD, Ärnlöv J, Lind L, Sundström J, Syvänen AC, Kivimaki M, Kähönen M, Mononen N, Raitakari OT, Viikari JS, Adamkova V, Kiechl S, Brion M, Nicolaides AN, Paulweber B, Haerting J, Dominiczak AF, Nyberg F, Whincup PH, Hingorani AD, Schott JJ, Bezzina CR, Ingelsson E, Ferrucci L, Gasparini P, Wilson JF, Rudan I, Franke A, Mühleisen TW, Pramstaller PP, Lehtimäki TJ, Paterson AD, Parsa A, Liu Y, van Duijn CM, Siscovick DS, Gudnason V, Jamshidi Y, Salomaa V, Felix SB, Sanna S, Ritchie MD, Stricker BH, Stefansson K, Boyer LA, Cappola TP, Olsen JV, Lage K, Schwartz PJ, Kääb S, Chakravarti A, Ackerman MJ, Pfeufer A, de Bakker PI, Newton-Cheh C. Genetic association study of QT interval highlights role for calcium signaling pathways in myocardial repolarization. Nat Genet. 2014;46(8):826–36.
10. Strauss DG, Vicente J, Johannesen L, Blinova K, Mason JW, Weeke P, Behr ER, Roden DM, Woosley R, Kosova G, Rosenberg MA, Newton-Cheh C. Common genetic variant risk score is associated with drug-induced QT prolongation and torsade de pointes risk: a pilot study. Circulation. 2017;135(14):1300–10.
11. Chockalingam P, Crotti L, Girardengo G, Johnson JN, Harris KM, van der Heijden JF, Hauer RN, Beckmann BM, Spazzolini C, Rordorf R, Rydberg A, Clur SA, Fischer M, van den Heuvel F, Kääb S, Blom NA, Ackerman MJ, Schwartz PJ, Wilde AA. Not all beta-blockers are equal in the management of long QT syndrome types 1 and 2: higher recurrence of events under metoprolol. J Am Coll Cardiol. 2012;60(20):2092–9.
12. Schwartz PJ, Priori SG, Cerrone M, Spazzolini C, Odero A, Napolitano C, Bloise R, De Ferrari GM, Klersy C, Moss AJ, Zareba W, Robinson JL, Hall WJ, Brink PA, Toivonen L, Epstein AE, Li C, Hu D. Left cardiac sympathetic denervation in the management of high-risk patients affected by the long-QT syndrome. Circulation. 2004;109(15):1826–33.
13. Schwartz PJ, Spazzolini C, Priori SG, Crotti L, Vicentini A, Landolina M, Gasparini M, Wilde AA, Knops RE, Denjoy I, Toivonen L, Mönnig G, Al-Fayyadh M, Jordaens L, Borggrefe M, Holmgren C, Brugada P, De Roy L, Hohnloser SH, Brink PA. Who are the long-QT syndrome patients who receive an implantable cardioverter-defibrillator and what happens to them?: data from the European long-QT syndrome implantable cardioverter-defibrillator (LQTS ICD) Registry. Circulation. 2010;122(13):1272–82.
14. Spazzolini C, Mullally J, Moss AJ, Schwartz PJ, McNitt S, Ouellet G, Fugate T, Goldenberg I, Jons C, Zareba W, Robinson JL, Ackerman MJ, Benhorin J, Crotti L, Kaufman ES, Locati EH, Qi M, Napolitano C, Priori SG, Towbin JA, Vincent GM. Clinical implications for patients with long QT syndrome who experience a cardiac event during infancy. J Am Coll Cardiol. 2009;54(9):832–7.

15. Wang Q, Curran ME, Splawski I, Burn TC, Millholland JM, VanRaay TJ, Shen J, Timothy KW, Vincent GM, de Jager T, Schwartz PJ, Toubin JA, Moss AJ, Atkinson DL, Landes GM, Connors TD, Keating MT. Positional cloning of a novel potassium channel gene: KVLQT1 mutations cause cardiac arrhythmias. Nat Genet. 1996;12:17–23.

16. Curran ME, Splawski I, Timothy KW, Vincent GM, Green ED, Keating MT. A molecular basis for cardiac arrhythmia: HERG mutations cause long QT syndrome. Cell. 1995;80:795–803.

17. Wang Q, Shen J, Splawski I, Atkinson D, Li Z, Robinson JL, Moss AJ, Towbin JA, Keating MT. SCN5A mutations associated with an inherited cardiac arrhythmia, long QT syndrome. Cell. 1995;80:805–11.

18. Kapplinger JD, Tester DJ, Salisbury BA, Carr JL, Harris-Kerr C, Pollevick GD, Wilde AA, Ackerman MJ. Spectrum and prevalence of mutations from the first 2,500 consecutive unrelated patients referred for the FAMILION long QT syndrome genetic test. Heart Rhythm. 2009;6:1297–303.

19. Moss AJ, Shimizu W, Wilde AA, Towbin JA, Zareba W, Robinson JL, Qi M, Vincent GM, Ackerman MJ, Kaufman ES, Hofman N, Seth R, Kamakura S, Miyamoto Y, Goldenberg I, Andrews ML, McNitt S. Clinical aspects of type-1 long-QT syndrome by location, coding type, and biophysical function of mutations involving the KCNQ1 gene. Circulation. 2007;115(19):2481–9.

20. Smith JL, Anderson CL, Burgess DE, Elayi CS, January CT, Delisle BP. Molecular pathogenesis of long QT syndrome type 2. J Arrhythm. 2016;32:373–80.

21. Makita N, Behr E, Shimizu W, Horie M, Sunami A, Crotti L, Schulze-Bahr E, Fukuhara S, Mochizuki N, Makiyama T, Itoh H, Christiansen M, McKeown P, Miyamoto K, Kamakura S, Tsutsui H, Schwartz PJ, George AL Jr, Roden DM. The E1784K mutation in SCN5A is associated with mixed clinical phenotype of type 3 long QT syndrome. J Clin Invest. 2008;118 (6):2219–29.

22. Schwartz PJ, Crotti L. Founder populations with channelopathies and church records reveal all sorts of interesting secrets: some are scientifically relevant. Heart Rhythm. 2017;14 (12):1882–3.

23. Schwartz PJ, Crotti L. Long QT and short QT syndromes. In: Zipes DP, Jalife J, editors. Cardiac electrophysiology: from cell to bedside. 7th ed. Philadelphia: Elsevier/Saunders; 2017. p. 893–904.

24. Schwartz PJ, Priori SG, Spazzolini C, Moss AJ, Vincent GM, Napolitano C, Denjoy I, Guicheney P, Breithardt G, Keating MT, Towbin JA, Beggs AH, Brink P, Wilde AA, Toivonen L, Zareba W, Robinson JL, Timothy KW, Corfield V, Wattanasirichaigoon D, Corbett C, Haverkamp W, Schulze-Bahr E, Lehmann MH, Schwartz K, Coumel P, Bloise R. Genotype-phenotype correlation in the long-QT syndrome: gene-specific triggers for life-threatening arrhythmias. Circulation. 2001;103(1):89–95.

25. Priori SG, Schwartz PJ, Napolitano C, Bloise R, Ronchetti E, Grillo M, Vicentini A, Spazzolini C, Nastoli J, Bottelli G, Folli R, Cappelletti D. Risk stratification in the long-QT syndrome. N Engl J Med. 2003;348(19):1866–74.

26. Sauer AJ, Moss AJ, McNitt S, Peterson DR, Zareba W, Robinson JL, Qi M, Goldenberg I, Hobbs JB, Ackerman MJ, Benhorin J, Hall WJ, Kaufman ES, Locati EH, Napolitano C, Priori SG, Schwartz PJ, Towbin JA, Vincent GM, Zhang L. Long QT syndrome in adults. J Am Coll Cardiol. 2007;49(3):329–37.

27. Moss AJ, Zareba W, Kaufman ES, Gartman E, Peterson DR, Benhorin J, Towbin JA, Keating MT, Priori SG, Schwartz PJ, Vincent GM, Robinson JL, Andrews ML, Feng C, Hall WJ, Medina A, Zhang L, Wang Z. Increased risk of arrhythmic events in long-QT syndrome with mutations in the pore region of the human ether-a-go-go-related gene potassium channel. Circulation. 2002;105:794–9.

28. Crotti L, Spazzolini C, Schwartz PJ, Shimizu W, Denjoy I, Schulze-Bahr E, Zaklyazminskaya EV, Swan H, Ackerman MJ, Moss AJ, Wilde AA, Horie M, Brink PA, Insolia R, De Ferrari GM, Crimi G. The common long-QT syndrome mutation KCNQ1/A341V causes unusually

severe clinical manifestations in patients with different ethnic backgrounds: toward a mutation-specific risk stratification. Circulation. 2007;116(21):2366–75.

29. Barhanin J, Lesage F, Guillemare E, Fink M, Lazdunski M, Romey G. K(V)LQT1 and lsK (minK) proteins associate to form the I(Ks) cardiac potassium current. Nature. 1996;384 (6604):78–80.

30. Schulze-Bahr E, Wang Q, Wedekind H, Haverkamp W, Chen Q, Sun Y, Rubie C, Hordt M, Towbin J, Borggrefe M. KCNE1 mutations cause Jervell and Lange-Nielsen syndrome. Nat Genet. 1997;17:267–8.

31. Abbott GW, Sesti F, Splawski I, Buck ME, Lehmann MH, Timothy KW, Keating MT, Goldstein SA. MiRP1 forms IKr potassium channels with HERG and is associated with cardiac arrhythmia. Cell. 1999;97(2):175–87.

32. Sesti F, Abbott GW, Wei J, Murray KT, Saksena S, Schwartz PJ, Priori SG, Roden DM, George ALJ, Goldstein SAN. A common polymorphism associated with antibiotic-induced cardiac arrhythmia. Proc Natl Acad Sci U S A. 2000;97:10613–8.

33. Medeiros-Domingo A, Kaku T, Tester DJ, Iturralde-Torres P, Itty A, Ye B, Valdivia C, Ueda K, Canizales-Quinteros S, Tusie-Luna MT, Makielski JC, Ackerman MJ. SCN4B-encoded sodium channel β4 subunit in congenital long-QT syndrome. Circulation. 2007;116:134–42.

34. Vatta M, Ackerman MJ, Ye B, Makielski JC, Ughanze EE, Taylor EW, Tester DJ, Balijepalli RC, Foell JD, Li Z, Kamp TJ, Towbin JA. Mutant caveolin-3 induces persistent late sodium current and is associated with long-QT syndrome. Circulation. 2006;114:2104–12.

35. Ueda K, Valdivia C, Medeiros-Domingo A, Tester DJ, Vatta M, Farrugia G, Ackerman MJ, Makielski JC. Syntrophin mutation associated with long QT syndrome through activation of the nNOS-SCN5A macromolecular complex. Proc Natl Acad Sci U S A. 2008;105:9355–60.

36. Chen L, Marquardt ML, Tester DJ, Sampson KJ, Ackerman MJ, Kass RS. Mutation of an A-kinase-anchoring protein causes long-QT syndrome. Proc Natl Acad Sci U S A. 2007;104:20990–5.

37. Yang Y, Yang Y, Liang B, Liu J, Li J, Grunnet M, Olesen SP, Rasmussen HB, Ellinor PT, Gao L, Lin X, Li L, Wang L, Xiao J, Liu Y, Liu Y, Zhang S, Liang D, Peng L, Jespersen T, Chen YH. Identification of a Kir3.4 mutation in congenital long QT syndrome. Am J Hum Genet. 2010;86:872–80.

38. Mohler PJ, Schott JJ, Gramolini AO, Dilly KW, Guatimosim S, duBell WH, Song LS, Haurogné K, Kyndt F, Ali ME, Rogers TB, Lederer WJ, Escande D, Le Marec H, Bennett V. Ankyrin-B mutation causes type 4 long-QT cardiac arrhythmia and sudden cardiac death. Nature. 2003;421(6923):634–9.

39. Plaster NM, Tawil R, Tristani-Firouzi M, Canún S, Bendahhou S, Tsunoda A, Donaldson MR, Iannaccone ST, Brunt E, Barohn R, Clark J, Deymeer F, George AL Jr, Fish FA, Hahn A, Nitu A, Ozdemir C, Serdaroglu P, Subramony SH, Wolfe G, Fu YH, Ptácek LJ. Mutations in Kir2.1 cause the developmental and episodic electrical phenotypes of Andersen's syndrome. Cell. 2001;105(4):511–9.

40. Marks ML, Whisler SL, Clericuzio C, Keating M. A new form of long QT syndrome associated with syndactyly. J Am Coll Cardiol. 1995;25(1):59–64.

41. Reichenbach H, Meister EM, Theile H. The heart-hand syndrome. A new variant of disorders of heart conduction and syndactylia including osseous changes in hands and feet. Kinderarztl Prax. 1992;60(2):54–6.

42. Splawski I, Timothy KW, Sharpe LM, Decher N, Kumar P, Bloise R, Napolitano C, Schwartz PJ, Joseph RM, Condouris K, Tager-Flusberg H, Priori SG, Sanguinetti MC, Keating MT. Ca (V)1.2 calcium channel dysfunction causes a multisystem disorder including arrhythmia and autism. Cell. 2004;119(1):19–31.

43. Betzenhauser MJ, Pitt GS, Antzelevitch C. Calcium channel mutations in cardiac arrhythmia syndromes. Curr Mol Pharmacol. 2015;8(2):133–42.

44. Crotti L, Johnson CN, Graf E, De Ferrari GM, Cuneo BF, Ovadia M, Papagiannis J, Feldkamp MD, Rathi SG, Kunic JD, Pedrazzini M, Wieland T, Lichtner P, Beckmann BM, Clark T,

Shaffer C, Benson DW, Kääb S, Meitinger T, Strom TM, Chazin WJ, Schwartz PJ, George AL Jr. Calmodulin mutations associated with recurrent cardiac arrest in infants. Circulation. 2013;127(9):1009–17.

45. Boczek NJ, Gomez-Hurtado N, Ye D, Calvert ML, Tester DJ, Kryshtal DO, Hwang HS, Johnson CN, Chazin WJ, Loporcaro CG, Shah M, Papez AL, Lau YR, Kanter R, Knollmann BC, Ackerman MJ. Spectrum and prevalence of CALM1-, CALM2-, and CALM3-encoded calmodulin variants in long QT syndrome and functional characterization of a novel long QT syndrome-associated calmodulin missense variant, E141G. Circ Cardiovasc Genet. 2016;9 (2):136–46.

46. Reed GJ, Boczek NJ, Etheridge SP, Ackerman MJ. CALM3 mutation associated with long QT syndrome. Heart Rhythm. 2015;12(2):419–22.

47. Pipilas DC, Johnson CN, Webster G, Schlaepfer J, Fellmann F, Sekarski N, Wren LM, Ogorodnik KV, Chazin DM, Chazin WJ, Crotti L, Bhuiyan ZA, George AL Jr. Novel calmodulin mutations associated with congenital long QT syndrome affect calcium current in human cardiomyocytes. Heart Rhythm. 2016;13(10):2012–9.

48. Chaix MA, Koopmann TT, Goyette P, Alikashani A, Latour F, Fatah M, Hamilton RM, Rioux JD. Novel CALM3 mutations in pediatric long QT syndrome patients support a CALM3-specific calmodulinopathy. HeartRhythm Case Rep. 2016;2(3):250–4.

49. Crotti L, Spazzolini C, Tester DJ, Ghidoni A, Baruteau A, Beckmann BM, Behr ER, Bennett JS, Bezzina CR, Bhuiyan ZA, Celiker A, Cerrone M, Dagradi F, De Ferrari GM, Etheridge SP, Fatah M, Garcia-Pavia P, Al-Ghamdi S, Hamilton RM, Al-Hassnan ZN, Horie M, Jimenez-Jaimez J, Kanter RJ, Kaski JP, Kotta MC, Lahrouchi N, Makita N, Norrish G, Odland HH, Ohno S, Papagiannis J, Parati G, Sekarski N, Tveten K, Vatta M, Webster G, Wilde AAM, Wojciak J, George AL, Ackerman MJ, Schwartz PJ. Calmodulin mutations and life-threatening cardiac arrhythmias: insights from the International Calmodulinopathy Registry. Eur Heart J. 2019. pii: ehz311.

50. Chin D, Means AR. Calmodulin: a prototypical calcium sensor. Trends Cell Biol. 2000;10:322–8.

51. Maier LS, Bers DM. Role of Ca2+/calmodulin-dependent protein kinase (CaMK) in excitation-contraction coupling in the heart. Cardiovasc Res. 2007;73(4):631–40.

52. Fischer R, Koller M, Flura M, Mathews S, Strehler-Page MA, Krebs J, Penniston JT, Carafoli E, Strehler EE. Multiple divergent mRNAs code for a single human calmodulin. J Biol Chem. 1988;263(32):17055–62.

53. Nyegaard M, Overgaard MT, Søndergaard MT, Vranas M, Behr ER, Hildebrandt LL, Lund J, Hedley PL, Camm AJ, Wettrell G, Fosdal I, Christiansen M, Børglum AD. Mutations in calmodulin cause ventricular tachycardia and sudden cardiac death. Am J Hum Genet. 2012;91 (4):703–12.

54. Roux-Buisson N, Cacheux M, Fourest-Lieuvin A, Fauconnier J, Brocard J, Denjoy I, Durand P, Guicheney P, Kyndt F, Leenhardt A, Le Marec H, Lucet V, Mabo P, Probst V, Monnier N, Ray PF, Santoni E, Trémeaux P, Lacampagne A, Fauré J, Lunardi J, Marty I. Absence of triadin, a protein of the calcium release complex, is responsible for cardiac arrhythmia with sudden death in human. Hum Mol Genet. 2012;21(12):2759–67.

55. Altmann HM, Tester DJ, Will ML, Middha S, Evans JM, Eckloff BW, Ackerman MJ. Homozygous/compound heterozygous triadin mutations associated with autosomal-recessive long-QT syndrome and pediatric sudden cardiac arrest: elucidation of the triadin knockout syndrome. Circulation. 2015;131(23):2051–60.

56. Priori SG, Napolitano C, Schwartz PJ. Low penetrance in the long-QT syndrome: clinical impact. Circulation. 1999;99:529–33.

57. Crotti L, Lundquist AL, Insolia R, Pedrazzini M, Ferrandi C, De Ferrari GM, Vicentini A, Yang P, Roden DM, George AL Jr, Schwartz PJ. KCNH2-K897T is a genetic modifier of latent congenital long-QT syndrome. Circulation. 2005;112(9):1251–8.

58. Nof E, Cordeiro JM, Pérez GJ, Scornik FS, Calloe K, Love B, Burashnikov E, Caceres G, Gunsburg M, Antzelevitch C. A common single nucleotide polymorphism can exacerbate

long-QT type 2 syndrome leading to sudden infant death. Circ Cardiovasc Genet. 2010;3 (2):199–206.

59. Paavonen KJ, Chapman H, Laitinen PJ, Fodstad H, Piippo K, Swan H, Toivonen L, Viitasalo M, Kontula K, Pasternack M. Functional characterization of the common amino acid 897 polymorphism of the cardiac potassium channel KCNH2 (HERG). Cardiovasc Res. 2003;59(3):603–11.

60. Pfeufer A, Jalilzadeh S, Perz S, Mueller JC, Hinterseer M, Illig T, Akyol M, Huth C, Schöpfer-Wendels A, Kuch B, Steinbeck G, Holle R, Näbauer M, Wichmann HE, Meitinger T, Kääb S. Common variants in myocardial ion channel genes modify the QT interval in the general population: results from the KORA study. Circ Res. 2005;96(6):693–701.

61. Pfeufer A, Sanna S, Arking DE, Müller M, Gateva V, Fuchsberger C, Ehret GB, Orrú M, Pattaro C, Köttgen A, Perz S, Usala G, Barbalic M, Li M, Pütz B, Scuteri A, Prineas RJ, Sinner MF, Gieger C, Najjar SS, Kao WH, Mühleisen TW, Dei M, Happle C, Möhlenkamp S, Crisponi L, Erbel R, Jöckel KH, Naitza S, Steinbeck G, Marroni F, Hicks AA, Lakatta E, Müller-Myhsok B, Pramstaller PP, Wichmann HE, Schlessinger D, Boerwinkle E, Meitinger T, Uda M, Coresh J, Kääb S, Abecasis GR, Chakravarti A. Common variants at ten loci modulate the QT interval duration in the QTSCD study. Nat Genet. 2009;41 (4):407–14.

62. Newton-Cheh C, Guo CY, Larson MG, Musone SL, Surti A, Camargo AL, Drake JA, Benjamin EJ, Levy D, D'Agostino RB Sr, Hirschhorn JN, O'Donnell CJ. Common genetic variation in KCNH2 is associated with QT interval duration: the Framingham Heart Study. Circulation. 2007;116(10):1128–36. Erratum in: Circulation. 2008 Jan 1/8;117(1):e9.

63. Bezzina CR, Verkerk AO, Busjahn A, Jeron A, Erdmann J, Koopmann TT, Bhuiyan ZA, Wilders R, Mannens MM, Tan HL, Luft FC, Schunkert H, Wilde AA. A common polymorphism in KCNH2 (HERG) hastens cardiac repolarization. Cardiovasc Res. 2003;59(1):27–36.

64. Ye B, Valdivia CR, Ackerman MJ, Makielski JC. A common human SCN5A polymorphism modifies expression of an arrhythmia causing mutation. Physiol Genomics. 2003;12:187–93.

65. Shinlapawittayatorn K, Du XX, Liu H, Ficker E, Kaufman ES, Deschênes I. A common SCN5A polymorphism modulates the biophysical defects of SCN5A mutations. Heart Rhythm. 2011;8:455–62.

66. Kapplinger JD, Erickson A, Asuri S, Tester DJ, McIntosh S, Kerr CR, Morrison J, Tang A, Sanatani S, Arbour L, Ackerman MJ. KCNQ1 p.L353L affects splicing and modifies the phenotype in a founder population with long QT syndrome type 1. J Med Genet. 2017;54 (6):390–8.

67. Paulussen AD, Gilissen RA, Armstrong M, Doevendans PA, Verhasselt P, Smeets HJ, Schulze-Bahr E, Haverkamp W, Breithardt G, Cohen N, Aerssens J. Genetic variations of KCNQ1, KCNH2, SCN5A, KCNE1 and KCNE2 in drug-induced long QT syndrome patients. J Mol Med. 2004;82:182–8.

68. Nishio Y, Makiyama T, Itoh H, Sakaguchi T, Ohno S, Gong YZ, Yamamoto S, Ozawa T, Ding WG, Toyoda F, Kawamura M, Akao M, Matsuura H, Kimura T, Kita T, Horie M. D85N, a KCNE1 polymorphism, is a disease causing gene variant in long QT syndrome. J Am Coll Cardiol. 2009,54.812–9.

69. Lahtinen AM, Marjamaa A, Swan H, Kontula K. KCNE1 D85N polymorphism—a sex-specific modifier in type 1 long QT syndrome? BMC Med Genet. 2011;12:11.

70. Kolder IC, Tanck MW, Postema PG, Barc J, Sinner MF, Zumhagen S, Husemann A, Stallmeyer B, Koopmann TT, Hofman N, Pfeufer A, Lichtner P, Meitinger T, Beckmann BM, Myerburg RJ, Bishopric NH, Roden DM, Kääb S, Wilde AA, Schott JJ, Schulze-Bahr E, Bezzina CR. Analysis for genetic modifiers of disease severity in patients with long-QT syndrome type 2. Circ Cardiovasc Genet. 2015;8:447–56.

71. Brink PA, Schwartz PJ. Of founder populations, long QT syndrome, and destiny. Heart Rhythm. 2009;6(11 Suppl):S25–33.

72. Crotti L, Monti MC, Insolia R, Peljto A, Goosen A, Brink PA, Greenberg DA, Schwartz PJ, George AL Jr. NOS1AP is a genetic modifier of the long-QT syndrome. Circulation. 2009;120 (17):1657–63.
73. Arking DE, Pfeufer A, Post W, Kao WH, Newton-Cheh C, Ikeda M, West K, Kashuk C, Akyol M, Perz S, Jalilzadeh S, Illig T, Gieger C, Guo CY, Larson MG, Wichmann HE, Marbán E, O'Donnell CJ, Hirschhorn JN, Kääb S, Spooner PM, Meitinger T, Chakravarti A. A common genetic variant in the NOS1 regulator NOS1AP modulates cardiac repolarization. Nat Genet. 2006;38(6):644–51.
74. Tomás M, Napolitano C, De Giuli L, Bloise R, Subirana I, Malovini A, Bellazzi R, Arking DE, Marban E, Chakravarti A, Spooner PM, Priori SG. Polymorphisms in the NOS1AP gene modulate QT interval duration and risk of arrhythmias in the long QT syndrome. J Am Coll Cardiol. 2010;55(24):2745–52.
75. Massion PB, Pelat M, Belge C, Balligand JL. Regulation of the mammalian heart function by nitric oxide. Comp Biochem Physiol A Mol Integr Physiol. 2005;142(2):144–50.
76. Khan SA, Lee K, Minhas KM, Gonzalez DR, Raju SV, Tejani AD, Li D, Berkowitz DE, Hare JM. Neuronal nitric oxide synthase negatively regulates xanthine oxidoreductase inhibition of cardiac excitation-contraction coupling. Proc Natl Acad Sci U S A. 2004;101(45):15944–8.
77. de Villiers CP, van der Merwe L, Crotti L, Goosen A, George AL Jr, Schwartz PJ, Brink PA, Moolman-Smook JC, Corfield VA. AKAP9 is a genetic modifier of congenital long-QT syndrome type 1. Circ Cardiovasc Genet. 2014;7(5):599–606.
78. Amin AS, Giudicessi JR, Tijsen AJ, Spanjaart AM, Reckman YJ, Klemens CA, Tanck MW, Kapplinger JD, Hofman N, Sinner MF, Müller M, Wijnen WJ, Tan HL, Bezzina CR, Creemers EE, Wilde AA, Ackerman MJ, Pinto YM. Variants in the 3′ untranslated region of the KCNQ1-encoded Kv7.1 potassium channel modify disease severity in patients with type 1 long QT syndrome in an allele-specific manner. Eur Heart J. 2012;33(6):714–23.
79. Crotti L, Lahtinen AM, Spazzolini C, Mastantuono E, Monti MC, Morassutto C, Parati G, Heradien M, Goosen A, Lichtner P, Meitinger T, Brink PA, Kontula K, Swan H, Schwartz PJ. Genetic modifiers for the long-QT syndrome: how important is the role of variants in the 3′ untranslated region of KCNQ1? Circ Cardiovasc Genet. 2016;9(4):330–9.
80. Duchatelet S, Crotti L, Peat RA, Denjoy I, Itoh H, Berthet M, Ohno S, Fressart V, Monti MC, Crocamo C, Pedrazzini M, Dagradi F, Vicentini A, Klug D, Brink PA, Goosen A, Swan H, Toivonen L, Lahtinen AM, Kontula K, Shimizu W, Horie M, George AL Jr, Trégouët DA, Guicheney P, Schwartz PJ. Identification of a KCNQ1 polymorphism acting as a protective modifier against arrhythmic risk in long-QT syndrome. Circ Cardiovasc Genet. 2013;6 (4):354–61.
81. Chai S, Wan X, Ramirez-Navarro A, Tesar PJ, Kaufman ES, Ficker E, George AL Jr, Deschênes I. Physiological genomics identifies genetic modifiers of long QT syndrome type 2 severity. J Clin Invest. 2018;128(3):1043–56.
82. Schwartz PJ, Crotti L, George AL Jr. Modifier genes for sudden cardiac death. Eur Heart J. 2018;39(44):3925–31.
83. Schwartz PJ, Garson A Jr, Paul T, Stramba-Badiale M, Vetter VL, Wren C, European Society of Cardiology. Guidelines for the interpretation of the neonatal electrocardiogram. A task force of the European Society of Cardiology. Eur Heart J. 2002;23(17):1329–44.
84. Quaglini S, Rognoni C, Spazzolini C, Priori SG, Mannarino S, Schwartz PJ. Cost-effectiveness of neonatal ECG screening for the long QT syndrome. Eur Heart J. 2006;27 (15):1824–32.
85. Ishikawa S, Yamada T, Kuwata T, Morikawa M, Yamada T, Matsubara S, Minakami H. Fetal presentation of long QT syndrome – evaluation of prenatal risk factors: a systematic review. Fetal Diagn Ther. 2013;33(1):1–7.
86. Mitchell JL, Cuneo BF, Etheridge SP, Horigome H, Weng HY, Benson DW. Fetal heart rate predictors of long QT syndrome. Circulation. 2012;126(23):2688–95.
87. Wacker-Gussmann A, Strasburger JF, Cuneo BF, Wakai RT. Diagnosis and treatment of fetal arrhythmia. Am J Perinatol. 2014;31(7):617–28.

88. Strasburger JF, Wakai RT. Fetal cardiac arrhythmia detection and in utero therapy. Nat Rev Cardiol. 2010;7(5):277–90.
89. Cuneo BF, Strasburger JF, Yu S, Horigome H, Hosono T, Kandori A, Wakai RT. In utero diagnosis of long QT syndrome by magnetocardiography. Circulation. 2013;128 (20):2183–91.
90. Horigome H, Nagashima M, Sumitomo N, Yoshinaga M, Ushinohama H, Iwamoto M, Shiono J, Ichihashi K, Hasegawa S, Yoshikawa T, Matsunaga T, Goto H, Waki K, Arima M, Takasugi H, Tanaka Y, Tauchi N, Ikoma M, Inamura N, Takahashi H, Shimizu W, Horie M. Clinical characteristics and genetic background of congenital long-QT syndrome diagnosed in fetal, neonatal, and infantile life: a nationwide questionnaire survey in Japan. Circ Arrhythm Electrophysiol. 2010;3(1):10–7.
91. Cuneo BF, Etheridge SP, Horigome H, Sallee D, Moon-Grady A, Weng HY, Ackerman MJ, Benson DW. Arrhythmia phenotype during fetal life suggests long-QT syndrome genotype: risk stratification of perinatal long-QT syndrome. Circ Arrhythm Electrophysiol. 2013;6 (5):946–51.
92. Lin MT, Wu MH, Chang CC, Chiu SN, Thériault O, Huang H, Christé G, Ficker E, Chahine M. In utero onset of long QT syndrome with atrioventricular block and spontaneous or lidocaine-induced ventricular tachycardia: compound effects of hERG pore region mutation and SCN5A N-terminus variant. Heart Rhythm. 2008;5(11):1567–74.
93. Murphy LL, Moon-Grady AJ, Cuneo BF, Wakai RT, Yu S, Kunic JD, Benson DW, George AL Jr. Developmentally regulated SCN5A splice variant potentiates dysfunction of a novel mutation associated with severe fetal arrhythmia. Heart Rhythm. 2012;9(4):590–7.
94. Wilde AA, Moss AJ, Kaufman ES, Shimizu W, Peterson DR, Benhorin J, Lopes C, Towbin JA, Spazzolini C, Crotti L, Zareba W, Goldenberg I, Kanters JK, Robinson JL, Qi M, Hofman N, Tester DJ, Bezzina CR, Alders M, Aiba T, Kamakura S, Miyamoto Y, Andrews ML, McNitt S, Polonsky B, Schwartz PJ, Ackerman MJ. Clinical aspects of type 3 long-QT syndrome: an international multicenter study. Circulation. 2016;134(12):872–82.
95. Schwartz PJ, Priori SG, Dumaine R, Napolitano C, Antzelevitch C, Stramba-Badiale M, Richard TA, Berti MR, Bloise R. A molecular link between the sudden infant death syndrome and the long-QT syndrome. N Engl J Med. 2000;343(4):262–7.
96. Wedekind H, Smits JP, Schulze-Bahr E, Arnold R, Veldkamp MW, Bajanowski T, Borggrefe M, Brinkmann B, Warnecke I, Funke H, Bhuiyan ZA, Wilde AA, Breithardt G, Haverkamp W. De novo mutation in the SCN5A gene associated with early onset of sudden infant death. Circulation. 2001;104(10):1158–64.
97. Schulze-Bahr E, Fenge H, Etzrodt D, Haverkamp W, Mönnig G, Wedekind H, Breithardt G, Kehl HG. Long QT syndrome and life threatening arrhythmia in a newborn: molecular diagnosis and treatment response. Heart. 2004;90(1):13–6.
98. Bankston JR, Yue M, Chung W, Spyres M, Pass RH, Silver E, Sampson KJ, Kass RS. A novel and lethal de novo LQT-3 mutation in a newborn with distinct molecular pharmacology and therapeutic response. PLoS One. 2007;2(12):e1258.
99. Yamamura K, Muneuchi J, Uike K, Ikeda K, Inoue H, Takahata Y, Shiokawa Y, Yoshikane Y, Makiyama T, Horie M, Hara T. A novel SCN5A mutation associated with the linker between III and IV domains of Nav1.5 in a neonate with fatal long QT syndrome. Int J Cardiol. 2010;145(1):61–4.
100. Ten Harkel AD, Witsenburg M, de Jong PL, Jordaens L, Wijman M, Wilde AA. Efficacy of an implantable cardioverter-defibrillator in a neonate with LQT3 associated arrhythmias. Europace. 2005;7(1):77–84.
101. Miller TE, Estrella E, Myerburg RJ, Garcia de Viera J, Moreno N, Rusconi P, Ahearn ME, Baumbach L, Kurlansky P, Wolff G, Bishopric NH. Recurrent third-trimester fetal loss and maternal mosaicism for long-QT syndrome. Circulation. 2004;109(24):3029–34.
102. Wang DW, Crotti L, Shimizu W, Pedrazzini M, Cantu F, De Filippo P, Kishiki K, Miyazaki A, Ikeda T, Schwartz PJ, George AL Jr. Malignant perinatal variant of long-QT syndrome caused

by a profoundly dysfunctional cardiac sodium channel. Circ Arrhythm Electrophysiol. 2008;1 (5):370–8.

103. Chang CC, Acharfi S, Wu MH, Chiang FT, Wang JK, Sung TC, Chahine M. A novel SCN5A mutation manifests as a malignant form of long QT syndrome with perinatal onset of tachycardia/bradycardia. Cardiovasc Res. 2004;64(2):268–78.

104. Paech C, Suchowerskyj P, Gebauer RA. Successful treatment of a newborn with genetically confirmed long QT syndrome 3 and repetitive Torsades De Pointes tachycardia. Pediatr Cardiol. 2011;32(7):1060–1.

105. Hoosien M, Ahearn ME, Myerburg RJ, Pham TV, Miller TE, Smets MJ, Baumbach-Reardon-L, Young ML, Farooq A, Bishopric NH. Dysfunctional potassium channel subunit interaction as a novel mechanism of long QT syndrome. Heart Rhythm. 2013;10(5):728–37.

106. Tester DJ, McCormack J, Ackerman MJ. Prenatal molecular genetic diagnosis of congenital long QT syndrome by strategic genotyping. Am J Cardiol. 2004;93(6):788–91.

107. Furushima H, Chinushi M, Sato A, Aizawa Y, Kikuchi A, Takakuwa K, Tanaka K. Fetal atrioventricular block and postpartum augmentative QT prolongation in a patient with long-QT syndrome with KCNQ1 mutation. J Cardiovasc Electrophysiol. 2010;21(10):1170–3.

108. Hoorntje T, Alders M, van Tintelen P, van der Lip K, Sreeram N, van der Wal A, Mannens M, Wilde A. Homozygous premature truncation of the HERG protein: the human HERG knock-out. Circulation. 1999;100(12):1264–7.

109. Beery TA, Shooner KA, Benson DW. Neonatal long QT syndrome due to a de novo dominant negative hERG mutation. Am J Crit Care. 2007;16(4):416, 412–5.

110. Priest JR, Ceresnak SR, Dewey FE, Malloy-Walton LE, Dunn K, Grove ME, Perez MV, Maeda K, Dubin AM, Ashley EA. Molecular diagnosis of long QT syndrome at 10 days of life by rapid whole genome sequencing. Heart Rhythm. 2014;11(10):1707–13.

111. Sy RW, Hamilton R, Klein GJ, Krahn AD. Neonatal heart block and long-QT syndrome. J Cardiovasc Electrophysiol. 2010;21(9):1059–60.

112. Simpson JM, Maxwell D, Rosenthal E, Gill H. Fetal ventricular tachycardia secondary to long QT syndrome treated with maternal intravenous magnesium: case report and review of the literature. Ultrasound Obstet Gynecol. 2009;34(4):475–80.

113. Johnson WH Jr, Yang P, Yang T, Lau YR, Mostella BA, Wolff DJ, Roden DM, Benson DW. Clinical, genetic, and biophysical characterization of a homozygous HERG mutation causing severe neonatal long QT syndrome. Pediatr Res. 2003;53(5):744–8.

114. Sauer CW, Marc-Aurele KL. A neonate with susceptibility to long QT syndrome type 6 who presented with ventricular fibrillation and sudden unexpected infant death. Am J Case Rep. 2016;17:544–8.

115. Schwartz PJ, Spazzolini C, Crotti L. All LQT3 patients need an ICD: true or false? Heart Rhythm. 2009;6(1):113–20.

116. Kapa S, Tester DJ, Salisbury BA, Harris-Kerr C, Pungliya MS, Alders M, Wilde AA, Ackerman MJ. Genetic testing for long-QT syndrome: distinguishing pathogenic mutations from benign variants. Circulation. 2009;120(18):1752–60.

117. Kambouris NG, Nuss HB, Johns DC, Tomaselli GF, Marban E, Balser JR. Phenotypic characterization of a novel long-QT syndrome mutation (R1623Q) in the cardiac sodium channel. Circulation. 1998;97(7):640–4.

118. Etheridge SP, Bowles NE, Arrington CB, Pilcher T, Rope A, Wilde AA, Alders M, Saarel EV, Tavernier R, Timothy KW, Tristani-Firouzi M. Somatic mosaicism contributes to phenotypic variation in Timothy syndrome. Am J Med Genet A. 2011;155A(10):2578–83.

119. Gillis J, Burashnikov E, Antzelevitch C, Blaser S, Gross G, Turner L, Babul-Hirji R, Chitayat D. Long QT, syndactyly, joint contractures, stroke and novel CACNA1C mutation: expanding the spectrum of Timothy syndrome. Am J Med Genet A. 2012;158A(1):182–7.

120. Dufendach KA, Giudicessi JR, Boczek NJ, Ackerman MJ. Maternal mosaicism confounds the neonatal diagnosis of type 1 Timothy syndrome. Pediatrics. 2013;131(6):e1991–5.

121. Diep V, Seaver LH. Long QT syndrome with craniofacial, digital, and neurologic features: is it useful to distinguish between Timothy syndrome types 1 and 2? Am J Med Genet A. 2015;167A(11):2780–5.
122. Corona-Rivera JR, Barrios-Prieto E, Nieto-García R, Bloise R, Priori S, Napolitano C, Bobadilla-Morales L, Corona-Rivera A, Zapata-Aldana E, Peña-Padilla C, Rivera-Vargas J, Chavana-Naranjo E. Unusual retrospective prenatal findings in a male newborn with Timothy syndrome type 1. Eur J Med Genet. 2015;58(6-7):332–5.
123. Landstrom AP, Boczek NJ, Ye D, Miyake CY, De la Uz CM, Allen HD, Ackerman MJ, Kim JJ. Novel long QT syndrome-associated missense mutation, L762F, in CACNA1C-encoded L-type calcium channel imparts a slower inactivation tau and increased sustained and window current. Int J Cardiol. 2016;220:290–8.
124. Splawski I, Timothy KW, Decher N, Kumar P, Sachse FB, Beggs AH, Sanguinetti MC, Keating MT. Severe arrhythmia disorder caused by cardiac L-type calcium channel mutations. Proc Natl Acad Sci U S A. 2005;102(23):8089–96; discussion 8086–8.
125. Philipp LR, Rodriguez FH 3rd. Cardiac arrest refractory to standard intervention in atypical Timothy syndrome (LQT8 type 2). Proc (Bayl Univ Med Cent). 2016;29(2):160–2.
126. Sepp R, Hategan L, Bácsi A, Cseklye J, Környei L, Borbás J, Széll M, Forster T, Nagy I, Hegedűs Z. Timothy syndrome 1 genotype without syndactyly and major extracardiac manifestations. Am J Med Genet A. 2017;173(3):784–9.
127. Dufendach KA, Timothy K, Ackerman MJ, Blevins B, Pflaumer A, Etheridge S, Perry J, Blom NA, Temple J, Chowdhury D, Skinner JR, Johnsrude C, Bratincsak A, Bos JM, Shah M. Clinical outcomes and modes of death in timothy syndrome: a multicenter international study of a rare disorder. JACC Clin Electrophysiol. 2018;4(4):459–66.
128. Findeisen F, Minor DL Jr. Disruption of the IS6-AID linker affects voltage-gated calcium channel inactivation and facilitation. J Gen Physiol. 2009;133(3):327–43.
129. Dick IE, Joshi-Mukherjee R, Yang W, Yue DT. Arrhythmogenesis in Timothy syndrome is associated with defects in Ca(2+)-dependent inactivation. Nat Commun. 2016;7:10370.
130. Drum BM, Dixon RE, Yuan C, Cheng EP, Santana LF. Cellular mechanisms of ventricular arrhythmias in a mouse model of Timothy syndrome (long QT syndrome 8). J Mol Cell Cardiol. 2014;66:63–71.
131. Badone B, Ronchi C, Kotta MC, Sala L, Ghidoni A, Crotti L, Zaza A. Calmodulinopathy: functional effects of CALM mutations and their relationship with clinical phenotypes. Front Cardiovasc Med. 2018;5:176.
132. Makita N, Yagihara N, Crotti L, Johnson CN, Beckmann BM, Roh MS, Shigemizu D, Lichtner P, Ishikawa T, Aiba T, Homfray T, Behr ER, Klug D, Denjoy I, Mastantuono E, Theisen D, Tsunoda T, Satake W, Toda T, Nakagawa H, Tsuji Y, Tsuchiya T, Yamamoto H, Miyamoto Y, Endo N, Kimura A, Ozaki K, Motomura H, Suda K, Tanaka T, Schwartz PJ, Meitinger T, Kääb S, Guicheney P, Shimizu W, Bhuiyan ZA, Watanabe H, Chazin WJ, George AL Jr. Novel calmodulin mutations associated with congenital arrhythmia susceptibility. Circ Cardiovasc Genet. 2014;7(4):466–74.
133. Marsman RF, Barc J, Beekman L, Alders M, Dooijes D, van den Wijngaard A, Ratbi I, Sefiani A, Bhuiyan ZA, Wilde AA, Bezzina CR. A mutation in CALM1 encoding calmodulin in familial idiopathic ventricular fibrillation in childhood and adolescence. J Am Coll Cardiol. 2014;63(3):259–66.
134. Limpitikul WB, Dick IE, Joshi-Mukherjee R, Overgaard MT, George AL Jr, Yue DT. Calmodulin mutations associated with long QT syndrome prevent inactivation of cardiac L-type Ca(2+) currents and promote proarrhythmic behavior in ventricular myocytes. J Mol Cell Cardiol. 2014;74:115–24.
135. Yin G, Hassan F, Haroun AR, Murphy LL, Crotti L, Schwartz PJ, George AL, Satin J. Arrhythmogenic calmodulin mutations disrupt intracellular cardiomyocyte Ca2+ regulation by distinct mechanisms. J Am Heart Assoc. 2014;3(3):e000996.
136. Limpitikul WB, Dick IE, Tester DJ, Boczek NJ, Limphong P, Yang W, Choi MH, Babich J, DiSilvestre D, Kanter RJ, Tomaselli GF, Ackerman MJ, Yue DT. A precision medicine

approach to the rescue of function on malignant calmodulinopathic long-QT syndrome. Circ Res. 2017;120(1):39–48.

137. Rocchetti M, Sala L, Dreizehnter L, Crotti L, Sinnecker D, Mura M, Pane LS, Altomare C, Torre E, Mostacciuolo G, Severi S, Porta A, De Ferrari GM, George AL Jr, Schwartz PJ, Gnecchi M, Moretti A, Zaza A. Elucidating arrhythmogenic mechanisms of long-QT syndrome CALM1-F142L mutation in patient-specific induced pluripotent stem cell-derived cardiomyocytes. Cardiovasc Res. 2017;113(5):531–41.

138. Gnecchi M, Stefanello M, Mura M. Induced pluripotent stem cell technology: Toward the future of cardiac arrhythmias. Int J Cardiol. 2017;237:49–52.

139. Wagner S, Dybkova N, Rasenack EC, Jacobshagen C, Fabritz L, Kirchhof P, Maier SK, Zhang T, Hasenfuss G, Brown JH, Bers DM, Maier LS. Ca2þ/calmodulin-dependent protein kinase II regulates cardiac Na+ channels. J Clin Invest. 2006;116:3127–38.

140. Koval OM, Snyder JS, Wolf RM, Pavlovicz RE, Glynn P, Curran J, Leymaster ND, Dun W, Wright PJ, Cardona N, Qian L, Mitchell CC, Boyden PA, Binkley PF, Li C, Anderson ME, Mohler PJ, Hund TJ. Ca2+/calmodulin-dependent protein kinase II-based regulation of voltage-gated Na+ channel in cardiac disease. Circulation. 2012;126(17):2084–94.

141. Shamgar L, Ma L, Schmitt N, Haitin Y, Peretz A, Wiener R, Hirsch J, Pongs O, Attali B. Calmodulin is essential for cardiac IKS channel gating and assembly: impaired function in long-QT mutations. Circ Res. 2006;98(8):1055–63.

142. Ghosh S, Nunziato DA, Pitt GS. KCNQ1 assembly and function is blocked by long-QT syndrome mutations that disrupt interaction with calmodulin. Circ Res. 2006;98(8):1048–54.

143. Kotta MC, Sala L, Ghidoni A, Badone B, Ronchi C, Parati G, Zaza A, Crotti L. Calmodulinopathy: a novel, life-threatening clinical entity affecting the young. Front Cardiovasc Med. 2018;5:175.

144. Chopra N, Yang T, Asghari P, Moore ED, Huke S, Akin B, Cattolica RA, Perez CF, Hlaing T, Knollmann-Ritschel BE, Jones LR, Pessah IN, Allen PD, Franzini-Armstrong C, Knollmann BC. Ablation of triadin causes loss of cardiac Ca2+ release units, impaired excitation-contraction coupling, and cardiac arrhythmias. Proc Natl Acad Sci U S A. 2009;106 (18):7636–41.

145. Knollmann BC. New roles of calsequestrin and triadin in cardiac muscle. J Physiol. 2009;587 (Pt 13):3081–7.

146. Walsh MA, Stuart AG, Schlecht HB, James AF, Hancox JC, Newbury-Ecob RA. Compound heterozygous triadin mutation causing cardiac arrest in two siblings. Pacing Clin Electrophysiol. 2016;39(5):497–501.

147. Byard RW, Krous HF. Sudden infant death syndrome: overview and update. Pediatr Dev Pathol. 2003;6(2):112–27.

148. Stillbirth Collaborative Research Network Writing Group. Causes of death among stillbirths. JAMA. 2011;306:2459–68.

149. Insolia R, Ghidoni A, Dossena C, Mastantuono E, Schwartz PJ. Sudden infant death syndrome and cardiac channelopathies: from mechanisms to prevention of avoidable tragedies. Cardiogenetics. 2011;1(s1):e6;28–34.

150. Schwartz PJ. Cardiac sympathetic innervation and the sudden infant death syndrome. A possible pathogenic link. Am J Med. 1976;60:167–72.

151. Arnestad M, Crotti L, Rognum TO, Insolia R, Pedrazzini M, Ferrandi C, Vege A, Wang DW, Rhodes TE, George AL Jr, Schwartz PJ. Prevalence of long-QT syndrome gene variants in sudden infant death syndrome. Circulation. 2007;115(3):361–7.

152. Otagiri T, Kijima K, Osawa M, Ishii K, Makita N, Matoba R, Umetsu K, Hayasaka K. Cardiac ion channel gene mutations in sudden infant death syndrome. Pediatr Res. 2008;64(5):482–7.

153. Klaver EC, Versluijs GM, Wilders R. Cardiac ion channel mutations in the sudden infant death syndrome. Int J Cardiol. 2011;152(2):162–70.

154. Wilders R. Cardiac ion channelopathies and the sudden infant death syndrome. ISRN Cardiol. 2012;2012:846171.

155. Crotti L, Ghidoni A, Insolia R, Schwartz PJ. The role of the cardiac sodium channel in perinatal early infant mortality. Card Electrophysiol Clin. 2014;6(4):749–59.
156. Filiano JJ, Kinney HC. A perspective on neuropathologic findings in victims of the sudden infant death syndrome: the triple-risk model. Biol Neonate. 1994;65:194–7.
157. Wang D, Shah KR, Um SY, Eng LS, Zhou B, Lin Y, Mitchell AA, Nicaj L, Prinz M, McDonald TV, Sampson BA, Tang Y. Cardiac channelopathy testing in 274 ethnically diverse sudden unexplained deaths. Forensic Sci Int. 2014;237:90–9.
158. Glengarry JM, Crawford J, Morrow PL, Stables SR, Love DR, Skinner JR. Long QT molecular autopsy in sudden infant death syndrome. Arch Dis Child. 2014;99(7):635–40.
159. Neubauer J, Lecca MR, Russo G, Bartsch C, Medeiros-Domingo A, Berger W, Haas C. Post-mortem whole-exome analysis in a large sudden infant death syndrome cohort with a focus on cardiovascular and metabolic genetic diseases. Eur J Hum Genet. 2017;25(4):404–9.
160. Tester DJ, Wong LCH, Chanana P, Jaye A, Evans JM, FitzPatrick DR, Evans MJ, Fleming P, Jeffrey I, Cohen MC, Tfelt-Hansen J, Simpson MA, Behr ER, Ackerman MJ. Cardiac genetic predisposition in sudden infant death syndrome. J Am Coll Cardiol. 2018;71(11):1217–27.
161. Andreasen C, Refsgaard L, Nielsen JB, Sajadieh A, Winkel BG, Tfelt-Hansen J, Haunsø S, Holst AG, Svendsen JH, Olesen MS. Mutations in genes encoding cardiac ion channels previously associated with sudden infant death syndrome (SIDS) are present with high frequency in new exome data. Can J Cardiol. 2013;29(9):1104–9.
162. Smith JL, Tester DJ, Hall AR, Burgess DE, Hsu CC, Claude Elayi S, Anderson CL, January CT, Luo JZ, Hartzel DN, Mirshahi UL, Murray MF, Mirshahi T, Ackerman MJ, Delisle BP. Functional invalidation of putative sudden infant death syndrome-associated variants in the KCNH2-encoded Kv11.1 channel. Circ Arrhythm Electrophysiol. 2018;11(5):e005859.
163. Schwartz PJ, Kotta MC. Sudden infant death syndrome and genetics: don't throw out the infant with the dirty water. J Am Coll Cardiol. 2018;71(11):1228–30.
164. MacDorman MF, Gregory EC. Fetal and perinatal mortality: United States, 2013. Natl Vital Stat Rep. 2015;64(8):1–24.
165. Bhuiyan ZA, Momenah TS, Gong Q, Amin AS, Ghamdi SA, Carvalho JS, Homfray T, Mannens MM, Zhou Z, Wilde AA. Recurrent intrauterine fetal loss due to near absence of HERG: clinical and functional characterization of a homozygous nonsense HERG Q1070X mutation. Heart Rhythm. 2008;5(4):553–61.
166. Crotti L, Tester DJ, White WM, Bartos DC, Insolia R, Besana A, Kunic JD, Will ML, Velasco EJ, Bair JJ, Ghidoni A, Cetin I, Van Dyke DL, Wick MJ, Brost B, Delisle BP, Facchinetti F, George AL, Schwartz PJ, Ackerman MJ. Long QT syndrome-associated mutations in intra-uterine fetal death. JAMA. 2013;309(14):1473–82.
167. Jones DK, Liu F, Dombrowski N, Joshi S, Robertson GA. Dominant negative consequences of a hERG 1b-specific mutation associated with intrauterine fetal death. Prog Biophys Mol Biol. 2016;120(1–3):67–76.

The Genetic Landscape of Cardiomyopathies

2

Brenda Gerull, Sabine Klaassen, and Andreas Brodehl

Contents

Abstract

Insights into genetic causes of cardiomyopathies have tremendously contributed to the understanding of the molecular basis and pathophysiology of hypertrophic, dilated, arrhythmogenic, restrictive and left ventricular noncompaction cardiomyopathy. More than thousand mutations in approximately 100 genes encoding proteins involved in many different subcellular systems have been identified indicating the diversity of pathways contributing to pathological cardiac remodeling. Moreover, the classical view based on morphology and physiology

B. Gerull (✉)
Department of Internal Medicine I, Comprehensive Heart Failure Center, University Hospital Würzburg, Würzburg, Germany
e-mail: gerull_b@ukw.de

S. Klaassen
Experimental and Clinical Research Center (ECRC), Max-Delbrück-Center for Molecular Medicine, Charité University Medicine Berlin, Berlin, Germany
e-mail: klaassen@mdc-berlin.de

A. Brodehl
Heart and Diabetes Centre NRW, Erich and Hanna Klessmann Institute, University Hospital of the Ruhr-University Bochum, Bad Oeynhausen, Germany
e-mail: abrodehl@hdz-nrw.de

© Springer Nature Switzerland AG 2019
J. Erdmann, A. Moretti (eds.), *Genetic Causes of Cardiac Disease*, Cardiac and Vascular Biology 7, https://doi.org/10.1007/978-3-030-27371-2_2

has been shifted toward genetic and molecular patterns defining the etiology of cardiomyopathies. Today, novel high-throughput genetic technologies provide an opportunity to diagnose individuals based on their genetic findings, sometimes before clinical signs of the disease occur. However, the challenge remains that rapid research developments and the complexity of genetic information are getting introduced into the clinical practice, which requires dedicated guidance in genetic counselling and interpretation of genetic test results for the management of families with inherited cardiomyopathies.

2.1   Introduction

Over the last three decades, genetic research has significantly contributed to the etiology of cardiomyopathies. In fact, most primary cardiomyopathies are influenced by genetic factors predisposing to the development of heart failure, one of the most common causes of death in industrialized countries. Cardiomyopathies (CMPs) are commonly grouped by their morphological signs leading to subtypes called hypertrophic cardiomyopathy (HCM), dilated cardiomyopathy (DCM), left ventricular noncompaction (LVNC), arrhythmogenic cardiomyopathy (ACM), and restrictive cardiomyopathy (RCM) (Fig. 2.1) [1].

Since, in 1990, the sarcomere gene beta-myosin heavy chain (*MYH7*) was identified as the first disease gene for HCM, it turned out that the sarcomere plays a major role in the genetic etiology of HCM but also of other CMPs [2, 3]. HCM can primarily be defined as a disease of the sarcomere, but the genetic basis for CMPs has evolved to be diverse. Genetic causes of ACM have been unraveled with disease genes involved in cell-cell contacts called desmosomes [4, 5]. But, in fact, today there are about hundred known disease genes associated with different subtypes of CMPs, which are expressed in various subcellular systems responsible for transcription, cardiac development, energy utilization, electrolyte imbalances, and others. The diversity is making it difficult to define a final common pathway leading from disturbed gene function to the disease phenotype.

Despite the genetic complexity, with the application of high-throughput genetic technology, we are able to analyze large amounts of data in a cost-effective and timely manner which makes it possible to use this technology not only for research but also for comprehensive cardiomyopathy diagnostics [6–8]. However, there are challenges that clinicians and their team of health-care providers face in translating research results and comprehensive data sets into the clinic in order to improve patient management. A major challenge is still the classification of variants into pathogenic or disease-causing versus benign. Often variants are defined as "uncertain," which makes the application for clinical purposes even more difficult [9]. Additionally, when a pathogenic variant has been identified, there can be substantial variation in penetrance, age of onset, and clinical expression of the phenotype, also within the same family. Here, modifying factors such as lifestyle,

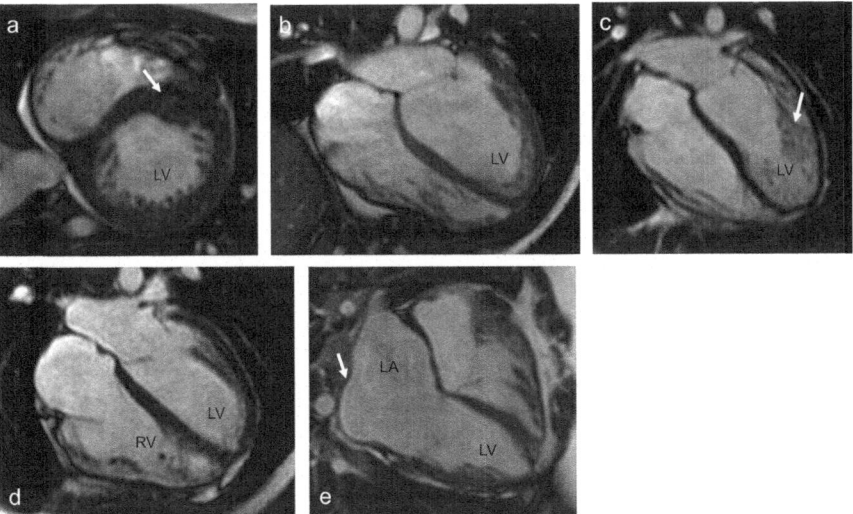

Fig. 2.1 Examples of cardiac magnetic resonance imaging (MRI) pictures of cardiomyopathy subtypes. (**a**) Short-axis two-chamber view of hypertrophic cardiomyopathy (HCM). Note the thick septal wall of the left ventricle (LV) (*white arrow*). (**b**) Long-axis four-chamber view of dilated cardiomyopathy (DCM). Note the enlarged left ventricle. (**c**) Long-axis four-chamber view of left ventricular noncompaction (LVNC). Note the noncompacted, trabeculated layer of the LV (*white arrow*). (**d**) Long-axis four-chamber view of arrhythmogenic cardiomyopathy (ACM). Note the enlarged right ventricle (RV). (**e**) Long-axis four-chamber view of restrictive cardiomyopathy (RCM). Note the massively enlarged left atrium (LA) (*white arrow*)

other genetic and epigenetic factors, pregnancy, and many still unknown entities are playing a role—an ongoing field of research. Moreover, different mutations in the same genes can cause variable clinical entities and even lead to the opposite phenotype (dilated vs. hypertrophic cardiomyopathy). Given the complexity, today the question arises how much a particular genetic diagnosis may help guiding diagnosis and clinical management? There are potential benefits such as differentiating a genetic etiology from other causes of CMPs or allowing the identification of at-risk individuals early, in a preclinical stage. But more importantly, the understanding of the molecular pathways in a translational setting will help to further understand these diseases and dissect more complex disease pathways toward targeted therapies.

This chapter summarizes current knowledge about clinical and molecular genetics of the five main subtypes of inherited cardiomyopathies in a bench-to-bedside approach.

2.2   Hypertrophic Cardiomyopathy (HCM)

Definition Hypertrophic cardiomyopathy (HCM) is a primary myocardial disorder affecting clinically about 1 in 500 individuals [10]. The pathological sign is an unexplained left ventricular hypertrophy which is not caused by abnormal loading conditions (Fig. 2.1a). Classically, HCM is characterized by asymmetric septal hypertrophy; however there can be any pattern of left ventricular hypertrophy associated with the disease [11–13].

Clinical Manifestations Clinically, the disease is highly variable in presentation and includes diastolic dysfunction, left ventricular outflow tract obstruction (in about 25%), ischemia, and atrial fibrillation. In about 5% of cases, end-stage HCM shows progression to systolic impairments. Many patients are asymptomatic and are diagnosed incidentally, and others may manifest shortness of breath, chest pain, palpitations, or syncope [11]. Disease-related mortality is most often attributable to sudden cardiac death, heart failure, and embolic stroke. However, early data about high mortality rates from tertiary centers from the 1970s and early 1980s, with annual death rates of 2–4% in adults and 4.2–5.9% in children, have been revised. Disease-related mortality in adults is today about 0.5% similar to that of the general population. Decreased mortality is an achievement of more effective risk stratification and the use of the implantable cardioverter-defibrillator for primary prevention of sudden death [14–16].

Inheritance HCM is mainly inherited as an autosomal dominant trait with incomplete penetrance and variable expression of the clinical phenotype. Some cases are explained by de novo mutations, and other apparently sporadic cases can arise due to autosomal recessive, compound heterozygous, or digenic mode of inheritance [17–19]. Furthermore, genetic modifiers and environmental and epigenetic factors are likely to influence the phenotype. Consequently, even identical mutations may show distinct degrees and morphology of hypertrophy, pattern of fibrosis, age of onset, and risk for arrhythmias and sudden death [20, 21]. On the other hand, population-based studies looking for variants in HCM-associated sarcomere genes indicated that changes in cardiac morphology and function may also occur without causing overt or typical HCM [22].

Disease Genes HCM was the first primary cardiomyopathy for which a genetic cause was identified back in 1990. Ever since over thousand mutations in numerous genes have been described. The most common eight genes encoding sarcomere proteins of the thick and thin filaments are beta-myosin heavy chain (*MYH7*) [2], alpha-tropomyosin (*TPM1*), cardiac troponin T (*TNNT2*) [23, 24], cardiac myosin-binding protein C (*MYBPC3*) [25], myosin regulatory light chain (*MYL2*), myosin essential light chain (*MYL3*) [26], cardiac troponin I (*TNNI3*) [27], and cardiac α-actin (*ACTC1*) [28]. Two major genes, *MYH7* and *MYBPC3*, account together for up to 75% of all identified mutations, whereas other genes including *TNNT2*, *TNNI3*, *TPM1*, *MYL2*, *MYL3*, and *ACTC1* each account for a small proportion of cases (1–5%) [29, 30]. Furthermore, rare variants have been described in additional genes encoding proteins of the sarcomere apparatus (MYH6, TNNC1, TTN) and the adjacent Z-disc (ACTN2, CSRP3, TCAP, VCL, NEXN, LDB3, MYOZ2, FLNC, MYPN, ANKRD1) or are involved in calcium homeostasis pathways (PLN, CASQ2, JPH2, CALR3). Some of these genes are less evident for direct pathogenicity and may function as modifiers (Table 2.1, Fig. 2.2a) [31–33].

Genotype-Phenotype Correlation The penetrance of expressing the phenotype when carrying a mutation appears to be highly variable and can be delayed or incomplete. Often left ventricular hypertrophy (LVH) does not develop until the third decade of life or even after. In addition, penetrance may also differ according to gender with males showing earlier clinical signs than females [21, 34]. Early genotype-phenotype studies described some genes (or particular mutations) with a higher degree of sudden cardiac death (*TNNT2*, *MYH7* p.Arg403Gln) and other genes (*MYBPC3*) with milder LVH and later age of onset [20, 24]. More recently, reports from the UK suggest a higher risk of sudden death in TNNT2 vs. MYBPC3 families (0.93% vs. 0.46% per year). Overall, currently available phenotypic data could not show clear correlations for specific genes and most of the mutations. Therefore, genotype data have only minor influence on current risk stratification [21, 35, 36].

Despite inherited as an autosomal dominant disease, complex phenotypes with patients carrying more than one mutation or variant exist, and they usually show more severe or earlier phenotypic expression. Such complexity has been reported in 5–7% of cases by Sanger sequencing and is even higher using next-generation sequencing (8–9%). Here, different scenarios have been seen including compound heterozygosity (different heterozygous mutation in each allele of the same gene), digenic cases (heterozygous mutations in two different genes), and also rare homozygosity [37, 38].

Lastly, mutations in classical HCM genes can result in divergent clinical features, also mimicking other forms of cardiomyopathy, in particular LVNC or RCM. Sometimes in the same family, different clinical manifestations may occur [39].

Genetic Counselling and Testing As indicated above, the majority of HCM cases is inherited as an autosomal dominant trait with a 50% risk for first-degree relatives. Current guidelines of the European Society of Cardiology (ESC) recommend genetic counselling for all patients with HCM, unless an acquired cause is demonstrated. Counselling should be performed by trained health-care professionals working within multidisciplinary teams to help patients understand and manage the psychological, social, professional, ethical, and legal implications of genetic disease [11].

If genetic testing is considered, genetic counselling should be performed to fully inform patients about the benefits and limitations of genetic testing and consequences for them and their families. In patients fulfilling the diagnostic criteria for HCM, genetic testing identifies a pathogenic mutation in up to 65% of cases. Although, nowadays, there is no direct benefit in most of the clinically affected cases having a genetic diagnosis in regard to treatment options and a dedicated prognosis,

Table 2.1 Overview about genes associated with HCM

Gene symbol	Encoded protein	Subcellular location/function	Frequency (%)	First description
Main disease-causing genes[a]				
MYBPC3	Cardiac myosin-binding protein C	Sarcomere, thick filament associated	30–40	1995
MYH7	β-myosin heavy chain	Sarcomere, thick filament	20–30	1990
MYL2	Regulatory myosin light chain	Sarcomere, thick filament	2–4	1996
MYL3	Essential myosin light chain	Sarcomere, thick filament	1–2	1996
TNNI3	Cardiac troponin I	Sarcomere, thin filament	5–7	1997
TNNT2	Cardiac troponin T	Sarcomere, thin filament	5–10	1994
TPM1	α-tropomyosin	Sarcomere, thin filament	<1	1994
ACTC1	Cardiac α-actin	Sarcomere, thin filament	<1	1999
Rare potentially associated genes[a,b]				
ACTN2	α-Actinin 2	Z-disc	Rare	2010
MYOZ2	Myozenin 2	Z-disc	Rare	2007
CSRP3	Cardiac LIM protein	Z-disc	Rare	2003
MYPN	Myopalladin	Z-disc	Rare	2012
FLNC	Filamin C	Z-disc	Rare	2014
TTN	Titin	Sarcomere	Rare	1999
TNNC1	Cardiac troponin C	Sarcomere, thin filament	Rare	2001
MYH6	α-Myosin heavy chain	Sarcomere, thick filament	Rare	2005
NEXN	Nexilin	Z-disc	Rare	2010
TCAP	Telethonin	Z-disc	Rare	2004
ANKRD1	Cardiac ankyrin repeat protein	Z-disc	Rare	2009
LDB3	LIM domain-binding protein 3	Z-disc	Rare	2006
VCL	Vinculin	Z-disc, intercalated disc, sarcolemma	Rare	2006
JPH2	Junctophilin 2	Sarcolemma, Ca^{2+} handling	Rare	2007
CAV3	Caveolin-3	Sarcolemma, repair	Rare	2004
PLN	Phospholamban	SR, Ca^{2+} handling	Rare	2007
CALR3	Calreticulin 3	SR, Ca^{2+} handling	Rare	2007
CASQ2	Calsequestrin	SR, Ca^{2+} handling	Rare	2007

[a]Autosomal dominant inheritance
[b]Evidence for pathogenicity unclear
SR sarcoplasmic reticulum

it is more important for pre-symptomatic testing in the family and reproductive advice. Generally speaking, predictive genetic testing can be performed in asymptomatic relatives of a patient with HCM when a definitive disease-causing mutation has been determined to initiate cascade family screening.

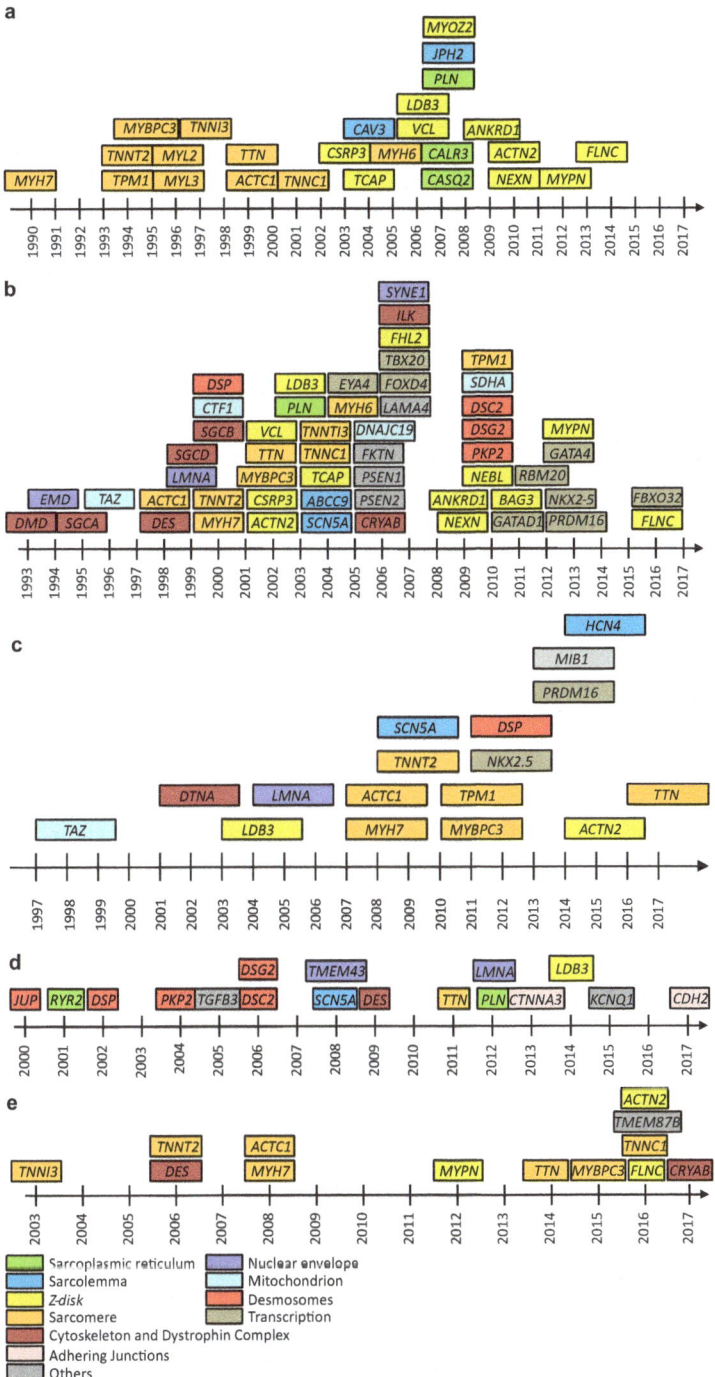

Fig. 2.2 Genes associated with different subtypes of cardiomyopathies according to the year of discovery. Colors indicate the subcellular location and/or functional association. (**a**) Hypertrophic cardiomyopathy. (**b**) Dilated cardiomyopathy. (**c**) Left ventricular noncompaction. (**d**) Arrhythmogenic cardiomyopathy. (**e**) Restrictive cardiomyopathy

However, in situations where predictive genetic testing is either not available or not considered, first-degree relatives should be offered clinical screening with an ECG and echocardiogram starting at an age of 10 years [40]. Individuals who have nondiagnostic clinical features that are consistent with early disease should be seen initially at intervals of 6–12 months and potentially less frequent depending on progression. As indicated above, age-dependent penetrance requires repeat screening even in individuals with no clinical features, unless a predictive genetic test has been performed and was "negative" for a proven disease-causing mutation [3, 11, 41].

Today, new high-throughput sequencing technologies are more and more used for genetic testing in diseases such as HCM by analyzing the whole exome with the potential identification of a large number of variants of unknown clinical significance (VUS). In contrast to other cardiomyopathies, broader genetic testing has shown only a modest increase in the diagnostic yield in HCM (45–72%) while producing more uncertainty. No matter what sequencing technology is used, for clinical practice, genetic testing should include the eight most commonly implicated sarcomere genes and may also consider genes of potential phenocopies of sarcomeric HCM, in particular when additional or atypical phenotypic signs are present [42, 43].

Other Genetic Etiologies There are phenocopies of HCM which can mimic sarcomeric HCM on the basis of cardiac imaging (Table 2.2). In adults, the main non-sarcomeric etiologies are metabolic storage disease such as Danon disease (LAMP2), LVH and Wolff-Parkinson-White syndrome (PRKAG2), Anderson-Fabry disease (GLA), familial TTR amyloidosis (TTR), and some mitochondrial cardiomyopathies. Others are more prevalent in children such as Pompe disease (GAA), Noonan syndrome and Leopard syndrome (PTPN11), or Friedreich ataxia (FXN). To distinguish these diseases from sarcomeric HCM is quite important as an early diagnosis ensures an appropriate management in particular in metabolic diseases [44, 45].

Functional Consequences Although the genetic basis for HCM is well established, the biochemical and biophysical mechanisms how sarcomere mutations lead to disease remain only partially understood. The majority of mutations described are missense mutations that result in mutant dominant-negative acting "poison" peptides, which are incorporated into the sarcomere and potentially have an adverse effect on sarcomere function [46]. The exception for the "poison" protein thesis is mutations in MYBPC3, which often result in truncated proteins. Here, mutant proteins are often cleared by cell surveillance mechanisms, which leads to a reduced amount of full-length protein resulting in haploinsufficiency and consequently an inadequate amount of functional proteins [47, 48].

Mechanistically, there is substantial evidence that myofilament mutations increase calcium sensitivity and Ca^{2+} affinity and thereby actin-dependent ATPase activity [49, 50]. Enhanced Ca^{2+} sensitivity and subsequent defects in calcium homeostasis such as intracellular calcium cycling, sarcoplasmic reticulum

Table 2.2 Other rare genes associated with cardiac hypertrophy

Gene symbol	Encoded protein	Subcellular location/ function	First description	Associated phenotypes/ disease	Inheritance
GLA	α-Galactosidase A	Lysosomes	1989	Fabry disease	X-linked
LAMP2	Lysosome-associated membrane protein 2	Lysosomes	2000	Danon disease	X-linked
PRKAG2	AMP-activated protein kinase, γ2-subunit (noncatalytic)	Energy sensor, cellular energy metabolism	2001	PRKAG2 cardiomyopathy	AD
GAA	α-Glucosidase	Lysosomes, extracellular	1991	Pompe disease	AR
TTR	Transthyretin	Lysosomes, extracellular	1989	Amyloidosis	AD
PTPN11	Protein tyrosine phosphatase, non-receptor 11	Cytosolic, nucleus, signaling molecule	2001	Noonan syndrome	AD
FXN	Frataxin	Mitochondria	1996	Friedreich ataxia	AR

AD autosomal dominant, *AR* autosomal recessive

[SR] Ca^{2+} reuptake, and CaMKII-mediated phosphorylation of proteins, including phospholamban, are likely to contribute to the pathogenesis of HCM [51]. A consequence of increased sarcomeric Ca^{2+} sensitivity is also an enhanced cross-bridge turnover and higher actin-activated ATPase activity. Subsequently, this generates higher energy need to produce a given tension. In accordance with this explanation, several studies have shown mitochondrial abnormalities and an impaired myocardial energy metabolism. Altered Ca^{2+} handling and energy deficiency appear to be major pathways leading to pathological changes such as hypertrophy and myofibrillar disarray as well as functional features such as impaired relaxation [52, 53].

More recently, understanding of fundamental mechanisms arises hope of more specific, substrate-modulating therapy ranging from metabolic modulators to small molecule effectors and gene therapy rescuing HCM at the mechanistic and biochemical level. Targets are energy utilization and oxidative stress, increased calcium sensitivity, sarcomere function (hypercontractility), and the action potential of the myocytes. However, the challenge exists that different mutations even at the same locus can lead to opposite functional consequences. For example, different mutations in MYH7 exert opposing effects on myofilament calcium sensitivity and contractility [54]. In silico models, such as motility assays using purified recombinant sarcomeric protein constructs containing dedicated mutations, are a way to investigate individual mutations [53]. Mutations in the myosin head region domain lead to hypercontractility and impaired relaxation by assessing the chemo-mechanical cycle in ex vivo preparations of mutated sarcomeric proteins. A small molecule, MYK-461, that decreases sarcomeric power due to inhibition of the myosin adenosine triphosphatase (ATPase) counteracts the effects of the mutation. In mice with hypercontractile mutations in myosin heavy chain, early chronic administration of MYK-461 prevented the development of LVH, myocyte disarray, and fibrosis and normalized expression of profibrotic and mitochondrial genes involved in energy utilization [55].

2.3   Dilated Cardiomyopathy (DCM)

Definition Dilated cardiomyopathy (DCM) is characterized by left or biventricular dilatation and systolic dysfunction that are not explained by abnormal loading conditions or coronary artery disease (Fig. 2.1b). The disease is a common cause of heart failure leading to heart transplantation and sudden death [1]. Causes of DCM can be genetic, estimated as high as 35–50% [56, 57]; however even non-genetic etiologies may predisposed by genetic factors and interact with extrinsic and environmental factors. Lately, there are suggestions for an updated definition of DCM recognizing a broader clinical spectrum of the disease, in particular in the setting of familial disease and before overt clinical manifestations are present. This includes criteria defining preclinical phases of the disease that may occur in mutation carriers without clinical expression, cases with isolated ventricular dilation, arrhythmic cardiomyopathy, or hypokinetic non-dilated cardiomyopathy [58].

Prevalence data based on echocardiography estimated unexplained DCM at 1:2500; more recently epidemiological and next-generation sequencing data suggest a much higher prevalence, possibly as high as 1:250. The disease incidence is 7 per 100,000, and males are more frequently affected than females (3:1) [59, 60].

Clinical Manifestations DCM can present as a clinical syndrome of systolic heart failure with reduced ejection fraction (HFrEF) but also with arrhythmias that may lead to sudden death or thromboembolic events. Management of HFrEF should be performed according to current heart failure guidelines; however, despite optimal therapy, the prognosis remains poor for patients showing heart failure symptoms with a 5-year mortality of up to 20%. Cardiac transplantation or other advanced therapies should be considered with progressive DCM, advanced heart failure, or otherwise refractory disease [61–63].

Another spectrum of the disease represents unidentified cases without clinical symptoms in early phases of the disease. Genetic approaches offer the possibility to identify individuals at risk in the setting of familial disease. Although not proven by placebo-controlled trials, early medical management may be beneficial, even in symptom-free mutation carriers with only minor cardiac abnormalities [58, 64, 65].

Inheritance DCM can be inherited as an autosomal dominant, recessive, or X-linked trait; however chromosomal abnormalities or even matrilineal inheritance may also be present.

Most cases of isolated DCM follow autosomal dominant inheritance with age-dependent penetrance and variable clinical expression, which can be delayed until the fifth or sixth decade. This indicates a role for potential modifying genetic, epigenetic, and environmental factors [66, 67]. However, these factors also contribute to the fact that familial DCM is often not recognized, as other more common etiologies are attributed. When compared to other cardiomyopathies, the genetic etiology of DCM is even more heterogenous, which became more obvious since the introduction of next-generation sequencing [68]. To date, more than 50 disease-related genes have been reported, although relatively few are supported by robust segregation analyses or experimental data (Table 2.3; Fig. 2.2b).

Genes Involved in DCM Pathogenic variants have been reported for a heterogenous group of genes encoding proteins with different functions involved in sarcomere integrity and force transmission, cytoskeletal architecture, cell-cell contacts, nuclear organization, transcription, and ion channel activity. Although extensive genetic heterogeneity is present, titin (*TTN*) is the most common disease gene accounting for 20–25% of the genetic causes [69–71]. The second most prevalent gene is lamin A/C (*LMNA*) with about 6%, followed by beta-myosin heavy chain (*MYH7*) [72–74]. Several other genes account for 1–2% of familial cases, while many others are less frequent or even reported once. Alterations, such as copy number variations (CNVs), have been reported in association with DCM but also with a low frequency. The most important genes and their role in subcellular systems will be discussed in more detail.

Titin and the Sarcomere Mutations in the titin gene (*TTN*) are the main cause of DCM accounting for about 20–25% in familial cases and even up to 18% of sporadic DCM cases. In particular, mutations that truncate the protein have been found to cause disease, although such variants have been also found in 1–2% of the control population [69–71]. Titin, the biggest protein in nature, provides the external scaffold interacting with the thin and thick filaments. It is important for sarcomere assembly and provides passive force and elasticity to maintain diastolic and systolic function, respectively. TTN is composed of different domains according to the sarcomere structure (Z-disc, I-band, A-band, M-band). In addition, *TTN* undergoes extensive alternative splicing and therefore produces many isoforms. Current data suggest that truncating variants in exons that are incorporated into all expressed isoforms, and are part of the final transcript in the heart, are more likely to cause disease (90%), whereas exons, e.g., of the I-band, which are not included in the mature transcript, may not be disease-causing and have been found more frequently in the control population [75].

Mutations in other proteins of the sarcomere are also associated with autosomal dominant DCM such as ACTC1, MYBPC3, MYH6, MYH7, TNNC1, TNNI3, TNNT2, and TPM1, with MYH7 as the most common one [73]. The frequency of

Table 2.3 Overview about genes associated with DCM

Gene symbol	Encoded protein	Subcellular localization/function	Frequency (%)	First description	Associated phenotypes/disease	Inheritance
Predominant DCM disease genes						
TTN	Titin	Sarcomere, molecular spring	20–25	2002	None	AD
LMNA	Lamin A/C	Nuclear envelope	6–10	1999	CD, EDMD2, EDMD3, LGMD1B	AD, AR
MYH7	β-myosin heavy chain	Sarcomere, thick filament	5	2000	None	AD
TNNT2	Cardiac troponin T	Sarcomere, thin filament	2	2000	None	AD
BAG3	Bcl2-associated athanogene 3	Z-disc, co-chaperone	2	2011	MFM	AD
RBM20	RNA-binding protein 20	Regulator of RNA splicing	2	2012	None	AD
SCN5A	Voltage-gated sodium channel type V, α-subunit	Ion channel/electrolyte imbalances	2	2004	CD, arrhythmias	AD
ACTC1	Cardiac α-actin	Sarcomere, thin filament	1	1998	None	AD
TPM1	Alpha-tropomyosin	Sarcomere, thin filament	1	2010	None	AD
TNNI3	Cardiac troponin I	Sarcomere, thin filament	1	2004	None	AR, AD
PLN	Phospholamban	SR, Ca^{2+} handling	1	2003	None	AD
TNNC1	Cardiac troponin C	Sarcomere, thin filament	1	2004	None	AD
Rare potentially associated genes						
MYBPC3	Cardiac myosin-binding protein C	Sarcomere, thick filament	Rare	2002	None	AD
MYPN	Myopalladin	Z-disc	Rare	2013	None	AD
DES	Desmin	IF protein, intercalated disc, Z-disc	Rare	1998	MFM	AD, AR
DMD	Dystrophin	Cytoskeleton, force transduction	Rare	1993	Duchenne, Becker muscular dystrophy	X-linked
DNAJC19	DNAJ (Hsp40) homolog	Mitochondrial chaperone	Rare	2006	Dilated cardiomyopathy with ataxia	X-linked

Gene	Protein	Function/location	Frequency	Year	Associated disease	Inheritance
EMD	Emerin	Inner nuclear membrane	Rare	1994	EDMD1	X-linked
TAZ	Tafazzin	Mitochondrial chaperone	Rare	1996	Barth syndrome	X-linked
ANKRD1	Cardiac ankyrin repeat protein	Transcription co-inhibitor	Rare	2009	None	AD
LDB3	LIM domain-binding protein 3	Z-disc, structural integrity	Rare	2003	Myofibrillar myopathy	AD
NEXN	Nexilin	Z-disc, actin-binding protein	Rare	2009	None	AD
SGCD	Delta-sarcoglycan	Transmembrane protein	Rare	1999	LGMD2F	AR, AD
EYA4	Eyes absent homolog 4	Transcriptional coactivator	Rare	2005	Sensorineural hearing loss	AD
PSEN1	Presenilin 1	Catalytic component of gamma-secretase	Rare	2006	None	AD
FBXO32	F-box protein 32/atrogin 1	Ubiquitin ligase	Rare	2016	None	AR
GATAD1	GATA zinc finger domain containing 1 protein	Transcription factor	Rare	2011	None	AR
GATA4	GATA-binding protein 4	Transcription factor	Rare	2013	Congenital heart disease	AD
MYH6	α-Myosin heavy chain	Sarcomere, thick filament	Rare	2005	None	AD
ACTN2	α-Actinin 2	Z-disc, actin-binding protein	Rare	2002	None	AD
CSRP3	Cardiac LIM protein	Z-disc, stretch sensing	Rare	2002	None	AD
TCAP	Telethonin	Z-disc, stretch sensing	Rare	2004	LGMD2G	AD
ABCC9	Sulfonylurea receptor 2	Subunit of the cardiac K(ATP) channel	Rare	2004	Cantú syndrome	AD
TBX20	T-box 20	Transcription factor	Rare	2007	Congenital heart disease	AD
NKX2-5	NK2 homeobox 5	Transcription factor	Rare	2013	Congenital heart disease	AD
CTF1	Cardiotrophin 1	Cytokine, hypertrophy	Rare	2000	None	AD
VCL	Vinculin	Cell-ECM and cell-cell adhesion	Rare	2002	None	AD
SGCB	β-Sarcoglycan	Transmembrane protein	Rare	2000	LGMD2E	AR
SGCA	α-Sarcoglycan	Transmembrane protein	Rare	1995	LGMD2D	AR
FHL2	Four and a half limb domains 2	Z-disc, focal adhesions	Rare	2007	Myopathy	X-linked
FKTN	Fukutin	Transmembrane protein, Golgi apparatus	Rare	2006	Fukuyama-type muscular dystrophy	AR

(continued)

Table 2.3 (continued)

Gene symbol	Encoded protein	Subcellular localization/function	Frequency (%)	First description	Associated phenotypes/disease	Inheritance
FOXD4	Forkhead box protein D4	Transcription factor	Rare	2007	Obsessive-compulsive disorder	AD
LAMA4	Laminin subunit alpha-4	ECM	Rare	2007	None	AD
ILK	Integrin-linked kinase	Focal adhesions, ECM connection	Rare	2007	None	AD
PSEN2	Presenilin 2	Catalytic component of γ-secretase	Rare	2006	None	AD
SDHA	Succinate dehydrogenase subunit A	Mitochondrial inner membrane	Rare	2010	None, neonatal DCM	AR
CRYAB	αB-Crystallin	Z-disc, small heat-shock protein	Rare	2006	MFM, cataract	AD, AR
SYNE1	Nesprin 1	Cytoskeleton, nuclear lamina	Rare	2007	EDMD4	AD
PRDM16	PR domain containing 16	Transcription factor	Rare	2013	1p36 deletion syndrome	AD
DSC2	Desmocollin 2	Desmosomes, intercalated disc, cell adhesion	Rare	2010	None	AD
DSG2	Desmoglein 2	Desmosomes, intercalated disc, cell adhesion	Rare	2010	None	AD, AR
DSP	Desmoplakin	Desmosomes, intercalated disc, cell adhesion	Rare	2000	Palmoplantar keratoderma	AD
PKP2	Plakophilin 2	Desmosomes, intercalated disc, cell adhesion	Rare	2010	None	AD
FLNC	Filamin C	Z-disc, actin cross-linking protein	Rare	2016	MFM	AD
NEBL	Nebulette	Z-disc	Rare	2010	10p14-p13 deletion	AD

AD autosomal dominant, *AR* autosomal recessive, *CD* conduction disorder, *ECM* extracellular matrix, *EDMD* Emery-Dreifuss muscular dystrophy, *LGMD* limb-girdle muscular dystrophy, *RNA* ribonucleic acid, *SR* sarcoplasmic reticulum, *IF* intermediate filament

sarcomere gene mutations to be associated with DCM is much lower compared to HCM and LVNC. Mechanistically, and in contrast to HCM, mutations are suggested to decrease Ca^{2+} sensitivity causing a hypocontractile state that leads to systolic dysfunction. For example, mice carrying a HCM-associated mutation (delGlu160) in TNNT2 showed increased Ca^{2+} sensitivity, whereas a DCM-associated mutation (delLys210) in the same gene showed the opposite, decreased Ca^{2+} sensitivity [53].

Nuclear Envelope Proteins DCM-associated mutations are also involving proteins of the nuclear envelope such as lamin A/C (LMNA) located in the nuclear lamina and emerin (EMD), a protein of the inner nuclear membrane. Dominant *LMNA* mutations account for 6–10% of genetic causes and are frequently associated with arrhythmias and conduction system disturbances [72]. In addition, limb-girdle myopathies are a variable feature. Lamins are involved in nuclear structure support, DNA repair, cell signaling pathway mediation, and chromatin organization. There are two major hypotheses how mutations in LMNA may lead to cardiac dysfunction: (1) disruption and uncoupling of structural proteins at the nuclear lamina that lead to mechanical instability and to an increased susceptibility to mechanical stress and (2) disrupted chromatin organization that impacts directly on gene transcription [76, 77].

EMD is a LEM-domain protein located in the nuclear lamina and has a role in assembly of the nuclear lamina and structural organization of the nuclear envelope. EMD mutations are X-linked and often associated with a skeletal muscle phenotype of Emery-Dreifuss muscular dystrophy [78].

Gene Expression Genes regulating transcription are also important in the pathogenesis of DCM. Rare mutations have been described in *TBX20, NKX2-5, GATA4, GATA6, FOXD4, PRDM16, EYA4, GATAD1,* and *RBM20.* Other associated phenotypes such as congenital heart defects, hearing loss, or more complex syndromic features are often observed (Table 2.3). Variants in most of those genes lead to decreased transcriptional activity influencing a set of genes important for cardiac development and structural remodeling [79–83].

A very interesting gene is *RBM20,* which encodes a spliceosome protein that regulates pre-mRNA splicing for many genes, including *TTN.* Interestingly, a mutational hotspot causing highly penetrant DCM alters an arginine-serine-rich region of RBM20, which influences binding with other splicing factors and changes transcript processing [84, 85].

Calcium Handling and Ion Imbalances Mutations in phospholamban (*PLN*) are predicted to alter Ca^{2+} homeostasis. The protein regulates Ca^{2+} uptake by the SR Ca^{2+} ATPase (SERCA2a) and inhibits Ca^{2+} cycling when dephosphorylated [86]. In its phosphorylated state by protein kinase A (PKA), muscle relaxation is enhanced, and beta-agonist activates PKA and enhances relaxation. Mutations in PLN blunt beta-adrenergic activation of PKA and control of Ca^{2+} cycling leading to decreased contractility [87].

Another important gene is the voltage-gated, type V alpha-subunit of the cardiac sodium channel (*SCN5A*) which is involved in the cardiac action potential. The channel is responsible for rapid depolarization of the myocardium and the mainte-nance of impulse conduction. Mutations lead to a variety of arrhythmic disease and also DCM by modifying electrical excitability of the channel leading to disturbances of currents and different ions which are also involved in contraction [88].

The ABCC9 protein is predicted to form ATP-sensitive potassium channels in cardiac and other muscles. Variants in this gene have been also associated with DCM [89].

Z-Disc-Associated Proteins The sarcomeric Z-disc defines the lateral borders of the sarcomere and consists of actin filaments coming from adjacent sarcomeres which are cross-linked by α-actinin molecules. The Z-disc serves as a nodal point with multiple functions such as intracellular signaling, mechanosensation and mechanotransduction. The Z-disc also links to the T-tubular system and the sarco-plasmic reticulum; moreover, several E3 ubiquitin ligases localized to the Z-disc link this structure to protein turnover and autophagy [32]. To date, mutations in multiple Z-disc proteins or Z-disc-related proteins have been reported to be associated with DCM including ACTN2, ANKRD1, BAG3, CRYAB, CSRP3, LDB3, MYPN, NEBL, NEXN, TCAP, FLH2, FLNC, and DES; however the evidence for clear pathogenicity is variable (Table 2.3) [90]. A typical sarcomeric Z-disc protein is α-actinin 2 (ACTN2), which is important for actin filament localization. The cardiac LIM protein CSRP3 is a protein with multiple interaction partners and is involved in various signal transduction cascades, including mechanosensation and mechanotransduction, calcium metabolism, and myofibrillogenesis. Mutations in *CSRP3* can either cause DCM or HCM [91, 92].

A protein, which is present at the sarcomeric Z-disc but also as an intermediate filament, is desmin (DES). It provides the link to the desmosomes and to other compartments such as the nucleus [93].

An interesting protein is the co-chaperone BAG3 (Bcl2-associated athanogene 3), which acts as a multifunctional adaptor protein responsible for cell survival, apopto-sis, migration, proliferation, macro-autophagy, and proteasomal processes. Mutations causing DCM may influence BAG3-interacting proteins such as HSP70 responsible for protein folding and quality control of newly synthetized proteins [94].

Overall, the Z-disc-associated proteins are diverse and are becoming a "hotspot" for cardiomyopathy-causing mutations; however the spectrum of functional disturbances is various and depending on their specific location and behavior.

Desmosomes Desmosomes are located at the intercalated discs and part of a junctional structure called "area composita." Mutations in desmosomal proteins can cause DCM-like phenotypes but are more frequently involved in ACM (see chapter ACM). There is a substantial overlap, in particular in phenotypic features of more left-sided ACM phenotypes and DCM with pronounced arrhythmic features, as seen mainly with mutations in desmoplakin (*DSP*), but also in other desmosomal proteins [95, 96].

DCM Associated with Other Pathologies As indicated in Table 2.3, DCM is sometimes associated with other non-cardiac phenotypic features, in particular myopathies. In addition, DCM can also occur as part of a syndrome that often affects children.

Interestingly, there are frequent associations of DCM with either myofibrillar myopathy (MFM), Emery-Dreifuss muscular dystrophy (EDMD), or limb-girdle muscular dystrophy (LGMD). Genetic variations in the same genes can cause isolated DCM and associations with those types of skeletal muscle diseases. Often, but not always, isolated DCM is inherited as an autosomal dominant disease, whereas associated myopathies are X-linked or autosomal recessive.

MFMs are chronic neuromuscular disorders. The morphologic changes in skeletal muscle and sometimes in cardiac muscle result from disintegration of the sarcomeric Z-disc and the myofibrils, followed by abnormal ectopic accumulation of multiple proteins (myofibrillar cytoplasmic inclusions). Mutations in genes encoding desmin (DES), alpha-B-crystallin (CRYAB), dystrophin (DMD), filamin C (FLNC), LIM domain-binding protein 3 (LDB3), and BAG3 [97] are causing MFM and/or DCM.

The two other common myopathies are limb-girdle muscular dystrophy (LGMD) and Emery-Dreifuss muscular dystrophy (EDMD), which are mainly inherited as recessive disorders. EDMD is characterized by myopathic changes in certain skeletal muscles and early contractures at the neck, elbows, and Achilles tendons, as well as cardiac conduction defects. Disease genes for EDMD are emerin (EMD), lamin A/C (LMNA), and nesprin-1 (SYNE1) [98]. Autosomal recessive LGMDs are a heterogenous group of disorders affecting primarily skeletal muscle in various forms; in some, DCM is part of the clinical spectrum such as in LGMD2F, LGMD2E, LGMD2D, and LGMD2G, whereas dominant mutations in LMNA can cause LGMD1B [99].

DCM is an important part of the disease spectrum in Duchenne and Becker muscular dystrophies caused by mainly truncating mutations in the dystrophin gene (DMD). In Duchenne or Becker with a later onset, cardiac features are present in about 95% of cases by the last years of life. Also, female carriers of these X-linked diseases develop DCM in about 15% of cases [100].

Lastly, DCM is also part of inherited syndromes. Some of them are listed in Table 2.3. They are usually associated with a congenital onset, inherited as a recessive or X-linked trait, and affect proteins that play a role in the heart and other tissues.

Genetic Testing and Counselling The increasing understanding of the genetic basis of DCM and the availability of comprehensive and cost-effective next-generation sequencing technologies have highlighted the importance of considering genetic testing in all patients with DCM, not just those with an obvious family history or a particular phenotype. Today, cardiomyopathy multi-gene panels or even whole-exome sequencing (WES) is available to perform genetic testing. In clinical settings, as opposed to research testing, the focus should be on known genes associated with DCM [101]. The major challenge occurs in interpreting genetic test results to be used in a clinical setting.

Moreover, not only the interpretation of genetic test results but also proper communication to the patients and their families require a clinician specifically trained in medical genetics and cardiology, who is embedded in a team of experts including genetic counsellors, psychologists, and specialized cardiologists. Expert knowledge of advances in testing, emerging data on newly identified variants and clinical correlations, and an understanding of the complex allelic heterogeneity and complicated scenarios such as multiple variants that may influence the disease are particularly important in DCM.

In the scenario that genetic testing results in a pathogenic/likely pathogenic mutation, the mutation can be considered to be tested in relatives at risk for cascade screening in order to facilitate prompt diagnosis, surveillance, and in selected cases preventative treatment. Furthermore, relatives who do not carry the mutation may be discharged from clinical surveillance which has long-range consequences for families and requires a correct review and classification of genetic variants identified by genetic testing.

Strict variant classification according to the ACMG criteria [9] can facilitate a highly accurate diagnostic yield in DCM, with a pathogenic/likely pathogenic variant detection rate of 35.2%, with 47.6% in familial DCM and 25.6% in sporadic cases [102]. However, even with these restrictions, many variants of uncertain significance (VUSs) are identified mainly in genes with weak evidence of being associated with DCM. For example, the interpretation of *TTN* variants is a big challenge, and pathogenicity depends on additional information and resources, as *TTN* truncating variants are apparently also present in healthy individuals. There are additional online tools available indicating details in the exon composition of the major *TTN* transcripts; information whether an exon is constitutively expressed, and other structural features for each exon; as well as the distribution of TTN variants in large published studies of cohorts of DCM patients and controls [71, 75].

Overall, genetic testing is challenging and takes an expert panel to draw the right conclusions. Likewise, there is no doubt about the importance of genetic counselling and clinical surveillance of first-degree relatives of individuals with DCM.

Genotype-Phenotype Correlation DCM is characterized by marked genetic heterogeneity and variable disease penetrance, which make direct applications to clinical management difficult. Additionally, clinical variability in the phenotype development with the influence of other contributing factors impedes the prediction for a certain genotype [8]. Only few genes such as *TTN* and *LMNA* as well as some founder mutations in *PLN* or *MYBPC3* are more common for comprehensive investigations to allow genotype-phenotype studies.

Mutations in *LMNA* are highly suggestive for progressive conduction disease and ventricular arrhythmias with an increased risk for sudden death. Risk assessment also involves a gender-specific risk with a higher mortality in men compared to women. LMNA-associated DCM demonstrates an age-related penetrance with onset in the third and fourth decades, so that by the seventh decade, penetrance is considered greater than 90–95%. The management implications are to assess for prophylactic implantable cardioverter-defibrillator placement to treat malignant ventricular arrhythmias for patients receiving a pacemaker for LMNA-associated conduction system disease, independent of ejection fraction [76, 103–105].

More recently, genotype-phenotype studies for *TTN* truncation variants suggested a milder phenotype compared to non-*TTN*-related cardiomyopathies, although the comparison was driven by a direct comparison to *LMNA* cardiomyopathy which has, as indicated above, a severe phenotype [106, 107]. First findings also suggest a gender-specific difference with regard to median age of onset and prognosis which was more preferable for woman compared to man. In terms of age-dependent penetrance, current data suggest a late age of onset, ranging between about 30% and 50% at the age of 50 and between 80% and 100% at the age of 70 [102, 106].

Severe phenotypes may also appear when additive genetic effects appear, for example, in cases carrying compound heterozygous variants [108].

2.4   Left Ventricular Noncompaction (LVNC)

Definition and Classification Left ventricular noncompaction (LVNC) is a rare disorder characterized by hypertrophic segments that consist of a thin compacted epicardial layer and a thick noncompacted endocardial layer. The noncompacted layer contains numerous prominent trabeculations and deep intertrabecular recesses [109, 110]. LVNC may be an isolated finding in the absence of any coexisting cardiac anomaly (isolated LVNC) or may be associated with other congenital heart anomalies such as complex congenital heart disease (non-isolated LVNC).

LVNC is a relatively new clinicopathologic condition, first described as such by Chin et al. in 1990 [109]. Based on the predominant myocardial involvement and genetic etiology, LVNC was classified by the American Heart Association (AHA) as a distinct primary cardiomyopathy [1]. The European Society of Cardiology (ESC) recognizes LVNC as an unclassified cardiomyopathy [41]. The ESC questions whether LVNC is a distinct cardiomyopathy or merely a congenital or acquired morphological trait shared by many phenotypically distinct cardiomyopathies [111, 112].

Clinical Manifestation LVNC presents with a unique congenital cardiac morphology, variable clinical features, and a diverse natural history. The heterogeneity of the clinical features includes both asymptomatic and symptomatic patients with progressive deterioration in cardiac function resulting in congestive heart failure, arrhythmias, thromboembolic events, and sudden cardiac death [113–115].

There was a prevalence of 0.014% in patients referred to an echocardiography laboratory in a tertiary referral center [115]. Lately, due to improved imaging techniques, increased awareness, and family screening, the rare entity LVNC is recognized with growing frequency [116, 117]. A substantial proportion of individuals is asymptomatic, suggesting that the true prevalence of LVNC may be higher. Greater extent of, and even excessive, LV trabeculation measured in end-diastole in asymptomatic population-representative individuals by cardiac magnetic resonance imaging (MRI) appeared benign and was not associated with deterioration in LV volumes or function during an almost 10-year period [MESA (Multi-Ethnic Study of Atherosclerosis)] [118]. LVNC was identified as the most frequent cardiomyopathy after DCM and HCM in childhood, with an estimated prevalence of 9% [119] in an Australian cohort and 5% in the US Pediatric Cardiomyopathy Registry [120].

The clinical diagnosis is performed with two-dimensional echocardiography and/or cardiac MRI on the basis of the prominent appearance of LV trabeculae and the ratio between the compacted and noncompacted LV wall (Fig. 2.1c). The diagnosis of LVNC in the symptomatic patient is made at any age, ranging from fetus to old age. Heart failure symptoms are the most common reason for hospital admission. Both systolic and diastolic cardiac dysfunctions have been described. Ventricular tachycardia with a significant risk of cardiac sudden death is frequently found (20–40%). The formation of thrombi in the extensive intertrabecular recesses leading to systemic thromboembolic events is another, although less common, presentation of patients with LVNC. The natural history of LVNC is largely unresolved. As in other cardiomyopathies, index cases represent the most severe spectrum of the disease [114]. LVNC has a high mortality rate and is strongly associated with arrhythmias in children. Preceding cardiac dysfunction or ventricular arrhythmias are associated with increased mortality. Children with normal cardiac dimensions and normal function are at low risk for sudden death [121].

Morphology and Pathogenesis The noncompacted endocardial layer of the myocardium comprises numerous, excessively prominent ventricular trabeculations with deep intertrabecular recesses. The recesses extend deeply into the trabecular meshwork and end at the thin compacted outer layer. In contrast to "persisting myocardial sinusoids" which were first observed in patients with left or right ventricular outflow tract obstruction, recesses in LVNC have no connection with the coronary circulation and are covered by endocardium throughout the ventricular cavity. The most widely used method for the morphologic diagnosis of LVNC is two-dimensional echocardiography and was established by Jenni et al. [110]. The diagnosis of LVNC is made irrespective of the presence of left ventricular systolic dysfunction or dilatation (DCM) due to hypokinetic segments in the altered LV myocardium. Given the common localization of the noncompacted areas in the apex and the common localization of septal hypertrophy in HCM, the two diagnoses of HCM and LVNC can coexist. LV dilation/dysfunction and LV hypertrophy can therefore be present or absent and do not influence LVNC diagnosis. There may be biventricular noncompaction, but criteria for the diagnosis of noncompacted myocardium involving the right ventricle have not been established. Imaging criteria for the assessment of LVNC are still evolving. A significant proportion of an asymptomatic population free from CVD satisfy all currently used cardiac magnetic resonance imaging diagnostic criteria for LVNC, suggesting that those criteria have poor specificity for LVNC [112].

There is evidence that LVNC can originate during embryonic development or be acquired later in life. The de novo LV trabeculations in a significant proportion (>25%) of pregnant women suggest that these may occur as a consequence of increased LV loading conditions or other physiological adaptation mechanisms related to pregnancy. Moreover, the general assumption that LVNC is caused by incomplete myocardial compaction during embryogenesis has been questioned [122, 123]. During fetal development, compaction of the ventricular myocardium normally progresses from epicardium to endocardium, from the base to the apex, and from the septum to the lateral wall. An arrest in this process of myocardial morphogenesis would explain the predominant localization of noncompacted myocardium in LVNC. If not reflecting compaction of pre-existing trabeculations, it is certainly possible that noncompaction relates to defects of proliferation of the compacted myocardium. Mutation of *PRDM16* causes LVNC in patients, and PRDM16 zebrafish mutants show impaired cardiomyocyte proliferation [79]. Genome editing of *PRDM16* leads to proliferation defects in iPSC-CMs, and abnormal TGF-β signaling was suggested as pathological mechanism [124]. Mouse embryos lacking Notch1, systemically or in the endocardium, show defective trabeculation. Molecular analysis of these mutants reveals that expression of endocardial and myocardial trabecular differentiation markers is impaired and ventricular cardiomyocyte proliferation is inhibited. These findings imply that endocardial Notch1 signaling is required for proliferation and differentiation of trabecular myocardium [125].

Disease Genes and Inheritance Among many sporadic cases, familial recurrence has already been observed in the first reports of isolated LVNC. In familial cases, autosomal dominant inheritance is more common than X-linked inheritance [126]. LVNC has been associated with mutations in almost 20 different genes (Table 2.4 and Fig. 2.2c). Defects in sarcomere genes are the most prevalent genetic cause occurring in 30% of adult patients with isolated LVNC. *MYH7* is the most frequent LVNC-associated gene in adult patients with isolated LVNC [127]. Mutations in the sarcomere genes encoding thick (MYH7) [128–131], intermediate/thick filament associated (MYBPC3) [127, 129, 132], and thin filaments (TNNT2, TPM1, ACTC1) [127, 129, 130, 133] have been described. *TTN*, encoding for the giant protein Titin that serves as a molecular spring of the sarcomere, has recently been added to the list of sarcomere genes in LVNC [134].

However, mutations in other genes are only rare causes of LVNC in single families. *TAZ* was the first gene shown to be associated with isolated LVNC by genetic linkage analysis in a family with X-linked inheritance [135]. *TAZ* encodes for taffazin, a protein involved in the biosynthesis of cardiolipin, an essential component of the inner mitochondrial membrane. A small proportion of familial autosomal dominant LVNC can be explained by mutations in genes encoding proteins of the Z line of the sarcomere, LDB3/ZASP [129, 136] and *ACTN2*/α-actinin 2 [137]. Two cardiac ion channels *HCN4*/hyperpolarization-activated cyclic nucleotide-gated potassium channel 4 [138, 139] and *SCN5A*/cardiac sodium channel alpha-subunit gene [140] lead to LVNC with sinus node dysfunction and arrhythmias, respectively. A component of the nuclear lamina,

Table 2.4 Overview about genes associated with LVNC

Gene symbol	Encoded protein	Subcellular location	Frequency (%)	First description	Associated phenotype/disease
Predominant LVNC disease genes					
MYH7	β-Myosin heavy chain	Sarcomere, thick filament	<15	2007	Congenital heart defects
MYBPC3	Cardiac myosin-binding protein C	Sarcomere, thick filament associated	<10	2010	Congenital heart defects
ACTC1	Cardiac α-actin	Sarcomere, thin filament	<5	2007	Congenital heart defects
TNNT2	Cardiac troponin T	Sarcomere, thin filament	<5	2008	None
TPM1	α-Tropomyosin	Sarcomere, thin filament	<5	2010	None
PRDM16	PR domain-containing protein 16	Transcription factor	<5	2013	1p36 deletion syndrome
TAZ[a]	Taffazin	Mitochondrial chaperone	Rare	1997	Barth syndrome
DTNA	α-Dystrobrevin	Cytoskeleton, contractile force transduction	Rare	2001	Congenital heart defects
LDB3	LIM domain-binding protein 3	Z-disc	Rare	2003	None
MIB1	Mindbomb drosophila homolog 1	Cell-cell signaling, ubiquitin ligase	Rare	2013	None
ACTN2	α-Actinin 2	Z-disc	Rare	2014	Ventricular fibrillation, sudden death
HCN4	Hyperpolarization-activated cyclic nucleotide-gated potassium channel 4	Ion channel	Rare	2014	Sinus node dysfunction/bradycardia
Rare, potentially associated genes					
LMNA	Lamin A/C	Nuclear envelope	Rare	2004	Cardiac conduction disease
SCN5A	Voltage-gated sodium channel type V, α-subunit	Ion channel	Rare	2008	Arrhythmia
NKX2-5	NK2 homeobox 5	Transcription factor	Rare	2011	Congenital heart defects, sudden death

DSP[b]	Desmoplakin	Desmosomes, intercalated disc, cell adhesion	Rare	2011	Acantholytic palmoplantar keratoderma
TTN	Titin	Sarcomere, molecular spring	Rare	2016	None

[a]X-linked inheritance
[b]Autosomal recessive; all other disorders are autosomal dominant

lamin A/C (*LMNA*) [141] causes LVNC with cardiac conduction disease. Mutations in the gene *MIB1* segregated with autosomal dominant LVNC in two families, and a conditional loss-of-function allele in a mouse model, also led to LVNC [142]. The hypertrabeculation and noncompaction seen in the MIb1 mouse was mimicked in a mouse with inactivation of Jagged1 in the myocardium or Notch1 in the endocardium, suggesting that the Notch1 signaling pathway was involved.

Genetic Syndromes and/or Other Disorders LVNC is present in a number of neuromuscular disorders, metabolic and mitochondrial disease, congenital malformations, and chromosomal syndromes. Some of these disorders may share pathogenetic mechanisms with LVNC. Alternatively, LVNC might be secondary to other cardiac malformations or even vice versa.

LVNC can occur as part of a syndrome in combination with dysmorphic features and other congenital malformations. Chromosomal deletions have been found on chromosomes 1p36, 1q43, and 5q35 [79]. Left ventricular noncompaction is known to be a part of various syndromes, including the Barth, Noonan, Roifman, Melnick-Needles, Nail-Patella, Toriello-Carey, and other uncommon syndromes [143]. *TAZ* mutation typically results in Barth syndrome, which is characterized by cardiomyopathy (frequently LVNC), skeletal myopathy, cyclic neutropenia, and 3-methylglutaconic aciduria (a marker of mitochondrial dysfunction) [144]. *DSP*, encoding for desmoplakin, a gene known for the first recessive human mutation that causes a generalized striate keratoderma particularly affecting the palmoplantar epidermis, woolly hair, and dilated left ventricular cardiomyopathy (Carvajal syndrome), also leads to LVNC with acantholytic palmoplantar keratoderma with autosomal recessive inheritance [145].

The co-occurrence of congenital heart defects (CHD) and noncompaction is seen in children and in adults [144, 146]. *DTNA*, encoding for α-dystrobrevin, a cytoskeletal component responsible for force transduction, is mutated in patients with hypoplastic left heart syndrome and LVNC [144]. Septal defects and Ebstein anomaly are the most prevalent congenital heart defects in LVNC. Mutations in the cardiac transcription factor *NKX2-5* were identified in children with LVNC and atrial septal defects [147]. Noncompaction associated with Ebstein anomaly, a rare type of CHD, has been shown to result from mutations in *MYH7* [131]. The association of sarcomere gene defects, cardiomyopathy, and structural CHD still requires further investigation. Very much like HCM and DCM, LVNC has been linked to neuromuscular disorders such as dystrophinopathies and with mitochondrial disease. In a number of muscular dystrophies, "left ventricular hypertrabeculation" was identified [148]. In a study of 113 pediatric patients with mitochondrial disease, LVNC was identified in 13% [149].

Genetic Counselling and Testing Since extensive family studies showed that the majority of affected relatives are asymptomatic, cardiologic evaluation should include all first-degree relatives irrespective of medical history. Other cardiomyopathies may co-occur within families, like HCM and DCM, so cardiac screening should aim at identifying all cardiomyopathies. Cardiac screening of relatives may show minor abnormalities not fulfilling LVNC criteria. Genetic testing, preferably with a targeted cardiomyopathy gene panel, may lead to the identification of a molecular defect in over 40% of isolated LVNC patients [129].

Molecular studies of LVNC have thus far shown that there are few recurrent mutations. Therefore, it is difficult to establish genotype-phenotype correlations. Additionally, intrafamilial phenotypic variability complicates predictions based on an identified mutation. Compound heterozygous or homozygous truncating sarcomere gene mutations appear to result in a more severe phenotype with childhood onset [132]. Whereas TAZ mutations have been reported in pediatric patients often diagnosed with LVNC during infancy, heterozygous mutations in autosomal dominant LVNC were mainly identified in adult patients.

Probst et al. [127] reported a cohort of 63 LVNC probands, previously studied by Klaassen et al. [130], in which 8 sarcomere genes were analyzed and heterozygous mutations found in 18 (29%) of the probands: 8 mutations were in the MYH7 gene, 5 in MYBPC3, 2 in ACTC1, 2 in TPM1, and 1 in TNNT2. There were no significant differences between mutation-positive and mutation-negative probands in terms of average age, myocardial function, or presence of heart failure or tachyarrhythmias at initial presentation or at follow-up. Probst et al. [127] noted that although 8 of the 15 distinct mutations were novel in this cohort, they were likely not specific to LVNC, because the other 7 mutations had previously been described in patients with other forms of cardiomyopathy, including hypertrophic and dilated forms. In conclusion, molecular analyses of LVNC support the concept of a shared molecular etiology of the different cardiomyopathic phenotypes.

For further reading: Left Ventricular Noncompaction [143] in Clinical Cardiogenetics, Springer International Publishing, H.F. Baars et al. (eds), 2016.

2.5 Arrhythmogenic Cardiomyopathy (ACM)

Definition Arrhythmogenic right ventricular cardiomyopathy (ARVC) is a primary cardiomyopathy characterized by progressive fibro-fatty replacement of the right and left ventricular myocardium (Fig. 2.1d) [150]. In the early eighteenth century, the clinical phenotype of ARVC was first reported by Giovanni Maria Lancisi in a four-generation family, where patients presented with right ventricular dilation or aneurysms and suffered from sudden cardiac death, both typical signs of the disease [151]. During the 1980s, clinical features were systematically described by Marcus and colleagues [152]. Particularly, the central involvement of the right ventricle was described. However, today there is growing evidence that the left ventricle is also affected leading to the term "arrhythmogenic cardiomyopathy" (ACM), which will be also used in this chapter. The estimated disease prevalence ranges from 1:2000 to 1:5000 considering ACM as a rare disease [153].

Clinical Manifestation Clinical features of the disease are highly variable ranging from unaffected mutation carriers to patients suffering from sudden cardiac death or requiring heart transplantation. Classical symptoms include palpitations, cardiac syncope, and aborted cardiac arrest due to ventricular arrhythmias. Heart failure is rare and may develop in later stages. ACM is a progressive disease which typically manifests in the second to fourth life decade. However, about 20% of cases develop the disease after the age of 50 [154]. Penetrance is age and gender dependent, and clinical manifestations and progression of the disease are highly variable. Men are clinically more often affected than women. The influence of sex hormones and higher physical activity of men may contribute to the disease expression. The clinical diagnosis is challenging, because of the lack of a specific diagnostic test. Hence, the diagnosis is based on fulfilling a set of major and minor criteria proposed by an international task force initially in 1994 [155]. This set of Task Force Criteria (TFC) was highly specific but lacked sensitivity for early forms of ACM and has been revised in 2010 to incorporate new findings and diagnostic modalities [150]. The TFC include evaluation of findings from six different diagnostic categories including cardiac imaging, assessment of electrical alterations, and family history. Management is individualized and focused on prevention of sudden death through use of antiarrhythmic medication and implantable cardioverter-defibrillators and rarely heart transplantation.

Inheritance An autosomal dominant trait is the most common inheritance pattern for non-syndromic forms of the disease. Today about 40–60% of the genetic causes are known, and plakophilin 2 (*PKP2*) is the most prevalent disease gene [4]. However, there have been reports of autosomal recessive patterns of almost all desmosomal disease genes; with and without additional syndromic features. For example, mutations in genes like plakoglobin (*JUP*) and desmoplakin (*DSP*) lead to ACM, palmoplantar keratoderma, and abnormalities of the hair structure. Carriers with homozygous mutations in desmocollin 2 (*DSC2*) show an isolated cardiac phenotype (Table 2.5). As seen in other cardiomyopathies, incomplete penetrance and variable clinical expression are also common; even in the same family, the severity of phenotypes demonstrates a wide range. In some cases, different family members develop DCM or ACM [156]. Besides cases with one mutation, there are several reports about compound heterozygous mutations or two or more variants in different genes [157]. Of note, there are also rare cases of de novo mutations [158]. Besides the main genetic drivers, also modifiers such as environmental (e.g., athletic activity) and epigenetic factors might influence the phenotype.

Disease Genes At the beginning of the 2000s, it was recognized that about 50% of ACM patients carry mutations in five different genes encoding desmosomal proteins: *JUP* [5], *DSP* [159], *PKP2* [4], *DSC2* [160, 161], and *DSG2* [162]. Desmosomes are cell-cell junctions, which have important functions for the nano-mechanical coupling of cells [96]. In organs exposed to high mechanical stress like the skin or the heart, cells are connected by these multi-protein complexes. Desmosomes are linked to the intermediate filament system, which is mainly formed in the heart by desmin (*DES*). Besides cardiac desmosomes, also adhering junctions are important for mechanical coupling of cardiomyocytes. Adhering junctions are linked to actin filaments [163]. Interestingly, ACM-associated mutations were also

Table 2.5 Overview about genes associated with ACM

Gene symbol	Encoded protein	Subcellular location	Frequency (%)	First description	Inheritance
PKP2	Plakophilin 2	Desmosomes, intercalated disc, cell adhesion	20–40	2004	AD, AR
DSP	Desmoplakin	Desmosomes, intercalated disc, cell adhesion	5–10	2002	AR[a], AD
DSC2	Desmocollin 2	Desmosomes, intercalated disc, cell adhesion	5–10	2006	AR, AD
DSG2	Desmoglein 2	Desmosomes, intercalated disc, cell adhesion	5–10	2006	AD; AR
JUP	Plakoglobin	Desmosomes, intercalated disc, cell adhesion	<5	2000	AR[b]
RYR2	Ryanodine receptor 2	SR, Ca^{2+} channel	Rare	2001	AD
TGFB3	Transforming growth factor β3	Cytokine	Rare	2005	AD
TMEM43	Luma	Nuclear envelope, endoplasmic reticulum	Rare	2008	AD
DES	Desmin	IF protein, intercalated disc, Z-disc,	Rare	2009	AD[c]
TTN	Titin	Sarcomere, molecular spring	Rare	2011	AD
PLN	Phospholamban	SR, Ca^{2+} handling	Rare	2012	AD
LMNA	Lamin A/C	IF protein, nuclear lamina	Rare	2012	AD
SCN5A	Voltage-gated sodium channel, type V, α-subunit	Intercalated disc, T-tubules system, sodium channel	Rare	2008	AD
CTNNA3	αT-Catenin	Intercalated disc, cell adhesion	Rare	2013	AD
LDB3	LIM domain-binding protein 3	Z-disc, structural integrity	Rare	2014	AD
KCNQ1	Voltage-gated potassium channel Q1	T-tubules system, potassium channel	Rare	2014	AD
CDH2	N-Cadherin	Intercalated disc, cell adhesion	Rare	2017	AD

[a]Carvajal syndrome
[b]Naxos disease
[c]Skeletal myopathy
AD autosomal dominant, *AR* autosomal recessive, *SR* sarcoplasmic reticulum, *IF* intermediate filament

identified in *CTNNA3* and *CDH2*, which encode two structural proteins of adhering junctions [164, 165]. Therefore, it is suggested that ACM is mainly a disease of the cardiac intercalated disc, as both desmosomes and adhering junctions are located there. In 40–60% of ACM patients, one or more mutations in genes encoding proteins involved in cell adhesion could be identified (Table 2.5; Fig. 2.2d) [158, 166]. In addition to genes, encoding structural proteins involved in cell adhesion of cardiomyocytes, there are some reports about mutations in other genes like *RYR2* [167], *TGFB3* [168], *TMEM43* [169], *SCN5A* [170], *TTN* (M [171]), *PLN* [156], *LMNA* [172], *LDB3* [173], and *KCNQ1* [174]. However, mutations in those genes are rare and account for less than 10% (Fig. 2.2d; Table 2.5).

Cell-Cell Adhesion McKoy et al. identified the first homozygous 2 bp deletion mutation in *JUP* in families from the Greek island Naxos [5]. Patients developed ACM in combination with keratosis and woolly hair. The typical triad of features is known in the literature as "Naxos disease" [175]. *JUP* encodes plakoglobin, a member of the armadillo protein family, which is a structural cytoplasmic component of the desmosome. Nonsense-mediated mRNA decay (NMD) is linked to the pre-mRNA splicing process, and the exon-junction complexes are necessary for the degradation of mRNA carrying a premature termination codon (PTC) [176]. Zhang et al. generated two *Jup* knock-in mouse models with the same deletion mutation. In one of them, the authors deleted additionally the introns to block NMD. Remarkably, the knock-in mice without the introns were completely healthy [177]. Those findings suggest that potentially "loss of function" mediated by NMD may explain the mechanism due to *JUP* mutations.

Similar to Naxos disease, Carvajal syndrome is characterized by striate keratoderma, woolly hair, and left ventricular arrhythmic cardiomyopathy. It is caused by mutations in *DSP* [178]. Rampazzo et al. described the first *DSP* mutation in dominant ACM [159]. Desmoplakin is a cytolinker protein and connects the cardiac desmosomes with the desmin intermediate filaments. The C-terminal tail domain is responsible for this protein-protein interaction. Desmoplakin forms dimers by the formation of a coiled coil of the Rod domains [179]. Autosomal recessive as well as autosomal dominant *DSP* mutations were reported in 5–10% of ACM patients (Table 2.5).

In 2004, Gerull et al. identified *PKP2* mutations in a large cohort of ACM patients [4]. *PKP2* is the most common gene for autosomal dominant ACM, and it accounts for about 20–40% of the known mutations. Most mutations are small deletions, insertions, nonsense, or splice site changes leading to the truncation of the protein. *PKP2* encodes plakophilin 2, which is also a member of the Armadillo protein family. Plakophilin 2 binds to the desmosomal cadherins and mediates the molecular interaction with the cytolinker protein desmoplakin [180]. Plakophilin 2 is necessary for the vesicle transport of the desmosomal cadherin desmocollin 2 to the plasma membrane [181]. It is suggested that haploinsufficiency could be the main genetic mode of action. Presumably, NMD and protein degradation pathways are involved in *PKP2* haploinsufficiency [182].

Different groups described also mutations in the genes *DSC2* [160, 161] and *DSG2* [162] encoding for desmosomal cadherins. The extracellular domains of both cadherins mediate the protein-protein interactions in a Ca^{2+}-dependent way by homophilic and presumably also by heterophilic trans-interaction [183]. Besides the description of heterozygous missense and nonsense variants in *DSC2* with, so far, lack of clear pathogenicity, there have also been homozygous truncation mutations reported causing a severe biventricular form of ACM without skin and hair features [184, 185]. Interestingly, Wong et al. investigated the clinical phenotype of a founder population with 28 heterozygous and 11 homozygous *DSC2*-p. Q554X mutation carriers. Almost all of the heterozygous carriers were healthy or presented only with mild clinical symptoms, whereas all homozygous mutation carriers developed ACM [186].

Besides the desmosomal genes, ACM-related mutations were also identified in desmin (*DES*) [158, 187]. Desmin is the major component of the intermediate filaments in cardiomyocytes, which provides the connection to different cellular organelles like the desmosomes and the costameres. Interestingly, ACM-associated *DES* mutations disturb the intermediate filament assembly [188].

Recently, mutations in αT-Catenin (*CTNNA3*) and in N-cadherin (*CDH2*) have been described [164, 165, 189]. Both proteins are structural components of the adhering junctions. Adhering junctions are cell-cell junctions which are connected to F-actin filaments. However, further studies are required to investigate the frequency and the pathogenic impact of variants/mutations in those genes.

Other Genetic Contributors Besides mutations affecting cell-cell adhesion proteins, there are also rare mutations reported in genes involved in Ca^{2+} handling like the *RYR2* [167] or *PLN* [156]. *RYR2* encodes the cardiac ryanodine receptor 2 which is mediating the Ca^{2+} release from the sarcoplasmic reticulum [190], whereas phospholamban (*PLN*) is a regulator of the sarco(endo)plasmic reticulum Ca^{2+} ATPase (SERCA), which transports Ca^{2+} into the sarcoplasmic reticulum [191]. Rare variants were also identified in the $5'$- and $3'$-untranslated regions of *TGFB3*, a cytokine from the transforming growth factor family [168]. TGF-β cytokines are involved in fibrotic remodeling process by upregulation of expression of extracellular matrix proteins like different collagens [192]. Merner et al. identified a missense mutation in *TMEM43* in a founder population from Newfoundland [169]. Remarkably, this specific mutation leads to nearly complete penetrance [169] and was also identified in ACM patients from different other countries [193]. *TMEM43* encodes the nuclear envelope protein luma [194]. The exact biochemical functions and the molecular structure of luma are widely unknown. Additionally, mutations were also identified in *LMNA* [172], which encodes the nuclear intermediate filament lamin A/C involved in the structural stabilization of the nuclei. Other rare variants were identified in *SCN5A* [170], encoding the α-subunit of the voltage-gated sodium channel 5, and in *KCNQ1* [174], encoding the voltage-gated potassium channel Q1. Mutations in both genes cause inherited arrhythmias like Brugada syndrome or long and short QT syndrome, indicating a genetic overlap of ACM with channelopathies.

Genetic Counselling and Testing Clinical genetic testing in ACM is in particular challenging and requires to be performed in dedicated cardio-genetic centers, with pre- and post-counselling possibilities. As indicated above, the inheritance pattern of ACM is more complex than previously thought, with frequent requirement for more than one "hit" for fully penetrant disease. More frequent than in other CMPs, patients carry homozygous or compound heterozygous variants in the same gene or digenic/oligogenic variants in a cluster of desmosomal genes. The broad use of large gene panels or whole-exome sequencing often results in a high number of variants of unknown clinical significance (VUSs), which should be carefully classified according to guidelines as suggested by the ACMG [9]. Additionally, genetic testing should be performed with the consideration that even a positive genetic test result assigned as a pathogenic or likely pathogenic mutation may not be the only genetic contributor to the phenotype in the patient and/or family.

However, genetic counselling should be done in all families with a confirmed case of ACM, and first-degree relatives should be screened according to the TFC by ECG, echocardiography, Holter ECG, and signal-averaged ECG starting at the age of 10 years, and this should be regularly repeated [40]. In case there is a reliable genetic test result (pathogenic or likely pathogenic mutation) available, this may be used for cascade family screening, although cardiologic evaluation may still continue at a lower level as a negative genetic result may not exclude any genetic predisposition. Lifestyle advice and reproductive counselling on the risk of transmission to offspring should be also provided. Restriction from competitive sports activity and strenuous physical exercise is strongly recommended to prevent disease onset and progression.

Despite some progress in the understanding of the functional impact of certain mutations, there is still a gap of knowledge of all contributing factors resulting in the clinical picture of ACM. Hopefully, the increasing genetic knowledge will also lead to more profound biochemical and cellular understanding of the pathomechanisms toward the development of more efficient molecular therapies.

2.6 Restrictive Cardiomyopathy (RCM)

Definition Restrictive cardiomyopathy (RCM) is characterized by an abnormal left ventricular filling pattern, diastolic dysfunction but normal wall thickness, and usually preserved systolic function. The restrictive filling pattern is caused by an increased muscle stiffness leading to atrial enlargement and an increase of ventricular end-diastolic blood pressure (Fig. 2.1e). RCM can be classified as a rare primary cardiomyopathy, often affecting children but also adults. However, there are also secondary causes known, such as storage or infiltrative disorders [41, 195].

Clinical Manifestation and Inheritance Patients typically develop heart failure symptoms such as dyspnea and fatigue [196]. Preserved LV ejection fraction, normal or mildly increased left and right ventricular wall thicknesses, restrictive diastolic filling patterns and severe atrial enlargement are common echocardiographic findings. The general prognosis is poor, patients often requiring heart

Table 2.6 Overview about genes associated with RCM

Gene symbol[a]	Encoded protein	Subcellular location	First description
TNNI3	Cardiac troponin I	Sarcomere, thin filament	2003
TNNT2	Cardiac troponin T	Sarcomere, thin filament	2006
DES	Desmin	IF protein, intercalated disc, Z-disc	2006
ACTC1	Cardiac actin	Sarcomere, thin filament	2008
MYH7	β-Myosin heavy chain	Sarcomere, thick filament	2008
TTN	Titin	Sarcomere	2014
MYPN	Myopalladin	Sarcomere, Z-band	2012
MYBPC3	Cardiac myosin-binding protein C	Sarcomere, thick filament associated	2015
TNNC1	Cardiac troponin C	Sarcomere, thin filament	2016
FLNC	Filamin C	Cytoskeleton, intercalated disc	2016
TMEM87B	Transmembrane protein 87 B	Membrane	2016
ACTN2	α-Actinin 2	Z-disc	2016
CRYAB	αB-Crystallin	IF protein, intercalated disc, Z-disc,	2017[b]

[a]Autosomal dominant inheritance
[b]Phenotype associated with skeletal myopathy
IF intermediate filament

transplantation. Clinically, the differentiation between RCM and constrictive pericarditis imposed by external pericardial constraint can be challenging. Secondary RCM can be part of systemic diseases, e.g., a mineralization disorder called pseudoxanthoma elasticum (PXE), caused by *ABCC6* mutations [197] or of cardiac amyloidosis associated with *TTR* mutations [198]. Additionally, metabolic diseases such as Pompe, Fabry, and Danon disease as well as Friedreich ataxia can be associated with LVH (Table 2.2) but can also present as RCM. Primary RCM is mainly a familial disease and inherited as an autosomal dominant trait (Table 2.6).

Disease Genes RCM is a rare disease and the proportion of genetic etiology remains unknown. However, Kostareva et al. investigated 24 non-related index patients using broad next-generation sequencing panels and identified in about 54% of them a pathogenic or likely pathogenic mutation [199]. Overall, mutations in 13 different genes have been reported (Fig. 2.2e, Table 2.6). Most of those genes encode for sarcomeric or cytoskeletal proteins. Of note, there is also a genetic overlap with other primary cardiomyopathies, in particular with HCM.

Sarcomeric Proteins Sarcomeres are the essential contracting units of striated muscle cells consisting of thick and thin filaments, which are connected by Z-bands. First mutations associated with familial RCM were identified in the *TNNI3* gene, encoding cardiac troponin I (cTnI) [200]. cTnI inhibits the ATPase function of the actin-myosin complex during the contraction cycle [201]. Mutations in cardiac troponin C (*TNNC1*, cTnC) and troponin T (*TNNT2*, cTnT), encoding the two other cardiac components of the troponin complex, were also identified in RCM patients [202, 203]. cTnC is the Ca^{2+}-binding and cTnT is the tropomyosin-binding subunit of the troponin complex, mediating the molecular interaction of thin and thick filaments during the contraction cycles [204]. Other sarcomeric disease genes are *ACTC1* and *MYH7* [205, 206]. Recently, Wu et al. identified a family with three affected patients carrying a nonsense mutation in *MYBPC3*, which encodes the myosin-binding protein C and binds to the thick filaments [207]. Peled and colleagues identified a de novo missense mutation (p.Y7621C) in titin (*TTN*) causing RCM [208]. Furthermore, other RCM-associated mutations were found in genes encoding myopalladin (*MYPN*) and α-actinin 2 (*ACTN2*) [199, 209]. α-Actinin 2 is the major structural component of the Z-bands, and myopalladin is a binding partner. Myopalladin is involved in the regulation and maintenance of sarcomeres close to the Z-disc [210].

Cytoskeletal Proteins The second group of RCM-associated disease genes includes different cytoskeletal structural proteins like desmin (*DES*) [211], filamin C (*FLNC*) [212], and αB-Crystallin (*CRYAB*) [213]. *DES* mutations cause besides RCM a wide spectrum of different other cardiac and skeletal myopathies including myofibrillar myopathy (MFM). Desmin is the major component of the cardiac intermediate filaments and connects different cell organelles like desmosomes, Z-bands, costameres, and presumably also the nuclei with the cytoskeleton [214]. αB-Crystallin is a member of the small heat-shock protein family and binds directly to the intermediate filaments suggesting a stabilizing function, and filamin C is an actin cross-linking protein, which is involved in the integrin-mediated cell-extracellular matrix adhesion [215]. Interestingly, mutations in those three genes cause also myofibrillar myopathy (MFM), characterized by toxic protein aggregation within the skeletal myocytes [216]. There is some evidence that an abnormal protein aggregation and accumulation in the myocardial tissue might contribute to the increased stiffness of the ventricular walls in RCM.

Genetic Counselling and Testing Because RCM is less common than other primary cardiomyopathies, the knowledge about prevalence and genetic associations are currently limited or even missing. Most reports about disease genes are based on single families or patients. As indicated above, there is substantial overlap in the genetic etiology, but also clinical presentation with other cardiomyopathies. Moreover, some genes are also involved in skeletal muscle disease, in particular MFM. Formal genetic counselling should be performed in all familial cases, and repeated clinical screenings of first-degree relatives are recommended using ECG, Holter ECG, and echocardiogram [40]. Secondary forms of RCM and constrictive pericarditis should be considered as a differential diagnosis as some metabolic and systemic diseases may lead to instituting specific therapy. As indicated for other cardiomyopathies, genetic testing should be performed for clinical purposes using NGS panels focusing on known genes. Broader approaches such as whole-exome analysis should be considered for research studies. However, in single patients, the classification of novel variants will be challenging and often leading to VUS. Over time more genetic and phenotypic data on bigger cohorts, genotype-phenotype studies, and further functional validation are required to provide more certain recommendations.

2.7   Translational Perspectives

Classical genetic approaches such as linkage analysis and positional cloning in combination with newer technologies such as next-generation sequencing have provided dramatic progress in unraveling the genetic causes of cardiomyopathies but have also demonstrated the complexity of human disease. What we previously thought are monogenic diseases are now much more complex diseases where the primary mutation appears to be a "more or less" important contributor. Despite making some progress in the understanding of disease mechanisms and the gain of knowledge about molecular changes of specific genes and mutations, there are still a lot of unknowns in the pathogenesis of cardiomyopathies—to understand how a mutated gene leads to clinical expression. Prospectively, this inspiring field of research provides remarkable opportunities to dissect what are the genetic and non-genetic contributors in the early and late disease process, how does different disease expression develop, and how to disrupt disease progression or even prevent full clinical expression.

From the clinical and translational perspective, many more questions should be answered to translate research findings to the benefit of patient care. We still have a "genetic gap"—overall about 50% of the genetic causes of CMPs remain unknown. How do we handle the increasing number of variants of unknown significance? Is there a way to translate research information about specific mutations and how they alter protein function directly to the benefit of patients to improve health care, rather than identifying just family members at risk?

Last but not least, someday there is the hope that genetic cardiomyopathies are not only treated but may also be cured. Given the current advances in genome editing technologies and other personalized genetic and molecular approaches, also outlined in other chapters, this day may be not so far away.

Compliance with Ethical Standards

Conflict of Interest   Brenda Gerull declares that she has no conflict of interest. Sabine Klaassen declares that she has no conflict of interest. Andreas Brodehl declares that he has no conflict of interest.

Ethical Approval   This article does not contain any studies with human participants or animals performed by any of the authors.

References

1. Maron BJ, Towbin JA, Thiene G, Antzelevitch C, Corrado D, Arnett D, et al. Contemporary definitions and classification of the cardiomyopathies: an American Heart Association Scientific Statement from the Council on Clinical Cardiology, Heart Failure and Transplantation Committee; Quality of Care and Outcomes Research and Functional Genomics and Translational Biology Interdisciplinary Working Groups; and Council on Epidemiology and Prevention. Circulation. 2006;113(14):1807–16. https://doi.org/10.1161/CIRCULATIONAHA.106. 174287.
2. Geisterfer-Lowrance AAT, Kass S, Tanigawa G, Vosberg H-P, McKenna W, Seidman CE, Seidman JG. A molecular basis for familial hypertrophic cardiomyopathy: a β cardiac myosin heavy chain gene missense mutation. Cell. 1990;62(5):999–1006. https://doi.org/10.1016/0092-8674(90)90274-I.
3. Ho CY, Charron P, Richard P, Girolami F, Van Spaendonck-Zwarts KY, Pinto Y. Genetic advances in sarcomeric cardiomyopathies: state of the art. Cardiovasc Res. 2015;105 (4):397–408. https://doi.org/10.1093/cvr/cvv025.
4. Gerull B, Heuser A, Wichter T, Paul M, Basson CT, McDermott DA, et al. Mutations in the desmosomal protein plakophilin-2 are common in arrhythmogenic right ventricular cardiomyopathy. Nat Genet. 2004;36(11):1162–4. https://doi.org/10.1038/ng1461.
5. McKoy G, Protonotarios N, Crosby A, Tsatsopoulou A, Anastasakis A, Coonar A, et al. Identification of a deletion in plakoglobin in arrhythmogenic right ventricular cardiomyopathy with palmoplantar keratoderma and woolly hair (Naxos disease). Lancet (London, England). 2000;355(9221):2119–24.
6. Genomes Project, C, Abecasis GR, Altshuler D, Auton A, Brooks LD, Durbin RM, et al. A map of human genome variation from population-scale sequencing. Nature. 2010;467 (7319):1061–73. https://doi.org/10.1038/nature09534.
7. Golbus JR, Puckelwartz MJ, Dellefave-Castillo L, Fahrenbach JP, Nelakuditi V, Pesce LL, et al. Targeted analysis of whole genome sequence data to diagnose genetic cardiomyopathy. Circ Cardiovasc Genet. 2014;7(6):751–9. https://doi.org/10.1161/CIRCGENETICS.113. 000578.
8. Kayvanpour E, Sedaghat-Hamedani F, Amr A, Lai A, Haas J, Holzer DB, et al. Genotype-phenotype associations in dilated cardiomyopathy: meta-analysis on more than 8000 individuals. Clin Res Cardiol. 2017;106(2):127–39. https://doi.org/10.1007/s00392-016-1033-6.
9. Richards S, Aziz N, Bale S, Bick D, Das S, Gastier-Foster J, et al. Standards and guidelines for the interpretation of sequence variants: a joint consensus recommendation of the American College of Medical Genetics and Genomics and the Association for Molecular Pathology. Genet Med. 2015;17(5):405–23. https://doi.org/10.1038/gim.2015.30.
10. Maron BJ, Gardin JM, Flack JM, Gidding SS, Kurosaki TT, Bild DE. Prevalence of hypertrophic cardiomyopathy in a general population of young adults. Echocardiographic Analysis of 4111 Subjects in the CARDIA Study. Coronary Artery Risk Development in (Young) Adults. Circulation. 1995;92(4):785–9. https://doi.org/10.1161/01.cir.92.4.785.
11. Elliott PM, Anastasakis A, Borger MA, Borggrefe M, Cecchi F, Charron P, et al. 2014 ESC Guidelines on diagnosis and management of hypertrophic cardiomyopathy: the Task Force for the Diagnosis and Management of Hypertrophic Cardiomyopathy of the European Society of Cardiology (ESC). Eur Heart J. 2014;35(39):2733–79. https://doi.org/10.1093/eurheartj/ehu284.
12. Gersh BJ, Maron BJ, Bonow RO, Dearani JA, Fifer MA, Link MS, et al. ACCF/AHA Guideline for the Diagnosis and Treatment of Hypertrophic Cardiomyopathy: A Report of the American College of Cardiology Foundation/American Heart Association Task Force on Practice Guidelines Developed in Collaboration With the American Association for Thoracic Surgery, American Society of Echocardiography, American Society of Nuclear Cardiology, Heart Failure Society of America, Heart Rhythm Society, Society for Cardiovascular

Angiography and Interventions, and Society of Thoracic Surgeons. J Am Coll Cardiol. 2011;58(25):e212–60. https://doi.org/10.1016/j.jacc.2011.06.011.

13. Nagueh SF, Mahmarian JJ. Noninvasive cardiac imaging in patients with hypertrophic cardiomyopathy. J Am Coll Cardiol. 2006;48(12):2410–22. https://doi.org/10.1016/j.jacc.2006.07.065.

14. Maron BJ, Rowin EJ, Casey SA, Garberich RF, Maron MS. What do patients with hypertrophic cardiomyopathy die from? Am J Cardiol. 2016a;117(3):434–5. https://doi.org/10.1016/j.amjcard.2015.11.013.

15. Maron BJ, Rowin EJ, Casey SA, Lesser JR, Garberich RF, McGriff DM, Maron MS. Hypertrophic cardiomyopathy in children, adolescents, and young adults associated with low cardiovascular mortality with contemporary management strategies. Circulation. 2016b;133(1):62–73. https://doi.org/10.1161/circulationaha.115.017633.

16. Maron BJ, Rowin EJ, Casey SA, Link MS, Lesser JR, Chan RH, et al. Hypertrophic cardiomyopathy in adulthood associated with low cardiovascular mortality with contemporary management strategies. J Am Coll Cardiol. 2015;65(18):1915–28. https://doi.org/10.1016/j.jacc.2015.02.061.

17. Charron P, Carrier L, Dubourg O, Tesson F, Desnos M, Richard P, et al. Penetrance of familial hypertrophic cardiomyopathy. Genet Couns. 1997;8(2):107–14.

18. Ingles J, Doolan A, Chiu C, Seidman J, Seidman C, Semsarian C. Compound and double mutations in patients with hypertrophic cardiomyopathy: implications for genetic testing and counselling. J Med Genet. 2005;42(10):e59. https://doi.org/10.1136/jmg.2005.033886.

19. Maron BJ. Hypertrophic cardiomyopathy: a systematic review. JAMA. 2002;287(10):1308–20.

20. Niimura H, Bachinski LL, Sangwatanaroj S, Watkins H, Chudley AE, McKenna W, et al. Mutations in the gene for cardiac myosin-binding protein c and late-onset familial hypertrophic cardiomyopathy. N Engl J Med. 1998;338(18):1248–57. https://doi.org/10.1056/nejm199804303381802.

21. Page SP, Kounas S, Syrris P, Christiansen M, Frank-Hansen R, Andersen PS, et al. Cardiac myosin binding protein-C mutations in families with hypertrophic cardiomyopathy: disease expression in relation to age, gender, and long term outcome. Circ Cardiovasc Genet. 2012;5(2):156–66. https://doi.org/10.1161/circgenetics.111.960831.

22. Golbus JR, Puckelwartz MJ, Fahrenbach JP, Dellefave-Castillo LM, Wolfgeher D, McNally EM. Population-based variation in cardiomyopathy genes. Circ Cardiovasc Genet. 2012;5(4):391–9. https://doi.org/10.1161/CIRCGENETICS.112.962928.

23. Thierfelder L, Watkins H, MacRae C, Lamas R, McKenna W, Vosberg H-P, et al. α-tropomyosin and cardiac troponin T mutations cause familial hypertrophic cardiomyopathy: a disease of the sarcomere. Cell. 1994;77(5):701–12. https://doi.org/10.1016/0092-8674(94)90054-X.

24. Watkins H, Conner D, Thierfelder L, Jarcho JA, MacRae C, McKenna WJ, et al. Mutations in the cardiac myosin binding protein-C gene on chromosome 11 cause familial hypertrophic cardiomyopathy. Nat Genet. 1995;11(4):434–7.

25. Bonne G, Carrier L, Bercovici J, Cruaud C, Richard P, Hainque B, et al. Cardiac myosin binding protein-C gene splice acceptor site mutation is associated with familial hypertrophic cardiomyopathy. Nat Genet. 1995;11(4):438–40.

26. Poetter K, Jiang H, Hassanzadeh S, Master SR, Chang A, Dalakas MC, et al. Mutations in either the essential or regulatory light chains of myosin are associated with a rare myopathy in human heart and skeletal muscle. Nat Genet. 1996;13(1):63–9.

27. Kimura A, Harada H, Park J-E, Nishi H, Satoh M, Takahashi M, et al. Mutations in the cardiac troponin I gene associated with hypertrophic cardiomyopathy. Nat Genet. 1997;16(4):379–82.

28. Mogensen J, Klausen IC, Pedersen AK, Egeblad H, Bross P, Kruse TA, et al. α-cardiac actin is a novel disease gene in familial hypertrophic cardiomyopathy. J Clin Invest. 1999;103(10):R39–43. https://doi.org/10.1172/JCI6460.

29. Andersen PS, Havndrup O, Hougs L, Sørensen KM, Jensen M, Larsen LA, et al. Diagnostic yield, interpretation, and clinical utility of mutation screening of sarcomere encoding genes in Danish hypertrophic cardiomyopathy patients and relatives. Hum Mutat. 2009;30(3):363–70. https://doi.org/10.1002/humu.20862.
30. Richard P, Charron P, Carrier L, Ledeuil C, Cheav T, Pichereau C, et al. Hypertrophic cardiomyopathy. Distribution of disease genes, spectrum of mutations, and implications for a molecular diagnosis strategy. Circulation. 2003;107(17):2227–32. https://doi.org/10.1161/01.cir.0000066323.15244.54.
31. Burke MA, Cook SA, Seidman JG, Seidman CE. Clinical and mechanistic insights into the genetics of cardiomyopathy. J Am Coll Cardiol. 2016;68(25):2871–86. https://doi.org/10.1016/j.jacc.2016.08.079.
32. Knoll R, Buyandelger B, Lab M. The sarcomeric Z-disc and Z-discopathies. J Biomed Biotechnol. 2011;2011:569628. https://doi.org/10.1155/2011/569628.
33. Landstrom AP, Ackerman MJ. Beyond the cardiac myofilament: hypertrophic cardiomyopathy-associated mutations in genes that encode calcium-handling proteins. Curr Mol Med. 2012;12(5):507–18.
34. Millat G, Bouvagnet P, Chevalier P, Dauphin C, Simon Jouk P, Da Costa A, et al. Prevalence and spectrum of mutations in a cohort of 192 unrelated patients with hypertrophic cardiomyopathy. Eur J Med Genet. 2010;53(5):261–7. https://doi.org/10.1016/j.ejmg.2010.07.007.
35. Anan R, Greve G, Thierfelder L, Watkins H, McKenna WJ, Solomon S, et al. Prognostic implications of novel beta cardiac myosin heavy chain gene mutations that cause familial hypertrophic cardiomyopathy. J Clin Invest. 1994;93(1):280–5. https://doi.org/10.1172/JCI116957.
36. Pasquale F, Syrris P, Kaski JP, Mogensen J, McKenna WJ, Elliott P. Long-term outcomes in hypertrophic cardiomyopathy caused by mutations in the cardiac troponin T gene. Circ Cardiovasc Genet. 2012;5(1):10–7. https://doi.org/10.1161/circgenetics.111.959973.
37. Alpert NR, Mohiddin SA, Tripodi D, Jacobson-Hatzell J, Vaughn-Whitley K, Brosseau C, et al. Molecular and phenotypic effects of heterozygous, homozygous, and compound hetero-zygote myosin heavy-chain mutations. Am J Physiol Heart Circ Physiol. 2005;288(3):H1097–102. https://doi.org/10.1152/ajpheart.00650.2004.
38. Fourey D, Care M, Siminovitch KA, Weissler-Snir A, Hindieh W, Chan RH, et al. Prevalence and clinical implication of double mutations in hypertrophic cardiomyopathy: revisiting the gene-dose effect. Circ Cardiovasc Genet. 2017;10(2):e001685. https://doi.org/10.1161/circgenetics.116.001685.
39. Olivotto I, d'Amati G, Basso C, Van Rossum A, Patten M, Emdin M, et al. Defining phenotypes and disease progression in sarcomeric cardiomyopathies: contemporary role of clinical investigations. Cardiovasc Res. 2015;105(4):409–23. https://doi.org/10.1093/cvr/cvv024.
40. Charron P, Arad M, Arbustini E, Basso C, Bilinska Z, Elliott P, et al. Genetic counselling and testing in cardiomyopathies: a position statement of the European Society of Cardiology Working Group on Myocardial and Pericardial Diseases. Eur Heart J. 2010;31(22):2715–26. https://doi.org/10.1093/eurheartj/ehq271.
41. Elliott P, Andersson B, Arbustini E, Bilinska Z, Cecchi F, Charron P, et al. Classification of the cardiomyopathies: a position statement from the European Society of Cardiology Working Group on Myocardial and Pericardial Diseases. Eur Heart J. 2008;29(2):270–6. https://doi.org/10.1093/eurheartj/ehm342.
42. Andreasen C, Nielsen JB, Refsgaard L, Holst AG, Christensen AH, Andreasen L, et al. New population-based exome data are questioning the pathogenicity of previously cardiomyopathy-associated genetic variants. Eur J Hum Genet. 2013;21(9):918–28. https://doi.org/10.1038/ejhg.2012.283.
43. Lopes LR, Zekavati A, Syrris P, Hubank M, Giambartolomei C, Dalageorgou C, et al. Genetic complexity in hypertrophic cardiomyopathy revealed by high-throughput sequencing. J Med Genet. 2013;50(4):228–39. https://doi.org/10.1136/jmedgenet-2012-101270.

44. Colan SD, Lipshultz SE, Lowe AM, Sleeper LA, Messere J, Cox GF, et al. Epidemiology and cause-specific outcome of hypertrophic cardiomyopathy in children: findings from the Pediatric Cardiomyopathy Registry. Circulation. 2007;115(6):773–81. https://doi.org/10.1161/circulationaha.106.621185.
45. Rapezzi C, Arbustini E, Caforio AL, Charron P, Gimeno-Blanes J, Helio T, et al. Diagnostic work-up in cardiomyopathies: bridging the gap between clinical phenotypes and final diagnosis. A position statement from the ESC Working Group on Myocardial and Pericardial Diseases. Eur Heart J. 2013;34(19):1448–58. https://doi.org/10.1093/eurheartj/ehs397.
46. Sata M, Ikebe M. Functional analysis of the mutations in the human cardiac beta-myosin that are responsible for familial hypertrophic cardiomyopathy. Implication for the clinical outcome. J Clin Invest. 1996;98(12):2866–73. https://doi.org/10.1172/jci119115.
47. Harris SP, Lyons RG, Bezold KL. In the thick of it: HCM-causing mutations in myosin binding proteins of the thick filament. Circ Res. 2011;108(6):751–64. https://doi.org/10.1161/circresaha.110.231670.
48. van Dijk SJ, Dooijes D, dos Remedios C, Michels M, Lamers JMJ, Winegrad S, et al. Cardiac myosin-binding protein C mutations and hypertrophic cardiomyopathy. Haploinsufficiency, deranged phosphorylation, and cardiomyocyte dysfunction. Circulation. 2009;119 (11):1473–83. https://doi.org/10.1161/circulationaha.108.838672.
49. Robinson P, Griffiths PJ, Watkins H, Redwood CS. Dilated and hypertrophic cardiomyopathy mutations in troponin and alpha-tropomyosin have opposing effects on the calcium affinity of cardiac thin filaments. Circ Res. 2007;101(12):1266–73. https://doi.org/10.1161/circresaha. 107.156380.
50. Robinson P, Mirza M, Knott A, Abdulrazzak H, Willott R, Marston S, et al. Alterations in thin filament regulation induced by a human cardiac troponin T mutant that causes dilated cardiomyopathy are distinct from those induced by troponin T mutants that cause hypertrophic cardiomyopathy. J Biol Chem. 2002;277(43):40710–6. https://doi.org/10.1074/jbc. M203446200.
51. Guinto PJ, Haim TE, Dowell-Martino CC, Sibinga N, Tardiff JC. Temporal and mutation-specific alterations in Ca2+ homeostasis differentially determine the progression of cTnT-related cardiomyopathies in murine models. Am J Physiol Heart Circ Physiol. 2009;297(2): H614–26. https://doi.org/10.1152/ajpheart.01143.2008.
52. Ashrafian H, McKenna WJ, Watkins H. Disease pathways and novel therapeutic targets in hypertrophic cardiomyopathy. Circ Res. 2011;109(1):86–96. https://doi.org/10.1161/circresaha.111.242974.
53. Spudich JA. Hypertrophic and dilated cardiomyopathy: four decades of basic research on muscle lead to potential therapeutic approaches to these devastating genetic diseases. Biophys J. 2014;106(6):1236–49. https://doi.org/10.1016/j.bpj.2014.02.011.
54. Kirschner SE, Becker E, Antognozzi M, Kubis HP, Francino A, Navarro-Lopez F, et al. Hypertrophic cardiomyopathy-related beta-myosin mutations cause highly variable calcium sensitivity with functional imbalances among individual muscle cells. Am J Physiol Heart Circ Physiol. 2005;288(3):H1242–51. https://doi.org/10.1152/ajpheart.00686.2004.
55. Green EM, Wakimoto H, Anderson RL, Evanchik MJ, Gorham JM, Harrison BC, et al. A small-molecule inhibitor of sarcomere contractility suppresses hypertrophic cardiomyopathy in mice. Science. 2016;351(6273):617–21. https://doi.org/10.1126/science.aad3456.
56. Grunig E, Tasman JA, Kucherer H, Franz W, Kubler W, Katus HA. Frequency and phenotypes of familial dilated cardiomyopathy. J Am Coll Cardiol. 1998;31(1):186–94.
57. Petretta M, Pirozzi F, Sasso L, Paglia A, Bonaduce D. Review and metaanalysis of the frequency of familial dilated cardiomyopathy. Am J Cardiol. 2011;108(8):1171–6. https://doi.org/10.1016/j.amjcard.2011.06.022.
58. Pinto YM, Elliott PM, Arbustini E, Adler Y, Anastasakis A, Bohm M, et al. Proposal for a revised definition of dilated cardiomyopathy, hypokinetic non-dilated cardiomyopathy, and its implications for clinical practice: a position statement of the ESC working group on

myocardial and pericardial diseases. Eur Heart J. 2016;37(23):1850–8. https://doi.org/10.1093/eurheartj/ehv727.

59. Codd MB, Sugrue DD, Gersh BJ, Melton LJ. Epidemiology of idiopathic dilated and hypertrophic cardiomyopathy. A population-based study in Olmsted County, Minnesota, 1975–1984. Circulation. 1989;80(3):564–72. https://doi.org/10.1161/01.cir.80.3.564.

60. Schafer S, de Marvao A, Adami E, Fiedler LR, Ng B, Khin E, et al. Titin-truncating variants affect heart function in disease cohorts and the general population. Nat Genet. 2017;49 (1):46–53. https://doi.org/10.1038/ng.3719.. http://www.nature.com/ng/journal/v49/n1/abs/ng.3719.html#supplementary-information

61. Gulati A, Jabbour A, Ismail TF, Guha K, Khwaja J, Raza S, et al. Association of fibrosis with mortality and sudden cardiac death in patients with nonischemic dilated cardiomyopathy. JAMA. 2013;309(9):896–908. https://doi.org/10.1001/jama.2013.1363.

62. Kober L, Thune JJ, Nielsen JC, Haarbo J, Videbaek L, Korup E, et al. Defibrillator implantation in patients with nonischemic systolic heart failure. N Engl J Med. 2016;375(13):1221–30. https://doi.org/10.1056/NEJMoa1608029.

63. McMurray JJ, Adamopoulos S, Anker SD, Auricchio A, Bohm M, Dickstein K, et al. ESC guidelines for the diagnosis and treatment of acute and chronic heart failure 2012: The Task Force for the Diagnosis and Treatment of Acute and Chronic Heart Failure 2012 of the European Society of Cardiology. Developed in collaboration with the Heart Failure Association (HFA) of the ESC. Eur J Heart Fail. 2012;14(8):803–69. https://doi.org/10.1093/eurjhf/hfs105.

64. Duboc D, Meune C, Pierre B, Wahbi K, Eymard B, Toutain A, et al. Perindopril preventive treatment on mortality in Duchenne muscular dystrophy: 10 years' follow-up. Am Heart J. 2007;154(3):596–602. https://doi.org/10.1016/j.ahj.2007.05.014.

65. Mahon NG, Murphy RT, MacRae CA, Caforio AL, Elliott PM, McKenna WJ. Echocardiographic evaluation in asymptomatic relatives of patients with dilated cardiomyopathy reveals preclinical disease. Ann Intern Med. 2005;143(2):108–15.

66. Hershberger RE, Hedges DJ, Morales A. Dilated cardiomyopathy: the complexity of a diverse genetic architecture. Nat Rev Cardiol. 2013;10(9):531–47. https://doi.org/10.1038/nrcardio.2013.105.

67. Mestroni L, Brun F, Spezzacatene A, Sinagra G, Taylor MR. Genetic causes of dilated cardiomyopathy. Prog Pediatr Cardiol. 2014;37(1–2):13–8. https://doi.org/10.1016/j.ppedcard.2014.10.003.

68. Haas J, Frese KS, Peil B, Kloos W, Keller A, Nietsch R, et al. Atlas of the clinical genetics of human dilated cardiomyopathy. Eur Heart J. 2015;36(18):1123–35. https://doi.org/10.1093/eurheartj/ehu301.

69. Gerull B, Gramlich M, Atherton J, McNabb M, Trombitas K, Sasse-Klaassen S, et al. Mutations of TTN, encoding the giant muscle filament titin, cause familial dilated cardiomyopathy. Nat Genet. 2002;30(2):201–4. https://doi.org/10.1038/ng815.

70. Herman DS, Lam L, Taylor MR, Wang L, Teekakirikul P, Christodoulou D, et al. Truncations of titin causing dilated cardiomyopathy. N Engl J Med. 2012;366(7):619–28. https://doi.org/10.1056/NEJMoa1110186.

71. Roberts AM, Ware JS, Herman DS, Schafer S, Baksi J, Bick AG, et al. Integrated allelic, transcriptional, and phenomic dissection of the cardiac effects of titin truncations in health and disease. Sci Transl Med. 2015;7(270):270ra276. https://doi.org/10.1126/scitranslmed.3010134.

72. Fatkin D, MacRae C, Sasaki T, Wolff MR, Porcu M, Frenneaux M, et al. Missense mutations in the rod domain of the lamin A/C gene as causes of dilated cardiomyopathy and conduction-system disease. N Engl J Med. 1999;341(23):1715–24. https://doi.org/10.1056/nejm199912023412302.

73. Kamisago M, Sharma SD, DePalma SR, Solomon S, Sharma P, McDonough B, et al. Mutations in sarcomere protein genes as a cause of dilated cardiomyopathy. N Engl J Med. 2000;343(23):1688–96. https://doi.org/10.1056/nejm200012073432304.

74. McNally EM, Puckelwartz MJ. Genetic variation in cardiomyopathy and cardiovascular disorders. Circ J. 2015;79(7):1409–15. https://doi.org/10.1253/circj.CJ-15-0536.
75. Akinrinade O, Alastalo TP, Koskenvuo JW. Relevance of truncating titin mutations in dilated cardiomyopathy. Clin Genet. 2016;90(1):49–54. https://doi.org/10.1111/cge.12741.
76. Brayson D, Shanahan CM. Current insights into LMNA cardiomyopathies: existing models and missing LINCs. Nucleus. 2017;8(1):17–33. https://doi.org/10.1080/19491034.2016.1260798.
77. Nikolova V, Leimena C, McMahon AC, Tan JC, Chandar S, Jogia D, et al. Defects in nuclear structure and function promote dilated cardiomyopathy in lamin A/C-deficient mice. J Clin Invest. 2004;113(3):357–69. https://doi.org/10.1172/JCI19448.
78. Bione S, Maestrini E, Rivella S, Mancini M, Regis S, Romeo G, Toniolo D. Identification of a novel X-linked gene responsible for Emery-Dreifuss muscular dystrophy. Nat Genet. 1994;8(4):323–7. https://doi.org/10.1038/ng1294-323.
79. Arndt AK, Schafer S, Drenckhahn JD, Sabeh MK, Plovie ER, Caliebe A, et al. Fine mapping of the 1p36 deletion syndrome identifies mutation of PRDM16 as a cause of cardiomyopathy. Am J Hum Genet. 2013;93(1):67–77. https://doi.org/10.1016/j.ajhg.2013.05.015.
80. Kirk EP, Sunde M, Costa MW, Rankin SA, Wolstein O, Castro ML, et al. Mutations in cardiac T-box factor gene TBX20 are associated with diverse cardiac pathologies, including defects of septation and valvulogenesis and cardiomyopathy. Am J Hum Genet. 2007;81(2):280–91. https://doi.org/10.1086/519530.
81. Williams T, Hundertmark M, Nordbeck P, Voll S, Arias-Loza PA, Oppelt D, et al. Eya4 induces hypertrophy via regulation of p27kip1. Circ Cardiovasc Genet. 2015;8(6):752–64. https://doi.org/10.1161/circgenetics.115.001134.
82. Xu L, Zhao L, Yuan F, Jiang WF, Liu H, Li RG, et al. GATA6 loss-of-function mutations contribute to familial dilated cardiomyopathy. Int J Mol Med. 2014;34(5):1315–22. https://doi.org/10.3892/ijmm.2014.1896.
83. Yuan F, Qiu XB, Li RG, Qu XK, Wang J, Xu YJ, et al. A novel NKX2-5 loss-of-function mutation predisposes to familial dilated cardiomyopathy and arrhythmias. Int J Mol Med. 2015;35(2):478–86. https://doi.org/10.3892/ijmm.2014.2029.
84. Guo W, Schafer S, Greaser ML, Radke MH, Liss M, Govindarajan T, et al. RBM20, a gene for hereditary cardiomyopathy, regulates titin splicing. Nat Med. 2012;18(5):766–73. https://doi.org/10.1038/nm.2693.
85. Maatz H, Jens M, Liss M, Schafer S, Heinig M, Kirchner M, et al. RNA-binding protein RBM20 represses splicing to orchestrate cardiac pre-mRNA processing. J Clin Invest. 2014;124(8):3419–30. https://doi.org/10.1172/jci74523.
86. Schmitt JP, Kamisago M, Asahi M, Li GH, Ahmad F, Mende U, et al. Dilated cardiomyopathy and heart failure caused by a mutation in phospholamban. Science. 2003;299(5611):1410–3. https://doi.org/10.1126/science.1081578.
87. Abrol N, de Tombe PP, Robia SL. Acute inotropic and lusitropic effects of cardiomyopathic R9C mutation of phospholamban. J Biol Chem. 2015;290(11):7130–40. https://doi.org/10.1074/jbc.M114.630319.
88. McNair WP, Sinagra G, Taylor MRG, Di Lenarda A, Ferguson DA, Salcedo EE, et al. SCN5A mutations associate with arrhythmic dilated cardiomyopathy and commonly localize to the voltage-sensing mechanism. J Am Coll Cardiol. 2011;57(21):2160–8. https://doi.org/10.1016/j.jacc.2010.09.084.
89. Bienengraeber M, Olson TM, Selivanov VA, Kathmann EC, O'Cochlain F, Gao F, et al. ABCC9 mutations identified in human dilated cardiomyopathy disrupt catalytic KATP channel gating. Nat Genet. 2004;36(4):382–7.
90. Hershberger RE, Parks SB, Kushner JD, Li D, Ludwigsen S. Coding sequence mutations identified in MYH7, TNNT2, SCN5A, CSRP3, LBD3, and TCAP from 313 patients with familial or idiopathic dilated cardiomyopathy. Clin Transl Sci. 2008;1:21–6. https://doi.org/10.1111/j.1752-8062.2008.00017.x.

91. Geier C, Gehmlich K, Ehler E, Hassfeld S, Perrot A, Hayess K, et al. Beyond the sarcomere: CSRP3 mutations cause hypertrophic cardiomyopathy. Hum Mol Genet. 2008;17 (18):2753–65. https://doi.org/10.1093/hmg/ddn160.
92. Mohapatra B, Jimenez S, Lin JH, Bowles KR, Coveler KJ, Marx JG, et al. Mutations in the muscle LIM protein and alpha-actinin-2 genes in dilated cardiomyopathy and endocardial fibroelastosis. Mol Genet Metab. 2003;80(1-2):207–15.
93. Dalakas MC, Park KY, Semino-Mora C, Lee HS, Sivakumar K, Goldfarb LG. Desmin myopathy, a skeletal myopathy with cardiomyopathy caused by mutations in the desmin gene. N Engl J Med. 2000;342(11):770–80. https://doi.org/10.1056/nejm200003163421104.
94. Arimura T, Ishikawa T, Nunoda S, Kawai S, Kimura A. Dilated cardiomyopathy-associated BAG3 mutations impair Z-disc assembly and enhance sensitivity to apoptosis in cardiomyocytes. Hum Mutat. 2011;32(12):1481–91. https://doi.org/10.1002/humu.21603.
95. Garcia-Pavia P, Syrris P, Salas C, Evans A, Mirelis JG, Cobo-Marcos M, et al. Desmosomal protein gene mutations in patients with idiopathic dilated cardiomyopathy undergoing cardiac transplantation: a clinicopathological study. Heart. 2011;97(21):1744–52. https://doi.org/10. 1136/hrt.2011.227967.
96. Patel DM, Green KJ. Desmosomes in the heart: a review of clinical and mechanistic analyses. Cell Commun Adhes. 2014;21(3):109–28. https://doi.org/10.3109/15419061.2014.906533.
97. Behin A, Salort-Campana E, Wahbi K, Richard P, Carlier RY, Carlier P, et al. Myofibrillar myopathies: state of the art, present and future challenges. Rev Neurol (Paris). 2015;171 (10):715–29. https://doi.org/10.1016/j.neurol.2015.06.002.
98. Sommerville RB, Vincenti MG, Winborn K, Casey A, Stitziel NO, Connolly AM, Mann DL. Diagnosis and management of adult hereditary cardio-neuromuscular disorders: a model for the multidisciplinary care of complex genetic disorders. Trends Cardiovasc Med. 2017;27 (1):51–8. https://doi.org/10.1016/j.tcm.2016.06.005.
99. Worman HJ, Bonne G. "Laminopathies": a wide spectrum of human diseases. Exp Cell Res. 2007;313(10):2121–33. https://doi.org/10.1016/j.yexcr.2007.03.028.
100. Kamdar F, Garry DJ. Dystrophin-deficient cardiomyopathy. J Am Coll Cardiol. 2016;67 (21):2533–46. https://doi.org/10.1016/j.jacc.2016.02.081.
101. Pua CJ, Bhalshankar J, Miao K, Walsh R, John S, Lim SQ, et al. Development of a comprehensive sequencing assay for inherited cardiac condition genes. J Cardiovasc Transl Res. 2016;9(1):3–11. https://doi.org/10.1007/s12265-016-9673-5.
102. Akinrinade O, Ollila L, Vattulainen S, Tallila J, Gentile M, Salmenpera P, et al. Genetics and genotype-phenotype correlations in Finnish patients with dilated cardiomyopathy. Eur Heart J. 2015;36(34):2327–37. https://doi.org/10.1093/eurheartj/ehv253.
103. Hershberger RE, Morales A. LMNA-related dilated cardiomyopathy. In: Pagon RA, Adam MP, Ardinger HH, Wallace SE, Amemiya A, Bean LJH, Bird TD, Ledbetter N, Mefford HC, Smith RJH, Stephens K, editors. GeneReviews(R). Seattle (WA): University of Washington, Seattle; 1993.
104. Pasotti M, Klersy C, Pilotto A, Marziliano N, Rapezzi C, Serio A, et al. Long-term outcome and risk stratification in dilated cardiolaminopathies. J Am Coll Cardiol. 2008;52 (15):1250–60. https://doi.org/10.1016/j.jacc.2008.06.044.
105. van Rijsingen IAW, Arbustini E, Elliott PM, Mogensen J, Hermans-van Ast JF, van der Kooi AJ, et al. Risk factors for malignant ventricular arrhythmias in lamin A/C mutation carriers: a European Cohort Study. J Am Coll Cardiol. 2012;59(5):493–500. https://doi.org/10.1016/j. jacc.2011.08.078.
106. Franaszczyk M, Chmielewski P, Truszkowska G, Stawinski P, Michalak E, Rydzanicz M, et al. Titin truncating variants in dilated cardiomyopathy – prevalence and genotype-phenotype correlations. PLoS One. 2017;12(1):e0169007. https://doi.org/10.1371/journal. pone.0169007.
107. Taylor MR, Fain PR, Sinagra G, Robinson ML, Robertson AD, Carniel E, et al. Natural history of dilated cardiomyopathy due to lamin A/C gene mutations. J Am Coll Cardiol. 2003;41 (5):771–80.

108. Roncarati R, Viviani Anselmi C, Krawitz P, Lattanzi G, von Kodolitsch Y, Perrot A, et al. Doubly heterozygous LMNA and TTN mutations revealed by exome sequencing in a severe form of dilated cardiomyopathy. Eur J Hum Genet. 2013;21(10):1105–11. https://doi.org/10.1038/ejhg.2013.16.
109. Chin TK, Perloff JK, Williams RG, Jue K, Mohrmann R. Isolated noncompaction of left ventricular myocardium. A study of eight cases. Circulation. 1990;82(2):507–13.
110. Jenni R, Oechslin E, Schneider J, Attenhofer Jost C, Kaufmann PA. Echocardiographic and pathoanatomical characteristics of isolated left ventricular non-compaction: a step towards classification as a distinct cardiomyopathy. Heart. 2001;86(6):666–71.
111. Oechslin E, Jenni R. Left ventricular non-compaction revisited: a distinct phenotype with genetic heterogeneity? Eur Heart J. 2011;32(12):1446–56. https://doi.org/10.1093/eurheartj/ehq508.
112. Weir-McCall JR, Yeap PM, Papagiorcopulo C, Fitzgerald K, Gandy SJ, Lambert M, et al. Left ventricular noncompaction: anatomical phenotype or distinct cardiomyopathy? J Am Coll Cardiol. 2016;68(20):2157–65. https://doi.org/10.1016/j.jacc.2016.08.054.
113. Ichida F, Hamamichi Y, Miyawaki T, Ono Y, Kamiya T, Akagi T, et al. Clinical features of isolated noncompaction of the ventricular myocardium: long-term clinical course, hemodynamic properties, and genetic background. J Am Coll Cardiol. 1999;34(1):233–40.
114. Murphy RT, Thaman R, Blanes JG, Ward D, Sevdalis E, Papra E, et al. Natural history and familial characteristics of isolated left ventricular non-compaction. Eur Heart J. 2005;26(2):187–92. https://doi.org/10.1093/eurheartj/ehi025.
115. Oechslin EN, Attenhofer Jost CH, Rojas JR, Kaufmann PA, Jenni R. Long-term follow-up of 34 adults with isolated left ventricular noncompaction: a distinct cardiomyopathy with poor prognosis. J Am Coll Cardiol. 2000;36(2):493–500.
116. Belanger AR, Miller MA, Donthireddi UR, Najovits AJ, Goldman ME. New classification scheme of left ventricular noncompaction and correlation with ventricular performance. Am J Cardiol. 2008;102(1):92–6. https://doi.org/10.1016/j.amjcard.2008.02.107.
117. Kohli SK, Pantazis AA, Shah JS, Adeyemi B, Jackson G, McKenna WJ, et al. Diagnosis of left-ventricular non-compaction in patients with left-ventricular systolic dysfunction: time for a reappraisal of diagnostic criteria? Eur Heart J. 2008;29(1):89–95. https://doi.org/10.1093/eurheartj/ehm481.
118. Zemrak F, Ahlman MA, Captur G, Mohiddin SA, Kawel-Boehm N, Prince MR, et al. The relationship of left ventricular trabeculation to ventricular function and structure over a 9.5-year follow-up: the MESA study. J Am Coll Cardiol. 2014;64(19):1971–80. https://doi.org/10.1016/j.jacc.2014.08.035.
119. Nugent AW, Daubeney PE, Chondros P, Carlin JB, Cheung M, Wilkinson LC, et al. The epidemiology of childhood cardiomyopathy in Australia. N Engl J Med. 2003;348(17):1639–46. https://doi.org/10.1056/NEJMoa021737.
120. Jefferies JL, Wilkinson JD, Sleeper LA, Colan SD, Lu M, Pahl E, et al. Cardiomyopathy phenotypes and outcomes for children with left ventricular myocardial noncompaction: results from the pediatric cardiomyopathy registry. J Card Fail. 2015;21(11):877–84. https://doi.org/10.1016/j.cardfail.2015.06.381.
121. Brescia ST, Rossano JW, Pignatelli R, Jefferies JL, Price JF, Decker JA, et al. Mortality and sudden death in pediatric left ventricular noncompaction in a tertiary referral center. Circulation. 2013;127(22):2202–8. https://doi.org/10.1161/CIRCULATIONAHA.113.002511.
122. Anderson RH, Jensen B, Mohun TJ, Petersen SE, Aung N, Zemrak F, et al. Key questions relating to left ventricular noncompaction cardiomyopathy: is the emperor still wearing any clothes? Can J Cardiol. 2017;33(6):747–57. https://doi.org/10.1016/j.cjca.2017.01.017.
123. Freedom RM, Yoo SJ, Perrin D, Taylor G, Petersen S, Anderson RH. The morphological spectrum of ventricular noncompaction. Cardiol Young. 2005;15(4):345–64. https://doi.org/10.1017/S1047951105000752.
124. Kodo K, Ong SG, Jahanbani F, Termglinchan V, Hirono K, InanlooRahatloo K, et al. iPSC-derived cardiomyocytes reveal abnormal TGF-beta signalling in left ventricular

non-compaction cardiomyopathy. Nat Cell Biol. 2016;18(10):1031–42. https://doi.org/10. 1038/ncb3411.

125. Grego-Bessa J, Luna-Zurita L, del Monte G, Bolos V, Melgar P, Arandilla A, et al. Notch signaling is essential for ventricular chamber development. Dev Cell. 2007;12(3):415–29. https://doi.org/10.1016/j.devcel.2006.12.011.

126. Sasse-Klaassen S, Gerull B, Oechslin E, Jenni R, Thierfelder L. Isolated noncompaction of the left ventricular myocardium in the adult is an autosomal dominant disorder in the majority of patients. Am J Med Genet A. 2003;119A(2):162–7. https://doi.org/10.1002/ajmg.a.20075.

127. Probst S, Oechslin E, Schuler P, Greutmann M, Boye P, Knirsch W, et al. Sarcomere gene mutations in isolated left ventricular noncompaction cardiomyopathy do not predict clinical phenotype. Circ Cardiovasc Genet. 2011;4(4):367–74. https://doi.org/10.1161/ CIRCGENETICS.110.959270.

128. Budde BS, Binner P, Waldmuller S, Hohne W, Blankenfeldt W, Hassfeld S, et al. Noncompaction of the ventricular myocardium is associated with a de novo mutation in the beta-myosin heavy chain gene. PLoS One. 2007;2(12):e1362. https://doi.org/10.1371/journal. pone.0001362.

129. Hoedemaekers YM, Caliskan K, Michels M, Frohn-Mulder I, van der Smagt JJ, Phefferkorn JE, et al. The importance of genetic counseling, DNA diagnostics, and cardiologic family screening in left ventricular noncompaction cardiomyopathy. Circ Cardiovasc Genet. 2010;3 (3):232–9. https://doi.org/10.1161/CIRCGENETICS.109.903898.

130. Klaassen S, Probst S, Oechslin E, Gerull B, Krings G, Schuler P, et al. Mutations in sarcomere protein genes in left ventricular noncompaction. Circulation. 2008;117(22):2893–901. https:// doi.org/10.1161/CIRCULATIONAHA.107.746164.

131. Postma AV, van Engelen K, van de Meerakker J, Rahman T, Probst S, Baars MJ, et al. Mutations in the sarcomere gene MYH7 in Ebstein anomaly. Circ Cardiovasc Genet. 2011;4 (1):43–50. https://doi.org/10.1161/CIRCGENETICS.110.957985.

132. Wessels MW, Herkert JC, Frohn-Mulder IM, Dalinghaus M, van den Wijngaard A, de Krijger RR, et al. Compound heterozygous or homozygous truncating MYBPC3 mutations cause lethal cardiomyopathy with features of noncompaction and septal defects. Eur J Hum Genet. 2015;23(7):922–8. https://doi.org/10.1038/ejhg.2014.211.

133. Monserrat L, Hermida-Prieto M, Fernandez X, Rodriguez I, Dumont C, Cazon L, et al. Mutation in the alpha-cardiac actin gene associated with apical hypertrophic cardiomyopathy, left ventricular non-compaction, and septal defects. Eur Heart J. 2007;28(16):1953–61. https:// doi.org/10.1093/eurheartj/ehm239.

134. Hastings R, de Villiers CP, Hooper C, Ormondroyd L, Pagnamenta A, Lise S, et al. Combina-tion of whole genome sequencing, linkage, and functional studies implicates a missense mutation in titin as a cause of autosomal dominant cardiomyopathy with features of left ventricular noncompaction. Circ Cardiovasc Genet. 2016;9(5):426–35. https://doi.org/10. 1161/CIRCGENETICS.116.001431.

135. Bleyl SB, Mumford BR, Thompson V, Carey JC, Pysher TJ, Chin TK, Ward K. Neonatal, lethal noncompaction of the left ventricular myocardium is allelic with Barth syndrome. Am J Hum Genet. 1997;61(4):868–72. https://doi.org/10.1086/514879.

136. Vatta M, Mohapatra B, Jimenez S, Sanchez X, Faulkner G, Perles Z, et al. Mutations in Cypher/ZASP in patients with dilated cardiomyopathy and left ventricular non-compaction. J Am Coll Cardiol. 2003;42(11):2014–27.

137. Bagnall RD, Molloy LK, Kalman JM, Semsarian C. Exome sequencing identifies a mutation in the ACTN2 gene in a family with idiopathic ventricular fibrillation, left ventricular noncompaction, and sudden death. BMC Med Genet. 2014;15:99. https://doi.org/10.1186/ s12881-014-0099-0.

138. Milano A, Vermeer AM, Lodder EM, Barc J, Verkerk AO, Postma AV, et al. HCN4 mutations in multiple families with bradycardia and left ventricular noncompaction cardiomyopathy. J Am Coll Cardiol. 2014;64(8):745–56. https://doi.org/10.1016/j.jacc.2014.05.045.

139. Schweizer PA, Schroter J, Greiner S, Haas J, Yampolsky P, Mereles D, et al. The symptom complex of familial sinus node dysfunction and myocardial noncompaction is associated with mutations in the HCN4 channel. J Am Coll Cardiol. 2014;64(8):757–67. https://doi.org/10.1016/j.jacc.2014.06.1155.

140. Shan L, Makita N, Xing Y, Watanabe S, Futatani T, Ye F, et al. SCN5A variants in Japanese patients with left ventricular noncompaction and arrhythmia. Mol Genet Metab. 2008;93 (4):468–74.

141. Hermida-Prieto M, Monserrat L, Castro-Beiras A, Laredo R, Soler R, Peteiro J, et al. Familial dilated cardiomyopathy and isolated left ventricular noncompaction associated with lamin A/C gene mutations. Am J Cardiol. 2004;94(1):50–4. https://doi.org/10.1016/j.amjcard.2004.03. 029.

142. Luxan G, Casanova JC, Martinez-Poveda B, Prados B, D'Amato G, MacGrogan D, et al. Mutations in the NOTCH pathway regulator MIB1 cause left ventricular noncompaction cardiomyopathy. Nat Med. 2013;19(2):193–201. https://doi.org/10.1038/nm.3046.

143. Hoedemaekers YM, Klaassen S. Left ventricular noncompaction. In: Baars HF, Doevendans PAFM, Houweling A, Tintelen JP, editors. Clinical cardiogenetics. Berlin: Springer; 2016. p. 113–35. https://doi.org/10.1007/978-3-319-44203-7.

144. Ichida F, Tsubata S, Bowles KR, Haneda N, Uese K, Miyawaki T, et al. Novel gene mutations in patients with left ventricular noncompaction or Barth syndrome. Circulation. 2001;103 (9):1256–63.

145. Williams T, Machann W, Kuhler L, Hamm H, Muller-Hocker J, Zimmer M, et al. Novel desmoplakin mutation: juvenile biventricular cardiomyopathy with left ventricular non-compaction and acantholytic palmoplantar keratoderma. Clin Res Cardiol. 2011;100 (12):1087–93. https://doi.org/10.1007/s00392-011-0345-9.

146. Stahli BE, Gebhard C, Biaggi P, Klaassen S, Valsangiacomo Buechel E, Attenhofer Jost CH, et al. Left ventricular non-compaction: prevalence in congenital heart disease. Int J Cardiol. 2013;167(6):2477–81. https://doi.org/10.1016/j.ijcard.2012.05.095.

147. Ouyang P, Saarel E, Bai Y, Luo C, Lv Q, Xu Y, et al. A de novo mutation in NKX2.5 associated with atrial septal defects, ventricular noncompaction, syncope and sudden death. Clin Chim Acta. 2011;412(1-2):170–5. https://doi.org/10.1016/j.cca.2010.09.035.

148. Finsterer J, Stollberger C, Towbin JA. Left ventricular noncompaction cardiomyopathy: cardiac, neuromuscular, and genetic factors. Nat Rev Cardiol. 2017;14(4):224–37. https://doi.org/10.1038/nrcardio.2016.207.

149. Scaglia F, Towbin JA, Craigen WJ, Belmont JW, Smith EO, Neish SR, et al. Clinical spectrum, morbidity, and mortality in 113 pediatric patients with mitochondrial disease. Pediatrics. 2004;114(4):925–31. https://doi.org/10.1542/peds.2004-0718.

150. Marcus FI, McKenna WJ, Sherrill D, Basso C, Bauce B, Bluemke DA, et al. Diagnosis of arrhythmogenic right ventricular cardiomyopathy/dysplasia: proposed modification of the task force criteria. Circulation. 2010;121(13):1533–41. https://doi.org/10.1161/CIRCULATIONAHA.108.840827.

151. Lancisi GM. De motu cordis et aneurysmatibus opus postumum. 1740.

152. Marcus FI, Fontaine GH, Guiraudon G, Frank R, Laurenceau JL, Malergue C, Grosgogeat Y. Right ventricular dysplasia: a report of 24 adult cases. Circulation. 1982;65(2):384–98.

153. Romero J, Mejia-Lopez E, Manrique C, Lucariello R. Arrhythmogenic right ventricular cardiomyopathy (ARVC/D): a systematic literature review. Clin Med Insights Cardiol. 2013;7:97–114. https://doi.org/10.4137/CMC.S10940.

154. Bhonsale A, Te Riele A, Sawant AC, Groeneweg JA, James CA, Murray B, et al. Cardiac phenotype and long-term prognosis of arrhythmogenic right ventricular cardiomyopathy/dysplasia patients with late presentation. Heart Rhythm. 2017;14(6):883–91. https://doi.org/10.1016/j.hrthm.2017.02.013.

155. McKenna WJ, Thiene G, Nava A, Fontaliran F, Blomstrom-Lundqvist C, Fontaine G, Camerini F. Diagnosis of arrhythmogenic right ventricular dysplasia/cardiomyopathy. Task Force of the Working Group Myocardial and Pericardial Disease of the European Society of

Cardiology and of the Scientific Council on Cardiomyopathies of the International Society and Federation of Cardiology. Br Heart J. 1994;71(3):215–8.

156. van der Zwaag PA, van Rijsingen IA, Asimaki A, Jongbloed JD, van Veldhuisen DJ, Wiesfeld AC, et al. Phospholamban R14del mutation in patients diagnosed with dilated cardiomyopathy or arrhythmogenic right ventricular cardiomyopathy: evidence supporting the concept of arrhythmogenic cardiomyopathy. Eur J Heart Fail. 2012;14(11):1199–207. https://doi.org/10.1093/eurjhf/hfs119.

157. Xu T, Yang Z, Vatta M, Rampazzo A, Beffagna G, Pilichou K, et al. Compound and digenic heterozygosity contributes to arrhythmogenic right ventricular cardiomyopathy. J Am Coll Cardiol. 2010;55(6):587–97. https://doi.org/10.1016/j.jacc.2009.11.020.

158. Klauke B, Kossmann S, Gaertner A, Brand K, Stork I, Brodehl A, et al. De novo desmin-mutation N116S is associated with arrhythmogenic right ventricular cardiomyopathy. Hum Mol Genet. 2010;19(23):4595–607. https://doi.org/10.1093/hmg/ddq387.

159. Rampazzo A, Nava A, Malacrida S, Beffagna G, Bauce B, Rossi V, et al. Mutation in human desmoplakin domain binding to plakoglobin causes a dominant form of arrhythmogenic right ventricular cardiomyopathy. Am J Hum Genet. 2002;71(5):1200–6. https://doi.org/10.1086/344208.

160. Heuser A, Plovie ER, Ellinor PT, Grossmann KS, Shin JT, Wichter T, et al. Mutant desmocollin-2 causes arrhythmogenic right ventricular cardiomyopathy. Am J Hum Genet. 2006;79(6):1081–8. https://doi.org/10.1086/509044.

161. Syrris P, Ward D, Evans A, Asimaki A, Gandjbakhch E, Sen-Chowdhry S, McKenna WJ. Arrhythmogenic right ventricular dysplasia/cardiomyopathy associated with mutations in the desmosomal gene desmocollin-2. Am J Hum Genet. 2006;79(5):978–84. https://doi.org/10.1086/509122.

162. Pilichou K, Nava A, Basso C, Beffagna G, Bauce B, Lorenzon A, et al. Mutations in desmoglein-2 gene are associated with arrhythmogenic right ventricular cardiomyopathy. Circulation. 2006;113(9):1171–9. https://doi.org/10.1161/CIRCULATIONAHA.105.583674.

163. Vermij SH, Abriel H, van Veen TA. Refining the molecular organization of the cardiac intercalated disc. Cardiovasc Res. 2017;113(3):259–75. https://doi.org/10.1093/cvr/cvw259.

164. Mayosi BM, Fish M, Shaboodien G, Mastantuono E, Kraus S, Wieland T, et al. Identification of cadherin 2 (CDH2) mutations in arrhythmogenic right ventricular cardiomyopathy. Circ Cardiovasc Genet. 2017;10(2):e001605. https://doi.org/10.1161/CIRCGENETICS.116.001605.

165. van Hengel J, Calore M, Bauce B, Dazzo E, Mazzotti E, De Bortoli M, et al. Mutations in the area composita protein alphaT-catenin are associated with arrhythmogenic right ventricular cardiomyopathy. Eur Heart J. 2013;34(3):201–10. https://doi.org/10.1093/eurheartj/ehs373.

166. Fressart V, Duthoit G, Donal E, Probst V, Deharo JC, Chevalier P, et al. Desmosomal gene analysis in arrhythmogenic right ventricular dysplasia/cardiomyopathy: spectrum of mutations and clinical impact in practice. Europace. 2010;12(6):861–8. https://doi.org/10.1093/europace/euq104.

167. Tiso N, Stephan DA, Nava A, Bagattin A, Devaney JM, Stanchi F, et al. Identification of mutations in the cardiac ryanodine receptor gene in families affected with arrhythmogenic right ventricular cardiomyopathy type 2 (ARVD2). Hum Mol Genet. 2001;10(3):189–94.

168. Beffagna G, Occhi G, Nava A, Vitiello L, Ditadi A, Basso C, et al. Regulatory mutations in transforming growth factor-beta3 gene cause arrhythmogenic right ventricular cardiomyopathy type 1. Cardiovasc Res. 2005;65(2):366–73. https://doi.org/10.1016/j.cardiores.2004.10.005.

169. Merner ND, Hodgkinson KA, Haywood AF, Connors S, French VM, Drenckhahn JD, et al. Arrhythmogenic right ventricular cardiomyopathy type 5 is a fully penetrant, lethal arrhythmic disorder caused by a missense mutation in the TMEM43 gene. Am J Hum Genet. 2008;82(4):809–21. https://doi.org/10.1016/j.ajhg.2008.01.010.

170. Erkapic D, Neumann T, Schmitt J, Sperzel J, Berkowitsch A, Kuniss M, et al. Electrical storm in a patient with arrhythmogenic right ventricular cardiomyopathy and SCN5A mutation. Europace. 2008;10(7):884–7. https://doi.org/10.1093/europace/eun065.
171. Taylor M, Graw S, Sinagra G, Barnes C, Slavov D, Brun F, et al. Genetic variation in titin in arrhythmogenic right ventricular cardiomyopathy-overlap syndromes. Circulation. 2011;124 (8):876–85. https://doi.org/10.1161/CIRCULATIONAHA.110.005405.
172. Quarta G, Syrris P, Ashworth M, Jenkins S, Zuborne Alapi K, Morgan J, et al. Mutations in the Lamin A/C gene mimic arrhythmogenic right ventricular cardiomyopathy. Eur Heart J. 2012;33(9):1128–36. https://doi.org/10.1093/eurheartj/ehr451.
173. Lopez-Ayala JM, Ortiz-Genga M, Gomez-Milanes I, Lopez-Cuenca D, Ruiz-Espejo F, Sanchez-Munoz JJ, et al. A mutation in the Z-line Cypher/ZASP protein is associated with arrhythmogenic right ventricular cardiomyopathy. Clin Genet. 2015;88(2):172–6. https://doi.org/10.1111/cge.12458.
174. Xiong Q, Cao Q, Zhou Q, Xie J, Shen Y, Wan R, et al. Arrhythmogenic cardiomyopathy in a patient with a rare loss-of-function KCNQ1 mutation. J Am Heart Assoc. 2015;4(1):e001526. https://doi.org/10.1161/JAHA.114.001526.
175. Protonotarios N, Tsatsopoulou A. Naxos disease: cardiocutaneous syndrome due to cell adhesion defect. Orphanet J Rare Dis. 2006;1:4. https://doi.org/10.1186/1750-1172-1-4.
176. Brogna S, Wen J. Nonsense-mediated mRNA decay (NMD) mechanisms. Nat Struct Mol Biol. 2009;16(2):107–13. https://doi.org/10.1038/nsmb.1550.
177. Zhang Z, Stroud MJ, Zhang J, Fang X, Ouyang K, Kimura K, et al. Normalization of Naxos plakoglobin levels restores cardiac function in mice. J Clin Invest. 2015;125(4):1708–12. https://doi.org/10.1172/JCI80335.
178. Norgett EE, Hatsell SJ, Carvajal-Huerta L, Cabezas JC, Common J, Purkis PE, et al. Recessive mutation in desmoplakin disrupts desmoplakin-intermediate filament interactions and causes dilated cardiomyopathy, woolly hair and keratoderma. Hum Mol Genet. 2000;9(18):2761–6.
179. Green KJ, Stappenbeck TS, Parry DA, Virata ML. Structure of desmoplakin and its association with intermediate filaments. J Dermatol. 1992;19(11):765–9.
180. Chen X, Bonne S, Hatzfeld M, van Roy F, Green KJ. Protein binding and functional characterization of plakophilin 2. Evidence for its diverse roles in desmosomes and beta-catenin signaling. J Biol Chem. 2002;277(12):10512–22. https://doi.org/10.1074/jbc.M108765200.
181. Nekrasova OE, Amargo EV, Smith WO, Chen J, Kreitzer GE, Green KJ. Desmosomal cadherins utilize distinct kinesins for assembly into desmosomes. J Cell Biol. 2011;195 (7):1185–203. https://doi.org/10.1083/jcb.201106057.
182. Kirchner F, Schuetz A, Boldt LH, Martens K, Dittmar G, Haverkamp W, et al. Molecular insights into arrhythmogenic right ventricular cardiomyopathy caused by plakophilin-2-missense mutations. Circ Cardiovasc Genet. 2012;5(4):400–11. https://doi.org/10.1161/CIRCGENETICS.111.961854.
183. Harrison OJ, Brasch J, Lasso G, Katsamba PS, Ahlsen G, Honig B, Shapiro L. Structural basis of adhesive binding by desmocollins and desmogleins. Proc Natl Acad Sci USA. 2016;113 (26):7160–5. https://doi.org/10.1073/pnas.1606272113.
184. Gerull B, Kirchner F, Chong JX, Tagoe J, Chandrasekharan K, Strohm O, et al. Homozygous founder mutation in desmocollin-2 (DSC2) causes arrhythmogenic cardiomyopathy in the Hutterite population. Circ Cardiovasc Genet. 2013;6(4):327–36. https://doi.org/10.1161/CIRCGENETICS.113.000097.
185. Simpson MA, Mansour S, Ahnood D, Kalidas K, Patton MA, McKenna WJ, et al. Homozygous mutation of desmocollin-2 in arrhythmogenic right ventricular cardiomyopathy with mild palmoplantar keratoderma and woolly hair. Cardiology. 2009;113(1):28–34. https://doi.org/10.1159/000165696.
186. Wong JA, Duff HJ, Yuen T, Kolman L, Exner DV, Weeks SG, Gerull B. Phenotypic analysis of arrhythmogenic cardiomyopathy in the Hutterite population: role of electrocardiogram in

identifying high-risk desmocollin-2 carriers. J Am Heart Assoc. 2014;3(6):e001407. https:// doi.org/10.1161/JAHA.114.001407.

187. van Tintelen JP, Van Gelder IC, Asimaki A, Suurmeijer AJ, Wiesfeld AC, Jongbloed JD, et al. Severe cardiac phenotype with right ventricular predominance in a large cohort of patients with a single missense mutation in the DES gene. Heart Rhythm. 2009;6(11):1574–83. https:// doi.org/10.1016/j.hrthm.2009.07.041.

188. Brodehl A, Hedde PN, Dieding M, Fatima A, Walhorn V, Gayda S, et al. Dual color photoactivation localization microscopy of cardiomyopathy-associated desmin mutants. J Biol Chem. 2012;287(19):16047–57. https://doi.org/10.1074/jbc.M111.313841.

189. Turkowski KL, Tester DJ, Bos JM, Haugaa KH, Ackerman MJ. Whole exome sequencing with genomic triangulation implicates CDH2-encoded N-cadherin as a novel pathogenic substrate for arrhythmogenic cardiomyopathy. Congenit Heart Dis. 2017;12(2):226–35. https://doi.org/10.1111/chd.12462.

190. Bers DM, Perez-Reyes E. Ca channels in cardiac myocytes: structure and function in Ca influx and intracellular Ca release. Cardiovasc Res. 1999;42(2):339–60.

191. Kranias EG, Bers DM. Calcium and cardiomyopathies. Subcell Biochem. 2007;45:523–37.

192. Ma Y, Zou H, Zhu XX, Pang J, Xu Q, Jin QY, et al. Transforming growth factor beta: a potential biomarker and therapeutic target of ventricular remodeling. Oncotarget. 2017;8 (32):53780–90. https://doi.org/10.18632/oncotarget.17255.

193. Milting H, Klauke B, Christensen AH, Musebeck J, Walhorn V, Grannemann S, et al. The TMEM43 Newfoundland mutation p.S358L causing ARVC-5 was imported from Europe and increases the stiffness of the cell nucleus. Eur Heart J. 2015;36(14):872–81. https://doi.org/10. 1093/eurheartj/ehu077.

194. Bengtsson L, Otto H. LUMA interacts with emerin and influences its distribution at the inner nuclear membrane. J Cell Sci. 2008;121(Pt 4):536–48. https://doi.org/10.1242/jcs.019281.

195. Garcia MJ. Constrictive pericarditis versus restrictive cardiomyopathy? J Am Coll Cardiol. 2016;67(17):2061–76. https://doi.org/10.1016/j.jacc.2016.01.076.

196. Kushwaha SS, Fallon JT, Fuster V. Restrictive cardiomyopathy. N Engl J Med. 1997;336 (4):267–76. https://doi.org/10.1056/NEJM199701233360407.

197. Schulz V, Hendig D, Szliska C, Gotting C, Kleesiek K. Novel mutations in the ABCC6 gene of German patients with pseudoxanthoma elasticum. Hum Biol. 2005;77(3):367–84.

198. Ruberg FL, Berk JL. Transthyretin (TTR) cardiac amyloidosis. Circulation. 2012;126 (10):1286–300. https://doi.org/10.1161/CIRCULATIONAHA.111.078915.

199. Kostareva A, Kiselev A, Gudkova A, Frishman G, Ruepp A, Frishman D, et al. Genetic spectrum of idiopathic restrictive cardiomyopathy uncovered by next-generation sequencing. PLoS One. 2016;11(9):e0163362. https://doi.org/10.1371/journal.pone.0163362.

200. Mogensen J, Kubo T, Duque M, Uribe W, Shaw A, Murphy R, et al. Idiopathic restrictive cardiomyopathy is part of the clinical expression of cardiac troponin I mutations. J Clin Invest. 2003;111(2):209–16. https://doi.org/10.1172/JCI16336.

201. Filatov VL, Katrukha AG, Bulargina TV, Gusev NB. Troponin: structure, properties, and mechanism of functioning. Biochemistry (Mosc). 1999;64(9):969–85.

202. Peddy SB, Vricella LA, Crosson JE, Oswald GL, Cohn RD, Cameron DE, et al. Infantile restrictive cardiomyopathy resulting from a mutation in the cardiac troponin T gene. Pediatrics. 2006;117(5):1830–3. https://doi.org/10.1542/peds.2005-2301.

203. Ploski R, Rydzanicz M, Ksiazczyk TM, Franaszczyk M, Pollak A, Kosinska J, et al. Evidence for troponin C (TNNC1) as a gene for autosomal recessive restrictive cardiomyopathy with fatal outcome in infancy. Am J Med Genet A. 2016;170(12):3241–8. https://doi.org/10.1002/ ajmg.a.37860.

204. Marques MA, de Oliveira GA. Cardiac troponin and tropomyosin: structural and cellular perspectives to unveil the hypertrophic cardiomyopathy phenotype. Front Physiol. 2016;7:429. https://doi.org/10.3389/fphys.2016.00429.

205. Karam S, Raboisson MJ, Ducreux C, Chalabreysse L, Millat G, Bozio A, Bouvagnet P. A de novo mutation of the beta cardiac myosin heavy chain gene in an infantile restrictive

cardiomyopathy. Congenit Heart Dis. 2008;3(2):138–43. https://doi.org/10.1111/j.1747-0803. 2008.00165.x.

206. Kaski JP, Syrris P, Burch M, Tome-Esteban MT, Fenton M, Christiansen M, et al. Idiopathic restrictive cardiomyopathy in children is caused by mutations in cardiac sarcomere protein genes. Heart. 2008;94(11):1478–84. https://doi.org/10.1136/hrt.2007.134684.

207. Wu W, Lu CX, Wang YN, Liu F, Chen W, Liu YT, et al. Novel phenotype-genotype correlations of restrictive cardiomyopathy with myosin-binding protein C (MYBPC3) gene mutations tested by next-generation sequencing. J Am Heart Assoc. 2015;4(7):e001879. https://doi.org/10.1161/JAHA.115.001879.

208. Peled Y, Gramlich M, Yoskovitz G, Feinberg MS, Afek A, Polak-Charcon S, et al. Titin mutation in familial restrictive cardiomyopathy. Int J Cardiol. 2014;171(1):24–30. https://doi. org/10.1016/j.ijcard.2013.11.037.

209. Purevjav E, Arimura T, Augustin S, Huby AC, Takagi K, Nunoda S, et al. Molecular basis for clinical heterogeneity in inherited cardiomyopathies due to myopalladin mutations. Hum Mol Genet. 2012;21(9):2039–53. https://doi.org/10.1093/hmg/dds022.

210. Bang ML, Mudry RE, McElhinny AS, Trombitas K, Geach AJ, Yamasaki R, et al. Myopalladin, a novel 145-kilodalton sarcomeric protein with multiple roles in Z-disc and I-band protein assemblies. J Cell Biol. 2001;153(2):413–27.

211. Arbustini E, Pasotti M, Pilotto A, Pellegrini C, Grasso M, Previtali S, et al. Desmin accumulation restrictive cardiomyopathy and atrioventricular block associated with desmin gene defects. Eur J Heart Fail. 2006;8(5):477–83. https://doi.org/10.1016/j.ejheart.2005.11.003.

212. Brodehl A, Ferrier RA, Hamilton SJ, Greenway SC, Brundler MA, Yu W, et al. Mutations in FLNC are associated with familial restrictive cardiomyopathy. Hum Mutat. 2016;37 (3):269–79. https://doi.org/10.1002/humu.22942.

213. Brodehl A, Gaertner-Rommel A, Klauke B, Grewe SA, Schirmer I, Peterschroder A, et al. The novel alphaB-crystallin (CRYAB) mutation p.D109G causes restrictive cardiomyopathy. Hum Mutat. 2017;38(8):947–52. https://doi.org/10.1002/humu.23248.

214. Goldfarb LG, Dalakas MC. Tragedy in a heartbeat: malfunctioning desmin causes skeletal and cardiac muscle disease. J Clin Invest. 2009;119(7):1806–13. https://doi.org/10.1172/ JCI38027.

215. Jahed Z, Shams H, Mehrbod M, Mofrad MR. Mechanotransduction pathways linking the extracellular matrix to the nucleus. Int Rev Cell Mol Biol. 2014;310:171–220. https://doi.org/ 10.1016/B978-0-12-800180-6.00005-0.

216. Olive M, Kley RA, Goldfarb LG. Myofibrillar myopathies: new developments. Curr Opin Neurol. 2013;26(5):527–35. https://doi.org/10.1097/WCO.0b013e328364d6b1.

Genetic Basis of Mitochondrial Cardiomyopathy

3

Elisa Mastantuono, Cordula Maria Wolf, and Holger Prokisch

Contents

E. Mastantuono
Institute of Human Genetics, Technische Universität München, Munich, Germany

Institute of Human Genetics, Helmholtz Zentrum München, Munich, Germany

DZHK (German Centre for Cardiovascular Research), Partner Site Munich Heart Alliance, Munich, Germany

C. M. Wolf
Department of Pediatric Cardiology and Congenital Heart Defects, German Heart Center Munich, Technical University Munich, Munich, Germany

H. Prokisch (✉)
Institute of Human Genetics, Technische Universität München, Munich, Germany

Institute of Human Genetics, Helmholtz Zentrum München, Munich, Germany
e-mail: prokisch@helmholtz-muenchen.de

© Springer Nature Switzerland AG 2019
J. Erdmann, A. Moretti (eds.), *Genetic Causes of Cardiac Disease*, Cardiac and Vascular Biology 7, https://doi.org/10.1007/978-3-030-27371-2_3

Abstract

Mitochondrial disorders are a clinically heterogeneous group of disorders that arise as a result of dysfunction of the mitochondrial energy metabolism, and they represent one of the largest groups of inborn errors of metabolism. Mitochondrial disorders can be caused by mutation of genes encoded by either nuclear DNA or mitochondrial DNA (mtDNA), with more than 300 disease-associated genes identified to date. Among these genes, around 100 have so far been associated with cardiac manifestations. Cardiomyopathy is estimated to occur in 20–40% of children with mitochondrial disorders. Genetic defects can affect a vast range of different mitochondrial functions including electron transport chain complex subunits and their assembly factors, mitochondrial transfer or ribosomal RNAs, factors involved in translation or mtDNA maintenance, and cofactor metabolism such as coenzyme Q10 synthesis. With collectively more than 1000 described cases, the most frequent mitochondrial cardiac diseases include Barth syndrome, Sengers syndrome, *ACAD9*- or *TMEM70*-related mitochondrial complex I or V deficiency, and Friedreich ataxia. Hypertrophic cardiomyopathy is the most common type of cardiomyopathy, but mitochondrial cardiomyopathies might also present as dilated, restrictive, left ventricular non-compaction, and histiocytoid cardiomyopathies. Mitochondrial cardiomyopathy can vary in severity from asymptomatic to severe manifestations including heart failure and sudden cardiac death. Congenital arrhythmias and congenital heart defects (CHDs) are also part of the clinical spectrum of mitochondrial disorders. In this chapter, we provide an overview of the constantly growing number of mitochondrial cardiac disorders and comment on the current practice in the diagnosis and treatment of patients with mitochondrial cardiomyopathy, including optimal therapeutic management and long-term monitoring.

3.1   Introduction

Mitochondrial disorders are a heterogeneous group of inborn errors of metabolism encompassing a wide range of clinical presentations, with > 300 disease-associated genes identified to date [1, 192]. Mitochondria are largely known as the powerhouse of the cell due to their crucial function in the generation of cellular energy. They exploit the energy stored in fats, carbohydrates, and proteins to produce ATP in a process called oxidative phosphorylation (OXPHOS). OXPHOS requires the transport of electrons to molecular oxygen via the mitochondrial respiratory chain, which involves four multi-subunit complexes (complex I–complex IV) and two mobile electron carriers, coenzyme Q10 (CoQ10) and cytochrome *c*. The respiratory chain generates a transmembrane proton gradient that is used by complex V (also known as ATP synthase) to synthesize ATP.

Major metabolic consequences of OXPHOS impairment include accumulation of intermediary metabolic products, increased generation of reactive oxygen species, and decreased energy production. As a consequence of energy deficiency, high-energy demand tissues such as muscle, brain, and liver can be impaired, with resulting multi-organ disease, often with a progression of symptoms developing over time. The variable and frequently systemic nature of mitochondrial disorders make molecular diagnosis difficult. Many different medical specialties are often involved in patient care (with individual physicians often becoming discouraged from solving such complex phenotypes). On the other hand, its very systemic nature aids in raising suspicion for the diagnosis of mitochondrial disorders. Specific combinations of the clinical features associated with mitochondrial disorders can subsequently be grouped into specific syndromes [2].

With an estimated prevalence of 1 in 5000 live births, mitochondrial disorders represent one of the largest groups of inborn errors of metabolism [3]. The onset of mitochondrial disorders has an apparent peak in early childhood (first 3 years of life) and a second broad peak beginning toward the second and the fourth decade of life (adult-onset diseases) [3]. Childhood-onset mitochondrial disorders are generally more severe, especially when clinical manifestation occurs in infancy (<1 year of age) or if they include cardiac involvement (mortality of 71% vs 26% without cardiac phenotype) [4].

Cardiac involvement, specifically cardiomyopathy and arrhythmias, are common features associated with both early- and late-onset forms of mitochondriopathy. Cardiomyopathy is estimated to occur in 20–40% of children with mitochondrial disorders [4, 5]. However, screening for cardiac involvement is not always performed, and this figure is therefore possibly underestimated. The most common form of cardiomyopathy is hypertrophic; however dilated cardiomyopathy and left ventricular (LV) non-compaction also occur relatively frequently [6, 7]. Conduction system and bradyarrhythmias, in addition to WPW syndrome and tachyarrhythmias, are the most commonly encountered arrhythmias in mitochondrial disorders [8, 9].

On the other hand, when a mitochondrial condition affects at the initial stage selectively the heart, the mitochondrial cardiomyopathy may be clinically indistinguishable from other genetic determined cardiomyopathies [147]. Hence, mitochondrial disorders should be suspected in idiopathic isolated forms of cardiomyopathies. Conventional cardiomyopathy gene panels often do not include all the mitochondrial cardiomyopathy-associated genes, and only the increasing use of whole-exome sequencing (WES) has led to unmasking of such cases. In general, the introduction of next-generation sequencing (NGS) has dramatically improved diagnostic yield for mitochondrial disorders and shifted the focus from mtDNA to the nuclear genome. Indeed, both mtDNA and nuclear genomes can be analyzed for pathogenic variants within a single experiment. Consequently, NGS has enabled the identification of a constantly growing list of new disease genes, thereby setting the foundation to uncover an increasing variety of pathophysiologic disease mechanisms [10–13].

Defects of the mitochondrial energy metabolism can be caused by mutations in a number of mitochondrial pathways and functions, including the ATP producing chain from pyruvate dehydrogenase via Krebs cycle to OXPHOS by affecting replication of mtDNA or transcription and translation of mitochondrially encoded OXPHOS subunits. Furthermore, there are numerous additional functions involved in the mitochondrial energy metabolism, such as transport processes of proteins and substrates through the mitochondrial membranes, quality control systems, essential cofactors, motility of the organelle, and membrane integrity.

Mitochondrial genetics is complex, as the workhorse of mitochondria, the mitochondrial proteome (around 1500 proteins), is under the control of two genomes, the nuclear genome and the mitochondrial genome (mtDNA). Although only a small fraction of their proteins are encoded by mtDNA, abnormalities in mtDNA affect more than half of the adult cases, whereas nuclear DNA defects account for up to 80% of mitochondrial disorders in children [14]. Therefore, the most common mode of inheritance of mitochondrial disorders in children is autosomal recessive. For mtDNA variants, the number of wild-type copies is a key factor that determines whether a cell expresses a biochemical defect. This is usually determined by the proportion of mutated copies versus wild-type copies of mtDNA. Homoplasmy describes the setting in which all mtDNA copies have the same mutation, while heteroplasmy, on the other hand, occurs if a mixture of different genotypes, e.g., wild-type and mutant mtDNA, coexist. Variants in mtDNA causing mitochondrial disorders can be further classified into three types: (a) mutations in genes encoding structural proteins or in genes involved in protein synthesis, (b) single or multiple mtDNA deletions, and (c) mtDNA depletion. Multiple mtDNA deletions and mtDNA depletion are caused by nuclear gene defects. Screening the literature, we found an cassociated cardiac phenotype in around 100 genes ($n = 1855$ cases) among the 300 mitochondrial disease-associated genes described to date. However, concerning the number of patients per gene defect, a non-equal long tail distribution was observed. There are disorders like Friedreich ataxia for which more than 1200 patients have a reported cardiac involvement, Barth syndrome with more than 200 patients, and *ACAD9* with over 50 patients, while there are around 70 disease entities with less than 10 published cases with cardiac involvement (Table 3.1). Many individuals display a cluster of clinical features which fall into defined mitochondrial syndromes. Among them, MELAS (mitochondrial encephalomyopathy, lactic acidosis, and stroke-like episodes) and MERRF (myoclonic epilepsy with ragged red fibers) represent the syndromes most frequently associated with cardiac manifestations.

3.2   Cardiac Manifestation in Mitochondrial Disorders

Cardiomyopathies represent a significant clinical manifestation associated with mitochondrial disorders that can result in sudden cardiac death. Neonatal cardiomyopathies are often characterized by hypertrophy of one (mostly left) or both ventricles [147]. Ventricular dysfunction may be progressive in utero and after birth. Neonatal cardiomyopathy is often followed by fatal heart failure or, in other

Table 3.1 List of mitochondrial genes associated with cardiac manifestations

Gene	Inheritance	HCM	DCM	LVNC	RCM	HICM	Arrhythmias	CHD	SCD/CA	HT	Extra-cardiac manifestations	Patients with cardiac phenotype[a]
OXPHOS												
Complex I												
NDUFA2	AR	Yes							Yes		Leigh syndrome	1 pt. (Ref. [15])
NDUFA10	AR	Yes							Yes		Leigh syndrome	1 pt. (Ref. [16])
NDUFA11	AR	Yes	Yes						Yes	Yes	Encephalocardiomyopathy with brain atrophy	4 pts. (Ref. [17])
NDUFB11	XLR/de novo		Yes			Yes			Yes	Yes	Linear skin defects, hypotonia, eye abnormalities, microcephaly, DD, thyroid abnormalities	6 pts. (Refs. [18–21])
NDUFS2	AR	Yes									Leigh syndrome	6 pts. (Refs. [22–24])
NDUFS4	AR	Yes									Leigh syndrome	1 pt. (Ref. [25])
NDUFS8	AR	Yes									Leigh syndrome	2 pts. (Refs. [23, 26])
NDUFV2	AR	Yes							Yes		Early-onset encephalopathy	5 pts. (Refs. [27, 28])
MT-ND1	mt	Yes		Yes					Yes		LHON, Alzheimer's disease, cancer of colon, sudden infant death syndrome, adult-onset dystonia, MELAS	7 pts. (Refs. [7, 29])

(continued)

Table 3.1 (continued)

Gene	Inheritance	HCM	DCM	LVNC	RCM	HICM	Arrhythmias	CHD	SCD/CA	HT	Extra-cardiac manifestations	Patients with cardiac phenotype[a]
MT-ND2	mt	Yes									LHON, Leigh syndrome	1 pt. (Ref. [30])
Assembly												
NDUFAF1	AR	Yes		Yes			Yes		Yes		Cardioencephalomyopathy	2 pts. (Refs. [31, 32])
NDUFAF4	AR	Yes									Infantile mitochondrial encephalopathy	1 pt. (Ref. [33])
ACAD9	AR	Yes	yes	Yes			Yes	Yes	Yes	Yes	Cardiorespiratory depression, encephalopathy, and severe lactic acidosis	56 pts. (Ref. [34])
FOXRED1	AR	Yes									Leigh syndrome	1 pt. (Ref. [35])
TMEM126B	AR	Yes						Yes			Myopathy	1 pt. (Ref. [36])
Complex II												
SDHA	AR	Yes	Yes	Yes					Yes		Leigh syndrome, paragangliomas	5 pts. (Refs. [37–39])
SDHD	AR		Yes	Yes				Yes	Yes	Yes	Encephalomyopathy; carcinoid tumors, intestinal; Cowden-like syndrome; Merkel cell carcinoma, somatic; paraganglioma and gastric stromal sarcoma; paragangliomas 1, with our without deafness; pheocytochromocytoma	1 pt. (Ref. [40])

Gene	Inheritance						Phenotype	Points (Ref.)
Complex III								
BCS1L	AR	Yes					Gracile syndrome	2 pts. (Ref. [41])
MT-CYB	mt	Yes	Yes	Yes			Complex III deficiency, LHON, colorectal cancer, exercise intolerance, mitochondrial encephalomyopathy, multisystem disorder, infantile histiocytoid cardiomyopathy, exercise intolerance+cardiomyopathy +septooptic dysplasia, Parkinsonism/MELAS overlap syndrome	3 pts. (Ref. [42])
Complex IV								
COX6B1	AR	Yes				Yes	Encephalomyopathy, hydrocephalus	1 pt. (Ref. [43])
MT-CO2	AR	Yes	Yes				Complex IV deficiency; pigmentary retinopathy; colorectal cancer	3 pts. (Ref. [44])
MT-CO3	AR		Yes				Complex IV deficiency; LHON, seizures+lactic acidosis	5 pts. (Ref. [45])
COX7B	AD/de novo	Yes			Yes		Microphthalmia with linear skin lesions	2 pts. (Ref. [46])
Assembly								
SURF1	AR	Yes					Leigh syndrome	2 pts. (Refs. [47, 48])
COX14	AR	Yes					Severe congenital lactic acidosis and dysmorphic features associated with a	1 pt. (Ref. [49])

(continued)

Table 3.1 (continued)

Gene	Inheritance	HCM	DCM	LVNC	RCM	HICM	Arrhythmias	CHD	SCD/CA	HT	Extra-cardiac manifestations	Patients with cardiac phenotype[a]
											complex IV assembly defect and a specific decrease in the synthesis of COX I; brain hypertrophy, altered myelination, numerous cavities throughout the brain, hepatomegaly, renal hypoplasia, and adrenal hyperplasia	
COA5	AR	Yes			Yes						Lethal neonatal hypertrophic cardiomyopathy	2 pts. (Ref. [50])
Complex V												
TMEM70	AR	Yes	Yes				Yes	Yes	Yes		Complex V (ATP synthase) deficiency, nuclear type 2	49 pts. (Refs. [51–54])
MT-ATP6	mt	Yes									Leigh syndrome, LHON, infantile mitochondrial bilateral striatal necrosis, seizures+lactic acidosis, NARP	2 pts. (Ref. [55, 191])
MT-ATP8	mt	Yes						Yes			Susceptibility to reversible valproate-induced brain pseudoatrophy, neuropathy	5 pts. (Refs. [55–57])
Homeostasis												
TAZ	XLR	Yes	Yes	Yes			Yes	Yes	Yes	Yes	Barth syndrome, cardioskeletal myopathy with neutropenia and abnormal mitochondria,	200 pts. (Ref. [58])

Gene	Inheritance						Phenotype	Points
						Yes	3-methylglutaconic aciduria, type III	
AGK	AR	Yes				Yes	Sengers syndrome, cardioskeletal myopathy, cataracts, failure to thrive, marked respiratory fatigue, marked respiratory distress, pulmonary hypertension, mitochondrial DNA depletion syndrome 10 (myopathic type)	25 pts. (Refs. [59–61])
DNAJC19	AR	Yes	Yes	Yes	Yes	Yes	3-Methylglutaconic aciduria, type V, ataxia	22 pts. (Refs. [62–65])
C1QBP	AR	Yes	Yes			Yes	Severe neonatal-, childhood-, or later-onset cardiomyopathy associated with combined respiratory chain deficiencies; severe neonatal-, childhood-, or later-onset cardiomyopathy associated with combined respiratory chain deficiencies	4 pts. (Ref. [66])
AIFM1	XLR		Yes				Combined oxidative phosphorylation deficiency 6	1 pt. (Ref. [67])
XPNPEP3	AR	Yes					Nephronophthisis-like nephropathy 1	2 pts. (Ref. [68])
FBXL4	AR	Yes		Yes			Mitochondrial DNA depletion syndrome	7 pts. (Ref. [69])

(continued)

Table 3.1 (continued)

Gene	Inheritance	HCM	DCM	LVNC	RCM	HICM	Arrhythmias	CHD	SCD/CA	HT	Extra-cardiac manifestations	Patients with cardiac phenotype[a]
											13 (encephalomyopathic type)	
mtDNA replication, transcription and translation												
POLG	AR		Yes								Mitochondrial DNA depletion syndrome 4A (Alpers); mitochondrial depletion syndrome 4B (MNGIE type); mitochondrial recessive ataxia syndrome (includes SANDO and SCAE); progressive external ophthalmoplegia, autosomal dominant; progressive external ophthalmoplegia, autosomal recessive	5 pt. (Refs. [70–72])
TWNK	AR		Yes				Yes		Yes		Mitochondrial DNA depletion syndrome 7 (hepatocerebral type), progressive external ophthalmoplegia	8 pts. (Refs. [73, 74])
MGME1	AR		Yes				Yes				External ophthalmoplegia, emaciation, and respiratory failure	2 pts. (Ref. [75])
SLC25A4	AR	Yes	Yes						Yes		Progressive external ophthalmoplegia with mitochondrial DNA deletions	4 pts. (Refs. [76–78])

Gene	Inheritance							Phenotype			Points (Refs)
ELAC2	AR	Yes	Yes					Encephalopathy	Yes	Yes	8 pts. (Refs. [79–81])
HSD17B10	AR	Yes						17-Beta-hydroxysteroid dehydrogenase X deficiency; mental retardation, X-linked 17/31, microduplication; mental retardation, X-linked syndrome 10			1 pt. (Ref. [82])
LRPPRC	AR	Yes						Leigh syndrome, French-Canadian type			5 pts. (Ref. [83])
GTPBP3	AR	Yes	Yes				Yes	Lactic acidosis and encephalopathy			6 pts. (Ref. [84])
MTO1	AR	Yes					Yes	Cardiomyopathy, variable clinical symptoms depending on type of mutation			16 pts. (Refs. [85–87])
TRMT5	AR	Yes						Hypertrophic cardiomyopathy muscular hypotonia, psychomotor retardation			1 pt. (Ref. [88])
TRMT10C	AR	Yes						Congenital lactic acidosis, hypotonia, feeding difficulties, and deafness, death at 5 months			1 pt. (Ref. [89])
TRMU	AR	Yes						Acute infantile liver failure; deafness, mitochondrial			1 pt. (Ref. [81])
GFM1	AR	Yes						Combined oxidative phosphorylation deficiency 1; progressive hepatoencephalopathy, early-onset Leigh syndrome			2 pts. (Refs. [90, 91])

(continued)

Table 3.1 (continued)

Gene	Inheritance	HCM	DCM	LVNC	RCM	HICM	Arrhythmias	CHD	SCD/CA	HT	Extra-cardiac manifestations	Patients with cardiac phenotype[a]
TSFM	AR	Yes	Yes								Combined oxidative phosphorylation deficiency 3	3 pts. (Refs. [92, 93])
RMND1	AR	Yes	Yes				Yes	Yes	Yes		Encephaloneuromyopathy, multiple respiratory chain deficiency	14 pts. (Refs. [81, 94])
PARS2	AR		Yes								Alpers syndrome; infantile-onset neurodegenerative disorders	2 pts. (Ref. [95])
MRPS22	AR	Yes					Yes	Yes			Combined oxidative phosphorylation deficiency 5; edema, tubulopathy, and hypotonia	5 pts. (Refs. [96–98])
MRPL3	AR	Yes							Yes		Psychomotor retardation and multiple respiratory chain deficiency	5 pts. (Refs. [99, 100])
MRPL44	AR	Yes									Infantile cardiomyopathy	2 pts. (Refs. [101, 102])
MTFMT	AR	Yes		Yes			Yes	Yes			Leigh syndrome and combined OXPHOS deficiency	8 pts. (Refs. [81, 103, 104])
AARS2	AR	Yes	Yes								Combined oxidative phosphorylation deficiency 8; fatal infantile hypertrophic mitochondrial cardiomyopathy; brain and skeletal muscles also showed milder respiratory chain defects	7 pts. (Refs. [81, 105])

Gene	Inheritance						Phenotype	Score (Refs.)
GARS	AR					Yes	Charcot-Marie-Tooth disease 2D (CMT2D) and distal hereditary motor neuropathy type V	2 pts. (Refs. [81, 106])
KARS	AR					Yes	Charcot-Marie-Tooth disease, recessive intermediate, B	2 pts. (Refs. [107, 108])
LARS2	AR	Yes	Yes			Yes	Premature ovarian failure and hearing loss in Perrault syndrome	1 pt. (Ref. [109])
IARS2	AR					Yes	Cataracts, growth hormone deficiency, sensory neuropathy, sensorineural hearing loss, and skeletal dysplasia	1 pt. (Ref. [110])
WARS2	AR					Yes	Mitochondrial encephalopathy	1 pt. (Ref. [13])
RRM2B	AR					Yes	Mitochondrial DNA depletion syndrome 8A and 8B	2 pts. (Ref. [111])
VARS2	AR	Yes	Yes	Yes	Yes	Yes	Neonatal encephalocardiomyopathy	9 pts (Ref [112])
YARS2	AR				Yes	Yes	Myopathy, lactic acidosis, and sideroblastic anemia 2	4 pts. (Refs. [110, 113])
SUCLG1	AR					Yes	Mitochondrial DNA depletion syndrome 9 (encephalomyopathic type with methylmalonic aciduria)	3 pts. (Ref. [114])
MT-TA	mt					Yes	Myotonic dystrophy-like myopathy, mitochondrial myopathy	1 pt. (Ref. [115])

(continued)

Table 3.1 (continued)

Gene	Inheritance	HCM	DCM	LVNC	RCM	HICM	Arrhythmias	CHD	SCD/CA	HT	Extra-cardiac manifestations	Patients with cardiac phenotype[a]
MT-TG	mt	Yes									Exercise intolerance, sudden death	2 pts. (Ref [115])
MT-TL1	mt	Yes	Yes				Yes		Yes		MELAS syndrome, MERRF syndrome, cardiomyopathy with or without skeletal myopathy, mitochondrial encephalomyopathy, PEO	15 pts. (Ref. [116])
MT-TP	mt	Yes									Myopathy, susceptibility to Parkinson disease, MERRF syndrome	1 pt. (Ref. [115])
Cofactors												
COQ2	AR	Yes	Yes						Yes		Coenzyme Q10 deficiency; infantile encephalomyopathy, nephropathy	2 pts. (Refs. [117, 118])
COQ4	AR	Yes									Haploinsufficiency, mental retardation, encephalomyopathy, and dysmorphic features.	7 pts. (Refs. [119, 120])
BOLA3	AR	Yes									Combined deficiency of the 2-oxoacid dehydrogenases, associated with a defect in lipoate synthesis and accompanied by defects in complexes I, II, and III of the mitochondrial respiratory chain	3 pts. (Ref. [121])
FXN	AR	Yes	Yes	Yes			Yes		Yes		Friedreich ataxia, Friedreich ataxia with retained reflexes	1200 pts. (Ref. [190])

Gene	Inheritance				Disease/phenotype	Points (Ref.)
COQ9	AR				Coenzyme Q10 deficiency	1 pt. (Ref. [122])
PDSS1	AR	Yes			Coenzyme Q10 deficiency; multisystem disease with early-onset deafness, mild mental retardation, optic atrophy, peripheral neuropathy, obesity, livedo reticularis, and cardiac valvulopathy	2 pts. (Ref. [123])
ISCU	AR	Yes			Myopathy with lactic acidosis, hereditary	2 pts. (Ref. [124])
NFU1	AR	Yes	Yes		Multiple mitochondrial dysfunctions syndrome 1	2 pts. (Refs. [121, 125])
LIAS	AR	Yes			Combined glycine cleavage and pyruvate dehydrogenase deficiency	3 pts. (Ref. [126])
COX10	AR	Yes			Complex IV deficiency; encephalopathy, progressive mitochondrial, with proximal renal tubulopathy	2 pts. (Ref. [127, 128])
COX15	AR	Yes			Complex IV deficiency; Leigh syndrome, early-onset fatal	1 pt. (Ref. [129])
SCO1	AR	Yes			Hepatic failure, early onset, and neurologic disorder	1 pt. (Ref. [130])
SCO2	AR	Yes		Yes	Cardioencephalomyopathy, fatal infantile, due to cytochrome c oxidase deficiency	38 pts. (Refs. [117, 128, 131–131])
COA6	AR	Yes			Hypertrophic cardiomyopathy	1 pt. (Ref. [134])

(continued)

Table 3.1 (continued)

Gene	Inheritance	HCM	DCM	LVNC	RCM	HICM	Arrhythmias	CHD	SCD/CA	HT	Extra-cardiac manifestations	Patients with cardiac phenotype[a]
SLC25A26	AR	Yes									Neonatal mortality resulting from respiratory insufficiency and hydrops to childhood acute episodes of cardiopulmonary failure and slowly progressive muscle weakness	3 pts. (Ref. [135])
Inhibitors												
ECHS1	AR	Yes	Yes				Yes	Yes			Leigh-like, neonatal onset, cardiomyopathy	8 pts. (Ref. [136])
Substrates												
SLC25A3	AR	Yes						Yes			Mitochondrial phosphate carrier deficiency	7 pts. (Ref. [137])
ACADVL	AR	Yes	Yes						Yes		VLCAD deficiency	13 pts. (Refs. [138, 139])
CPT1A	AR	Yes									CPT deficiency, hepatic, type IA	1 pt. (Ref. [140])
CPT2	AR	Yes									Hepatic CPT deficiency type II, lethal neonatal CPT II deficiency; myopathy due to CPT II deficiency	1 pt. (Ref. [141])
HADH	AR	Yes									3-Hydroxyacyl-CoA dehydrogenase deficiency; hyperinsulinemic hypoglycemia, familial, 4	1 pt. (Ref. [142])
HADHA	AR	Yes									Fatty liver, acute, of pregnancy; HELLP syndrome, maternal, of	1 pt. (Ref. [143])

Gene	Inheritance							Cardiac/clinical manifestation	No. of patients (Refs.)[a]
								pregnancy; LCHAD deficiency; trifunctional protein deficiency	
HADHB	AR	Yes						Trifunctional protein deficiency	1 pt. (Ref. [143])
SLC22A5	AR	Yes			Yes			Cardiomyopathy, systemic carnitine deficiency	8 pts. (Refs. [132, 144])
SLC25A20	AR	Yes		Yes				Carnitine-acylcarnitine translocase deficiency	1 pt. (Ref. [145])
PPA2	AR	Yes			Yes			Early death, alcohol sensitivity	18 pts. (Ref. [146])

HCM hypertrophic cardiomyopathy, *DCM* dilated cardiomyopathy, *LVNC* left ventricular non-compaction, *RCM* restrictive cardiomyopathy, *HICM* histiocytoid cardiomyopathy, *CHD* congenital heart defect, *SCD/CA* sudden cardiac death/cardiac arrest, *HT* heart transplantation

[a]Minimal number of patients described in literature with cardiac manifestations

cases, can progress into a dilated form, non-compaction of the left ventricle, improve, or even regress completely [147].

Hypertrophic cardiomyopathy (HCM) is the most frequent manifestation in mitochondrial disorders. HCM is characterized by progressive myocardial thickening; diastolic and systolic ventricular dysfunction; histopathologic changes, such as myocyte disarray and fibrosis; and arrhythmias which may cause sudden cardiac death. In adulthood, hypertrophic cardiomyopathy has a prevalence of 1 in 500 and is typically caused by mutations in sarcomere genes [148]. Pediatric HCM is most frequently associated with Noonan syndrome (1 in 10,000), but it is also frequently found with inborn errors of metabolism, including mitochondrial disorder [149]. HCM was reported to be associated with about 90 mitochondrial disease genes out of 100 with a cardiac phenotype (Table 3.1).

Dilated cardiomyopathy (DCM) is characterized by progressive myocardial dilatation and thinning. Both diastolic and systolic ventricular dysfunctions can occur, and DCM is often associated with the occurrence of cardiac conduction system diseases, arrhythmias, and sudden arrhythmic death. DCM as a sequela of myocarditis is the most common cardiomyopathy in childhood, but it can also be inherited or part of systemic diseases, such as a mitochondrial disorder [150]. In mitochondrial disorders, DCM is often a consequence of a progressed HCM. It has been found in association with 30 disease genes, most frequently with Barth and DCMA syndromes [151].

Mitochondrial disorders can present with cardiac conduction defects typically involving sinus node dysfunction, atrioventricular block, ventricular conduction delay, or WPW syndrome. Progression in intraventricular conduction defect is unpredictable and responsible for sudden cardiac death. Overall, 17 mitochondrial disease genes were found in association with arrhythmias. Prevalence of Wolff-Parkinson-White syndrome among patients with mtDNA mutations is high (up to 15%), specifically in patients with MELAS and MERRF [152]. Also in the clinical presentation of Kearns-Sayre syndrome are cardiac conduction defects, together with progressive external ophthalmoplegia and pigmentary retinopathy [153].

Congenital heart defects (CHDs) are rarely recognized to be linked to mitochondrial disorders. CHDs have been identified in patients with mitochondrial disorders due to 16 different genetic defects; however the pathomechanism remains largely unknown (Table 3.1). The CHDs include patent ductus arteriosus (PDA), ventricular and septal defects (VSD and ASD), or more complex CHD defects (tetralogy of Fallot, transpositions of great arteries). Among them, if we consider three of the most frequent causes of cardiac disorder, we found CHDs to be reported in 30% of *TMEM70* cases and 10% of *ACAD9* and *TAZ* cases ([34, 53]; https://www. barthsyndrome.org;). This figure is clearly above the 1% of cases with CHDs identified in the general population. The observation of CHD is present in only a fraction of patients with the clinical presentation of mitochondrial disorders. Usually not all patients with a certain gene defect develop a cardiac phenotype, but the mitochondrial dysfunction predisposes patients to cardiac abnormalities [154]. Sengers syndrome or *ELAC2* patients seemed to be an exception, but for most gene defects which were tightly linked to cardiac manifestations, NGS diagnostics extended the clinical spectrum to patients without cardiac phenotypes [59, 79, 80, 125].

Left ventricular non-compaction (LVNC) is a morphological abnormality of excessive trabeculation of the left ventricular myocardium, which is often complicated by ventricular dysfunction, arrhythmias, and cardioembolism. LVNC can be part of a structural heart disease. Additionally, mutations in a variety of genes, such as transcription factors, structural, nuclear, or ion channel genes, may cause LVNC. In general, the etiology of this disease entity is still poorly understood and object of current investigations. In mitochondrial disorders, Barth syndrome is most often associated with LVNC, but it is also typical for *DNAJC19* defects and reported in a total of 11 mitochondrial genes [62, 153, 155].

3.3 Genetic Causes of Mitochondrial Cardiac Manifestation

With about 300 gene defects reported in association with human mitochondrial disorders, the spectrum of respiratory chain defects is rather intricate. To break down the complexity, we have grouped the genes here into (1) disorders of oxidative phosphorylation (OXPHOS) subunits and their assembly factors; (2) defects of mitochondrial DNA, RNA, and protein synthesis; (3) defects in the substrate-generating upstream reactions of OXPHOS; (4) defects in cofactor metabolism; (5) defects in mitochondrial homeostasis; and (6) defects in relevant inhibitors (Fig. 3.1). In the following text, we will describe these groups in more details.

3.3.1 OXPHOS Subunits and Their Assembling Factors

Complex I deficiency is the most common biochemical phenotype observed in individuals with mitochondrial disorder and is responsible for approximately 30% of childhood-onset cases [156]. Complex I is composed of 44 subunits (7 encoded by mtDNA and 37 by nDNA), with genetic defects identified in 28 structural genes, in 11 assembly factors, and in a number of factors involved in mitochondrial translation (Fig. 3.1) [154].

Complex I deficiency is clinically heterogeneous, with a diverse spectrum of clinical presentations, such as Leigh syndrome, fatal infantile lactic acidosis (FILA), leukoencephalopathy, MELAS, or also hypertrophic cardiomyopathy. HCM, in particular, can be isolated or associated with multi-organ disease. Isolated HCM has been reported with mutations in nuclear-encoded subunits (*NDUFS2, NDUFV2*) and assembly factors (*ACAD9,* and less commonly *NDUFAF1*). Less frequently patients manifest with dilated cardiomyopathy, left ventricle non-compaction, or conduction defects (such as Wolff-Parkinson-White) [157]. Among 95 cases with complex I deficiency and cardiac manifestations described in the literature (Table 3.1), *ACAD9* deficiency, with 56 cases, represents the most frequent genetic cause observed [34]. Considering the number of *ACAD9* patients and the minor allele frequency of deleterious variants described so far, an

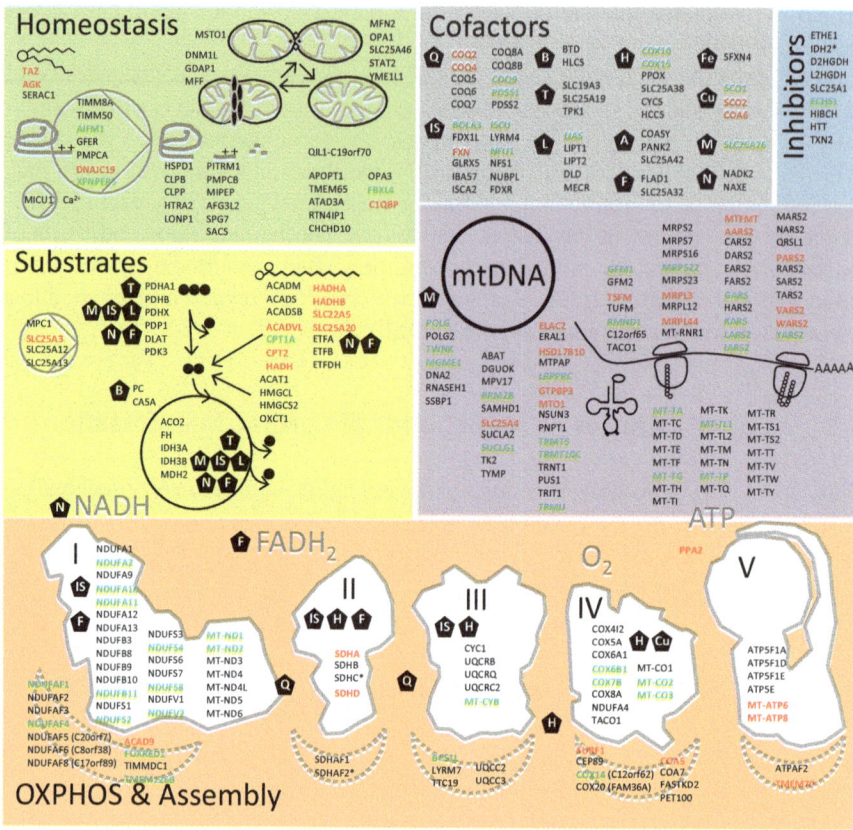

Fig. 3.1 Known disease genes ($n = 309$) in different parts of the mitochondrial energy metabolism. Q (coenzyme Q10), IS (iron-sulfur clusters), B (biotin), T (thiamine pyrophosphate), L (lipoic acid), H (heme), A (coenzyme A), F (riboflavin/FMN/FAD), Fe (iron), Cu (copper), M (s-adenosyl methionine), N (NADH/NADPH). Summary of mitochondrial genes with cardiac manifestations: in red are highlighted genes in which the cardiac involvement represents a major symptom among the other clinical features associated with the gene defect; in green are highlighted genes in which the cardiac involvement represents a minor symptom among the other clinical features associated with the gene defect

incidence of 59 new *ACAD9* patients born every year in Europe has been recently estimated [34]. The clinical spectrum of ACAD9 deficiency is characterized by predominant heart involvement. Other clinical recurrent symptoms include lactic acidosis and muscular weakness [158]. Age of onset, severity of the disease, and progression can vary significantly [34]. On a biochemical level, residual ACAD9 enzyme activity, and not complex I activity, seems to correlate with the severity of clinical symptoms in ACAD9-deficient patients [159]. Regarding the clinical presentation, the typical age of onset of patients with *ACAD9* deficiency is in the neonatal period or in early childhood, with the majority of the patients presenting in the first year of life. Looking at the survival data, for this subgroup with early onset, the survival was poor (50% not surviving the first 2 years) and the outcome extremely severe compared to patients with a later presentation (more than 90% surviving 10 years) [34]. The patients with severe developmental delay or intellectual disability also showed an early disease onset; however neurological manifestations seem to be a rare presentation of the disease [34]. Furthermore, most patients currently alive are able to perform routine activities of daily living. This aspect could also influence the management of *ACAD9* patients, for example, for the decision of performing a cardiac transplant. Four patients with *ACAD9* defects underwent a cardiac transplant. Unfortunately, the two patients who presented within the first year died despite all efforts. In contrast, the two patients presenting after the age of 1 year developed normally, and their clinical status remains stable after years of follow-up (currently aged 15 and 35 years, respectively). However, additional longitudinal studies are warranted to better identify which patients with *ACAD9* deficiency are appropriate heart transplant candidates [160]. Supplementation with riboflavin showed improvement in complex I activity in the majority of patient-derived fibroblasts, and most patients similarly were reported to have clinical benefit with treatment. Most notably, patients presenting within the first year of life show a significantly better survival when treated with riboflavin [34]. However, detailed data about the starting point of riboflavin treatment, the dosage, etc. in large cohorts are needed [34, 161].

Isolated **complex II deficiency** is a rare cause of mitochondrial disorder. Complex II, or succinate dehydrogenase, is formed of four subunits, and it is the only respiratory chain complex entirely encoded by nDNA. Genetic defects were identified in four structural genes and two assembly factors. Clinically, neurological symptoms represent the most common presentation (especially Leigh syndrome or leukoencephalopathy), but hypertrophic and dilated cardiomyopathy, as well as left ventricular non-compaction, were identified in six of the reported cases [162]. In particular, early-onset hypertrophic cardiomyopathy with left ventricular non-compaction and severe complex II deficiency has been observed in association with SDHD defects [40]. Isolated and severe neonatal dilated cardiomyopathy has been associated with *SDHA* mutations [163]. Conduction abnormalities were present in the same patients and thus should be investigated in patients with complex II deficiency [162].

Complex III (coenzyme Q or cytochrome reductase) is a multi-subunit transmembrane protein encoded by both the mitochondrial (cytochrome *b*) and the nuclear genomes (ten subunits). Gene defects have been identified in five structural genes and five assembly factors [164]. Complex III or combined complex I and III deficiency typically manifests in infancy as a severe, multisystem disorder that includes features such as hypotonia, seizures, lactic acidosis, hypoglycemia, and intellectual disability. Variants in *BCS1L* can result in Björnstad syndrome and GRACILE syndrome, severe neonatal syndromes with multisystem and neurological manifestations. However, milder cases with survival into adulthood have also been described. In such cases, HCM has been reported in two patients [41]. In addition, different forms of cardiomyopathy (hypertrophic, dilated, and histiocytoid cardiomyopathy), either isolated or accompanied by multisystem mitochondrial disorder, have been described in three patients with complex III deficiency associated with mutations in mitochondrial DNA-encoded cytochrome *b* (*MT-CYB*) [42, 165].

Complex IV—also called cytochrome *c* oxidase—is the terminal enzyme of the respiratory chain and consists of 14 subunits, 3 of which (named *COX1*, *COX2*, and *COX3*) are encoded by mitochondrial DNA. Complex IV deficiency has been associated with mutations in 11 structural genes and 8 assembly factors.

There are four types of complex IV deficiency differentiated by symptoms and age of onset: benign infantile mitochondrial type, French-Canadian type, infantile mitochondrial myopathy type, and Leigh syndrome. The clinical spectrum can vary among affected individuals, even within the same family. In mildly affected individuals can occur muscle weakness and hypotonia. Whereas in more severely affected individuals neurological dysfunction, heart and liver manifestations, lactic acidosis, and/or Leigh syndrome may also be present [166]. Cardiac involvement has been described in 16 published patients with complex IV deficiency (Table 3.1). Isolated dilated cardiomyopathy has been reported in individuals with complex IV deficiency and carrying *MT-CO2* and *MT-CO3* variants [167]. *SURF1*, a nuclear gene encoding a complex IV assembly factor, represents the most common cause of Leigh syndrome due to complex IV deficiency and is sometimes accompanied by cardiomyopathy [48]. In addition, there are a number of cofactor deficiencies which result in complex IV deficiency with cardiac manifestations (see Sect. 3.4 and Table 3.1).

Mitochondrial ATP synthase (**complex V**) synthesizes ATP from ADP and inorganic phosphate using the energy provided by the proton electrochemical gradient (proton-motive force) across the inner mitochondrial membrane. It consists of two functional domains, F1 and Fo. F1 comprises five different subunits, while the Fo region includes three main subunits a, b, and c, and six additional subunits. Gene defects have been identified in six structural genes and two assembly factors.

Most cases present with neonatal-onset hypotonia, lactic acidosis, hyperammonemia, hypertrophic cardiomyopathy, and 3-methylglutaconic aciduria. Among 55 cases with cardiac manifestations and complex V deficiency present in literature, 49 have been associated with variants in *TMEM70*. *TMEM70* deficiency is the most common genetic defect affecting the ATP synthase [51]. Frequent symptoms at onset are poor feeding, hypotonia, lethargy, and respiratory and heart failure, accompanied by lactic acidosis, 3-methylglutaconic aciduria, and hyperammonemia. In children with *TMEM70* deficiency, the most common heart problem is non-obstructive concentric HCM with preserved systolic function [52, 54]. With the exception of neonates with heart failure, the prognosis of cardiomyopathy in *TMEM70* patients is favorable, because HCM is mostly non-progressive or even regressive during long-term follow-up. Conduction defects (Wolff-Parkinson-White syndrome) have been found in 13% of *TMEM70* patients. Cardiomyopathy associated with systemic mitochondrial disorder has also been described with mutations in mitochondrial genes encoding complex V subunits, including *MT-ATP6* [191] and *MT-ATP8* [57, 168, 169].

3.3.2 Defects of Mitochondrial DNA, RNA, and Protein Synthesis

Mitochondrial protein translation defects typically cause multiple OXPHOS abnormalities and severe mitochondrial disorder. Gene defects have been identified in genes encoding elongation factors, aminoacyl-tRNA synthetases, tRNA-modifying enzymes, a mitochondrial peptide release factor, and an RNase that processes mitochondrial RNA. Mitochondrial disorders linked to protein translation defects manifest with neurological involvement and hypertrophic cardiomyopathy, as well as other multisystem abnormalities. Conditions with cardiac manifestations include defects in mitochondrial ribosomal proteins (*MRPS22*, *MRPL3*, *MRPL44*), mitochondrial tyrosine (*YARS2*) or alanine (*AARS2*) tRNA aminoacylation, and other enzymes involved in mitochondrial RNA metabolism [186, 188]. Cardiomyopathy resulting from mutations in *ELAC2* are usually associated with early severe forms of hypertrophic cardiomyopathy in the context of a multisystem disorder. However, isolated forms of hypertrophic and dilated cardiomyopathy may also be present.

Posttranscriptional modification of mitochondrial tRNAs is necessary for their stability and function. Abnormalities in these processes are illustrative of recently identified protein translation disorders that lead to cardiomyopathy. Hypertrophic cardiomyopathy and cardiac conduction defects in combination with psychomotor delay, encephalopathy, hypotonia, and lactic acidosis were found in children with variants in *MTO1* [86]. Mutations in *GTPBP3* have been described in association with neonatal hypertrophic or dilated cardiomyopathy and conduction defects. Again, most of the patients also showed extra-cardiac symptoms, such as encephalopathy, lactic acidemia, hypoglycemia, and hyperammonemia [84]. Another posttranscriptional modification defect of mitochondrial tRNAs was found in tRNA methyltransferase 5 (*TRMT5*) deficiency. Mutations in *TRMT5* have been described in association with hypertrophic cardiomyopathy, exercise intolerance, global developmental delay, hypotonia, peripheral neuropathy, renal tubulopathy, and lactic acidosis [88].

3.3.3 Defects in the Substrate-Generating Upstream Reactions of OXPHOS

Deficiencies of the mitochondrial phosphate carrier (*SLC25A3*) have been described in cases with hypertrophic cardiomyopathy accompanied by myopathy and lactic acidosis [137]. Lahrouchi et al. have recently reported a loss of function of the carnitine transport SLC22A5 in a family with history of pediatric cardiomyopathy and sudden cardiac death. The patients showed a reduction in the degree of cardiac hypertrophy, and her exercise tolerance improved markedly after L-carnitine supplementation [144]. Biallelic missense variants in the nuclear-encoded mitochondrial inorganic pyrophosphatase (PPA2) were identified in 17 individuals from 7 unrelated pedigrees presenting with seizures, lactic acidosis, cardiac arrhythmia, and exquisite sensitivity to alcohol, leading to sudden cardiac death [146].

3.3.4   Defects in Relevant Cofactors

Many cofactors have a key role in mitochondrial energy metabolism. Among them, some are required for the respiratory chain enzymes like coenzyme Q, iron-sulfur clusters, riboflavin, and heme. Their deficiency typically results in defects of more than one respiratory enzyme. Primary CoQ10 deficiency is a phenotypically and genetically heterogeneous condition, with various clinical presentations including encephalomyopathy, myopathy, cerebellar ataxia, nephrotic syndrome, and severe infantile multisystem mitochondrial disorder. Hypertrophic cardiomyopathy has been reported in cases with mutations in *COQ2*, *COQ4*, and *COQ9* [118, 119, 170, 187]. Fatal infantile cardioencephalomyopathy due to cytochrome c oxidase (COX) deficiency 1 is caused by biallelic variants in the *SCO2* gene. *SCO2* is a mitochondrial copper-binding protein involved in the biogenesis of the Cu(A) site in the cytochrome c oxidase (CcO) subunit Cox2 and in the maintenance of cellular copper homeostasis. Mutations in both *SCO1* and *SCO2* are associated with distinct clinical phenotypes in addition to tissue-specific cytochrome c oxidase deficiency. *SCO2* is highly expressed in the muscle, whereas *SCO1* is expressed at higher levels in the liver. This reflects the different clinical presentation of *SCO1*, mostly associated with hepatic liver failure, and *SCO2*, predominantly associated with severe early-onset cardiac failure. The onset of cardiomyopathy is either in utero or in the first days of life [132]. Copper-histidine supplementation in cell culture, and also the treatment in one patient, was reported to be beneficial for the cardiac phenotype [131].

3.3.5   Defects in Mitochondrial Homeostasis

Mitochondrial homeostasis involves several essential aspects of mitochondrial biogenesis, lipid synthesis, protein import, fission and fusion, quality control, and targeted degradation.

Barth syndrome is due to mutations in the X-linked *TAZ* gene, which codes for Tafazzin, a phospholipid transacylase involved in the remodelling of cardiolipin. Barth syndrome is characterized by cardiomyopathy, skeletal myopathy, distinctive facial features, developmental delay, neutropenia, and increased urinary levels of 3-methylglutaconic acid. Cardiomyopathy is the presenting manifestation in more than 70% of affected males and usually appears in infancy. Interestingly, in these patients, left ventricular non-compaction and dilated cardiomyopathies are more frequent cardiological findings than hypertrophic cardiomyopathy [171]. Patients can manifest with supraventricular and ventricular arrhythmias, sometimes related to sudden cardiac death [172]. Recently, a systematic mutation screening of *TAZ* in a large cohort of pediatric patients with primary cardiomyopathy identified pathogenic variants in 3.5% of the male patients [173]. In a mouse model of Barth syndrome, cardiac-specific loss of succinate dehydrogenase (complex II) activity has been described as a key event in the pathogenesis of cardiomyopathy [174]. Sengers syndrome is caused by the deficiency of the acylglycerol kinase (AGK) which is also involved in the mitochondrial protein import. The clinical spectrum is characterized by the presence of hypertrophic cardiomyopathy, cataracts, myopathy, exercise intolerance, and lactic acidosis [60]. 3-Methylglutaconic aciduria associated with *DNAJC19* mutations (DCMA syndrome) is mainly associated with dilated cardiomyopathy or left ventricular non-compaction and non-progressive cerebellar ataxia [63]. Variants in *C1QBP*, encoding a complement component 1 Q subcomponent-binding protein, have been recently associated with severe forms of neonatal or later-onset cardiomyopathy associated with combined respiratory chain deficiencies [66].

3.3.6 Defects in Relevant Inhibitors

Mitochondrial short-chain enoyl-CoA hydratase-1 deficiency (ECHS1) is an autosomal recessive inborn error of metabolism caused by compound heterozygous mutations in the *ECHS1* gene and characterized by severely delayed psychomotor development, neurodegeneration, increased lactic acid, and cardiomyopathy due to the accumulation of toxic metabolites. Usually these patients are affected by hypertrophic cardiomyopathy and, more rarely, by the dilated form [136, 189].

3.3.7 Mitochondrial Syndromes

Mitochondrial syndromes represent a spectrum of symptoms typically associated with specific abnormalities of mitochondrial DNA (Table 3.2). They can be associated with cardiomyopathy, conduction abnormalities, or both. Kearns-Sayre syndrome is characterized by extra-cardiac manifestations such as pigmentary retinopathy, progressive external ophthalmoplegia, increased cerebrospinal fluid protein concentration, and cerebellar ataxia. Cardiac conduction block is one of the cardinal manifestations; however hypertrophic cardiomyopathy may also be observed. Sudden cardiac death occurs in up to 20% of cases [175].

Cardiomyopathy and/or conduction abnormalities may also accompany MELAS, MERRF, and mtDNA-associated Leigh syndrome/NARP (neuropathy, ataxia, and retinitis pigmentosa) syndromes. Cardiomyopathy occurs in 18–30% of individuals with MELAS syndrome [176]. Both dilated and hypertrophic cardiomyopathies have been observed in MELAS syndrome; however, more typical is non-obstructive concentric hypertrophy [177]. Cardiac conduction abnormalities including Wolff-Parkinson-White syndrome has been reported in 13–27% of individuals with MELAS syndrome [176]. Cardiomyopathy has been identified in 30% of MERRF cases and Wolff-Parkinson-White in 22% [178]. In NARP syndrome, the cardiological manifestations described thus far are hypertrophic cardiomyopathy and atrioventricular block [179].

Table 3.2 Genetic defects and clinical manifestations of mitochondrial syndromes

Mitochondrial syndrome	Genetic defect	Cardiac manifestation	Extra-cardiac manifestation
Kearns-Sayre	mtDNA deletion	A/C (heart block); DCM, HCM, PMVT	Progressive external ophthalmoplegia, pigmentary retinopathy, cerebellar ataxia, short stature, deafness, dementia, limb weakness, diabetes mellitus, renal tubulopathy
MELAS	mtDNA mut.: MT-TL1 enc. tRNALeu (m.3243A > G (80%), m.3271T > C (7.5%)), also MD-ND1 and others	HCM, DCM, LVNC, RCM; A/C (heart block, WPW syndrome)	Encephalomyopathy, lactic acidosis, stroke-like episodes, deafness, myopathy
MERRF	mtDNA mut.: MT-TK enc. tRNALys (m.8344A > G) also MT-TF, MT-TL1, MT-TI, MT-TP	HCM, DCM, HICM; A/C (WPW syndrome, others)	Myoclonic epilepsy, ataxia, myopathy, intellectual disability, deafness, short stature, optic atrophy
NARP/ mtDNA-associated LS	mtDNA mut.: LS, MT-ATP6, MT-TL1, MT-TK, MTTW, MT-TV, MT-ND1, MT-ND2, MT-ND3, MT-ND4, MT-ND5, MT-ND6, MT-CO3; NARP, MT-ATP6(m.8993T > G, m.8993T > G)	HCM; A/C (heart block)	Leigh syndrome, neuropathy, ataxia, pigmentary retinopathy

DCM dilated cardiomyopathy, *HCM* hypertrophic cardiomyopathy, *HICM* histiocytoid cardiomyopathy, *LVNC* left ventricular non-compaction, *MELAS* mitochondrial encephalomyopathy, lactic acidosis, and stroke-like episodes, *MERRF* myoclonic epilepsy with ragged red fibers, *PMVT* polymorphic ventricular tachycardia, *RCM* restrictive cardiomyopathy, *WPW* Wolff-Parkinson-White syndrome

3.4   Characterization of the Cardiac Involvement

The clinical presentation in patients with cardiomyopathy varies from asymptomatic to severe heart failure with asphyxia. In some cases, a sudden cardiac death can even be the first manifestation of the underlying cardiac dysfunction in an apparently healthy individual. Therefore, the assessment of the cardiac status represents a fundamental step for the management, follow-up, and prognosis of these patients. History and physical examination might reveal multisystem involvement, such as failure to thrive, global developmental delay, muscle weakness, respiratory abnormalities, epilepsy, vision problems, and others [180]. Family history might be positive for the occurrence of multisystem disorders in ancestors (Fig. 3.2). If a significant ventricular dysfunction is present, the cardiac examination might show

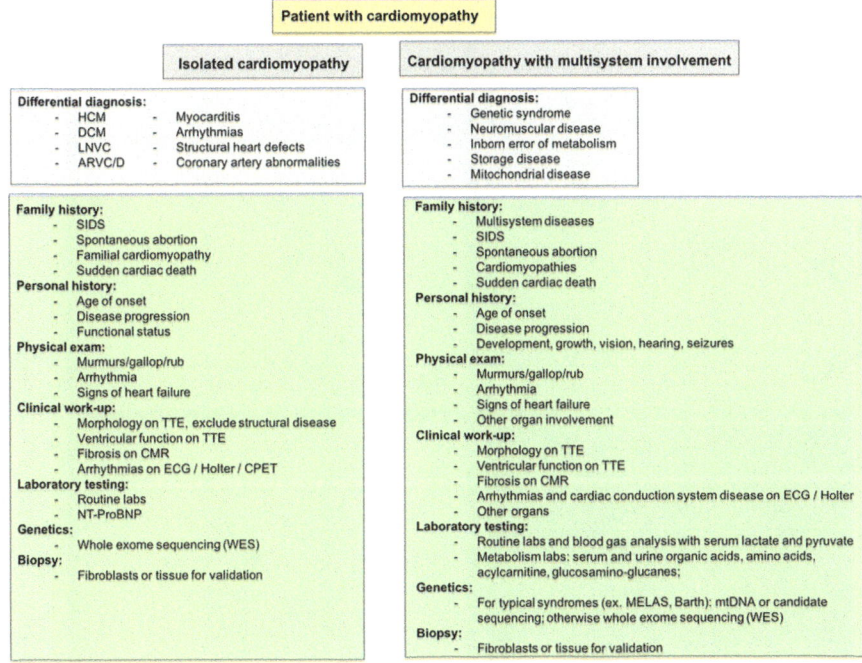

Fig. 3.2 Clinical work-up of patient with suspected mitochondrial cardiomyopathy

signs of heart failure, such as dyspnea, sweating, failure to thrive, bilateral lung crackles, pitting edema, hepatomegaly, and signs of hypoperfusion. Severe cardiac manifestations including heart failure, ventricular tachyarrhythmias, and sudden cardiac death can occur during a metabolic crisis often precipitated by physiologic stressors such as febrile illness or surgery. The cardiac investigation should include chest X-ray, transthoracic echocardiography (TTE), electrocardiography (ECG), Holter ECG, cardiac magnetic resonance (CMR), and, in older children, cardiopulmonary exercise testing (CPET) (Fig. 3.2). TTE is performed to assess ventricular chamber sizes, myocardial thickness, and systolic and diastolic function [181]. If hypertrophy is present, it is mostly concentric [182] (Fig. 3.3), but asymmetric septal hypertrophy with or without left ventricular outflow obstruction has been described. ECG and Holter recordings are performed to screen for the presence of arrhythmias and cardiac conduction system disorders, such as sinus node dysfunction, atrioventricular block, WPW syndrome, or intraventricular conduction delay [181] (Fig. 3.3e-f). CPET is performed to assess exercise capacity as reflected by the maximum rate of oxygen consumption (peak VO2) [181] and to evaluate the risk for exercise-triggered arrhythmias or conduction system disease. CMR should be performed to fully evaluate cardiac morphology and ventricular function, to screen for storage diseases, and to assess the presence of interstitial or focal fibrosis. The indication for cardiac catheterization is on an individual basis and should be

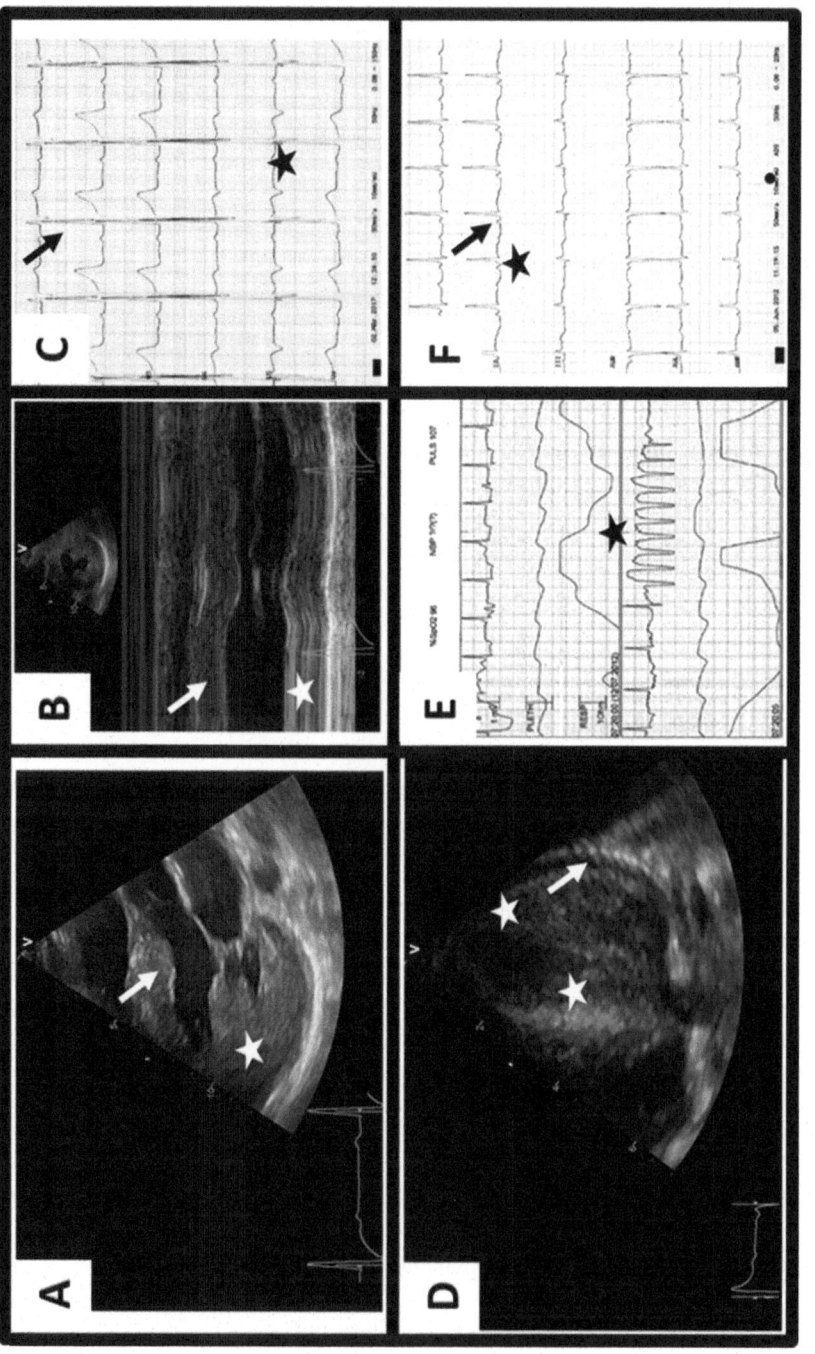

Fig. 3.3 Case presentation: cardiac manifestation of mitochondrial disorder. (**a**, **b**, **c**) A 15-year-old girl with complex I and IV deficiency and concentric hypertrophic cardiomyopathy presenting with metabolic crisis and seizures, but preserved cardiac function. Parasternal long axis (**a**) and parasternal short axis

(b) on transthoracic echocardiography show thickening of the left ventricular posterior wall (star) and of the interventricular septum (arrowhead); there is left ventricular hypertrophy on electrocardiography with deep S waves in the right (star) and tall R waves in the left (arrowhead) precordial leads. (d, e, f) A 2-month-old boy with complex I deficiency and severe concentric hypertrophic cardiomyopathy presenting with heart failure. Four-chamber view reveals massive thickening of the left ventricular myocardium (star) and pericardial effusion (arrowhead) on transthoracic echocardiography (d); telemetric monitoring shows brief runs of non-sustained self-limiting polymorphic ventricular tachycardia (star) (e); there is WPW syndrome with short PR interval (star) and delta-waves (arrowhead) on 12-lead surface electrocardiogram

performed if ventricular function is very poor in order to assess hemodynamics for possible heart transplantation or if the underlying diagnosis is not clear, so that endomyocardial biopsy results are made available. The conventional clinical approach of patients with mitochondrial cardiomyopathy includes detailed family and personal history and a clinical work-up to evaluate the cardiac defect, followed by histologic, enzymologic, molecular, and metabolic findings [153] (Fig. 3.2). If the cardiac phenotype is suspected to be based on a mitochondrial disorder, laboratory tests should include basic measurements, as well as detailed metabolic tests including ammonia, serum pyruvate, serum lactate, creatine kinase, quantitative amino acids, plasma acyl carnitine profile, and quantitative urine organic acids [153]. Care should be taken for proper processing of blood samples, as many metabolic tests require specific handling for accurate results. However, in clinical practice can frequently occur a mitochondrial cardiomyopathy without alteration of such classical metabolic findings.

3.5   Diagnosis of Potential Mitochondrial Cardiac Disease

The clinical spectrum of mitochondrial disorders often overlaps with common cardiological and extra-cardiac diseases. The diagnostic approach of these disorders is complex, time-consuming, and expensive. As mentioned above, although a positive biochemical result can substantiate a clinical diagnosis, the results are often inconclusive [11]. In this scenario, today's use of genetic testing represents a fundamental step in the diagnosis of mitochondrial disorders [12, 13] (Fig. 3.2). Whole-exome sequencing enables rapid, cost-effective, genome-wide screening and dramatically increases the diagnostic yield to greater than 60%, revealing a remarkable heterogeneity of underlying gene defects [11, 192]. With a genome-wide approach, missing genotype-phenotype correlations are revealed, and subsequently the list of new disease genes is ever growing. Indeed, in clinical practice, many mitochondrial cardiomyopathies would have been clinically misdiagnosed without the identification of the causal gene defects by NGS techniques. On the other hand, early-onset forms of hypertrophic cardiomyopathy related to specific disorders, such as Noonan syndrome, could also be erroneously suspected to be due to mitochondrial defects without a genetic confirmation. In case of identification of variants of uncertain significance (VUS), a further valuable diagnostic step to validate the genetic finding is represented by a targeted biochemical analysis in fibroblasts or tissue biopsies (endomyocardial biopsy) (Fig. 3.2). Fibroblasts should be stored in parallel to be available for functional confirmation without further delaying diagnosis [193].

3.6   Management of Patients with Mitochondrial Cardiomyopathy

During the past 6 years, more than 100 novel mitochondrial disorders have been identified, affecting diverse mitochondrial pathways. The application of NGS technologies facilitates our understanding of the pathophysiology of mitochondrial disorders in general and potentially treatable disease subgroups; but the clinical management of affected individuals is challenging, and diagnostic strategies are in flux [102]. Unfortunately, in current clinical practice, therapeutic options for the majority of classical mitochondrial syndromes are limited to supportive care. Nevertheless, there are several mitochondrial defects, especially related to cofactor metabolism, that can be corrected by specific treatment strategies [102]. If the patient is severely and progressively ill (e.g., neonatal lactic acidosis), a more rapid diagnostic procedure is important, and a muscle biopsy with functional investigations is necessitated in parallel with starting genetic testing. This holds true especially if the suspected diagnosis influences the disease management (e.g., PDHc deficiency: initiation of ketogenic diet) or if it can guide the end of life decisions (e.g., Leigh syndrome). In this regard, empirical therapy with thiamine (20 mg/kg/day), biotin (5 mg/kg/day), riboflavin (20 mg/kg/day), and coenzyme Q10 (15 mg/kg/day) might be considered in patients with rapidly progressive or potential life-threatening course of disease. The treatment effect depends on the underlying mitochondrial disorders, for example, L-carnitine supplementation is highly effective in patients who have dilated cardiomyopathy secondary to primary systemic carnitine deficiency and in Barth syndrome. On the other hand, it has low effect on other types of mitochondrial cardiomyopathy. L-Arginine is effective in MELAS syndrome. Riboflavin is highly effective in complex I deficiencies and in particular in *ACAD9* patients [34]. A low-fat, high-protein diet is important for long-chain fatty acid metabolism disorders [102]. Patients with mitochondrial disorders should avoid certain medications that interfere with mitochondrial function and can precipitate a crisis state, such as metformin, propofol, statins, valproic acid, macrolide antibiotics, or tetracyclines [153]. Regular clinical checkups of all organs affected are necessitated depending on the severity of multisystem involvement. Cardiac involvement is treated by conventional heart failure therapy which includes diuretics, angiotensin-converting enzyme inhibitors, beta-blockers, and calcium antagonists. Oral anticoagulation may be indicated with poor systolic ventricular function according to the clinical guidelines. Implantation of a pacemaker might be necessary in severe cardiac conduction system disease. An implantable cardioverter-defibrillator device might be indicated if there is a significant risk for sudden arrhythmic death or for secondary prevention after survived sudden death. Prophylactic cardiac pacing device implantation is thus generally proposed to prevent cardiac death. In the absence of consistent data regarding the incidence of ventricular arrhythmia in patients with mitochondrial cardiomyopathy, an implantable cardiac defibrillator should be proposed according to current clinical guidelines. A metabolic crisis with acute or subacute multi-organ failure secondary to physiologic stressors, such as infection, toxic triggers, medications, psychological stress, heat, or dehydration, is an emergency and needs to be managed appropriately. Cardiac complications during a crisis include cardiogenic shock, arrhythmias, and sudden cardiac death. Management should be focused to the underlying cause of the crisis and on treatment that can improve mitochondrial function. Fever needs to be treated and empiric antibiotics administered if an infection is suspected. Mechanical ventilation is required in case of respiratory failure and care needs to be taken with oxygenation, as it worsens the crisis by increasing free radical production. Correction of acid-base and electrolyte disturbances should be gradual. Continuous infusion of 10% arginine and dextrose-containing intravenous fluids should be best administered via central venous access given the risk for phlebitis and necrosis. Hemodialysis might be necessary for treatment-resistant hyperammonemia, hyperkalemia, or lactic acidosis [183, 184].

3.7   Conclusion and Perspectives

Cardiomyopathy and conduction defects are the most frequent manifestations of mitochondrial cardiac disease. Mitochondrial cardiomyopathy is associated with an ever-growing list of genes, currently around 100. This list outmatches the number of non-mitochondrial cardiomyopathy genes. Therefore, screening for cardiomyopathy should be a routine part of the management of individuals with known or suspected mitochondrial disorders. The diagnosis of mitochondrial disorders, however, remains challenging in many cases due to the myriad of different symptoms in overlap with other clinical conditions. NGS techniques of leukocyte-derived DNA are the state-of-the-art tool to diagnose mitochondrial disorders, enabling rapid, cost-effective, genome-wide screening with a diagnostic yield of greater than 60% [12, 13, 191]. During the past 6 years, more than 100 novel mitochondrial disorders have been identified and have facilitated our understanding of the pathophysiology of mitochondrial disorders. For the remaining undiagnosed cases, the role of complementary NGS approaches, such as genome sequencing and RNA sequencing, is gaining importance [185, 193]. These genomic approaches are still used on a research-oriented base, but we predict that in the future, they will become a valuable tool also in a diagnostic setting. Despite the scientific progress discussed thus far, it is important to underline the fact that most of these disorders do not have a treatment. This is with the exception of the group of cofactor metabolism deficiencies for which there are promising treatment options, highlighting the need for molecular diagnostics. There are still more than 70 diseases with less than 10 patients described. With the increasing number of patients, we will better understand the genotype-phenotype correlations which will help in counseling families. The better we understand the pathomechanism, the greater the chance we will identify new treatment options.

References

1. Mayr JA, Haack TB, Freisinger P, Karall D, Makowski C, Koch J, Feichtinger RG, Zimmermann FA, Rolinski B, Ahting U, Meitinger T, Prokisch H, Sperl W. Spectrum of combined respiratory chain defects. J Inherit Metab Dis. 2015;38:629–40.
2. Gorman GS, Chinnery PF, DiMauro S, Hirano M, Koga Y, McFarland R, Suomalainen A, Thorburn DR, Zeviani M, Turnbull DM. Mitochondrial diseases. Nat Rev Dis Primers. 2016;2:16080.
3. Gorman GS, Schaefer AM, Ng Y, Gomez N, Blakely EL, Alston CL, Feeney C, Horvath R, Yu-Wai-Man P, Chinnery PF, Taylor RW, Turnbull DM, McFarland R. Prevalence of nuclear and mitochondrial DNA mutations related to adult mitochondrial disease. Ann Neurol. 2015;77:753–9.
4. Holmgren D, Wåhlander H, Eriksson BO, Oldfors A, Holme E, Tulinius M. Cardiomyopathy in children with mitochondrial disease; clinical course and cardiological findings. Eur Heart J. 2003;24:280–8.
5. Scaglia F, Towbin JA, Craigen WJ, Belmont JW, Smith EO, Neish SR, Ware SM, Hunter JV, Fernbach SD, Vladutiu GD, Wong L-JC, Vogel H. Clinical spectrum, morbidity, and mortality in 113 pediatric patients with mitochondrial disease. Pediatrics. 2004;114:925–31.
6. Limongelli G, Masarone D, Pacileo G. Mitochondrial disease and the heart. Heart. 2017;103:390–8.
7. Tang S, Batra A, Zhang Y, Ebenroth ES, Huang T. Left ventricular noncompaction is associated with mutations in the mitochondrial genome. Mitochondrion. 2010;10:350–7.
8. Bates MGD, Bourke JP, Giordano C, d'Amati G, Turnbull DM, Taylor RW. Cardiac involvement in mitochondrial DNA disease: clinical spectrum, diagnosis, and management. Eur Heart J. 2012;33:3023–33.
9. Finsterer J, Kothari S. Cardiac manifestations of primary mitochondrial disorders. Int J Cardiol. 2014;177:754–63.
10. Alston CL, Rocha MC, Lax NZ, Turnbull DM, Taylor RW. The genetics and pathology of mitochondrial disease. J Pathol. 2017;241:236–50.
11. Horvath R, Chinnery PF. The effect of neurological genomics and personalized mitochondrial medicine. JAMA Neurol. 2017;74:11.
12. Wortmann S, Mayr J, Nuoffer J, Prokisch H, Sperl W. A guideline for the diagnosis of pediatric mitochondrial disease: the value of muscle and skin biopsies in the genetics era. Neuropediatrics. 2017a;48:309–14.
13. Wortmann SB, Timal S, Venselaar H, Wintjes LT, Kopajtich R, Feichtinger RG, Onnekink C, Mühlmeister M, Brandt U, Smeitink JA, Veltman JA, Sperl W, Lefeber D, Pruijn G, Stojanovic V, Freisinger P, v Spronsen F, Derks TG, Veenstra-Knol HE, Mayr JA, Rötig A, Tarnopolsky M, Prokisch H, Rodenburg RJ. Biallelic variants in WARS2 encoding mitochondrial tryptophanyl-tRNA synthase in six individuals with mitochondrial encephalopathy. Hum Mutat. 2017b;38:1786–95.
14. Goldstein AC, Bhatia P, Vento JM. Mitochondrial disease in childhood: nuclear encoded. Neurotherapeutics. 2013;10:212–26.
15. Hoefs SJG, Dieteren CEJ, Distelmaier F, Janssen RJRJ, Epplen A, Swarts HGP, Forkink M, Rodenburg RJ, Nijtmans LG, Willems PH, Smeitink JAM, van den Heuvel LP. NDUFA2 complex I mutation leads to Leigh disease. Am J Hum Genet. 2008;82:1306–15.
16. Hoefs SJG, van Spronsen FJ, Lenssen EWH, Nijtmans LG, Rodenburg RJ, Smeitink JAM, van den Heuvel LP. NDUFA10 mutations cause complex I deficiency in a patient with Leigh disease. Eur J Hum Genet. 2011;19:270 4.
17. Berger I, Hershkovitz E, Shaag A, Edvardson S, Saada A, Elpeleg O. Mitochondrial complex I deficiency caused by a deleterious NDUFA11 mutation. Ann Neurol. 2008;63:405–8.
18. Rea G, Homfray T, Till J, Roses-Noguer F, Buchan RJ, Wilkinson S, Wilk A, Walsh R, John S, McKee S, Stewart FJ, Murday V, Taylor RW, Ashworth M, Baksi AJ, Daubeney P, Prasad S, Barton PJR, Cook SA, Ware JS. Histiocytoid cardiomyopathy and microphthalmia

with linear skin defects syndrome: phenotypes linked by truncating variants in NDUFB11. Cold Spring Harb Mol Case Stud. 2017;3:a001271.

19. Shehata BM, Cundiff CA, Lee K, Sabharwal A, Lalwani MK, Davis AK, Agrawal V, Sivasubbu S, Iannucci GJ, Gibson G. Exome sequencing of patients with histiocytoid cardiomyopathy reveals a de novo NDUFB11 mutation that plays a role in the pathogenesis of histiocytoid cardiomyopathy. Am J Med Genet A. 2015;167A:2114–21.

20. Torraco A, Bianchi M, Verrigni D, Gelmetti V, Riley L, Niceta M, Martinelli D, Montanari A, Guo Y, Rizza T, Diodato D, Di Nottia M, Lucarelli B, Sorrentino F, Piemonte F, Francisci S, Tartaglia M, Valente EM, Dionisi-Vici C, Christodoulou J, Bertini E, Carrozzo R. A novel mutation in *NDUFB11* unveils a new clinical phenotype associated with lactic acidosis and sideroblastic anemia. Clin Genet. 2017;91:441–7.

21. van Rahden VA, Fernandez-Vizarra E, Alawi M, Brand K, Fellmann F, Horn D, Zeviani M, Kutsche K. Mutations in NDUFB11, encoding a complex I component of the mitochondrial respiratory chain, cause microphthalmia with linear skin defects syndrome. Am J Hum Genet. 2015;96:640–50.

22. Loeffen J, Elpeleg O, Smeitink J, Smeets R, Stöckler-Ipsiroglu S, Mandel H, Sengers R, Trijbels F, van den Heuvel L. Mutations in the complex I NDUFS2 gene of patients with cardiomyopathy and encephalomyopathy. Ann Neurol. 2001;49:195–201.

23. Loeffen J, Smeitink J, Triepels R, Smeets R, Schuelke M, Sengers R, Trijbels F, Hamel B, Mullaart R, van den Heuvel L. The first nuclear-encoded complex I mutation in a patient with Leigh syndrome. Am J Hum Genet. 1998;63:1598–608.

24. Ngu LH, Nijtmans LG, Distelmaier F, Venselaar H, van Emst-de Vries SE, van den Brand MAM, Stoltenborg BJM, Wintjes LT, Willems PH, van den Heuvel LP, Smeitink JA, Rodenburg RJT. A catalytic defect in mitochondrial respiratory chain complex I due to a mutation in NDUFS2 in a patient with Leigh syndrome. Biochim Biophys Acta Mol basis Dis. 2012;1822:168–75.

25. Haack TB, Madignier F, Herzer M, Lamantea E, Danhauser K, Invernizzi F, Koch J, Freitag M, Drost R, Hillier I, Haberberger B, Mayr JA, Ahting U, Tiranti V, Rötig A, Iuso A, Horvath R, Tesarova M, Baric I, Uziel G, Rolinski B, Sperl W, Meitinger T, Zeviani M, Freisinger P, Prokisch H. Mutation screening of 75 candidate genes in 152 complex I deficiency cases identifies pathogenic variants in 16 genes including *NDUFB9*. J Med Genet. 2012b;49:83–9.

26. Haack TB, Haberberger B, Frisch E-M, Wieland T, Iuso A, Gorza M, Strecker V, Graf E, Mayr JA, Herberg U, Hennermann JB, Klopstock T, Kuhn KA, Ahting U, Sperl W, Wilichowski E, Hoffmann GF, Tesarova M, Hansikova H, Zeman J, Plecko B, Zeviani M, Wittig I, Strom TM, Schuelke M, Freisinger P, Meitinger T, Prokisch H. Molecular diagnosis in mitochondrial complex I deficiency using exome sequencing. J Med Genet. 2012a;49:277–83.

27. Bénit P, Beugnot R, Chretien D, Giurgea I, De Lonlay-Debeney P, Issartel J-P, Corral-Debrinski M, Kerscher S, Rustin P, Rötig A, Munnich A. Mutant NDUFV2 subunit of mitochondrial complex I causes early onset hypertrophic cardiomyopathy and encephalopathy. Hum Mutat. 2003;21:582–6.

28. Cameron JM, MacKay N, Feigenbaum A, Tarnopolsky M, Blaser S, Robinson BH, Schulze A. Exome sequencing identifies complex I NDUFV2 mutations as a novel cause of Leigh syndrome. Eur J Paediatr Neurol. 2015;19:525–32.

29. Zifa E, Theotokis P, Kaminari A, Maridaki H, Leze H, Petsiava E, Mamuris Z, Stathopoulos C. A novel G3337A mitochondrial ND1 mutation related to cardiomyopathy co-segregates with tRNALeu(CUN) A12308G and tRNAThr C15946T mutations. Mitochondrion. 2008;8:229–36.

30. Alila OF, Rebai EM, Tabebi M, Tej A, Chamkha I, Tlili A, Bouguila J, Tilouche S, Soyah N, Boughamoura L, Fakhfakh F. Whole mitochondrial genome analysis in two families with dilated mitochondrial cardiomyopathy: detection of mutations in *MT-ND2* and *MT-TL1* genes. Mitochondrial DNA A. 2016;27:2873–80.

31. Dunning CJR, McKenzie M, Sugiana C, Lazarou M, Silke J, Connelly A, Fletcher JM, Kirby DM, Thorburn DR, Ryan MT. Human CIA30 is involved in the early assembly of mitochondrial complex I and mutations in its gene cause disease. EMBO J. 2007;26:3227–37.
32. Fassone E, Taanman J-W, Hargreaves IP, Sebire NJ, Cleary MA, Burch M, Rahman S. Mutations in the mitochondrial complex I assembly factor NDUFAF1 cause fatal infantile hypertrophic cardiomyopathy. J Med Genet. 2011;48:691–7.
33. Saada A, Edvardson S, Rapoport M, Shaag A, Amry K, Miller C, Lorberboum-Galski H, Elpeleg O. C6ORF66 is an assembly factor of mitochondrial complex I. Am J Hum Genet. 2008;82:32–8.
34. Repp BM, Mastantuono E, Alston CL, Schiff M, Haack TB, Rötig A, Ardissone A, Lombes A, Catarino CB, Diodato D, Schottmann G, Poulton J, Burlina A, Jonckheere A, Munnich A, Ghezzi D, Rokicki D, Wellesley D, Martinelli D, Lamantea E, et al. Clinical, biochemical and genetic spectrum of 70 patients with ACAD9 deficiency: is riboflavin supplementation effective? Orphanet J Rare Dis. 2018;13(1):120.
35. Fassone E, Duncan AJ, Taanman J-W, Pagnamenta AT, Sadowski MI, Holand T, Qasim W, Rutland P, Calvo SE, Mootha VK, Bitner-Glindzicz M, Rahman S. FOXRED1, encoding an FAD-dependent oxidoreductase complex-I-specific molecular chaperone, is mutated in infantile-onset mitochondrial encephalopathy. Hum Mol Genet. 2015;24:4183.
36. Alston CL, Compton AG, Formosa LE, Strecker V, Oláhová M, Haack TB, Smet J, Stouffs K, Diakumis P, Ciara E, Cassiman D, Romain N, Yarham JW, He L, De Paepe B, Vanlander AV, Seneca S, Feichtinger RG, Płoski R, Rokicki D, Pronicka E, Haller RG, Van Hove JLK, Bahlo M, Mayr JA, Van Coster R, Prokisch H, Wittig I, Ryan MT, Thorburn DR, Taylor RW. Biallelic mutations in TMEM126B cause severe complex I deficiency with a variable clinical phenotype. Am J Hum Genet. 2016;99:217–27.
37. Alston CL, Davison JE, Meloni F, van der Westhuizen FH, He L, Hornig-Do H-T, Peet AC, Gissen P, Goffrini P, Ferrero I, Wassmer E, McFarland R, Taylor RW. Recessive germline SDHA and SDHB mutations causing leukodystrophy and isolated mitochondrial complex II deficiency. J Med Genet. 2012;49:569–77.
38. Van Coster R, Seneca S, Smet J, Van Hecke R, Gerlo E, Devreese B, Van Beeumen J, Leroy JG, De Meirleir L, Lissens W. Homozygous Gly555Glu mutation in the nuclear-encoded 70 kDa flavoprotein gene causes instability of the respiratory chain complex II. Am J Med Genet A. 2003;120A:13–8.
39. Courage C, Jackson CB, Hahn D, Euro L, Nuoffer J-M, Gallati S, Schaller A. SDHA mutation with dominant transmission results in complex II deficiency with ocular, cardiac, and neurologic involvement. Am J Med Genet A. 2017;173:225–30.
40. Alston CL, Ceccatelli Berti C, Blakely EL, Oláhová M, He L, McMahon CJ, Olpin SE, Hargreaves IP, Nolli C, McFarland R, Goffrini P, O'Sullivan MJ, Taylor RW. A recessive homozygous p.Asp92Gly SDHD mutation causes prenatal cardiomyopathy and a severe mitochondrial complex II deficiency. Hum Genet. 2015;134:869–79.
41. Al-Owain M, Colak D, Albakheet A, Al-Younes B, Al-Humaidi Z, Al-Sayed M, Al-Hindi H, Al-Sugair A, Al-Muhaideb A, Rahbeeni Z, Al-Sehli A, Al-Fadhli F, Ozand PT, Taylor RW, Kaya N. Clinical and biochemical features associated with BCS1L mutation. J Inherit Metab Dis. 2013;36:813–20.
42. Hagen CM, Aidt FH, Havndrup O, Hedley PL, Jespersgaard C, Jensen M, Kanters JK, Moolman-Smook JC, Møller DV, Bundgaard H, Christiansen M. MT-CYB mutations in hypertrophic cardiomyopathy. Mol Genet Genomic Med. 2013;1:54–65.
43. Abdulhag UN, Soiferman D, Schueler-Furman O, Miller C, Shaag A, Elpeleg O, Edvardson S, Saada A. Mitochondrial complex IV deficiency, caused by mutated COX6B1, is associated with encephalomyopathy, hydrocephalus and cardiomyopathy. Eur J Hum Genet. 2015;23:159–64.
44. Campos Y, García-Redondo A, Fernández-Moreno MA, Martínez-Pardo M, Goda G, Rubio JC, Martín MA, del Hoyo P, Cabello A, Bornstein B, Garesse R, Arenas J. Early-onset

multisystem mitochondrial disorder caused by a nonsense mutation in the mitochondrial DNA cytochrome C oxidase II gene. Ann Neurol. 2001;50:409–13.

45. Arbustini E, Favalli V, Narula N, Serio A, Grasso M. Left ventricular noncompaction: a distinct genetic cardiomyopathy? J Am Coll Cardiol. 2016;68:949–66.

46. Indrieri A, van Rahden VA, Tiranti V, Morleo M, Iaconis D, Tammaro R, D'Amato I, Conte I, Maystadt I, Demuth S, Zvulunov A, Kutsche K, Zeviani M, Franco B. Mutations in COX7B cause microphthalmia with linear skin lesions, an unconventional mitochondrial disease. Am J Hum Genet. 2012;91:942–9.

47. Piekutowska-Abramczuk D, Magner M, Popowska E, Pronicki M, Karczmarewicz E, Sykut-Cegielska J, Kmiec T, Jurkiewicz E, Szymanska-Debinska T, Bielecka L, Krajewska-Walasek M, Vesela K, Zeman J, Pronicka E. *SURF1* missense mutations promote a mild Leigh phenotype. Clin Genet. 2009;76:195–204.

48. Wedatilake Y, Brown RM, McFarland R, Yaplito-Lee J, Morris AAM, Champion M, Jardine PE, Clarke A, Thorburn DR, Taylor RW, Land JM, Forrest K, Dobbie A, Simmons L, Aasheim ET, Ketteridge D, Hanrahan D, Chakrapani A, Brown GK, Rahman S. SURF1 deficiency: a multi-centre natural history study. Orphanet J Rare Dis. 2013;8:96.

49. Weraarpachai W, Sasarman F, Nishimura T, Antonicka H, Auré K, Rötig A, Lombès A, Shoubridge EA. Mutations in C12orf62, a factor that couples COX I synthesis with cytochrome c oxidase assembly, cause fatal neonatal lactic acidosis. Am J Hum Genet. 2012;90:142–51.

50. Huigsloot M, Nijtmans LG, Szklarczyk R, Baars MJH, van den Brand MAM, Hendriksfranssen MGM, van den Heuvel LP, Smeitink JAM, Huynen MA, Rodenburg RJT. A mutation in C2orf64 causes impaired cytochrome c oxidase assembly and mitochondrial cardiomyopathy. Am J Hum Genet. 2011;88:488–93.

51. Čížková A, Stránecký V, Mayr JA, Tesařová M, Havlíčková V, Paul J, Ivánek R, Kuss AW, Hansíková H, Kaplanová V, Vrbacký M, Hartmannová H, Nosková L, Honzík T, Drahota Z, Magner M, Hejzlarová K, Sperl W, Zeman J, Houštěk J, Kmoch S. TMEM70 mutations cause isolated ATP synthase deficiency and neonatal mitochondrial encephalocardiomyopathy. Nat Genet. 2008;40:1288–90.

52. Honzík T, Tesarová M, Mayr JA, Hansíková H, Jesina P, Bodamer O, Koch J, Magner M, Freisinger P, Huemer M, Kostková O, van Coster R, Kmoch S, Houstêk J, Sperl W, Zeman J. Mitochondrial encephalocardio-myopathy with early neonatal onset due to TMEM70 mutation. Arch Dis Child. 2010;95:296–301.

53. Magner M, Dvorakova V, Tesarova M, Mazurova S, Hansikova H, Zahorec M, Brennerova K, Bzduch V, Spiegel R, Horovitz Y, Mandel H, Eminoğlu FT, Mayr JA, Koch J, Martinelli D, Bertini E, Konstantopoulou V, Smet J, Rahman S, Broomfield A, Stojanović V, Dionisi-Vici C, van Coster R, Morava-Kozicz E, Sperl W, Zeman J, Honzik T, Honzik T. TMEM70 deficiency: long-term outcome of 48 patients. J Inherit Metab Dis. 2015;38:417–26.

54. Spiegel R, Khayat M, Shalev SA, Horovitz Y, Mandel H, Hershkovitz E, Barghuti F, Shaag A, Saada A, Korman SH, Elpeleg O, Yatsiv I. TMEM70 mutations are a common cause of nuclear encoded ATP synthase assembly defect: further delineation of a new syndrome. J Med Genet. 2011;48:177–82.

55. Imai A, Fujita S, Kishita Y, Kohda M, Tokuzawa Y, Hirata T, Mizuno Y, Harashima H, Nakaya A, Sakata Y, Takeda A, Mori M, Murayama K, Ohtake A, Okazaki Y. Rapidly progressive infantile cardiomyopathy with mitochondrial respiratory chain complex V deficiency due to loss of ATPase 6 and 8 protein. Int J Cardiol. 2016;207:203–5.

56. Jonckheere AI, Hogeveen M, Nijtmans LGJ, van den Brand MAM, Janssen AJM, Diepstra JHS, van den Brandt FCA, van den Heuvel LP, Hol FA, Hofste TGJ, Kapusta L, Dillmann U, Shamdeen MG, Smeitink JAM, Rodenburg RJT. A novel mitochondrial ATP8 gene mutation in a patient with apical hypertrophic cardiomyopathy and neuropathy. J Med Genet. 2007;45:129–33.

57. Ware SM, El-Hassan N, Kahler SG, Zhang Q, Ma Y-W, Miller E, Wong B, Spicer RL, Craigen WJ, Kozel BA, Grange DK, Wong L-J. Infantile cardiomyopathy caused by a

mutation in the overlapping region of mitochondrial ATPase 6 and 8 genes. J Med Genet. 2009;46:308–14.

58. Barth Syndrome Foundation: Home (n.d.).

59. Haghighi A, Haack TB, Atiq M, Mottaghi H, Haghighi-Kakhki H, Bashir RA, Ahting U, Feichtinger RG, Mayr JA, Rötig A, Lebre A-S, Klopstock T, Dworschak A, Pulido N, Saeed MA, Saleh-Gohari N, Holzerova E, Chinnery PF, Taylor RW, Prokisch H. Sengers syndrome: six novel AGK mutations in seven new families and review of the phenotypic and mutational spectrum of 29 patients. Orphanet J Rare Dis. 2014;9:119.

60. Mayr JA, Haack TB, Graf E, Zimmermann FA, Wieland T, Haberberger B, Superti-Furga A, Kirschner J, Steinmann B, Baumgartner MR, Moroni I, Lamantea E, Zeviani M, Rodenburg RJ, Smeitink J, Strom TM, Meitinger T, Sperl W, Prokisch H. Lack of the mitochondrial protein acylglycerol kinase causes Sengers syndrome. Am J Hum Genet. 2012;90:314–20.

61. Vukotic M, Nolte H, König T, Saita S, Ananjew M, Krüger M, Tatsuta T, Langer T. Acylglycerol kinase mutated in sengers syndrome is a subunit of the TIM22 protein translocase in mitochondria. Mol Cell. 2017;67:471–483.e7.

62. Davey KM, Parboosingh JS, McLeod DR, Chan A, Casey R, Ferreira P, Snyder FF, Bridge PJ, Bernier FP. Mutation of DNAJC19, a human homologue of yeast inner mitochondrial membrane co-chaperones, causes DCMA syndrome, a novel autosomal recessive Barth syndrome-like condition. J Med Genet. 2006;43:385–93.

63. Ojala T, Polinati P, Manninen T, Hiippala A, Rajantie J, Karikoski R, Suomalainen A, Tyni T. New mutation of mitochondrial DNAJC19 causing dilated and noncompaction cardiomyopathy, anemia, ataxia, and male genital anomalies. Pediatr Res. 2012;72:432–7.

64. Al Teneiji A, Siriwardena K, George K, Mital S, Mercimek-Mahmutoglu S. Progressive cerebellar atrophy and a novel homozygous pathogenic DNAJC19 variant as a cause of dilated cardiomyopathy ataxia syndrome. Pediatr Neurol. 2016;62:58–61.

65. Ucar SK, Mayr JA, Feichtinger RG, Canda E, Çoker M, Wortmann SB. Previously unreported biallelic mutation in DNAJC19: are sensorineural hearing loss and basal ganglia lesions additional features of dilated cardiomyopathy and ataxia (DCMA) syndrome? JIMD Rep. 2017;35:39–45.

66. Feichtinger RG, Oláhová M, Kishita Y, Garone C, Kremer LS, Yagi M, Uchiumi T, Jourdain AA, Thompson K, D'Souza AR, Kopajtich R, Alston CL, Koch J, Sperl W, Mastantuono E, Strom TM, Wortmann SB, Meitinger T, Pierre G, Chinnery PF, Chrzanowska-Lightowlers ZM, Lightowlers RN, DiMauro S, Calvo SE, Mootha VK, Moggio M, Sciacco M, Comi GP, Ronchi D, Murayama K, Ohtake A, Rebelo-Guiomar P, Kohda M, Kang D, Mayr JA, Taylor RW, Okazaki Y, Minczuk M, Prokisch H. Biallelic C1QBP mutations cause severe neonatal-, childhood-, or later-onset cardiomyopathy associated with combined respiratory-chain deficiencies. Am J Hum Genet. 2017b;101:525–38.

67. Heimer G, Eyal E, Zhu X, Ruzzo EK, Marek-Yagel D, Sagiv D, Anikster Y, Reznik-Wolf H, Pras E, Oz Levi D, Lancet D, Ben-Zeev B, Nissenkorn A. Mutations in AIFM1 cause an X-linked childhood cerebellar ataxia partially responsive to riboflavin. Eur J Paediatr Neurol. 2018;22:93–101.

68. O'Toole JF, Liu Y, Davis EE, Westlake CJ, Attanasio M, Otto EA, Seelow D, Nurnberg G, Becker C, Nuutinen M, Kärppä M, Ignatius J, Uusimaa J, Pakanen S, Jaakkola E, van den Heuvel LP, Fehrenbach H, Wiggins R, Goyal M, Zhou W, Wolf MTF, Wise E, Helou J, Allen SJ, Murga-Zamalloa CA, Ashraf S, Chaki M, Heeringa S, Chernin G, Hoskins BE, Chaib H, Gleeson J, Kusakabe T, Suzuki T, Isaac RE, Quarmby LM, Tennant B, Fujioka H, Tuominen H, Hassinen I, Lohi H, van Houten JL, Rotig A, Sayer JA, Rolinski B, Freisinger P, Madhavan SM, Herzer M, Madignier F, Prokisch H, Nurnberg P, Jackson P, Khanna H, Katsanis N, Hildebrandt F, Hildebrandt F. Individuals with mutations in XPNPEP3, which encodes a mitochondrial protein, develop a nephronophthisis-like nephropathy. J Clin Invest. 2010;120:791–802.

69. Bonnen PE, Yarham JW, Besse A, Wu P, Faqeih EA, Al-Asmari AM, Saleh MAM, Eyaid W, Hadeel A, He L, Smith F, Yau S, Simcox EM, Miwa S, Donti T, Abu-Amero KK, Wong L-J,

Craigen WJ, Graham BH, Scott KL, McFarland R, Taylor RW. Mutations in FBXL4 cause mitochondrial encephalopathy and a disorder of mitochondrial DNA maintenance. Am J Hum Genet. 2013;93:471–81.

70. Van Goethem G, Luoma P, Rantamäki M, Al Memar A, Kaakkola S, Hackman P, Krahe R, Löfgren A, Martin JJ, De Jonghe P, Suomalainen A, Udd B, Van Broeckhoven C. POLG mutations in neurodegenerative disorders with ataxia but no muscle involvement. Neurology. 2004;63:1251–7.

71. Horvath R, Hudson G, Ferrari G, Fütterer N, Ahola S, Lamantea E, Prokisch H, Lochmüller H, Mcfarland R, Ramesh V, Klopstock T, Freisinger P, Salvi F, Mayr JA, Taylor RW, Turnbull D, Hanna M, Fialho D, Suomalainen A, Zeviani M, Chinnery PF. Phenotypic spectrum associated with mutations of the mitochondrial polymerase g gene. Brain. 2006;129:1674–84.

72. Verhoeven WM, Egger JI, Kremer BP, de Pont BJ, Marcelis CL. Recurrent major depression, ataxia, and cardiomyopathy: association with a novel POLG mutation? Neuropsychiatr Dis Treat. 2011;7:293–6.

73. Van Hove JLK, Cunningham V, Rice C, Ringel SP, Zhang Q, Chou P-C, Truong CK, Wong L-JC. Finding twinkle in the eyes of a 71-year-old lady: a case report and review of the genotypic and phenotypic spectrum of TWINKLE-related dominant disease. Am J Med Genet A. 2009;149A:861–7.

74. Fratter C, Gorman GS, Stewart JD, Buddles M, Smith C, Evans J, Seller A, Poulton J, Roberts M, Hanna MG, Rahman S, Omer SE, Klopstock T, Schoser B, Kornblum C, Czermin B, Lecky B, Blakely EL, Craig K, Chinnery PF, Turnbull DM, Horvath R, Taylor RW. The clinical, histochemical, and molecular spectrum of PEO1 (Twinkle)-linked adPEO. Neurology. 2010;74:1619–26.

75. Kornblum C, Nicholls TJ, Haack TB, Schöler S, Peeva V, Danhauser K, Hallmann K, Zsurka G, Rorbach J, Iuso A, Wieland T, Sciacco M, Ronchi D, Comi GP, Moggio M, Quinzii CM, DiMauro S, Calvo SE, Mootha VK, Klopstock T, Strom TM, Meitinger T, Minczuk M, Kunz WS, Prokisch H. Loss-of-function mutations in MGME1 impair mtDNA replication and cause multisystemic mitochondrial disease. Nat Genet. 2013;45:214–9.

76. Thompson K, Majd H, Dallabona C, Reinson K, King MS, Alston CL, He L, Lodi T, Jones SA, Fattal-Valevski A, Fraenkel ND, Saada A, Haham A, Isohanni P, Vara R, Barbosa IA, Simpson MA, Deshpande C, Puusepp S, Bonnen PE, Rodenburg RJ, Suomalainen A, Õunap K, Elpeleg O, Ferrero I, McFarland R, Kunji ERS, Taylor RW. Recurrent de novo dominant mutations in SLC25A4 cause severe early-onset mitochondrial disease and loss of mitochondrial DNA copy number. Am J Hum Genet. 2016;99:860–76.

77. Körver-Keularts IMLW, de Visser M, Bakker HD, Wanders RJA, Vansenne F, Scholte HR, Dorland L, Nicolaes GAF, Spaapen LMJ, Smeets HJM, Hendrickx ATM, van den Bosch BJC. Two novel mutations in the SLC25A4 gene in a patient with mitochondrial myopathy. JIMD Rep. 2015;22:39–45.

78. Palmieri L, Alberio S, Pisano I, Lodi T, Meznaric-Petrusa M, Zidar J, Santoro A, Scarcia P, Fontanesi F, Lamantea E, Ferrero I, Zeviani M. Complete loss-of-function of the heart/muscle-specific adenine nucleotide translocator is associated with mitochondrial myopathy and cardiomyopathy. Hum Mol Genet. 2005;14:3079–88.

79. Haack TB, Kopajtich R, Freisinger P, Wieland T, Rorbach J, Nicholls TJ, Baruffini E, Walther A, Danhauser K, Zimmermann FA, Husain RA, Schum J, Mundy H, Ferrero I, Strom TM, Meitinger T, Taylor RW, Minczuk M, Mayr JA, Prokisch H. ELAC2 mutations cause a mitochondrial RNA processing defect associated with hypertrophic cardiomyopathy. Am J Hum Genet. 2013a;93:211–23.

80. Akawi NA, Ben-Salem S, Hertecant J, John A, Pramathan T, Kizhakkedath P, Ali BR, Al-Gazali L. A homozygous splicing mutation in ELAC2 suggests phenotypic variability including intellectual disability with minimal cardiac involvement. Orphanet J Rare Dis. 2016;11:139.

81. Taylor RW, Pyle A, Griffin H, Blakely EL, Duff J, He L, Smertenko T, Alston CL, Neeve VC, Best A, Yarham JW, Kirschner J, Schara U, Talim B, Topaloglu H, Baric I, Holinski-Feder E, Abicht A, Czermin B, Kleinle S, Morris AAM, Vassallo G, Gorman GS, Ramesh V, Turnbull DM, Santibanez-Koref M, McFarland R, Horvath R, Chinnery PF. Use of whole-exome sequencing to determine the genetic basis of multiple mitochondrial respiratory chain complex deficiencies. JAMA. 2014;312:68.

82. Rauschenberger K, Schöler K, Sass JO, Sauer S, Djuric Z, Rumig C, Wolf NI, Okun JG, Kölker S, Schwarz H, Fischer C, Grziwa B, Runz H, Nümann A, Shafqat N, Kavanagh KL, Hämmerling G, Wanders RJA, Shield JPH, Wendel U, Stern D, Nawroth P, Hoffmann GF, Bartram CR, Arnold B, Bierhaus A, Oppermann U, Steinbeisser H, Zschocke J. A non-enzymatic function of 17β-hydroxysteroid dehydrogenase type 10 is required for mitochondrial integrity and cell survival. EMBO Mol Med. 2010;2:51–62.

83. Oláhová M, Hardy SA, Hall J, Yarham JW, Haack TB, Wilson WC, Alston CL, He L, Aznauryan E, Brown RM, Brown GK, Morris AAM, Mundy H, Broomfield A, Barbosa IA, Simpson MA, Deshpande C, Moeslinger D, Koch J, Stettner GM, Bonnen PE, Prokisch H, Lightowlers RN, McFarland R, Chrzanowska-Lightowlers ZMA, Taylor RW. *LRPPRC* mutations cause early-onset multisystem mitochondrial disease outside of the French-Canadian population. Brain. 2015;138:3503–19.

84. Kopajtich R, Nicholls TJ, Rorbach J, Metodiev MD, Freisinger P, Mandel H, Vanlander A, Ghezzi D, Carrozzo R, Taylor RW, Marquard K, Murayama K, Wieland T, Schwarzmayr T, Mayr JA, Pearce SF, Powell CA, Saada A, Ohtake A, Invernizzi F, Lamantea E, Sommerville EW, Pyle A, Chinnery PF, Crushell E, Okazaki Y, Kohda M, Kishita Y, Tokuzawa Y, Assouline Z, Rio M, Feillet F, de Camaret BM, Chretien D, Munnich A, Menten B, Sante T, Smet J, Régal L, Lorber A, Khoury A, Zeviani M, Strom TM, Meitinger T, Bertini ES, Van Coster R, Klopstock T, Rötig A, Haack TB, Minczuk M, Prokisch H. Mutations in GTPBP3 cause a mitochondrial translation defect associated with hypertrophic cardiomyopathy, lactic acidosis, and encephalopathy. Am J Hum Genet. 2014;95:708–20.

85. Baruffini E, Dallabona C, Invernizzi F, Yarham JW, Melchionda L, Blakely EL, Lamantea E, Donnini C, Santra S, Vijayaraghavan S, Roper HP, Burlina A, Kopajtich R, Walther A, Strom TM, Haack TB, Prokisch H, Taylor RW, Ferrero I, Zeviani M, Ghezzi D. MTO1 mutations are associated with hypertrophic cardiomyopathy and lactic acidosis and cause respiratory chain deficiency in humans and yeast. Hum Mutat. 2013;34:1501–9.

86. Ghezzi D, Baruffini E, Haack TB, Invernizzi F, Melchionda L, Dallabona C, Strom TM, Parini R, Burlina AB, Meitinger T, Prokisch H, Ferrero I, Zeviani M. Mutations of the mitochondrial-tRNA modifier MTO1 cause hypertrophic cardiomyopathy and lactic acidosis. Am J Hum Genet. 2012;90:1079–87.

87. Charif M, Titah SMC, Roubertie A, Desquiret-Dumas V, Gueguen N, Meunier I, Leid J, Massal F, Zanlonghi X, Mercier J, Raynaud de Mauverger E, Procaccio V, de Camaret BM, Lenaers G, Hamel CP. Optic neuropathy, cardiomyopathy, cognitive disability in patients with a homozygous mutation in the nuclear *MTO1* and a mitochondrial *MT-TF* variant. Am J Med Genet A. 2015;167:2366–74.

88. Powell CA, Kopajtich R, D'Souza AR, Rorbach J, Kremer LS, Husain RA, Dallabona C, Donnini C, Alston CL, Griffin H, Pyle A, Chinnery PF, Strom TM, Meitinger T, Rodenburg RJ, Schottmann G, Schuelke M, Romain N, Haller RG, Ferrero I, Haack TB, Taylor RW, Prokisch H, Minczuk M. TRMT5 mutations cause a defect in post-transcriptional modification of mitochondrial tRNA associated with multiple respiratory-chain deficiencies. Am J Hum Genet. 2015;97:319–28.

89. Metodiev MD, Thompson K, Alston CL, Morris AAM, He L, Assouline Z, Rio M, Bahi-Buisson N, Pyle A, Griffin H, Siira S, Filipovska A, Munnich A, Chinnery PF, McFarland R, Rötig A, Taylor RW. Recessive mutations in TRMT10C cause defects in mitochondrial RNA processing and multiple respiratory chain deficiencies. Am J Hum Genet. 2016;98:993–1000.

90. Simon MT, Ng BG, Friederich MW, Wang RY, Boyer M, Kircher M, Collard R, Buckingham KJ, Chang R, Shendure J, Nickerson DA, Bamshad MJ, University of Washington Center for

Mendelian Genomics, Van Hove JLK, Freeze HH, Abdenur JE. Activation of a cryptic splice site in the mitochondrial elongation factor GFM1 causes combined OXPHOS deficiency. Mitochondrion. 2017;34:84–90.

91. Smits P, Antonicka H, van Hasselt PM, Weraarpachai W, Haller W, Schreurs M, Venselaar H, Rodenburg RJ, Smeitink JA, van den Heuvel LP. Mutation in subdomain G' of mitochondrial elongation factor G1 is associated with combined OXPHOS deficiency in fibroblasts but not in muscle. Eur J Hum Genet. 2011a;19:275–9.

92. Ahola S, Isohanni P, Euro L, Brilhante V, Palotie A, Pihko H, Lönnqvist T, Lehtonen T, Laine J, Tyynismaa H, Suomalainen A. Mitochondrial EFTs defects in juvenile-onset Leigh disease, ataxia, neuropathy, and optic atrophy. Neurology. 2014;83:743–51.

93. Emperador S, Bayona-Bafaluy MP, Fernández-Marmiesse A, Pineda M, Felgueroso B, López-Gallardo E, Artuch R, Roca I, Ruiz-Pesini E, Couce ML, Montoya J. Molecular-genetic characterization and rescue of a TSFM mutation causing childhood-onset ataxia and nonobstructive cardiomyopathy. Eur J Hum Genet. 2016;25:153–6.

94. Ng YS, Alston CL, Diodato D, Morris AA, Ulrick N, Kmoch S, Houštěk J, Martinelli D, Haghighi A, Atiq M, Gamero MA, Garcia-Martinez E, Kratochvílová H, Santra S, Brown RM, Brown GK, Ragge N, Monavari A, Pysden K, Ravn K, Casey JP, Khan A, Chakrapani A, Vassallo G, Simons C, McKeever K, O'Sullivan S, Childs A-M, Østergaard E, Vanderver A, Goldstein A, Vogt J, Taylor RW, McFarland R. The clinical, biochemical and genetic features associated with RMND1-related mitochondrial disease. J Med Genet. 2016;53(11):768–75. https://doi.org/10.1136/jmedgenet-2016-103910.

95. Yin X, Tang B, Mao X, Peng J, Zeng S, Wang Y, Jiang H, Li N. The genotypic and phenotypic spectrum of PARS2-related infantile-onset encephalopathy. J Hum Genet. 2018;63:971–80.

96. Saada A, Shaag A, Arnon S, Dolfin T, Miller C, Fuchs-Telem D, Lombes A, Elpeleg O. Antenatal mitochondrial disease caused by mitochondrial ribosomal protein (MRPS22) mutation. J Med Genet. 2007;44:784–6.

97. Smits P, Saada A, Wortmann SB, Heister AJ, Brink M, Pfundt R, Miller C, Haas D, Hantschmann R, Rodenburg RJT, Smeitink JAM, van den Heuvel LP. Mutation in mitochondrial ribosomal protein MRPS22 leads to Cornelia de Lange-like phenotype, brain abnormalities and hypertrophic cardiomyopathy. Eur J Hum Genet. 2011b;19:394–9.

98. Baertling F, Haack TB, Rodenburg RJ, Schaper J, Seibt A, Strom TM, Meitinger T, Mayatepek E, Hadzik B, Selcan G, Prokisch H, Distelmaier F. MRPS22 mutation causes fatal neonatal lactic acidosis with brain and heart abnormalities. Neurogenetics. 2015b;16:237–40.

99. Galmiche L, Serre V, Beinat M, Assouline Z, Lebre A-S, Chretien D, Nietschke P, Benes V, Boddaert N, Sidi D, Brunelle F, Rio M, Munnich A, Rötig A. Exome sequencing identifies MRPL3 mutation in mitochondrial cardiomyopathy. Hum Mutat. 2011;32:1225–31.

100. Bursle C, Narendra A, Chuk R, Cardinal J, Justo R, Lewis B, Coman D. COXPD9 an evolving multisystem disease; congenital lactic acidosis, sensorineural hearing loss, hypertrophic cardiomyopathy, cirrhosis and interstitial nephritis. JIMD Rep. 2017;34:105–9.

101. Carroll CJ, Isohanni P, Pöyhönen R, Euro L, Richter U, Brilhante V, Götz A, Lahtinen T, Paetau A, Pihko H, Battersby BJ, Tyynismaa H, Suomalainen A. Whole-exome sequencing identifies a mutation in the mitochondrial ribosome protein MRPL44 to underlie mitochondrial infantile cardiomyopathy. J Med Genet. 2013;50:151–9.

102. Distelmaier F, Haack TB, Wortmann SB, Mayr JA, Prokisch H. Treatable mitochondrial diseases: cofactor metabolism and beyond. Brain. 2017;140:e11.

103. Haack TB, Gorza M, Danhauser K, Mayr JA, Haberberger B, Wieland T, Kremer L, Strecker V, Graf E, Memari Y, Ahting U, Kopajtich R, Wortmann SB, Rodenburg RJ, Kotzaeridou U, Hoffmann GF, Sperl W, Wittig I, Wilichowski E, Schottmann G, Schuelke M, Plecko B, Stephani U, Strom TM, Meitinger T, Prokisch H, Freisinger P. Phenotypic spectrum of eleven patients and five novel MTFMT mutations identified by exome sequencing and candidate gene screening. Mol Genet Metab. 2014;111:342–52.

104. Tucker EJ, Hershman SG, Köhrer C, Belcher-Timme CA, Patel J, Goldberger OA, Christodoulou J, Silberstein JM, McKenzie M, Ryan MT, Compton AG, Jaffe JD, Carr SA,

Calvo SE, RajBhandary UL, Thorburn DR, Mootha VK. Mutations in MTFMT underlie a human disorder of formylation causing impaired mitochondrial translation. Cell Metab. 2011;14:428–34.

105. Götz A, Tyynismaa H, Euro L, Ellonen P, Hyötyläinen T, Ojala T, Hämäläinen RH, Tommiska J, Raivio T, Oresic M, Karikoski R, Tammela O, Simola KOJ, Paetau A, Tyni T, Suomalainen A. Exome sequencing identifies mitochondrial alanyl-tRNA synthetase mutations in infantile mitochondrial cardiomyopathy. Am J Hum Genet. 2011;88:635–42.

106. McMillan HJ, Schwartzentruber J, Smith A, Lee S, Chakraborty P, Bulman DE, Beaulieu CL, Majewski J, Boycott KM, Geraghty MT. Compound heterozygous mutations in glycyl-tRNA synthetase are a proposed cause of systemic mitochondrial disease. BMC Med Genet. 2014;15:36.

107. Kohda M, Tokuzawa Y, Kishita Y, Nyuzuki H, Moriyama Y, Mizuno Y, Hirata T, Yatsuka Y, Yamashita-Sugahara Y, Nakachi Y, Kato H, Okuda A, Tamaru S, Borna NN, Banshoya K, Aigaki T, Sato-Miyata Y, Ohnuma K, Suzuki T, Nagao A, Maehata H, Matsuda F, Higasa K, Nagasaki M, Yasuda J, Yamamoto M, Fushimi T, Shimura M, Kaiho-Ichimoto K, Harashima H, Yamazaki T, Mori M, Murayama K, Ohtake A, Okazaki Y. A comprehensive genomic analysis reveals the genetic landscape of mitochondrial respiratory chain complex deficiencies. PLoS Genet. 2016;12:e1005679.

108. Verrigni D, Diodato D, Di Nottia M, Torraco A, Bellacchio E, Rizza T, Tozzi G, Verardo M, Piemonte F, Tasca G, D'Amico A, Bertini E, Carrozzo R. Novel mutations in *KARS* cause hypertrophic cardiomyopathy and combined mitochondrial respiratory chain defect. Clin Genet. 2017;91:918–23.

109. Riley LG, Rudinger-Thirion J, Schmitz-Abe K, Thorburn DR, Davis RL, Teo J, Arbuckle S, Cooper ST, Campagna DR, Frugier M, Markianos K, Sue CM, Fleming MD, Christodoulou J. LARS2 variants associated with hydrops, lactic acidosis, sideroblastic anemia, and multisystem failure. JIMD Rep. 2016;28:49–57.

110. Ardissone A, Lamantea E, Quartararo J, Dallabona C, Carrara F, Moroni I, Donnini C, Garavaglia B, Zeviani M, Uziel G. A novel homozygous YARS2 mutation in two Italian siblings and a review of literature. JIMD Rep. 2015;20:95–101.

111. Pitceathly RDS, Smith C, Fratter C, Alston CL, He L, Craig K, Blakely EL, Evans JC, Taylor J, Shabbir Z, Deschauer M, Pohl U, Roberts ME, Jackson MC, Halfpenny CA, Turnpenny PD, Lunt PW, Hanna MG, Schaefer AM, McFarland R, Horvath R, Chinnery PF, Turnbull DM, Poulton J, Taylor RW, Gorman GS. Adults with RRM2B-related mitochondrial disease have distinct clinical and molecular characteristics. Brain. 2012;135:3392–403.

112. Bruni F, Di Meo I, Bellacchio E, Webb BD, Mcfarland R, Chrzanowska-Lightowlers ZMA, He L, Skorupa E, Moroni I, Ardissone A, Walczak A, Tyynismaa H, Isohanni P, Mandel H, Prokisch H, Haack T, Bonnen PE, Enrico B, Pronicka E, Ghezzi D, Taylor RW, Diodato D. Clinical, biochemical, and genetic features associated with VARS2-related mitochondrial disease. Hum Mutat. 2018;39(4):563–78.

113. Riley LG, Menezes MJ, Rudinger-Thirion J, Duff R, de Lonlay P, Rotig A, Tchan MC, Davis M, Cooper ST, Christodoulou J. Phenotypic variability and identification of novel YARS2 mutations in YARS2 mitochondrial myopathy, lactic acidosis and sideroblastic anaemia. Orphanet J Rare Dis. 2013;8:193.

114. Carrozzo R, Verrigni D, Rasmussen M, de Coo R, Amartino H, Bianchi M, Buhas D, Mesli S, Naess K, Born AP, Woldseth B, Prontera P, Batbayli M, Ravn K, Joensen F, Cordelli DM, Santorelli FM, Tulinius M, Darin N, Duno M, Jouvencel P, Burlina A, Stangoni G, Bertini E, Redonnet-Vernhet I, Wibrand F, Dionisi-Vici C, Uusimaa J, Vieira P, Osorio AN, McFarland R, Taylor RW, Holme E, Ostergaard E. Succinate-CoA ligase deficiency due to mutations in SUCLA2 and SUCLG1: phenotype and genotype correlations in 71 patients. J Inherit Metab Dis. 2016;39:243–52.

115. WebHome < MITOMAP < Foswiki, (n.d.).

116. Bannwarth S, Procaccio V, Lebre AS, Jardel C, Chaussenot A, Hoarau C, Maoulida H, Charrier N, Gai X, Xie HM, Ferre M, Fragaki K, Hardy G, de Camaret BM, Marlin S,

Dhaenens CM, Slama A, Rocher C, Bonnefont JP, Rötig A, Aoutil N, Gilleron M, Desquiret-Dumas V, Reynier P, Ceresuela J, Jonard L, Devos A, Espil-Taris C, Martinez D, Gaignard P, Sang K-HLQ, Amati-Bonneau P, Falk MJ, Florentz C, Chabrol B, Durand-Zaleski I, Paquis-Flucklinger V. Prevalence of rare mitochondrial DNA mutations in mitochondrial disorders. J Med Genet. 2013;50:704–14.

117. Scalais E, Chafai R, Van Coster R, Bindl L, Nuttin C, Panagiotaraki C, Seneca S, Lissens W, Ribes A, Geers C, Smet J, De Meirleir L. Early myoclonic epilepsy, hypertrophic cardiomy-opathy and subsequently a nephrotic syndrome in a patient with CoQ10 deficiency caused by mutations in para-hydroxybenzoate-polyprenyl transferase (COQ2). Eur J Paediatr Neurol. 2013;17:625–30.

118. Desbats MA, Lunardi G, Doimo M, Trevisson E, Salviati L. Genetic bases and clinical manifestations of coenzyme Q10 (CoQ10) deficiency. J Inherit Metab Dis. 2015a;38:145–56.

119. Brea-Calvo G, Haack TB, Karall D, Ohtake A, Invernizzi F, Carrozzo R, Kremer L, Dusi S, Fauth C, Scholl-Bürgi S, Graf E, Ahting U, Resta N, Laforgia N, Verrigni D, Okazaki Y, Kohda M, Martinelli D, Freisinger P, Strom TM, Meitinger T, Lamperti C, Lacson A, Navas P, Mayr JA, Bertini E, Murayama K, Zeviani M, Prokisch H, Ghezzi D. COQ4 mutations cause a broad spectrum of mitochondrial disorders associated with CoQ10 deficiency. Am J Hum Genet. 2015;96:309–17.

120. Chung WK, Martin K, Jalas C, Braddock SR, Juusola J, Monaghan KG, Warner B, Franks S, Yudkoff M, Lulis L, Rhodes RH, Prasad V, Torti E, Cho MT, Shinawi M. Mutations in COQ4, an essential component of coenzyme Q biosynthesis, cause lethal neonatal mitochondrial encephalomyopathy. J Med Genet. 2015;52:627–35.

121. Cameron JM, Janer A, Levandovskiy V, Mackay N, Rouault TA, Tong W-H, Ogilvie I, Shoubridge EA, Robinson BH. Mutations in iron-sulfur cluster scaffold genes NFU1 and BOLA3 cause a fatal deficiency of multiple respiratory chain and 2-oxoacid dehydrogenase enzymes. Am J Hum Genet. 2011;89:486–95.

122. Lek M, Karczewski KJ, Minikel EV, Samocha KE, Banks E, Fennell T, O'Donnell-Luria AH, Ware JS, Hill AJ, Cummings BB, Tukiainen T, Birnbaum DP, Kosmicki JA, Duncan LE, Estrada K, Zhao F, Zou J, Pierce-Hoffman E, Berghout J, Cooper DN, Deflaux N, DePristo M, Do R, Flannick J, Fromer M, Gauthier L, Goldstein J, Gupta N, Howrigan D, Kiezun A, Kurki MI, Moonshine AL, Natarajan P, Orozco L, Peloso GM, Poplin R, Rivas MA, Ruano-Rubio V, Rose SA, Ruderfer DM, Shakir K, Stenson PD, Stevens C, Thomas BP, Tiao G, Tusie-Luna MT, Weisburd B, Won H-H, Yu D, Altshuler DM, Ardissino D, Boehnke M, Danesh J, Donnelly S, Elosua R, Florez JC, Gabriel SB, Getz G, Glatt SJ, Hultman CM, Kathiresan S, Laakso M, McCarroll S, McCarthy MI, McGovern D, McPherson R, Neale BM, Palotie A, Purcell SM, Saleheen D, Scharf JM, Sklar P, Sullivan PF, Tuomilehto J, Tsuang MT, Watkins HC, Wilson JG, Daly MJ, MacArthur DG. Exome aggregation consortium, analysis of protein-coding genetic variation in 60,706 humans. Nature. 2016;536:285–91.

123. Mollet J, Giurgea I, Schlemmer D, Dallner G, Chretien D, Delahodde A, Bacq D, de Lonlay P, Munnich A, Rötig A. Prenyldiphosphate synthase, subunit 1 (PDSS1) and OH-benzoate polyprenyltransferase (COQ2) mutations in ubiquinone deficiency and oxidative phosphory-lation disorders. J Clin Invest. 2007;117:765–72.

124. Kollberg G, Tulinius M, Melberg A, Darin N, Andersen O, Holmgren D, Oldfors A, Holme E. Clinical manifestation and a new ISCU mutation in iron–sulphur cluster deficiency myopa-thy. Brain. 2009;132:2170–9.

125. Haack TB, Rolinski B, Haberberger B, Zimmermann F, Schum J, Strecker V, Graf E, Athing U, Hoppen T, Wittig I, Sperl W, Freisinger P, Mayr JA, Strom TM, Meitinger T, Prokisch H. Homozygous missense mutation in BOLA3 causes multiple mitochondrial dysfunctions syndrome in two siblings. J Inherit Metab Dis. 2013b;36:55–62.

126. Mayr JA, Feichtinger RG, Tort F, Ribes A, Sperl W. Lipoic acid biosynthesis defects. J Inherit Metab Dis. 2014;37:553–63.

127. Antonicka H, Leary SC, Guercin G-H, Agar JN, Horvath R, Kennaway NG, Harding CO, Jaksch M, Shoubridge EA. Mutations in COX10 result in a defect in mitochondrial heme A

biosynthesis and account for multiple, early-onset clinical phenotypes associated with isolated COX deficiency. Hum Mol Genet. 2003;12:2693–702.

128. Vondrackova A, Vesela K, Hansikova H, Docekalova DZ, Rozsypalova E, Zeman J, Tesarova M. High-resolution melting analysis of 15 genes in 60 patients with cytochrome-c oxidase deficiency. J Hum Genet. 2012;57:442–8.

129. Alfadhel M, Lillquist YP, Waters PJ, Sinclair G, Struys E, McFadden D, Hendson G, Hyams L, Shoffner J, Vallance HD. Infantile cardioencephalopathy due to a COX15 gene defect: Report and review. Am J Med Genet A. 2011;155:840–4.

130. Stiburek L, Vesela K, Hansikova H, Hulkova H, Zeman J. Loss of function of Sco1 and its interaction with cytochrome c oxidase. AJP Cell Physiol. 2009;296:C1218–26.

131. Freisinger P, Horvath R, Macmillan C, Peters J, Jaksch M. Reversion of hypertrophic cardiomyopathy in a patient with deficiency of the mitochondrial copper binding protein Sco2: is there a potential effect of copper? J Inherit Metab Dis. 2004;27:67–79.

132. Papadopoulou LC, Sue CM, Davidson MM, Tanji K, Nishino I, Sadlock JE, Krishna S, Walker W, Selby J, Glerum DM, Coster RV, Lyon G, Scalais E, Lebel R, Kaplan P, Shanske S, De Vivo DC, Bonilla E, Hirano M, DiMauro S, Schon EA. Fatal infantile cardioencephalomyopathy with COX deficiency and mutations in SCO2, a COX assembly gene. Nat Genet. 1999;23:333–7.

133. Tran-Viet K-N, Powell C, Barathi VA, Klemm T, Maurer-Stroh S, Limviphuvadh V, Soler V, Ho C, Yanovitch T, Schneider G, Li Y-J, Nading E, Metlapally R, Saw S-M, Goh L, Rozen S, Young TL. Mutations in SCO2 are associated with autosomal-dominant high-grade myopia. Am J Hum Genet. 2013;92:820–6.

134. Baertling F, van den Brand MAM, Hertecant JL, Al-Shamsi A, van den Heuvel LP, Distelmaier F, Mayatepek E, Smeitink JA, Nijtmans LGJ, Rodenburg RJT. Mutations in COA6 cause cytochrome c oxidase deficiency and neonatal hypertrophic cardiomyopathy. Hum Mutat. 2015a;36:34–8.

135. Kishita Y, Pajak A, Bolar NA, Marobbio CMT, Maffezzini C, Miniero DV, Monné M, Kohda M, Stranneheim H, Murayama K, Naess K, Lesko N, Bruhn H, Mourier A, Wibom R, Nennesmo I, Jespers A, Govaert P, Ohtake A, Van Laer L, Loeys BL, Freyer C, Palmieri F, Wredenberg A, Okazaki Y, Wedell A. Intra-mitochondrial methylation deficiency due to mutations in SLC25A26. Am J Hum Genet. 2015;97:761–8.

136. Haack TB, Jackson CB, Murayama K, Kremer LS, Schaller A, Kotzaeridou U, de Vries MC, Schottmann G, Santra S, Büchner B, Wieland T, Graf E, Freisinger P, Eggimann S, Ohtake A, Okazaki Y, Kohda M, Kishita Y, Tokuzawa Y, Sauer S, Memari Y, Kolb-Kokocinski A, Durbin R, Hasselmann O, Cremer K, Albrecht B, Wieczorek D, Engels H, Hahn D, Zink AM, Alston CL, Taylor RW, Rodenburg RJ, Trollmann R, Sperl W, Strom TM, Hoffmann GF, Mayr JA, Meitinger T, Bolognini R, Schuelke M, Nuoffer J-M, Kölker S, Prokisch H, Klopstock T. Deficiency of ECHS1 causes mitochondrial encephalopathy with cardiac involvement. Ann Clin Transl Neurol. 2015;2:492–509.

137. Bhoj EJ, Li M, Ahrens-Nicklas R, Pyle LC, Wang J, Zhang VW, Clarke C, Wong LJ, Sondheimer N, Ficicioglu C, Yudkoff M. Pathologic variants of the mitochondrial phosphate carrier SLC25A3: two new patients and expansion of the cardiomyopathy/skeletal myopathy phenotype with and without lactic acidosis. JIMD Rep. 2015;19:59–66.

138. Cox GF, Souri M, Aoyama T, Rockenmacher S, Varvogli L, Rohr F, Hashimoto T, Korson MS. Reversal of severe hypertrophic cardiomyopathy and excellent neuropsychologic outcome in very-long-chain acyl-coenzyme A dehydrogenase deficiency. J Pediatr. 1998;133:247–53.

139. Mathur A, Sims HF, Gopalakrishnan D, Gibson B, Rinaldo P, Vockley J, Hug G, Strauss AW. Molecular heterogeneity in very-long-chain acyl-CoA dehydrogenase deficiency causing pediatric cardiomyopathy and sudden death. Circulation. 1999;99:1337–43.

140. Longo N, di San Filippo CA, Pasquali M. Disorders of carnitine transport and the carnitine cycle. Am J Med Genet C Semin Med Genet. 2006;142C:77–85.

141. Orngreen MC, Madsen KL, Preisler N, Andersen G, Vissing J, Laforet P. Bezafibrate in skeletal muscle fatty acid oxidation disorders: a randomized clinical trial. Neurology. 2014;82:607–13.
142. Martins E, Cardoso ML, Rodrigues E, Barbot C, Ramos A, Bennett MJ, Teles EL, Vilarinho L. Short-chain 3-hydroxyacyl-CoA dehydrogenase deficiency: the clinical relevance of an early diagnosis and report of four new cases. J Inherit Metab Dis. 2011;34:835–42.
143. Choi J-H, Yoon H-R, Kim G-H, Park S-J, Shin Y-L, Yoo H-W. Identification of novel mutations of the HADHA and HADHB genes in patients with mitochondrial trifunctional protein deficiency. Int J Mol Med. 2007;19(1):81–7.
144. Lahrouchi N, Lodder EM, Mansouri M, Tadros R, Zniber L, Adadi N, Clur S-AB, van Spaendonck-Zwarts KY, Postma AV, Sefiani A, Ratbi I, Bezzina CR. Exome sequencing identifies primary carnitine deficiency in a family with cardiomyopathy and sudden death. Eur J Hum Genet. 2017;25:783–7.
145. Iacobazzi V, Invernizzi F, Baratta S, Pons R, Chung W, Garavaglia B, Dionisi-Vici C, Ribes A, Parini R, Huertas MD, Roldan S, Lauria G, Palmieri F, Taroni F. Molecular and functional analysis of SLC25A20 mutations causing carnitine-acylcarnitine translocase deficiency. Hum Mutat. 2004;24:312–20.
146. Kennedy H, Haack TB, Hartill V, Mataković L, Baumgartner ER, Potter H, Mackay R, Alston CL, O'Sullivan S, McFarland R, Connolly G, Gannon C, King R, Mead S, Crozier I, Chan W, Florkowski CM, Sage M, Höfken T, Alhaddad B, Kremer LS, Kopajtich R, Feichtinger RG, Sperl W, Rodenburg RJ, Minet JC, Dobbie A, Strom TM, Meitinger T, George PM, Johnson CA, Taylor RW, Prokisch H, Doudney K, Mayr JA. Sudden cardiac death due to deficiency of the mitochondrial inorganic pyrophosphatase PPA2. Am J Hum Genet. 2016;99:674–82.
147. El-Hattab AW, Scaglia F. Mitochondrial cardiomyopathies. Front Cardiovasc Med. 2016;3:25.
148. Maron BJ, Gardin JM, Flack JM, Gidding SS, Kurosaki TT, Bild DE. Prevalence of hypertrophic cardiomyopathy in a general population of young adults. Echocardiographic analysis of 4111 subjects in the CARDIA study. Coronary artery risk development in (young) adults. Circulation. 1995;92:785–9.
149. Elliott P, Charron P, Blanes JRG, Tavazzi L, Tendera M, Konté M, Laroche C, Maggioni AP, Anastasakis A, Arbustini E, Asselbergs FW, Axelsson A, Brito D, Caforio ALP, Carr-White G, Czekaj A, Damy T, Devoto E, Favalli V, Findlay I, Garcia-Pavia P, Hagège A, Heliö T, Iliceto S, Isnard R, Jansweijer JA, Limongelli G, Linhart A, Cuenca DL, Mansencal N, McKeown P, Mogensen J, Mohiddin SA, Monserrat L, Olivotto I, Rapezzi C, Rigopoulos AG, Rosmini S, Pfeiffer B, Wicks E, Podzimkova J, Kuchynka P, Palecek T, Bundgaard H, Thune JJ, Kumme A, Vestergaard LD, Hey T, Ollila L, Kaartinen M, Dubourg O, Arslan M, Tsieu MS, Guellich A, Tissot CM, Guendouz S, Thevenin S, Khelifa RC, Gandjbakhch E, Komajda M, Neugebauer A, Pfeiffer B, Steriotis A, Ritsatos K, Vlagkouli V, Biagini E, Gentile N, Longhi S, Arretini A, Fornaro A, Cecchi F, Spirito P, Formisano F, Masarone D, Valente F, Pacileo G, Schiavo A, Testolina M, Serio A, Grasso M, Wilde A, Pinto Y, Klöpping C, Van Der Heijden JF, De Jonge N, Sikora-Puz A, Wybraniec M, Czekaj A, Francisco AR, Brito D, Madeira H, Ortiz-Genga M, Barriales-Villa R, Fernandez X, Lopez-Cuenca D, Gomez-Milanes I, Lopez-Ayala JM, Guzzo-Merello G, Gallego-Delgado M, Muir A, McOsker J, Jardine T, Iqbal H, Sekhri N, Rajani R, Bueser T, Watkinson O. European cardiomyopathy pilot registry: EURObservational research programme of the European society of cardiology. Eur Heart J. 2016;37(2):164–73.
150. Colan SD, Lipshultz SE, Lowe AM, Sleeper LA, Messere J, Cox GF, Lurie PR, Orav EJ, Towbin JA. Epidemiology and cause-specific outcome of hypertrophic cardiomyopathy in children. Circulation. 2007;115:773–81.
151. Nugent AW, Daubeney PEF, Chondros P, Carlin JB, Colan SD, Cheung M, Davis AM, Chow CW, Weintraub RG. National Australian childhood cardiomyopathy study, clinical features and outcomes of childhood hypertrophic cardiomyopathy: results from a national population-based study. Circulation. 2005;112:1332–8.

152. Montaigne D, Pentiah AD. Mitochondrial cardiomyopathy and related arrhythmias. Card Electrophysiol Clin. 2015;7:293–301.
153. Meyers DE, Basha HI, Koenig MK. Mitochondrial cardiomyopathy: pathophysiology, diagnosis, and management. Texas Hear Inst J. 2013;40:385–94.
154. Enns GM. Pediatric mitochondrial diseases and the heart. Curr Opin Pediatr. 2017;29 (5):541–51.
155. Spencer CT, Bryant RM, Day J, Gonzalez IL, Colan SD, Thompson WR, Berthy J, Redfearn SP, Byrne BJ. Cardiac and clinical phenotype in Barth syndrome. Pediatrics. 2006;118: e337–46.
156. Brunel-Guitton C, Levtova A, Sasarman F. Mitochondrial diseases and cardiomyopathies. Can J Cardiol. 2015;31:1360–76.
157. Fassone E, Rahman S. Complex I deficiency: clinical features, biochemistry and molecular genetics. J Med Genet. 2012;49:578–90.
158. Haack TB, Danhauser K, Haberberger B, Hoser J, Strecker V, Boehm D, Uziel G, Lamantea E, Invernizzi F, Poulton J, Rolinski B, Iuso A, Biskup S, Schmidt T, Mewes H-W, Wittig I, Meitinger T, Zeviani M, Prokisch H. Exome sequencing identifies ACAD9 mutations as a cause of complex I deficiency. Nat Genet. 2010;42:1131–4.
159. Schiff M, Haberberger B, Xia C, Mohsen A-W, Goetzman ES, Wang Y, Uppala R, Zhang Y, Karunanidhi A, Prabhu D, Alharbi H, Prochownik EV, Haack T, Häberle J, Munnich A, Rötig A, Taylor RW, Nicholls RD, Kim J-J, Prokisch H, Vockley J. Complex I assembly function and fatty acid oxidation enzyme activity of ACAD9 both contribute to disease severity in ACAD9 deficiency. Hum Mol Genet. 2015;24:3238–47.
160. Collet M, Assouline Z, Bonnet D, Rio M, Iserin F, Sidi D, Goldenberg A, Lardennois C, Metodiev MD, Haberberger B, Haack T, Munnich A, Prokisch H, Rötig A. High incidence and variable clinical outcome of cardiac hypertrophy due to ACAD9 mutations in childhood. Eur J Hum Genet. 2016;24:1112–6.
161. Gerards M, van den Bosch BJC, Danhauser K, Serre V, van Weeghel M, Wanders RJA, Nicolaes GAF, Sluiter W, Schoonderwoerd K, Scholte HR, Prokisch H, Rötig A, de Coo IFM, Smeets HJM. Riboflavin-responsive oxidative phosphorylation complex I deficiency caused by defective ACAD9: new function for an old gene. Brain. 2011;134:210–9.
162. Jain-Ghai S, Cameron JM, Al Maawali A, Blaser S, MacKay N, Robinson B, Raiman J. Complex II deficiency – a case report and review of the literature. Am J Med Genet A. 2013;161A:285–94.
163. Levitas A, Muhammad E, Harel G, Saada A, Caspi VC, Manor E, Beck JC, Sheffield V, Parvari R. Familial neonatal isolated cardiomyopathy caused by a mutation in the flavoprotein subunit of succinate dehydrogenase. Eur J Hum Genet. 2010;18:1160–5.
164. Feichtinger RG, Brunner-Krainz M, Alhaddad B, Wortmann SB, Kovacs-Nagy R, Stojakovic T, Erwa W, Resch B, Windischhofer W, Verheyen S, Uhrig S, Windpassinger C, Locker F, Makowski C, Strom TM, Meitinger T, Prokisch H, Sperl W, Haack TB, Mayr JA. Combined respiratory chain deficiency and UQCC2 mutations in neonatal encephalomyopathy: defective supercomplex assembly in complex III deficiencies. Oxidative Med Cell Longev. 2017a;2017:1–11.
165. Andreu AL, Checcarelli N, Iwata S, Shanske S, Dimauro S. A missense mutation in the mitochondrial cytochrome b gene in a revisited case with histiocytoid cardiomyopathy. Pediatr Res. 2000;48:311–4.
166. Rak M, Bénit P, Chrétien D, Bouchereau J, Schiff M, El-Khoury R, Tzagoloff A, Rustin P. Mitochondrial cytochrome c oxidase deficiency. Clin Sci (Lond). 2016;130:393–407.
167. Marin-Garcia J, Goldenthal MJ, Ananthakrishnan R, Pierpont ME. The complete sequence of mtDNA genes in idiopathic dilated cardiomyopathy shows novel missense and tRNA mutations. J Card Fail. 2000;6:321–9.
168. Finsterer J, Stöllberger C, Schubert B. Acquired left ventricular hypertrabeculation/ noncompaction in mitochondriopathy. Cardiology. 2004;102:228–30.

169. Mayr JA, Havlíčková V, Zimmermann F, Magler I, Kaplanová V, Jesina P, Pecinová A, Nusková H, Koch J, Sperl W, Houstek J. Mitochondrial ATP synthase deficiency due to a mutation in the ATP5E gene for the F1 epsilon subunit. Hum Mol Genet. 2010;19:3430–9.
170. Desbats MA, Vetro A, Limongelli I, Lunardi G, Casarin A, Doimo M, Spinazzi M, Angelini C, Cenacchi G, Burlina A, Rodriguez Hernandez MA, Chiandetti L, Clementi M, Trevisson E, Navas P, Zuffardi O, Salviati L. Primary coenzyme Q10 deficiency presenting as fatal neonatal multiorgan failure. Eur J Hum Genet. 2015b;23:1254–8.
171. Vernon HJ, Sandlers Y, McClellan R, Kelley RI. Clinical laboratory studies in Barth syndrome. Mol Genet Metab. 2014;112:143–7.
172. Jefferies JL. Barth syndrome. Am J Med Genet C Semin Med Genet. 2013;163C:198–205.
173. Wang J, Guo Y, Huang M, Zhang Z, Zhu J, Liu T, Shi L, Li F, Huang H, Fu L. Identification of TAZ mutations in pediatric patients with cardiomyopathy by targeted next-generation sequencing in a Chinese cohort. Orphanet J Rare Dis. 2017;12:26.
174. Dudek J, Cheng I-F, Chowdhury A, Wozny K, Balleininger M, Reinhold R, Grunau S, Callegari S, Toischer K, Wanders RJ, Hasenfuß G, Brügger B, Guan K, Rehling P. Cardiac-specific succinate dehydrogenase deficiency in Barth syndrome. EMBO Mol Med. 2016;8:139–54.
175. Kabunga P, Lau AK, Phan K, Puranik R, Liang C, Davis RL, Sue CM, Sy RW. Systematic review of cardiac electrical disease in Kearns-Sayre syndrome and mitochondrial cytopathy. Int J Cardiol. 2015;181:303–10.
176. El-Hattab AW, Adesina AM, Jones J, Scaglia F. MELAS syndrome: clinical manifestations, pathogenesis, and treatment options. Mol Genet Metab. 2015;116:4–12.
177. Thorburn DR. Mitochondrial disorders: prevalence, myths and advances. J Inherit Metab Dis. 2004;27:349–62.
178. DiMauro S, Hirano M. MERRF. Seattle: University of Washington; 1993.
179. Santorelli FM, Tanji K, Shanske S, DiMauro S. Heterogeneous clinical presentation of the mtDNA NARP/T8993G mutation. Neurology. 1997;49:270–3.
180. Rapezzi C, Arbustini E, Caforio ALP, Charron P, Gimeno-Blanes J, Helio T, Linhart A, Mogensen J, Pinto Y, Ristic A, Seggewiss H, Sinagra G, Tavazzi L, Elliott PM. Diagnostic work-up in cardiomyopathies: bridging the gap between clinical phenotypes and final diagnosis. A position statement from the ESC working group on myocardial and pericardial diseases. Eur Heart J. 2013;34:1448–58.
181. Limongelli G, Tome-Esteban M, Dejthevaporn C, Rahman S, Hanna MG, Elliott PM. Prevalence and natural history of heart disease in adults with primary mitochondrial respiratory chain disease. Eur J Heart Fail. 2010;12:114–21.
182. Guenthard J, Wyler F, Fowler B, Baumgartner R. Cardiomyopathy in respiratory chain disorders. Arch Dis Child. 1995;72:223–6.
183. Golden AS, Law YM, Shurtleff H, Warner M, Saneto RP. Mitochondrial electron transport chain deficiency, cardiomyopathy, and long-term cardiac transplant outcome. Pediatr Transplant. 2012;16:265–8.
184. Koga Y, Akita Y, Nishioka J, Yatsuga S, Povalko N, Tanabe Y, Fujimoto S, Matsuishi T. L-Arginine improves the symptoms of strokelike episodes in MELAS. Neurology. 2005;64:710–2.
185. Kremer LS, Bader DM, Mertes C, Kopajtich R, Pichler G, Iuso A, Haack TB, Graf E, Schwarzmayr T, Terrile C, Koňaříková E, Repp B, Kastenmüller G, Adamski J, Lichtner P, Leonhardt C, Funalot B, Donati A, Tiranti V, Lombes A, Jardel C, Gläser D, Taylor RW, Ghezzi D, Mayr JA, Rötig A, Freisinger P, Distelmaier F, Strom TM, Meitinger T, Gagneur J, Prokisch H. Genetic diagnosis of Mendelian disorders via RNA sequencing. Nat Commun. 2017;8:15824.
186. Distelmaier F, Haack TB, Catarino CB, Gallenmüller C, Rodenburg RJ, Strom TM, Baertling F, Meitinger T, Mayatepek E, Prokisch H, Klopstock T. MRPL44 mutations cause a slowly progressive multisystem disease with childhood-onset hypertrophic cardiomyopathy. Neurogenetics. 2015;16:319–23.

187. Duncan AJ, Bitner-Glindzicz M, Meunier B, Costello H, Hargreaves IP, López LC, Hirano M, Quinzii CM, Sadowski MI, Hardy J, Singleton A, Clayton PT, Rahman S. A nonsense mutation in COQ9 causes autosomal-recessive neonatal-onset primary coenzyme Q10 deficiency: a potentially treatable form of mitochondrial disease. Am J Hum Genet. 2009;84:558–66.
188. Van Haute L, Pearce SF, Powell CA, D'Souza AR, Nicholls TJ, Minczuk M. Mitochondrial transcript maturation and its disorders. J Inherit Metab Dis. 2015;38:655–80.
189. Peters H, Buck N, Wanders R, Ruiter J, Waterham H, Koster J, Yaplito-Lee J, Ferdinandusse S, Pitt J. ECHS1 mutations in Leigh disease: a new inborn error of metabolism affecting valine metabolism. Brain. 2014;137:2903–8.
190. Weidemann F, Rummey C, Bijnens B, Störk S, Jasaityte R, Dhooge J, Baltabaeva A, Sutherland G, Schulz JB, Meier T. Mitochondrial protection with idebenone in cardiac or neurological outcome (MICONOS) study group, the heart in Friedreich ataxia: definition of cardiomyopathy, disease severity, and correlation with neurological symptoms. Circulation. 2012;125:1626–34.
191. Ganetzky RD, Stendel C, McCormick EM, Zolkipli-Cunningham Z, Goldstein AC, Klopstock T, Falk MJ. MT-ATP6 mitochondrial disease variants: phenotypic and biochemical features analysis in 218 published cases and cohort of 14 new cases. Hum Mutat. 2019;40(5):499–515.
192. Stenton SL, Prokisch H. Advancing genomic approaches to the molecular diagnosis of mitochondrial disease. Essays Biochem. 2018;62(3):399–408.
193. Stenton SL, Kremer LS, Kopajtich R, Ludwig C, Prokisch H. The diagnosis of inborn errors of metabolism by an integrative "multi-omics" approach: a perspective encompassing genomics, transcriptomics, and proteomics. J Inherit Metab Dis. 2019.

The Genetics of Coronary Heart Disease

Jeanette Erdmann and Maria Loreto Muñoz Venegas

Contents

J. Erdmann (✉) · M. L. Muñoz Venegas
Institute for Cardiogenetics, University of Lübeck, Lübeck, Germany
e-mail: jeanette.erdmann@uni-luebeck.de

© Springer Nature Switzerland AG 2019
J. Erdmann, A. Moretti (eds.), *Genetic Causes of Cardiac Disease*, Cardiac and Vascular Biology 7, https://doi.org/10.1007/978-3-030-27371-2_4

4.1   Introduction

Cardiovascular disease (CVD) is a class of diseases that encompasses heart diseases, brain vascular diseases and blood vessel diseases. CVDs and their risk factors are a leading cause of death and morbidity in the world [1–3]. According to World Health Organization estimates, CVDs are responsible for 151,377 million disability-adjusted life years (DALY). DALY is the number of years lost because of ill health, disability or premature death. Coronary heart disease (CHD) accounts for 41.35% (or 62,587 million) of these years, while cerebrovascular diseases account for another 30.78% (or 46,591 million) of these years [1].

CVD is subdivided into two major groups: (1) CVD due to atherosclerosis and (2) other CVDs. In atherosclerosis, fatty material and cholesterol are deposited inside blood vessels, making it harder for blood to efficiently supply oxygen and nutrients to cells. Such deposits are also known as atherosclerotic plaques. As a result, blood vessels become less pliable. The build-up of atherosclerotic plaques is referred to as coronary artery disease (CAD) (also known as CHD). Over time, the atherosclerotic plaques can rupture, triggering the formation of a blood clot that deprives blood flow. In acute cases, obstruction of the coronary artery to the heart will lead to a heart attack (i.e. myocardial infarction, MI) [1, 4, 5].

Epidemiologic studies have revealed a variety of risk factors for CAD that can be broadly subdivided into behavioural (e.g. sedentary lifestyle, smoking, unhealthy diet), metabolic (e.g. hypertension, diabetes, cholesterol) and other factors (e.g. age, gender, genetic disposition) [1, 5, 6]. The interplay between lifestyle and genetic risk factors is characteristic of complex or multifactorial diseases, such as CAD [5, 7]. This chapter focuses on disentangling the genetics of CAD.

4.2   The Role of Genetic Studies in Deciphering CAD Mechanisms

As a complex trait, the mode of CAD inheritance follows Fisher's 1918 'infinitesimal model' [8]. In this model, discrete and continuous traits are consistent if quantitative trait variation is caused by a combination of many segregating genes, each with a small (infinitesimal) effect on the trait. This mechanism leads to a normal distribution of genetic values and, together with normally distributed environmental effects, results in a normal distribution of phenotypes in the population. This theory implies that the genetic and non-genetic sources of variation can be estimated by quantifying the correlation between relatives, without any knowledge of specific genes that potentially affect the trait [9]. In 1938, the first familial CAD risk was described [10], followed by clinical observations in the 1950s and subsequent familial and twin studies that supported this theory [11]. For example, the Framingham Heart Study found that there was a 29% increased risk of CAD for an individual with a family history of CAD [12, 13]. In addition, the first large-scale prospective Swedish twin study ($N = 21,004$ twins) [14] identified that the risk of death from CHD was greater in monozygotic (MZ) twins than in dizygotic

(DZ) twins in both men and women independent of CHD risk factors. A follow-up study showed that CHD heritability was 0.57 (95% CI, 0.45–0.69) and 0.38 (0.26–0.50) for male and female twins, respectively, with heritable effects most evident in younger individuals [15].

Such heritability estimates originate from family-based study designs. Briefly, family studies can be subdivided into three main groups: (1) single affected family member, (2) relative pairs and (3) extended families. Examples of study types with a single affected family member are case-control studies, trios (case and both parents) and case-only designs. However, the disadvantage of collecting single affected family members is that for complex diseases, multiple affected individuals are required to determine identity-by-descent (IBD) sharing. Therefore, relative pairs and/or extended family study designs are used. Examples of relative (affected and non-affected) pairs would be sib-pairs, twins or avuncular (e.g. aunt-nephew) pedigrees. Finally, extended family groups are large families with multiple affected individuals across many generations [16].

Heritability estimates are generally more precise using close relatives, whereas distant relatives are less precise and less biased [9]. Therefore, the use of twin (pedigree design) and full sibling (within-family design) data has been an important starting point for understanding CAD/MI genetics. In the proceeding paragraphs, we will briefly highlight a couple of studies that have paved the way for our understanding of CAD/MI genetics.

4.2.1 Twin Studies

Twin studies are a special case of pedigree studies consisting of six different types depending on the researcher's aim [17]. The first evidence of a genetic basis of CAD/MI was provided from a 'classical' twin design [18]. This study design used the phenotypic resemblance of MZ (genetically identical as a result of the division of a single fertilised egg) and DZ (non-identical twins that are formed from the separate fertilisation of two eggs) twins to estimate the contribution of genetic and environmental variation to phenotypic variation. As MZ and DZ twin pairs are exposed to similar pre- and postnatal environmental factors, the genetic origin of a trait can be determined [18].

The advantage of twin studies for complex or multifactorial traits such as CAD is the distinct characteristics of a twin pair, i.e. twins are the same age and exhibit a higher degree of shared family environment (e.g. lifestyle) compared to sib-pair, thereby 'controlling' the influence of environmental risk factors into a study model and attributing phenotypic differences to twin genetics. Furthermore, errors caused by non-paternity (i.e. different fathers) are reduced or nullified in comparison to sib-pair studies [18]. As previously mentioned, the twin pairs used to investigate the genetic basis of CAD mortality [14, 15] were the first to highlight the utility of twin studies in understanding CAD/MI genetics. In 2001, Wienk et al. used a Danish twin study to report that heritability estimates of frailty in CHD were within the range of 0.53–0.58 for males and females, respectively [19]. Subsequently, in 2005, Wienk et al. reported heritability estimates of 0.45 for both sexes in an additive genetics-unique environment (also known as AE, where A represents the additive genetic factors and E the unique environmental factors) model without covariates [20]. The lower heritability estimates and discrepancies between these two Wienk et al. studies could be due to different sample selection methods and overall age differences between the cohorts, as the Danish twin cohort is much older than the Swedish twin cohort.

Over the last few decades, twin studies have emphasised the genetic component of numerous CAD/MI risk factors such as smoking [21], plasma lipids, lipoproteins, and apolipoproteins [22]. Although genetic methods and molecular technologies continue to evolve more 'sophisticated' study designs (e.g. such as improved genotype chips), twin studies remain an important resource considering the unique features of this type of study design. For example, the advantage of studying the effects of epigenetic factors through DNA methylation or histone modification between twins with different lifestyles may help elucidate the environmental effect of genome expression or within-pair epigenetic drift over time [23, 24]. These epigenetic factors may help to explain why most identical twins do not contract CHD and may die of different causes and hint towards a structural gene variant mechanism. For example, Gordon and colleagues [25] investigated a cohort of 250 mothers and their newborn twins focusing on two cell types: human umbilical vein endothelial cells and cord blood mononuclear cells. They found that birthweight—a known predisposing factor for cardiovascular disease—was associated with gene expression involved in cardiovascular function. Subsequent studies using twin data will enhance our understanding of CAD pathomechanisms.

4.2.2 Full Sibling Studies

Similar to twin studies, full sibling study designs have been seminal in the ongoing search for CAD-/MI-causing genes. Murabito and colleagues [26], using population-based offspring cohort data from the Framingham Heart Study ($N = 2475$), found that middle-aged adult siblings display an increased risk for CVD events with an odds ratio (OR) 1.55 (95% CI: 1.19–2.03). Moreover, the OR for sibling CVD risk (OR = 1.99, 95% CI: 1.32–3.00) exceeded that for the parental CVD (OR = 1.45, 95% CI: 1.02–2.05). This implies that sibling CVD prevalence conferred an increased risk of future CVD events beyond the established risk factors and parental CVD. Interestingly, a 2003 review by the same authors reported that on average, there is a two- to threefold increase in CAD risk in first-degree relatives of cases, and having two or more first-degree relatives with CAD is associated with a three- to sixfold increased risk in developing CAD [27].

4.2.3 Linkage-Based Family Studies

Another type of genetic study design that is used to map the chromosomal locations of genes is linkage analyses. Briefly, a family (or families) is genotyped with polymorphic markers that span their genome, and the genotyping data is then analysed. This technique results in a logarithm of odds (LOD) score for each marker. A significant LOD score (LOD $\geq$ 3.0) indicates that in the family there is co-segregation with the disease and is identified as linkage.

There are two different types of linkage analyses that are used to map the chromosomal locations of genes for CAD and MI: (1) a model-based linkage analysis using large families in which the inheritance pattern in the families is clearly defined and (2) a model-free analysis using hundreds of small nuclear families with at least two affected siblings in each family [28]. Each of these types of linkage analyses will be discussed in detail.

(1) Model-based linkage analysis

Wang and colleagues [29] carried out a genome-wide linkage scan of a large Caucasian family (13 patients with CAD, of which nine were also affected with MI) that showed an autosomal dominant pattern of CAD/MI. The authors identified that there is a significant linkage score (LOD = 4.19) on chromosome 15q26.3 that contained approximately 93 genes. Of the known genes, myocyte enhancer factor 2 (*MEF2A*), which encodes a transcription factor, was a strong candidate for CAD/MI susceptibility due to its role in vasculogenesis and its potential role in controlling vascular morphogenesis [30, 31]. Subsequent in vitro studies showed that mice deficient in MEF2A because of a seven amino acid deletion had an effect on gene function [29]. A follow-up mutational study found 3 new mutations in exon 7 of MEF2A in 4 of the 207 independent CAD/MI patients [32]. However, follow-up efforts to re-sequence the coding sequence and splice sites of MEF2A in 300 patients with premature CAD failed to detect a MI-causing mutation or mutation co-segregation [30]. These negative findings were echoed in a later study of Iranian families [33] and in a separate Caucasian family with a history of CAD [34].

(2) Model-free linkage analysis

Using this linkage analysis approach, Helgadottir et al. [35] observed a suggestive linkage on chromosome 13q12-13 that they successfully mapped to arachidonate 5-lipoxygenase-activating protein (*ALOX5AP*) gene encoding a 5-lipoxygenase-activating protein (FLAP). This gene was associated with a twofold increase in MI risk in 296 multiplex Icelandic families. Furthermore, they observed that the gain-of-function mutation was largely attributed to male carriers of the at-risk haplotype who also had the strongest associations with the ALOX5AP haplotype. However, they did not find association between an at-risk haplotype called HapA and MI in a British cohort.

Some success with using linkage analysis to map chromosomal positions associated with CAD and MI has been reported in other studies. Three chromosomal positions have been mapped for CAD, namely, 2q21.1-22 [36], 3q13 [37] and Xq23-26 [36]. Two chromosomal positions have been mapped for MI, namely, 1p34-36 [38] and chromosome 14 (map position 123–130) [39]. These types of studies have been both a success and failure. In some circumstances, contradictory results may be due to the shortfall of analytical tools used. Referring back to the previously mentioned role of *MEF2A* variant in CAD susceptibility, two studies [29, 32] identified a 21 bp deletion of *MEF2A*. However, subsequent studies did not reach the same conclusion [33, 40]. In 2016, Xu et al. [41] conducted both exome and Sanger sequencing on a four-generation Chinese Han family with familial CAD and found a novel deletion in exon 11 of *MEF2A* that co-segregated with CAD/MI cases.

It is important to note that the downside of linkage analyses for complex traits such as CAD is that the effect sizes (or penetrance) of the individual causal variants may be too small to allow detection via co-segregation [42]. Therefore, the power to detect genes may be minimal [43] and mapping resolution may be low [44]. In this case, an alternative solution for gene identification may be to use unrelated individuals in hypothesis and hypothesis-free-based association studies.

4.2.4   Candidate Gene Studies

One strategy to identify risk variants associated with a particular disease is candidate gene studies. Briefly, these studies test whether selected genes are related to a disease based on prior knowledge about the gene function or pathophysiology of the disease. Succinctly, the following steps are used: (1) select candidate genes based on prior knowledge; (2) select the gene variant (also known as single nucleotide polymorphisms, SNPs) that is tagged by affecting gene regulation and/or its protein product; and then (3) confirm SNP association with a disease by detecting its occurrence in random cases versus controls [45].

Since the early 1990s, almost 5000 studies have analysed candidate genes in relation to CAD and MI with only 58% of variants showing consistent results in replication studies. Possible explanations for this disparity could be small study populations, false-positive associations and ethnic variations among studies [46]. Other reasons could be due to the inherent disadvantage of using candidate gene studies.

4.2.5   Genome-Wide Association Studies

There are three main disadvantages of the candidate gene approach: (1) the reliance on prior knowledge of the function of the studied gene(s), (2) the inherent bias towards choosing a candidate gene that is geared towards the researcher's specific study or interest and (3) causative variants outside the region of study that may be missed [47]. Considering the limitations of candidate gene studies, in addition to the improvement of genotyping chip designs, their continuous lowering of costs and the development of improved statistical methods (e.g. imputation and haplotype tagging), there has been an increased popularity of genome-wide association studies (GWAS).

One such imperative technological breakthrough that aided in GWAS rapid success rate is the 2007 completion and public availability of the International HapMap (short for haplotype map) Project [48], which allowed the mapping of haplotype landscapes to SNPs in three continental populations. Several years later, in 2015, the 1000 Genome Project [49] took advantage of the development of sequencing technology and released freely available human genetic variation data based on low-coverage whole-genome sequencing that reached its pinnacle with a reference panel called 1000GP3. Recently, the Haplotype Reference Consortium (HRC) [50] combined all whole-genome sequencing data sets into a single haplotype reference panel to facilitate genotype imputation. Promisingly, the HRC reference panel has been used in the imputation stage of an endophenotype of glaucoma meta-analysis GWAS. This was shown to improve the concordance between assayed and imputed genotypes, markedly in cases of low-frequency variants. In turn, this technique significantly improved p-values, particularly for suggestive variants [51], thereby outperforming 1000GP-based imputation concordance and final p-value results.

Unlike linkage studies, GWAS use unrelated subjects to detect associations between genetic variants and disease/traits, making it easier to obtain large sample sizes. The foundation of GWAS is the 'common disease, common variant' (CDCV) hypothesis that was first put forward by Lander [52]. This hypothesis implies that common genetic variants in the population with low penetrance (by common we mean allelic variants present in more than 1–5% of the population [53]) contribute to the genetic susceptibility to common complex traits and disease [52].

Exploiting GWAS analyses has led to significant progress in understanding the genetics of CAD/MI. For CAD-GWAS studies, 2007 was an important year when three independent GWAS studies discovered an association of variants on chromosome 9p21.3 and CAD in European ancestry population [54–56] and multiple races [57–61], apart from African Americans [58]. This result may be consistent with the 'Out Of Africa' hypothesis which states that all present population groups of *Homo sapiens* have evolved from a primitive African population [62] (Fig. 4.1).

Over the past decade, GWAS for CAD/MI has seen much success, which was catapulted in 2007 with the discovery of 9p21, by independent research groups. By 2009, twelve other genetic risk variants were discovered through GWAS to be associated with CAD [63]. Subsequently in 2013, with a larger sample size, the CARDIoGRAMplusC4D Consortium reported 46 loci associated with CAD, both confirming previously published and finding new variants [64]. This was followed by the identification of ten additional new loci in 2015 [65]. Currently, in 2017 and 2018, CARDIoGRAMplusC4D data together with the UK Biobank [66] data have been proven to be a wealthy resource of genetic data reflected by the increase in new CAD-associated loci [67–69].

In spite of these new CAD loci findings, less focus has been given to the X- and Y-chromosomes with fewer in studies on the X-chromosome [70]. It is common knowledge that there exists sexual dimorphism regarding the incidence, prevalence,

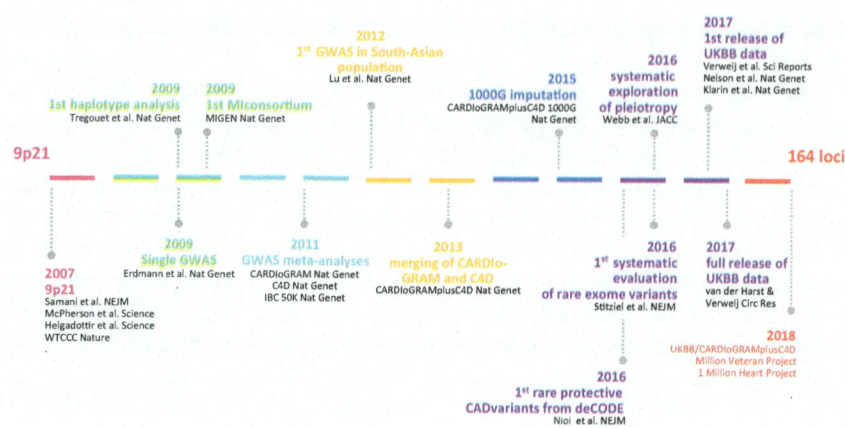

Fig. 4.1 Milestones in cardiovascular genetics from 2007 and beyond

morbidity and mortality of CVD and/or MI with men having an increased risk compared to age-matched women [71–75]. This is due to two barriers: (a) markers on genotyping chips and (b) statistical methods [70]. These are acutely present when analysing the X-chromosome, due to its unique properties such as X-inactivation, and the presence of two X-chromosome copies in females compared to males [70, 75, 76]. There are publications [76, 77] that do offer possible solutions and recommendations for incorporating X-chromosome. However, the results have been mixed with a 2016 meta-analysis study [78] reporting no association of CAD and X-chromosome variants, compared to a recent 2017 American Heart Association/ American Stroke Association (AHA/ASA) conference abstract [79] which found three novel CAD susceptibility loci on the X-chromosome. Much work remains for the inclusion of the sex chromosomes. Nevertheless, we can say that for the autosomal chromosomes, there are currently a total number of 163 loci associated with CAD.

Using the CAD/MI GWAS results thus far, we can tentatively say (a) the majority of common variants found show modest CAD risk increase; (b) most of the variants found are situated outside protein-coding regions; and (c) we have improved our understanding of CAD risk with the loci we have found so far [5] (Fig. 4.2).

4.3   9p21 Locus and Its Role in CAD/MI

Carried by 75% of the global population (excluding black Africans), 9p21 risk alleles are associated with coronary atherosclerosis risk [81]. Moreover, a prior study [82] showed that 9p21 is significantly associated with the risk of first CHD events (1.19 hazard ratio of first event; 95% CI: 1.17–1.22) compared to subsequent CHD events (1.01 hazard ratio of first event; 95% CI: 0.97–1.06). These

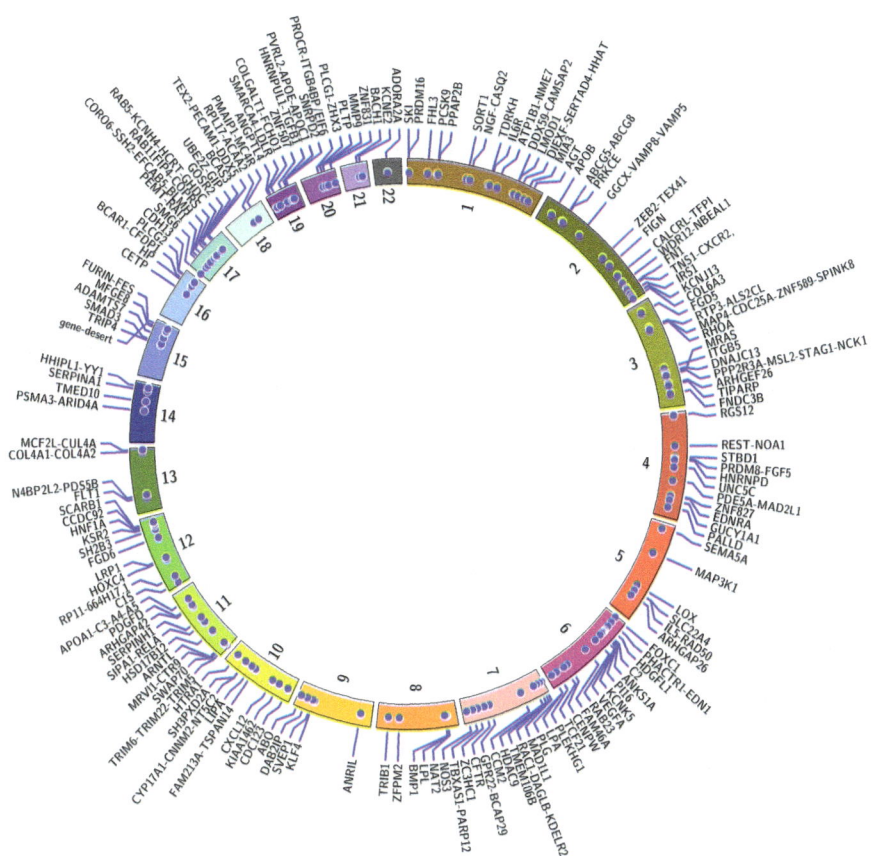

Fig. 4.2 Circos plot [80] with 163 risk loci identified by January 2018. Figure provided by Syed M. Ijlal Haider, Institute for Cardiogenetics

observations suggest that 9p21.3 stimulates coronary atherosclerosis (i.e. CAD) rather than MI [81, 83]. The 9p21.3 locus has been characterised as a 'gene desert' containing dispersed haplotype blocks [84]. It is thought that the CVD-associated region is adjacent to the last exons of a long non-coding RNA (lncRNA), specifically the antisense non-coding RNA in the *INK4* locus (*ANRIL*; also known as *CDKN2BAS*) [85]. The closest protein-coding (candidate) genes include the cyclin-dependent kinase (CDK) inhibitors *CDKN2A* and *CDKN2B* [83]. Furthermore, Holdt et al. [86] reported that *CDKN2A/B* gene protein products (p16^{INK4a}, p14^{ARF} and p15^{INK4b}) that are expressed in smooth muscle cell layers in both normal arteries and atherosclerotic plaques participate in atherosclerotic lesions. As covered in a review by Hannou and colleagues [83], several studies have failed to decipher the exact mechanism through which CDKN2A/B gene products work or which pathways are involved. Using unbiased genomic techniques based on chromosome conformation capture (3C) [87], Harismendy, O and colleagues [88] detected long-

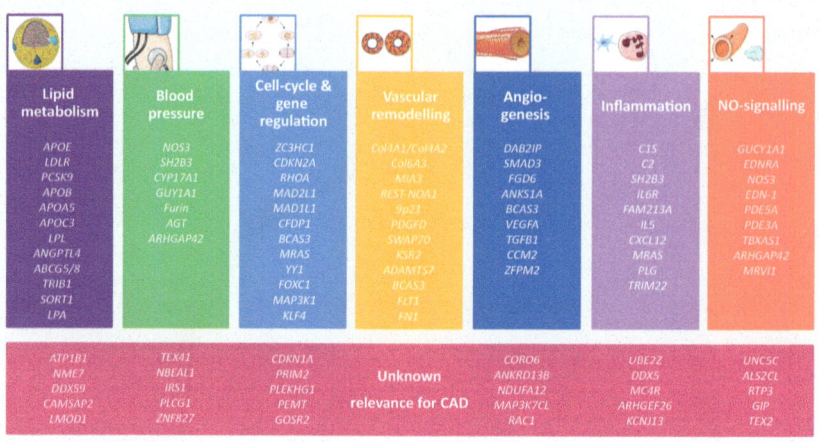

Fig. 4.3 Selection of CAD risk genes and their proven and/or predicted involvement in pathways related to CAD and MI

distance interactions between the enhancer interval containing the CAD locus and *CDKN2A/B*. Considering the effects of interactions across large distances, this observation tentatively (and excitingly) points to the possibility that 9p21.3 disease-associated SNPs interact and modify other distant genes. Very recently, Holdt and colleagues identified circANRIL as a prototype of a circRNA regulating ribosome biogenesis and conferring atheroprotection, thereby showing that circularisation of long non-coding RNAs may alter RNA function and in general might protect from human disease [89].

Broadly summarising, there are seven categories with their respective identified risk loci that underlie the pathways to CAD. Of the total number of loci (Fig. 4.3), 70% of loci are known to be involved in lipid metabolism (12%), blood pressure (7%), cell cycle and gene regulation (12%), vascular remodelling (12%), angiogenesis (9%), inflammation (10%) and nitric oxide signalling (9%). Unsurprisingly, due to the laborious and extensive experimental follow-up required, the vast majority (30%) of identified loci pathways have yet to be explained.

4.4   Does Sample Size Matter to Identify CAD/MI Risk Loci?

To detect significant SNP contributions in GWAS, large numbers of cohort participants are required, ranging from the tens to hundreds of thousands of subjects. Large GWAS consortia are formed to reach such numbers, usually through meta-analysis of GWAS. For example, a large Coronary ARtery DIsease Genome-wide Replication and Meta-analysis (CARDIoGRAM) Consortium [90] ($N_{cases} = 22{,}233$ and $N_{controls} = 64{,}762$) was constituted and identified 13 new CAD loci by a meta-analysis of GWAS (sometimes abbreviated to meta-GWAS). In parallel, the

Coronary Artery Disease (C4D) Genetics Consortium [91] (N_{total_cases} = 15,420 and $N_{total\_controls}$ = 15,062) found five newly associated CAD loci. In an effort to increase sample size, these two consortiums merged and are now known as the CARDIoGRAMplusC4D Consortium [64]. The merging, CARDIoGRAMplusC4D Consortium achieved a sample size of 63,746 CAD cases and 130,681 controls and identified an additional 15 loci to the already known CAD loci. Proceeding, the first interim 150,000 genotyped individuals from the UK Biobank have recently been made available [92] with the intention to release the remaining 350,000 participants in the near future. This would mean a publically available health resource of an unprecedented 500,000 individuals. Interestingly, this year, 3 studies [67, 68, 93] each found 13, 14 and 15 new loci associated with CAD, of which 7 loci overlapped between the 3 studies. Subtle differences in study design (e.g. phenome-wide association scan [67] vs. false discovery rate approach [93]) and phenotype definitions (e.g. CAD cases defined as multiple International Classification of Disease (ICD) 10 code [68] vs. subdivision of ICD10-coded CAD cases with/ without angina [93]) could be a minor contributing reason as to why the loci overlap between studies is low; however, further studies are merited. These three studies exemplify the gain of a large sample size for GWASs within the last decade: from ≈60 common genetic variants to >95 total number of CAD-associated loci [94].

A larger sample size could help low-frequency variant studies to reach the required level of significance. A prior large-scale exome-wide study of more than 120,000 participants only had 80% power to detect an OR of ±2.0 for CAD-associated variants with a minor allele frequency (MAF) of 0.1% [95].

Even with relatively common genetic variants (MAF > 0.01%), large cohorts like the CARDIoGRAMplusC4D [64] and the recent UK Biobank could facilitate the identification of single variant statistical analyses for common variants across the exome and thereby reach exome-wide significance, such as in the case of atherosclerosis lesions in young participants [96].

4.5 Rare/Low-Frequency Variants of CAD: 'Can "In-Betweeners" Explain the Missing Heritability?'

In classical genetics, narrow-sense heritability represents the joint distribution of allele frequency and effect sizes [97], meaning that dominant or epistatic effects are not considered.

In GWASs (and the CDCV hypothesis), the quantification of the proportion of additive genetic variance because of LD between the genotyped and imputed SNPs with the unknown causal variants (i.e. 'SNP heritability') implies that genetic variations can be tagged by common SNPs via LD [97]. It has been found that between one-third and two-thirds of the additive genetic variation in a population is tagged and is often referred to as the 'missing heritability' [42, 92, 97, 98]. This 'missingness' is clearly observed when comparing common variants (by common we mean MAF, >0.05%) of small effects (i.e. GWAS) and very rare variants (by rare we mean MAF ≤ 0.05%) with large effects (i.e. whole exome studies), even after

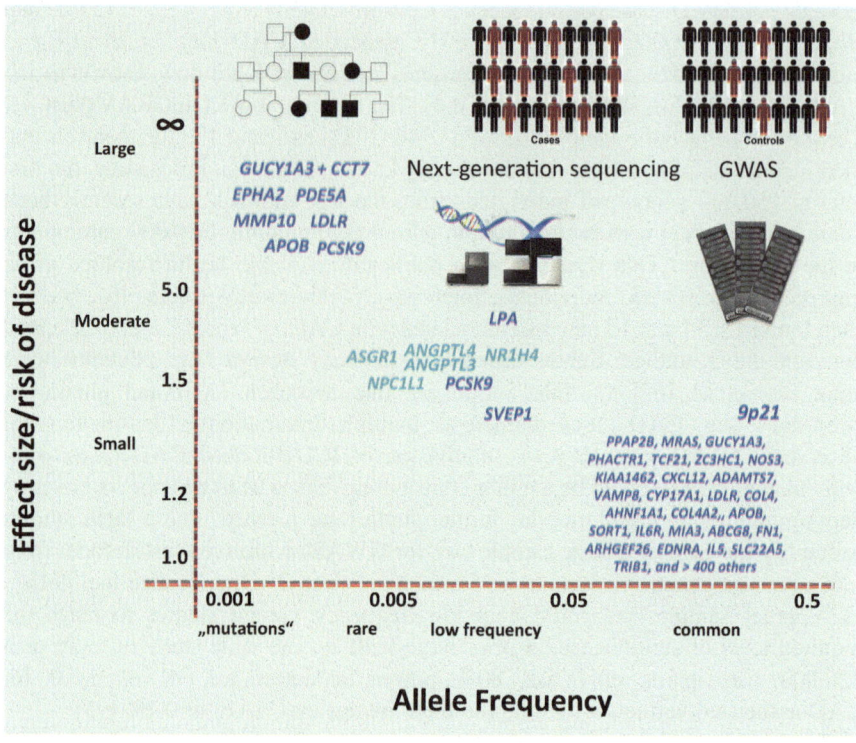

Fig. 4.4 Allelic spectrum of CAD ranging from private variants with monogenic effect on disease to common variants conferring only low disease risk (adapted from Manolio et al. 2009)

controlling for other effects such as environment [92]. One early example was the haplotype association of a rare CAD variant and *SLC2A-LPAL2-LPA* gene cluster [99]. More recently, the Myocardial Infarction Genetics and CARDIoGRAM Exome Consortia Investigators [95] investigated the effects of loss-of-function mutations in 72,868 CAD patients and 120,770 controls. They identified low-frequency loss-of-function missense variants in *ANGPTL4* gene (that was also associated with a protection against CAD) and an association with increased CAD risk in low-frequency coding variants in *SVEP1* gene.

Challenges remain in the search for rare CAD variants. For example, there are difficulties in detecting and replicating rare variants that are restricted to specific population groups. To overcome this issue, it has been suggested that combining summary GWAS association statistics by using 'local SNP heritability' could provide replication at the locus level rather than the SNP variant level [92]. Another issue for successful GWAS is the requirement for very large sample sizes. However, using large cohorts such as the UK Biobank would overcome this particular issue (Fig. 4.4).

4.6   A Note: Missing Does Not Imply Absence

Considering that the current CAD-associated SNPs only explain 10.6% of CAD variability and 40% heritability, it is natural to ask whether the 'missing' heritability is actually absent. However, this is not the case with most complex traits that present missing heritability. As we mentioned above, other explanations (e.g. rare SNPs that present large effects) or new analytical approaches (e.g. Generalized Compound Double Heterozygosity [GCDH] that can detect genetic associations by a relaxed form of CH) help to explain this gap. Another proposed explanation for the missing variability and heritability would be that some common risk alleles might have (very) small effects that are too small to pass the traditional GWAS significance level. This leads to a question of how many of these common risk alleles with small effects would a study need to explain 100% of CAD heritability? During the peak of GWAS studies, Wray and colleagues [100] addressed this question and reported that causal risk variants can be estimated as a function of the disease heritability and prevalence in the scenario that all risk alleles have the same relative risk and frequency [100, 101]. Another enticing idea put forward to help explain CAD 'missing heritability' is that heritable risk for CAD and common complex diseases could be partly attributed to interactions among diseases and traits [102, 103]. Making the case for the pleiotropic effect among complex disease, a recent publication by Webb et al. [103] identified six new CAD loci. In addition, they showed that 47% of the loci were association with another disease/trait with several loci showing multiple associations.

4.7   Adding Biological Meaning to GWAS Findings

An inherent limitation of association studies, such as GWASs, is that they do not provide biological meaning of the casual variants that were tagged via genome-wide significant SNPs (tagSNPs). To translate GWAS results to biological function, post-GWAS analysis is necessary. Usually, post-GWAS starts with inexpensive in silico (bioinformatics) analysis of GWAS-found tagSNPs that is then followed up with more time and costly in vitro and/or in vivo studies. There is an increasing number of in silico tools and statistical techniques used to give meaning and prioritisation to the resulting associated loci that are subsequently investigated with experiments. In the following paragraphs, we will briefly cover some of the most popular or recent methods.

4.8   Mendelian Randomisation

Although randomised controlled trial (RCT) studies are the optimum way to establish the causal relationship between risk factors, exposure and disease of interest (in our case, CAD), sometimes this is not possible. An alternative is to use Mendelian randomisation (MR) analysis of GWAS data. The rationale is to use genetic

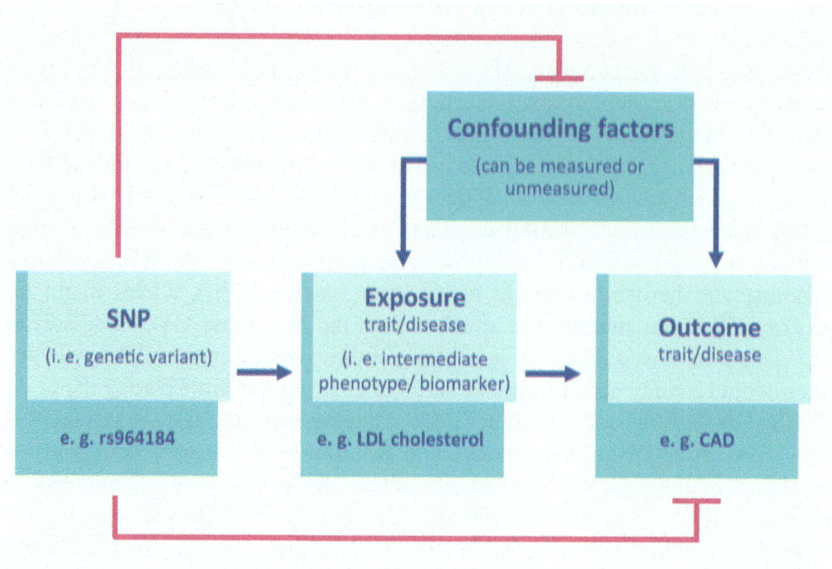

Fig. 4.5 Mendelian randomization basic principles used to find causal pathways. Compliance to three MR assumptions regarding the genetic variant must be kept, namely, the genetic variant must be robustly associated with the exposure variable (shown by the thick blue arrow) and should not be directly related to the outcome or have any confounding factors (shown by the two pink lines)

variants as a proxy for a potentially modifiable exposure to identify causal effects for risk of CAD [104], thus making MR analysis resistant to confounding factors, an advantage over randomised control trials [105]. Other useful characteristics of MR studies are that multiple genetic variants can be used to increase power and investigate pleiotropic influence on the trait of interest, bidirectional MR studies can be used to determine the direction of causal effects in more complex networks, gene-by-environment interactions can be studied and epigenetic profiles can be used as an intermediate phenotype [104] (Fig. 4.5).

One of the main utilities of MR results is their applicability in drug development. For example, MR studies of loss-of-function mutations in proprotein convertase subtilisin/kexin type 9 (*PCSK9*) have led to the development of several PCSK9 monoclonal antibodies. These antibodies are currently under study in four phase 3 trials to test whether such drugs reduce cardiovascular events. Various genetic studies have reported that gain- and loss-of-function mutations in *PCSK9* increase low-density lipoprotein-cholesterol (LDL-C) concentration and premature atherosclerosis and reduce LDL-C with low rates of CHD, respectively. Succinctly, this is achieved via hepatocyte endocytosis where circulating LDL-C is cleared from the blood by the binding of LDL-C to low-density lipoprotein receptor (LDL-R) that is situated on the hepatocyte cell membrane. PCSK9 regulates LDL-R metabolism by binding to it, thus leading it to be destroyed by lysosomes and thereby decreasing its recirculation [106].

4.9   Next-Generation Sequencing (NGS) Approaches for CAD and MI

We would be remiss if we didn't concisely highlight next-generation sequencing (NGS)—with whole-genome sequencing being the most widely used NGS technology [107]—and also MR, studies in providing strong CAD pathophysiological insight [108]. Following the lipoprotein example, Emdin et al. [109] used the NGS approach (via LPA gene sequencing) to show that one standard deviation (SD) genetically lowered lipoprotein(a) level is associated with a lower risk of 29% for CHD, of 31% for peripheral vascular disease, of 13% for stroke, of 17% for heart failure and of 37% for aortic stenosis.

4.10   Expression Quantitative Trait Loci (eQTL): CAD/MI

EQTLs are SNPs, previously found to be associated with the phenotype of interest (i.e. CAD), which have either a local effect (*cis*) from where the associated variant was found or distant effect (*trans*; e.g. more than 5 Mb away) from the associated variant [110, 111]. This method uses quantitative trait loci (QTLs) by looking at mRNA levels—the primary genome product—and correlating gene/protein/methylation level (i.e. intermediate molecular quantitative traits) with the genetic variants found [110, 112]. RNA samples from patients and healthy subjects are recruited and converted into microarray data creating a RNA expression data set. Publicly available multi-tissue databases usually accompany association studies and are used to add strong evidence that the associated SNP has a functional effect [113], in turn adding a targeted approach of candidate genes to follow up with experimental work. A variety of tissue disease-specific databases are available such as ENCODE [114], GTEx [115], Epigenome RoadMap [116] and STARNET [117], some of which have been used to focus on CAD-relevant tissues. For example, a study [118] using STARNET showed that *cis*- and *trans*-genes could act as a mechanism for multiple risk loci to contribute to cardiometabolic diseases (precursors of CAD) heritability (Fig. 4.6).

4.11   Network Analysis

There are a few tools [119–122] available that either prioritise variants found during association analysis or perform tissue enrichment analysis. An interesting study published last year [123] used various tissue-specific regulatory networks and protein-protein interaction networks that do not solely rely on a priori knowledge. They were able to detect well-implicated CAD genes in the prevalence of CAD and also unravelled novel key regulators (*LUM, HGD, F2, ANXA3* and *STAT3*) for CAD. Recently, Vilne et al. demonstrated how hypercholesterolaemia can hinder mitochondrial activity during atherosclerosis progression and identified oestrogen-related

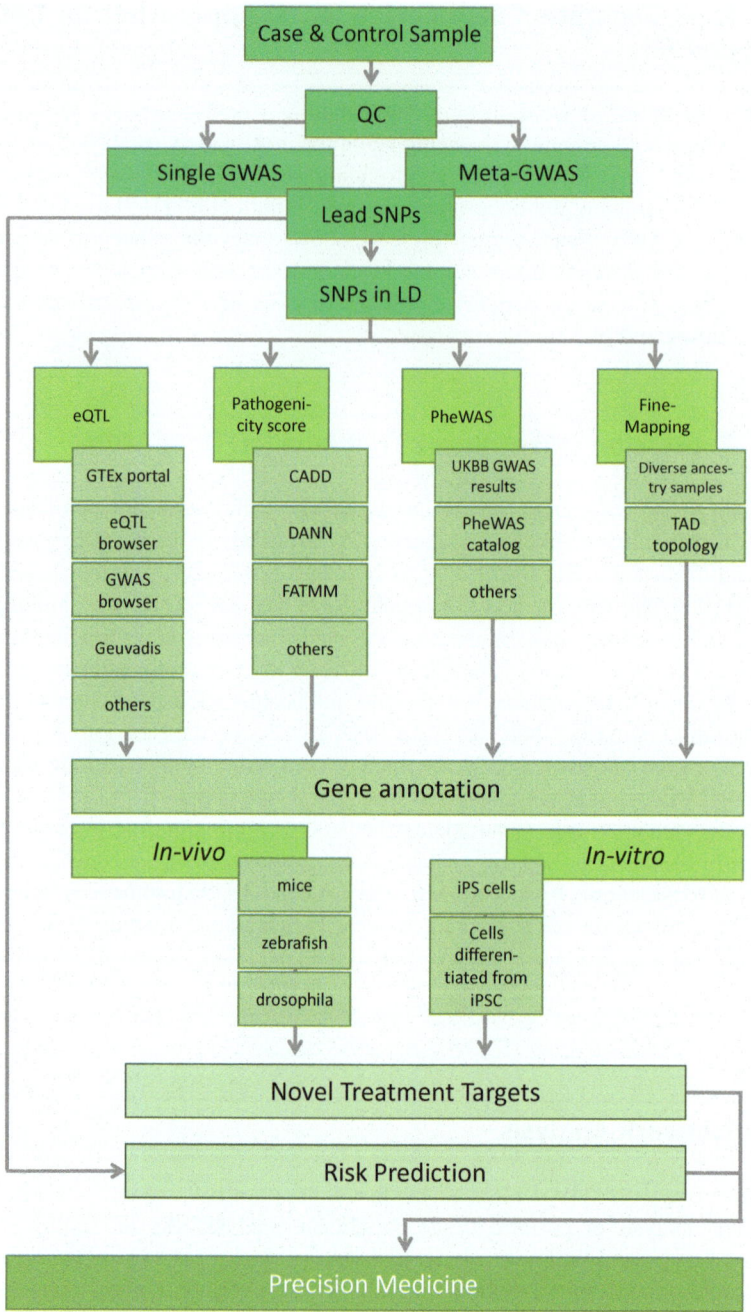

Fig. 4.6 Flow chart describing GWAS and post-GWAS analyses

receptor-α and its cofactors PGC1-α and PGC1-β as potential therapeutic targets to counteract these processes using a network approach [124].

4.12   The 'Omics Era'

One of the fields that has seen an almost exponentially rapid rise in popularity due to lowered cost and high-throughput analysis has been the '-omics' studies (namely, genomics, epigenomics, transcriptomics, proteomics, metabolomics and microbiomics), with a vital participatory role of cloud [125] and/or Web [126] computing that facilitates the handling of huge '-omics' study data volumes. As we become more aware that (1) identified CAD loci only explain a small percentage of heritability; (2) common diseases, such as CAD, tend to occur because of gene regulation changes; and (3) similar genetic variants contribute to different final outcomes, it may come as no surprise that systems genetics evolved to integrate the various '-omics' studies. In our case, this enabled the methodology to help explain the complexity of the underlying molecular patterns that are associated with CAD.

4.13   Systems Genetics Approaches in CAD/MI

As defined by Björkegren et al. [111], systems genetics uses molecular mechanisms to define disease-driving molecular processes that underlie GWAS, whole exome sequencing (WES) or whole-genome sequencing (WGS) and to integrate such processes with functional genomic data. One such example of associated SNPs exerting a tissue-dependent effect on gene expression is exemplified by Musunuru et al. [127]. They integrated eQTL and protein QTL (pQTL) information and found that a MI risk variant alters the expression of *SORT1* gene in the liver (via a lipoprotein metabolism-regulated pathway) and not in the blood. This observation is dissimilar to prior GWASs that identified a strong association between 1p13 locus and plasma LDL-C levels in MI patients and, moreover, that this locus influenced the risk of MI by conferring changes to plasma LDL-C. Another recent study [128] used a systems genetics approach to integrate DNA genotypes and gene expression profiles from seven CAD-relevant tissues with CAD CARDIoGRAM GWAS information. With this analytical perspective, they showed that RNA-processing genes play a pivotal role in causing CAD and furthermore identify several strongly inherited, evolutionarily conserved, risk-enriched CAD genes that cause regulatory gene network changes across vascular and metabolic tissues.

4.14   Multi-omics Approaches

Compared to looking at an individualised -omics approach, multi-omics provides a greater understanding of the flow of information from the disease initiator to its functional consequence or interaction. Multifactorial diseases, such as CAD, prove to be extremely entangled, which may be a contributing factor as to the lack of multi-omics studies of CAD. However, a preprint study by Santolini et al. [129] used >100 genetically diverse mice strains to investigate a multi-omics approach to cardiac hypertrophy and heart failure. Interestingly, they developed a personalised strategy to investigate stressor-induced heart failure and identified 36-fold change genes that were enriched in human cardiac disease genes and hypertrophic pathways and were missed by the traditional population-wide differentially expressed gene method. Additionally, the genes that they found were linked to both upstream regulators and signalling networks, providing insight into cardiac hypertrophy severity and resistance. Finally, they validated *Hes1* as a novel regulator of cardiac hypertrophy (Fig. 4.7).

4.15   Exciting Times Ahead

4.15.1 Reverse Genetics

An interesting topic that has appeared in the literature recently is the concept of 'human knockouts'. It is based on the idea that the accumulation of rare homozygous mutations is most likely in highly consanguineous populations, as is the case with the Pakistan Risk of Myocardial Infarction Study (PROMIS) [130]. It is known that heterozygous deficiency of the *APOC3* gene confers protection against CHD. PROMIS participants who were homozygous for *APOC3* loss-of-function mutations were challenged with an oral fat load. These same individuals were then compared with family members lacking the mutation, and they showed significantly improved clearance of the usual post-prandial rise in plasma triglycerides from their circulation. This study highlights the potential and impact of reverse genetics and functional research and the relevance of drug targets prior to their costly development.

4.15.2 Leveraging Genomic Data to Identify Novel CAD/MI Drug Targets

As has been published [131] and commented [132], looking at these naturally occurring 'human knockout' population groups, researchers can directly (and non-invasively) see what would happen when a protein's function or regulatory mechanism is completely removed thereafter derive subsequent dose-response curves in drug development, for example, if we follow the starting path of the *GUCY1A3* and *CCT7* genes. Prior GWAS studies [90, 133] had found a CAD/MI association with a common variant on chromosome 4q32.1 which overlapped with

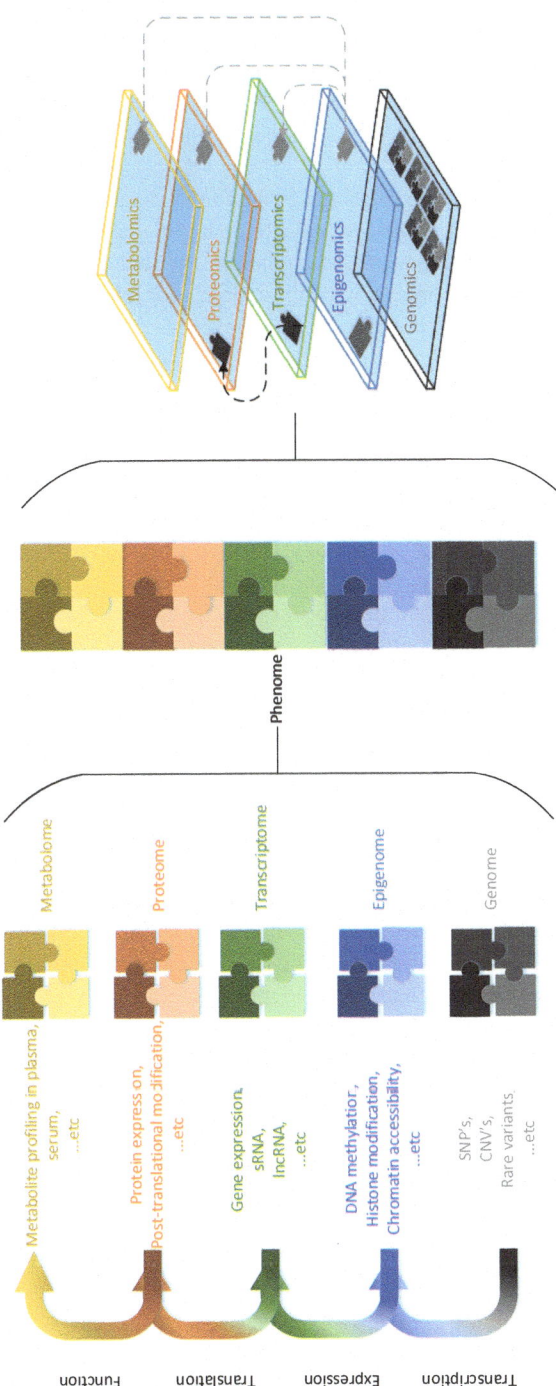

Fig. 4.7 Overview of the multi-omics approach. From a micro-level approach, we are able to consecutively combine information to form a picture at the '-ome' level, which, when aggregated together, results in a phenome informational level. Furthermore, different informational pieces can affect between multiple -omics layers. SNPs, single nucleotide polymorphisms; CNVs, copy number variants; sRNA, small RNA; lncRNA, long non-coding RNA

the *GUCY1A3* gene. Subsequently in 2013, Erdmann and colleagues [134] studied two German families and unrelated MI/CAD cases and found two rare variants (both heterozygous mutations) in the *GUCY1A3* and *CCT7* genes. With confirmation through in vitro/in vivo experimental work, they evidenced that mutations in the soluble guanylyl cyclase-dependent nitric oxide signalling pathway could be linked to MI. Moving on to today, Kessler et al. [135] elucidated via human samples and cell lines (vascular smooth muscle cell migration and platelet function experiments) that *GUCY1A3* affects the expression of soluble guanylyl cyclase, smooth muscle cell migration and platelet function. Further studies are merited to bring this pathway to its destination, which would hopefully be drug targets for individuals carrying the *GUCY1A3* risk allele (Table 4.1).

4.16   Precision Medicine

GWAS data has been used as a stepping stone to identify risk genes and, with subsequent post-GWAS analysis strategies, may help to elucidate disease-associated pathways that would then be used in drug development or selection. This is the case with *PCSK9* (mentioned previously). Although GWAS studies have been useful, it must be emphasised that these studies do not provide the necessary information required to stratify individuals according to severity, prognosis and responsiveness that is required for drug development and/or selection. To obtain such information, Morita and Komuro [143] suggested stratifying large-scale prospective studies according to clinically affected subphenotypes in patients with similar clinical presentations and then adding a second layer independent of the variant associated with the disease onset that would look for a variant associated (or driving pathways) with the subphenotype within the disease [143, 144]. One example of such stratification would be to discriminate between dyslipidemic patients with CAD and patients without CAD. In this example, a subpopulation analysis to identify genetic variants associated with CAD susceptibility could help in the selection of individuals susceptible to CAD who should receive proactive, perhaps intensive, cardiometabolic abnormality management to prevent CAD [143]. On a macro level, current pharmacogenomics research is following these lines. Contrary to the subphenotype method, some argue that in CAD patients, it is more useful to highlight the blend of genetic and environmental causal factors (or pathways) that underlie CAD patients in large-population-scale cohorts [144]. Such discussion can only promote this type of research approach and can be used as a thoroughfare towards a future of individualised precision medicine.

4.17   Closing Remarks

Indeed, much has happened since the journey towards unravelling the genetics underlying CAD/MI began. It will most likely be decades before we fully grasp the pathomechanisms that result in CAD/MI. However, considering the global

Table 4.1 Examples of genes protective of CAD and MI identified by large-scale array-based and deep-sequencing projects

Gene	PCSK9	NPC1L1	LPL	APOC3	ANGPTL4	ANGPTL3	ASGR1	NR1H4
Frequency	1 in 50 blacks	1 in 150	1 in 10	1 in 150	1 in 500	1 in 300	1 in 120 Icelanders	~1 in 450 Icelanders
Phenotype	LDL	LDL	TG	TG	TG	TG, LDL	non-HDL	non-HDL
Risk	80% lower risk	53% lower risk	17% lower risk	40% lower risk	57% lower risk	34% lower risk	34% lower risk	–
Therapy	Evolocumab Ecocizumab Alirocumab	Ezetimibe	–	Antisense in development	Monoclonal antibodies in development	Monoclonal antibodies in development	–	–
References	[136]	[137]	[95]	[138]	[95, 139]	[140]	[141]	[142]

TG = triglycerides

burden that is CAD/MI, we owe it to ourselves to continue our efforts in unravelling the genetic basis of these diseases. Luckily, we have allies in the continued lowering of genetic sequencing costs, increased computational facilities and strategies (such as machine learning and data mining) that can handle the influx of large data and new research approaches. Exciting times lie ahead of us.

Disclosure Statement The authors have nothing to disclose.

Conflict of Interest None declared.

Funding This work was supported by a grant from the Fondation Leducq [PlaqOmics], the German Federal Ministry of Education and Research (BMBF) within the framework of ERA-NET on Cardiovascular Disease, Joint Transnational Call 2017 [ERA-CVD: ENDLESS]. This work was also supported by the Deutsche Forschungsgemeinschaft (DFG, German Research Foundation) under Germany's Excellence Strategy—EXC 22167-390884018.

References

1. Mendis S, Puska P, Norrving B, editors. Global atlas on cardiovascular disease prevention and control. Geneva: World Health Organization (WHO) in collaboration with World Heart Foundation and the World Stroke Organization; 2011.
2. Nichols M, Townsend N, Scarborough P, Rayner M. Cardiovascular disease in Europe 2014: epidemiological update. Eur Heart J. 2014;35:2950–9.
3. Mozaffarian D, et al. Heart disease and stroke statistics—2016 update. Circulation. 2015;133: e38–e360.
4. Buja LM. Coronary artery disease: pathological anatomy and pathogenesis. In: Willerson JT, Holmes Jr DR, editors. Coronary artery disease. London: Springer; 2015.
5. Khera AV, Kathiresan S. Genetics of coronary artery disease: discovery, biology and clinical translation. Nat Rev Genet. 2017;18(6):331–44.
6. Ozaki K, Tanaka T. Molecular genetics of coronary artery disease. J Hum Genet. 2016;61:71–7.
7. Khera AV, et al. Genetic risk, adherence to a healthy lifestyle, and coronary disease. N Engl J Med. 2016;375:2349–58.
8. Fisher RA. The correlation between relatives on the supposition of mendelian inheritance. Philos Trans R Soc Edinb. 1918;52:399–433.
9. Vinkhuyzen AAE, Wray NR, Yang J, Goddard ME, Visscher PM. Estimation and partitioning of heritability in human populations using whole genome analysis methods. Annu Rev Genet. 2013;47:75–95.
10. Müller C. Xanthomata, hypercholesterolemia, angina pectoris. Acta Med Scand. 1938;95 (S89):75–84.
11. Gertler MM, Garn SM, White PD. Young candidates for coronary heart disease. J Am Med Assoc. 1951;147(7):621–5.
12. Schildkraut JM, Myers RH, Cupples LA, Kiely DK, Kannel WB. Coronary risk associated with age and sex of parental heart disease in the Framingham Study. Am J Cardiol. 1989;64 (10):555–9.
13. Myers RH, Kiely DK, Cupples LA, Kannel WB. Parental history is independent risk factor for coronary artery disease: The Framingham Study. Am Heart J. 1990;120(4):963–9.
14. Marenberg ME, Risch N, Berkman LF, Floderus B, de Faire U. Genetic susceptibility to death from coronary heart disease in a study of twins. N Engl J Med. 1994;330:1041–6.

15. Zdravkovic S, Wienke A, Pedersen NL, Marenberg ME, Yashin AI, de Faire U. Heritability of death from coronary heart disease: a 36-year follow-up of 20 966 Swedish twins. J Intern Med. 2002;252:247–54.
16. Banerjee A. A review of family history of cardiovascular disease: risk factor and research tool. Int J Clin Pract. 2012;66(6):536–43.
17. Boomsma D, Busjahn A, Peltonen L. Classical twin studies and beyond. Nat Rev Genet. 2002;3:872–82.
18. Mangino M, Spector T. Understanding coronary artery disease using twin studies. Heart. 2013;99(6):373–5.
19. Wienke A, Holm N, Skytthe A, Yashin AI. The heritability of mortality due to heart disease: a correlated frailty model applied to Danish twins. Twin Res. 2001;4(4):266–74.
20. Wienke A, Herskind AM, Christensen K, Skytthe A, Yashin AI. The heritability of CHD mortality in Danish Twins after controlling for smoking and BMI. Twin Res Hum Genet. 2005;8(1):53–9.
21. Koopmans JR, Slutske WS, Heath AC. The genetics of smoking initiation and quantity smoked in Dutch adolescent and young adult twins. Behav Genet. 1999;29(6):383–93.
22. Snieder H, van Doornen LJP, Boomsma DI. The age dependency of gene expression for plasma lipids, lipoproteins, and apolipoproteins. Am J Hum Genet. 1997;60(3):638–50.
23. Tamoki AD, Tamoki DL, Molnar AA. Past, present and future of cardiovascular twin studies. Cor Vasa. 2014;56(6):e486–93.
24. Fraga MF, et al. Epigenetic differences arise during the lifetime of monozygotic twins. Proc Natl Acad Sci USA. 2005;102(30):10604–9.
25. Gordon L, et al. Expression discordance of monozygotic twins at birth: Effect of intrauterine environment and a possible mechanism for fetal programming. Epigenetics. 2011;6:579–92.
26. Murabito JM, et al. Sibling cardiovascular disease as a risk factor for cardiovascular disease in middle-aged adults. J Am Med Assoc. 2005;294(24):3117–23.
27. Scheuner MT. Genetic evaluation for coronary artery disease. Genet Med. 2003;5(4):269–85.
28. Wang Q. Advances in the genetic basis of coronary artery disease. Curr Atheroscler Rep. 2005;7(3):235–41.
29. Wang L, Fan C, Topol SE, Topol EJ, Wang Q. Mutation of MEF2A in an inherited disorder with features of coronary artery disease. Science. 2003;302(5650):1578–81.
30. Edmondson DG, Lyons GE, Martin JF, Olson EN. Mef2 gene expression marks the cardiac and skeletal muscle lineage during mouse embryogenesis. Development. 1994;120:1251–63.
31. Subramanian SV, Nadal-Ginard B. Early expression of the different isoforms of the myocyte enhancer factor-2 (MEF2) protein in myogenic as well as non-myogenic cell lineages during mouse embryogenesis. Mech Dev. 1996;57(1):103–12.
32. Bhagavatula MRK, et al. Transcription factor MEF2A mutations in patients with coronary artery disease. Hum Mol Genet. 2004;13(24):3181–8.
33. Inanloo Rahatloo K, Davaran S, Elahi E. Lack of association between the MEF2A gene and coronary artery disease in Iranian families. Iran J Basic Med Sci. 2013;16(8):950–4.
34. Lieb W, et al. Lack of association between the MEF2A gene and myocardial infarction. Circulation. 2008;117:185–91.
35. Helgadottir A, et al. The gene encoding 5-lipoxygenase activating protein confers risk of myocardial infarction and stroke. Nat Genet. 2004;36(3):233–9.
36. Pajukanta P, et al. Two loci on chromosome 2 and X for premature coronary heart disease identified in early- and late-settlement populations of Finland. Am J Hum Genet. 2000;67 (6):1481–93.
37. Hauser ER, et al. A genome-wide scan for early-onset coronary artery disease in 438 families: the GENECARD study. Am J Hum Genet. 2004;75(3):436–47.
38. Wang Q, et al. Premature myocardial infarction novel susceptibility locus on chromosome 1P34-36 identified by genomewide linkage analysis. Am J Hum Genet. 2004;74:262–71.
39. Broeckel U, et al. A comprehensive linkage analysis for myocardial infarction and its related risk factors. Nat Genet. 2002;30(2):210–4.

40. Weng L, et al. Lack of MEF2A mutations in coronary artery disease. J Clin Invest. 2005;115 (4):1016–20.
41. Xu D-L, et al. Novel 6-bp deletion in MEF2A linked to premature coronary artery disease in a large Chinese family. Mol Med Rep. 2016;14(1):649–54.
42. Visscher PM, Brown MA, McCarthy MI, Yang J. Five years of GWAS discovery. Am J Hum Genet. 2012;90(1):7–24.
43. Risch N, Merikangas K. The future of genetic studies of complex human diseases. Science. 1996;273:1516–7.
44. Wang X, Prins BP, Siim S, Laan M, Snieder H. Beyond genome-wide association studies: new strategies for identifying genetic determinants of hypertension. Curr Hypertens Rep. 2011;13:442–51.
45. Patnala R, Clements J, Batra J. Candidate gene association studies: a comprehensive guide to useful in silico tools. BMC Genet. 2013;14:39.
46. Mayer B, Erdmann J, Schunkert H. Genetics and heritability of coronary artery disease and myocardial infarction. Clin Res Cardiol. 2007;96(1):1–7.
47. Zhu M, Zhao S. Candidate gene identification approach: progress and challenges. Int J Biol Sci. 2007;3:420–7.
48. International HapMap Consortium. The international HapMap project. Nature. 2003;426 (6968):789–96.
49. The 1000 Genomes Project Consortium. A global reference for human genetic variation. Nature. 2015;526(7571):68–74.
50. McCarthy S, et al. A reference panel of 64,976 haplotypes for genotype imputation. Nat Genet. 2016;48:1279–83.
51. Iglesias AI, et al. Haplotype reference consortium panel: practical implications of imputations with large reference panels. Hum Mutat. 2017;38(8):1025–32.
52. Lander ES. The new genomics: global views of biology. Science. 1996;274:536–9.
53. Manolio TA. Genomewide association studies and assessment for the risk of disease. N Engl J Med. 2010;363:166–1676.
54. Helgadottir A, et al. A common variant on chromosome 9p21 affects the risk of myocardial infarction. Science. 2007;316(5830):1491–3.
55. McPherson R, et al. A common allele on chromosome 9 associated with coronary heart disease. Science. 2007;316(5830):1488–91.
56. The Wellcome Trust Case Control Consortium. Genome-wide association study of 14 000 cases of seven common diseases and 3 000 shared controls. Nature. 2007;447:661–78.
57. Hinohara K, et al. Replication of the association between a chromosome 9p21 polymorphism and coronary artery disease in Japanese and Korean populations. J Hum Genet. 2008;53:357–9.
58. Assimes TL, et al. Susceptibility locus for clinical and subclinical coronary artery disease at chromosome 9p21 in the multi-ethnic ADVANCE study. Hum Mol Genet. 2008;17 (15):2320–8.
59. Ding H, et al. 9p21 is a shared susceptibility locus strongly for coronary artery disease and weakly for ischemic stroke in Chinese Han population. Circ Cardiovasc Genet. 2009;2:338–46.
60. Kumar P, Henikoff S, Ng PC. Predicting the effects of coding non-synonymous variants on protein function using the SIFT algorithm. Nat Protoc. 2009;4:1073–81.
61. Pignataro P, et al. Association study between coronary artery disease and rs1333049 polymorphism at 9p21.3 locus in Italian population. J Cardiovasc Transl Res. 2017;10(5–6):455–8.
62. Tattersall I. Human origin: out of Africa. Proc Natl Acad Sci USA. 2009;106(38):16018–21.
63. Roberts R. A genetic basis for coronary artery disease. Trends Cardiovasc Med. 2015;25 (3):171–8.
64. CARDIoGRAMplusC4D Consortium, et al. Large-scale association analysis identifies new risk loci for coronary artery disease. Nat Genet. 2013;45:25–33.

65. the CARDIoGRAMplusC4D Consortium. A comprehensive 1000 Genomes–based genome-wide association meta-analysis of coronary artery disease. Nat Genet. 2015;47:1121.
66. Bycroft C, et al. Genome-wide genetic data on ~500,000 UK Biobank participants. bioRxiv. 2017.
67. Klarin D, et al. Genetic analysis in UK Biobank links insulin resistance and transendothelial migration pathways to coronary artery disease. Nat Genet. 2017;49(9):1392–7.
68. Verweij N, Eppinga RN, Hagemeijer Y, van der Harst P. Identification of 15 novel risk loci for coronary artery disease and genetic risk of recurrent events, atrial fibrillation and heart failure. Sci Rep. 2017;7:2761.
69. van der Harst P, Verweij N. Identification of 64 novel genetic loci provides an expanded view on the genetic architecture of coronary artery disease. Circ Res. 2018;122(3):433–43.
70. Editorial. Accounting for sex in the genome. Nat Med. 2017;23:1243.
71. Miller VM. Family matters: sexual dimorphism in cardiovascular disease. Lancet. 2012;379 (9819):873–5.
72. Charchar FJ, et al. Inheritance of coronary artery disease in men: an analysis of the role of the Y chromosome. Lancet. 2012;379(9819):915–22.
73. Papakonstantinou NA, Stamou MI, Baikoussis NG, Goudevenos J, Apostolakis E. Sex differentiation with regard to coronary artery disease. J Cardiol. 2013;62(1):4–11.
74. Chiha J, Mitchell P, Gopinath B, Plant AJH, Kovoor P, Thiagalingam A. Gender differences in the severity and extent of coronary artery disease. IJC Heart Vasc. 2015;8:161–6.
75. Winham SJ, de Andrade M, Miller VM. Genetics of cardiovascular disease: Importance of sex and ethnicity. Atherosclerosis. 2015;241(1):219–28.
76. König IR, Loley C, Erdmann J, Ziegler A. How to include chromosome X in your genome-wide association study. Genet Epidemiol. 2014;38(2):97–103.
77. Gao F, et al. XWAS: a software toolset for genetic data analysis and association studies of the X chromosome. J Hered. 2015;106(5):666–71.
78. Loley C, et al. No association of coronary artery disease with X-chromosomal variants in comprehensive international meta-analysis. Sci Rep. 2016;6:35278.
79. Assimes TL, et al. Abstract 16167: A GWAS of EHR-Defined CAD Identifies Multiple Novel Loci Including the First 3 Loci on the X-Chromosome: The Million Veteran Program. Circulation. 2017;136(Suppl 1):A16167.
80. Krzywinski MI, et al. Circos: an information aesthetic for comparative genomics. Genome Res. 2009;19(9):1639–45.
81. Chen H-H, Almontashiri NAM, Antoine D, Stewart AFR. Functional genomics of the 9p21.3 locus for atherosclerosis: clarity or confusion? Curr Cardiol Rep. 2014;16(7):502.
82. Patel RS, et al. Genetic variants at chromosome 9p21 and risk of first versus subsequent coronary heart disease events. J Am Coll Cardiol. 2014;63(21):2234–45.
83. Hannou SA, Wouters K, Paumelle R, Staels B. Functional genomics of the CDKN2A/B locus in cardiovascular and metabolic disease: what have we learned from GWASs? Trends Endocrinol Metab. 2015;26(4):176–84.
84. Paquette M, Chong M, Luna Saavedra YG, Pare G, Dufour R, Baass A. The 9p21.3 locus and cardiovascular risk in familial hypercholesterolemia. J Clin Lipidol. 2017;11:406 12.
85. Congrains A, et al. Genetic variants at the 9p21 locus contribute to atherosclerosis through modulation of ANRIL and CDKN2A/B. Atherosclerosis. 2012;220(2):449–55.
86. Holdt LM, Sass K, Gäbel G, Bergert H, Thiery J, Tuepser D. Expression of Chr9p21genes CDKN2B (p15INK4b), CDKN2A (p16INK4a, p14ARF) and MTAP in human atherosclerotic plaque. Atherosclerosis. 2011;214:264–70.
87. Nagano T, et al. Single-cell Hi-C reveals cell-to-cell variability in chromosome structure. Nature. 2013;502(7469):59–64.
88. Harismendy O, et al. 9p21 DNA variants associated with coronary artery disease impair interferon-γ signalling response. Nature. 2011;470(7333):264–8.
89. Holdt LM, et al. Circular non-coding RNA ANRIL modulates ribosomal RNA maturation and atherosclerosis in humans. Nat Commun. 2016;7:12429.

90. Schunkert H, et al. Large-scale association analysis identifies 13 new susceptibility loci for coronary artery disease. Nat Genet. 2011;43(4):333–8.
91. The Coronary Artery Disease (C4D) Genetics Consortium. A genome-wide association study in Europeans and South Asians identifies five new loci for coronary artery disease. Nat Genet. 2011;43(4):339–44.
92. Haley CS. Ten years of the genomics of common diseases: 'The end of the beginning'. Genome Biol. 2016;17(1):254.
93. Nelson CP, et al. Association analyses based on false discovery rate implicate new loci for coronary artery disease. Nat Genet. 2017;49:1385–91.
94. van der Harst P, Verweij N. The identification of 64 novel genetic loci provides an expanded view on the genetic architecture of coronary artery disease. Circ Res. 2018;122(3):433–43.
95. Myocardial Infarction Genetics and CARDIoGRAM Exome Consortia Investigators. Coding variation in ANGPTL4, LPL, and SVEP1 and the risk of coronary disease. N Engl J Med. 2016;374:1134–44.
96. Hixson JE, et al. Whole exome sequencing to identify genetic variants associated with raised atherosclerotic lesions in young persons. Sci Rep. 2017;7(4091):2045–322.
97. Visscher PM, et al. 10 years of GWAS discover: biology, function, and translation. Am J Hum Genet. 2017;101(1):5–22.
98. Manolio TA, et al. Finding the missing heritability of complex diseases. Nature. 2009;461 (7265):747–53.
99. Tregouet D-A, et al. Genome-wide haplotype association study identifies the SLC22A3-LPAL2-LPA gene cluster as a risk locus for coronary artery disease. Nat Genet. 2009;41 (3):283–5.
100. Wray NR, Goddard ME, Visscher PM. Prediction of individual genetic risk to disease from genome-wide association studies. Genome Res. 2007;17(10):1520–8.
101. Prins BP, Lagou V, Asselbergs FW, Snieder H, Fu J. Genetics of coronary artery disease: genome-wide association studies and beyond. Atherosclerosis. 2012;225:1–10.
102. Kovacic JC. Unraveling the complex genetics of coronary artery disease. J Am Coll Cardiol. 2017;69(7):837–40.
103. Webb TR, et al. Systematic evaluation of pleiotropy identifies 6 further loci associated with coronary artery disease. J Am Coll Cardiol. 2017;69(7):823–36.
104. Burgess S, Timpson NJ, Ebrahim S, Davey Smith G. Mendelian randomization: where are we now and where are we going? Int J Epidemiol. 2015;44(2):379–88.
105. Jansen H, Lieb W, Schunkert H. Mendelian randomization for the identification of causal pathways in atherosclerotic vascular disease. Cardiovasc Drugs Ther. 2016;30(1):41–9.
106. Giugliano RP, Sabatine MS. Are PCSK9 inhibitors the next breakthrough in the cardiovascular field? J Am Coll Cardiol. 2015;65(24):2638–51.
107. Goodwin S, McPherson JD, McCombie WR. Coming of age: ten years of next-generation sequencing technologies. Nat Rev Genet. 2016;17(6):333–51.
108. Domenico G, Chiara P, Martinelli N, Corrocher R, Olivieri O. A decade of progress on the genetic basis of coronary artery disease. Practical insights for the internist. Eur J Intern Med. 2017;41:10–7.
109. Emdin CA, et al. Phenotypic characterization of genetically lowered human lipoprotein (a) levels. J Am Coll Cardiol. 2016;68(25):2761–72.
110. Westra H-J, Franke L. From genome to function by studying eQTLs. Biochim Biophys Acta. 2014;1842(10):1896–902.
111. Björkegren JL, Kovacic JC, Dudley JT, Schadt EE. Genome-wide significant loci: How important are they?: Systems genetics to understand heritability of coronary artery disease and other common complex disorders. J Am Coll Cardiol. 2015;65(8):830–45.
112. Jansen RC, Nap J-P. Genetical genomics: the added value from segregation. Trends Genet. 2001;17(7):388–91.

113. Hartiala J, Schwarzman WS, Gabbay J, Ghayalpour A, Bennett BJ, Allayee H. The genetic architecture of coronary artery disease: current knowledge and future opportunities. Curr Atheroscler Rep. 2017;19(2):6.
114. The ENCODE Project Consortium. An integrated encyclopedia of DNA elements in the human genome. Nature. 2012;489(7414):57–74.
115. Lonsdale J, et al. The Genotype-Tissue Expression (GTEx) project. Nat Genet. 2013;45 (6):580–5.
116. Roadmap Epigenomics Consortium, et al. Integrative analysis of 111 reference human epigenomes. Nature. 2015;518(7539):317–30.
117. Hägg S, et al. Multi-organ expression profiling uncovers a gene module in coronary artery disease involving transendothelial migration of leukocytes and LIM domain binding 2: The Stockholm Atherosclerosis Gene Expression (STAGE) Study. PLOS Genet. 2009;5(12): e1000754.
118. Franzén O, et al. Cardiometabolic risk loci share downstream cis- and trans-gene regulation across tissues and diseases. Science. 2016;353(6301):827–30.
119. Tranchevent L-C, et al. Endeavour update: a web resource for gene prioritization in multiple species. Nucleic Acids Res. 2008;36:W377–84.
120. Chen J, Bardes EE, Aronow BJ, Jegga AG. ToppGene Suite for gene list enrichment analysis and candidate gene prioritization. Nucleic Acids Res. 2009;37:W305–11.
121. Pers TH, Dworzynski P, Thomas CE, Lage K, Brunak S. MetaRanker 2.0: a web server for prioritization of genetic variation data. Nucleic Acids Res. 2013;41:W104–8.
122. Pers TH, et al. Biological interpretation of genome-wide association studies using predicted gene functions. Nat Commun. 2014;6:5890.
123. Zhao Y, Chen J, Freudenberg JM, Meng Q, Rajpal DK, Yang X. Network-based identification and prioritization of key regulators of coronary artery disease loci. Arterioscler Thromb Vasc Biol. 2016;36:928–41.
124. Vilne B, et al. Network analysis reveals a causal role of mitochondrial gene activity in atherosclerotic lesion formation. Atherosclerosis. 2017;267:39–48.
125. Wall DP, Kudtarkar P, Fusaro VA, Pivovarov R, Patil P, Tonellato PJ. Cloud computing for comparative genomics. BMC Bioinformatics. 2010;11:259.
126. Pavlovich M. Computing in biotechnology: omics and beyond. Trends Biotechnol. 2017;35 (6):479–80.
127. Musunuru K, et al. From noncoding variant to phenotype via SORT1 at the 1p13 cholesterol locus. Nature. 2010;466:714–9.
128. Talukdar HA, et al. Cross-tissue regulatory gene networks in coronary artery disease. Cell Syst. 2016;2(3):196–208.
129. Santolini M, et al. A personalized, multi-omics approach identifies genes involved in cardiac hypertrophy and heart failure. bioRxiv. 2017.
130. Saleheen D, et al. Human knockouts and phenotypic analysis in a cohort with a high rate of consanguinity. Nature. 2017;544(7649):235–9.
131. Plenge RM, Scolnick EM, Altshuler D. Validating therapeutic targets through human genetics. Nat Rev Drug Discov. 2013;12(8).581–94.
132. Mullard A. Calls grow to tap the gold mine of human genetic knockouts. Nat Rev Drug Discov. 2017;16(8):515–8.
133. Lu X, et al. Genome-wide association study in Han Chinese identifies four new susceptibility loci for coronary artery disease. Nat Genet. 2012;44(8):890–4.
134. Erdmann J, et al. Dysfunctional nitric oxide signalling increases risk of myocardial infarction. Nature. 2013;504(7480):432–6.
135. Kessler T, et al. Functional characterization of the GUCY1A3 coronary artery disease risk locus. Circulation. 2017;136(5):476.
136. Kathiresan S. A PCSK9 missense variant associated with a reduced risk of early-onset myocardial infarction. N Engl J Med. 2008;358(21):2299–300.

137. Inactivating Mutations in NPC1L1 and Protection from Coronary Heart Disease. N Engl J Med. 2014;371(22):2072–82.
138. The TG and HDL Working Group of the Exome Sequencing Project, National Heart, Lung, and Blood Institute. Loss-of-function mutations in APOC3, triglycerides, and coronary disease. N Engl J Med. 2014;371(1):22–31.
139. Dewey FE, et al. Inactivating variants in ANGPTL4 and risk of coronary artery disease. N Engl J Med. 2016;374(12):1123–33.
140. Ajufo E, Rader DJ. New therapeutic approaches for familial hypercholesterolemia. Annu Rev Med. 2018;69(1):113–31.
141. Callaway E. Protective gene offers hope for next blockbuster heart drug. Nat News. 2016.
142. Deaton AM, et al. A rare missense variant in NR1H4 associates with lower cholesterol levels. Commun Biol. 2018;1(1):14.
143. Morita H, Komuro I. A strategy for genomic research on common cardiovascular diseases aiming at the realization of precision medicine. Circ Res. 2016;119:900–9003.
144. Khera AV, Sekar K. Is coronary atherosclerosis one disease or many? Setting realistic expectations for precision medicine. Circulation. 2017;135(11):1005–7.

Complex Genetics and the Etiology of Human Congenital Heart Disease

5

Richard W. Kim and Peter J. Gruber

Contents

Abstract

The genetic architecture of human congenital heart disease (CHD) is as complex as the phenotypes it produces. The objective of this chapter is to review recent findings on the genetic basis and inheritance patterns of CHD. Rather than provide lists of identified genes, instead we offer a conceptual framework to understand the relationship between genetic variation and CHD. We review recent studies utilizing contemporary techniques, some of which may be difficult to interpret for the nonspecialist. This overview aims to educate students and clinicians, providing a background to understand pertinent genetic literature as it relates to human CHD.

R. W. Kim
Children's Hospital of Los Angeles, Keck School of Medicine, University of Southern California, Los Angeles, CA, USA

P. J. Gruber (✉)
Yale University School of Medicine, New Haven, CT, USA
e-mail: peter.gruber@yale.edu

© Springer Nature Switzerland AG 2019   169
J. Erdmann, A. Moretti (eds.), *Genetic Causes of Cardiac Disease*, Cardiac and
Vascular Biology 7, https://doi.org/10.1007/978-3-030-27371-2_5

5.1   Introduction

Despite the fact that congenital heart disease is the most common birth defect, the genetic architecture is incompletely understood. The initial sequencing of a reference human genome was completed in 2004, facilitating a period of unprecedented growth in our understanding of the genetic underpinnings of human disease [1–3]. Recent advances have improved our knowledge of the genetic architecture of congenital heart disease (CHD). Importantly, the genetic loci newly associated with CHD have accelerated mechanistic studies, allowing deep understanding and the development of rational diagnostics and therapies [4–6]. The wealth of new putatively causative genes has had important implications for cardiac development [7, 8]. However, with these new discoveries came the growing realization of the enormous complexity of the human genome, especially as it relates to human congenital heart disease [9–11], emphasizing the importance of incorporating nuanced genetic features beyond coding sequence alterations in analyses [12–14]. Much of what has been reported focuses on alterations in DNA sequence data that can be categorized and understood in terms of the size, character, location, or frequency of sequence variants. Given what we know of the extreme phenotypic (both anatomic and physiologic) variability of CHD, even within narrow CHD anatomic subtypes, it is not surprising that the genetic underpinnings of congenital heart disease are complex and incompletely understood. This overview aims to educate students and clinicians providing a background to understand pertinent genetic literature. An updated list of commonly associated chromosomal aneuploidies, copy number variants, and putative causative genes are presented separately in Tables 5.1, 5.2, and 5.3, but will not be discussed individually in significant depth.

5.2   Common Variants

A single, reference genome exists only in theory as the 6 billion base diploid genome is characterized by tremendous diversity and ongoing polymorphic variation through each generation. Typically, any two unrelated genomes vary at millions of loci (a genetic position) totaling more than 25 million base pairs of DNA (1000 Genomes Project Consortium, 2015) [15, 16]. These genetic differences can be categorized as either small-scale, intermediate-scale, or large-scale structural variants (Tables 5.1, 5.2, and 5.3). Small-scale structural variants are composed of single-nucleotide changes and short insertions or deletions, called "indels." Intermediate-scale sequence variants refer to copy number variants (with gain or loss) that impact hundreds of thousands to millions of base pairs. Large-scale structural variants refer to chromosomal abnormalities that can sometimes be evaluated microscopically. Each of these types of genetic variation will be described in turn below. It is in this genetic variation that lies the key to both individuality as well as disease pathogenesis.

Table 5.1 Syndromes associated with CHD

Syndrome	Defect	Locus	Causal Gene(s)	Most common CHD	% with CHD
Chromosomal abnormalities					
Down	Trisomy	Chr21	Unknown	AVC	40–50%
Turner	Monosomy	ChrX	Unknown	COA; BAV; dilation of ascending aorta; HLH; PAPVD without ASD	20–50%
Patau	Trisomy	Chr13	Unknown	ASD, VSD, PDA, polyvalvular disease	80–100%
Edwards	Trisomy	Chr18	Unknown	ASD, VSD, PDA, polyvalvular disease	80–100%
Chromosomal structural syndromes					
22q11 Deletion	Deletion	22q11.2	*TBX1*	TOF; IAA type B; TA; VSD; aortic arch abnormalities	80–100%
Williams-Beuren	Deletion	7q11.23	*ELN*	SVAS; PAS: multiple arterial stenoses; AV and MV defects	80–100%
Cri-Du-Chat	Deletion	5p15.2	*CTNND2*	VSD, PDA, ASD, TOF	10–55%
Cat Eye	Inversion duplication	22q11	Unknown	TAPVR, TOF	>50%
Jacobsen	Deletion	11q23	Unknown, *JAM-3*	HLH, LVOT defects	>50%
1p36 Deletion	Deletion	1p36	*DVL1*	PDA, noncompaction cardiomyopathy,	43–70%
Single gene mutation syndromes					
Alagille	Single gene	20p12; 1p12	*JAG1; NOTCH2*	Peripheral pulmonary hypoplasia; PS; TOF	>90%

(continued)

Table 5.1 (continued)

Syndrome	Defect	Locus	Causal Gene(s)	Most common CHD	% with CHD
Noonan	Single gene	12q24; 12p1.21; 2p21; 3p25.2; 7q34; 15q22.31; 11p15.5; 1p13.2; 10q25.2; 11q23.3; 17q11.2	*PTPN11; KRAS; SOS1; RAF1; BRAF; MEK1; HRAS; NRAS; SHOC2; CBL; NF1*	PS; ASD; VSD; PDA	80%
Holt-Oram	Single gene	12q24	*TBX5*	ASD; VSD; PDA	85%
Char	Single gene	6p12	*TFAP2B*	PDA	100%
Ellis-van Creveld	Single gene	4p16	*EVC; EVC2*	ASD/single atrium	60%
Costello	Single gene	11p15.5	*HRAS*	PS; other structural heart disease; hypertrophy; rhythm disturbances	63%
Cardiofaciocutaneous	Single gene	12p12.1; 7q34; 15q22.31; 19p13.3	*KRAS; BRAF; MAP2K1; MAP2K2*	PS; ASD; HCM	71%
CHARGE	Single gene	8p12; 7q21.11	*CHD7; SEMA3E*	TOF; ASD; VSD	85%
Duane-radial Ray Syndrome DDRS (Okihiro Syndrome)	Single gene	20q13.2	*SALL4*	VSD, PFO, TOF	…
Kabuki Syndrome	Single gene	12q13.12	*MLL2*	VSD, ASD, TOF, SV, COA, PDA, TGA, RBBB	31–55%

Table 5.2 Copy number variants associated with nonsyndromic CHD

Locus	Size (Kbp)	CNV	No of genes	Genes	Phenotype
1q21.1	418–3981	Gain, loss	3–45	*PRKAB2, FM05, CHD1L, BCL9, ACP6, GJA5, CD160, PDZK1, NBPF11, FMO5, GJA8*	TOF, AS, CoA, PA, VSD
3p25.1	175–12,380	Gain	2	*RAF J, TMEM40*	TOF
3q22.1–3q26.1	680–32,134	Gain, loss	0–300	*FOXL2, NPHP3, FAM62C, CEP70, FAIM, PIK3CB, FOXL2, BPESC1*	DORV, TAPVR, AVSD
4q22.1	45	Gain	1	*PPM1K*	TOF
5q14.1-q14.3	4937–5454	Gain	41,103	*EDIL3, VCAN, SSBP2, TMEM167A*	TOF
5q35.3	264–1777	Gain	19–38	*CNOT6, GFPT2, FLT4, ZNF879, ZNF345C, ADAMTS2, NSD1*	TOF
7q11.23	330–348	Gain	5–8	*FKBP6*	HLHS, Ebstein's
8p23.1	67–12,000	Gain, loss	4	*GATA4, NEIL2, FDFT1, CSTB, SOX7*	AVSD, VSD, TOF, ASD, BAV
9q34.3	190–263	Loss	2–9	*NOTCH1, EHMT1*	TOF, CoA, HLHS
11p15.5	256–271	Gain	13	*HRAS*	DILV, AS
13q14.11	555–1430	Gain	7	*TNFSF11*	TOF, TAPVR, VSD, BAV
15q11.2	238–2285	Loss	4	*TUBGCP5, CYFIP1, NIPA2, NIPA1*	CoA, ASD, VSD, TAPVD, Complex left-sided
16p13.11	1414–2903	Gain	11–14	*MYH11*	Malformations HLHS
18q11.1–18q11.2	308–6118	Gain	1–28	*GATA6*	VSD
19p13.3	52–805	Gain, Loss	1–34	*MIER2, CNN2, FSTL3, PTBP1, WDR18, GNA11, S1PR4*	TOF
Xp22.2	509–615	Gain	2–4	*MID1*	TOF, AVSD

Nearly all genetic discovery in the past decade is based upon the simple concept of identifying the genetic differences between patients and controls. One searches for either sequence or structural variation, identifying candidate genes to examine or comparing to the entire genome. Classically, two types of approaches are described: "forward" and "reverse." The "forward genetic" approach begins with the identification of a phenotype, followed by various molecular and genetic techniques to map

Table 5.3 Single genes associated with CHD

Gene	Protein	Phenotypes
ANKRD1	Ankyrin repeat domain	TAPVR
CITED2	c-AMP responsive element-binding protein	ASD; VSD
FOG2/ ZFPM2	Friend of GATA	TOF, DORV
GATA4	GATA4 transcription factor	ASD, PS, VSD, TOF, AVSD, PAPVR
GATA6	GATA6 transcription factor	ASD, TOF, PS, AVSD, PDA, OFT defects, VSD
HAND2	Helix-loop-helix transcription factor	TOF
IRX4	Iroquois homeobox 4	VSD
MED13L	Mediator complex subunit 13-like	TGA
NKX2-5/ NKX2.5	Homeobox-containing transcription factor	ASD, VSD, TOF, HLH, CoA, TGA, DORV, IAA, OFT defects
NKX2-6	Homeobox-containing transcription factor	PTA
TBX1	T-Box 1 transcription factor	TOF, (22q11 deletion syndromes)
TBX5	T-Box 5 transcription factor	AVSD, ASD, VSD, (Holt Oram syndrome)
TBX20	T-Box 20 transcription factor	ASD, MS, VSD
TFAP2B	Transcription factor AP-2 beta	PDA, (Char syndrome)
ZIC3	Zinc finger transcription factor	TGA, PS, DORV, TAPVR, ASD, HLH, VSD, Dextrocardia, L-R axis defects
ACVR1/ ALK2	BMP receptor	AVSD
ACVR2B	Activin receptor	PS, DORV, TGA, Dextrocardia,
ALDH1A2	Retinaldehyde dehydrogenase	TOF
CFC1/ CRYPTIC	Cryptic protein	TOF; TGA; AVSD; ASD; VSD; IAA; DORV
CRELD1	Epidermal growth factor-related proteins	ASD; AVSD
FOXH1	Forkhead activin signal transducer	TOF, TGA
GDF1	Growth differentiation factor-1	Heterotaxy, TOF, TGA, DORV
GJA1	Connexin 43	ASD, HLH, TAPVR, (Oculodentodigital dysplasia)
JAG1	Jagged-1 ligand	PAS, TOF, (Alagille syndrome)
LEFTY2	Left-right determination factor	TGA, AVSD, IAA, CoA, L-R axis defects, IVC defects
NODAL	Nodal homolog (TGF-beta superfamily)	TGA, PA, TOF, DORV, Dextrocardia, IVC defect, TAPVR, AVSD
NOTCH1	NOTCH1 (ligand of JAG1)	BAV, AS, CoA, HLH

(continued)

Table 5.3 (continued)

Gene	Protein	Phenotypes
PDGFRA	Platelet-derived growth factor receptor alpha	TAPVR
SMAD6	MAD-related protein	BAV, CoA, AS
TAB2	TGF-beta activated kinase	OFT defects
TDGF1	Teratocarcinoma-derived growth factor 1	TOF, VSD
VEGF	Vascular endothelial growth factor	CoA, OFT defects
ACTC	Alpha-cardiac actin	ASD
ELN	Elastin	SVAS, PAS, PS, AS, (Williams-Beuren syndrome)
MYH11	Myosin heavy chain 11	PDA, aortic neurysm
MYH6	Alpha-myosin heavy chain	ASD, TA, AS, PFO, TGA
MYH7	Beta-myosin heavy chain	Ebstein anomaly, ASD, NVM

associated genetic loci. "Reverse genetics" takes the opposite approach, in which a gene of interest is mutated, and the associated phenotype is interrogated. Both techniques provide insight into causality—both have limitations. Most prominently, forward genetic approaches rely on statistical associations that fail to provide mechanistic insights. Reverse genetic approaches provide a more robust association of gene function and phenotype, but until recently, were not experiments that could be performed in humans, and therefore lacked the complexity of other approaches. In general, if one wants to understand humans, it is necessary to study humans.

Prior to the sequencing of the human genome and the subsequent development of the International HapMap project, very little was known about the underlying contribution of genetic variation to CHD [17]. Although associations between large chromosomal aneuploidies such as Trisomy 21 and CHD were well described (Table 5.1), the pedigrees of multigenerational families necessary to determine genetic linkage were simply not available for CHD. As explained in more depth below, except for relatively minor phenotypes such as atrial septal defects and familial patent ductus arteriosus, cardiac-related complications invariably led to death during childhood, preventing the accumulation of affected individuals in families [18, 19]. When data from the HapMap project became available in 2005, for the first time, scientists had the tools to begin to understand the underlying genetic architecture of CHD.

The HapMap was the first attempt to categorize the genetic diversity of man using the millions of single-nucleotide variants found throughout the human genome [17]. At the time, sequencing technology was limited to microarrays, so maps of genetic variants could only be based on sequence variation found commonly throughout the human population—resolution was limited. These single-nucleotide polymorphisms (SNPs) were usually found at a population frequency of at least 5%.

Although 5% may seem infrequent, in genetic language, it was the operational definition of a common variant at the time; now common variants are more often defined as >1%. Genome-wide association studies (GWAS) identified the association of common variation to complex traits or diseases and were the first significant validation of the human genome project [20]. GWAS specifically refer to studies involving common variants in contrast to the study of rare variants which at the time were technologically limited. SNPs and common single-nucleotide variants are one in the same, whereas rare single-nucleotide variants that occur at very low frequencies, usually well below 1%, are considered mutations rather than polymorphisms or SNPs. In both candidate gene association studies and in GWAS, the association of common variants and CHD is not strongly associated with CHD [21–24].

5.3   Common Variants as Genetic Modifiers

The 1000 Genomes Project is a further iteration of the HapMap, and confirmed that >95% of variation within an individual genome is common. This does not mean individual genomes are overwhelmingly similar, but rather that most human genetic variants are repeated. Population frequency has important implications for the effects of the variant on human health. Unlike rare and highly detrimental genetic variants that likely reduce reproductive fitness, benign polymorphisms survive through generations and can accumulate within a population, becoming "common." If a gene mutation leads to a serious congenital abnormality, as are many CHD phenotypes, the affected individual is unlikely to live long enough to reproduce and pass the mutation on to his or her offspring. Only genetic variants that permit an individual to live long enough to reproduce or are otherwise advantageous are likely to accumulate within human populations. Thus, although they contribute >95% of the variability between individual genomes, the majority of common polymorphisms are unlikely to directly lead to or contribute strongly to CHD.

Though common variants are unlikely to be a major cause of CHD phenotypes, background common variation is still relevant to CHD. Both human and animal studies suggest that common variants are important modifiers of CHD phenotype expression. They are critical contributors to the variable expressivity and incomplete penetrance that are hallmarks of CHD [5, 25]. Variable expression occurs when identical genetic variants are associated with different disease phenotypes. Incomplete penetrance is a form of the carrier state, or the extreme version of expressivity in which there is no overt phenotype despite the presence of the causative mutation [26, 27]. One example of variable expression are inactivating mutations of the highly conserved cardiac transcription factor *NKX2-5*. Identical mutations can result in highly pleomorphic phenotypes including a spectrum of left ventricular outflow tract obstruction (interrupted aortic arch vs. hypoplastic aortic arch), while at other times carriers display structurally normal hearts. Other examples exist where a variety of disparate heart defects including complete heart block, atrial septal defects (ASD), ventricular septal defects (VSD), and tetralogy of Fallot (TOF) are all associated with mutations in *NKX2-5* [28, 29]. Polymorphic alleles within individual genetic backgrounds likely modified the phenotypic expression of the causative mutant. Experimental work in transgenic mouse models further corroborate this observation as *NKX2-5* mutations may harbor distinct anatomic phenotypes when bred into different inbred mouse backgrounds [25, 30].

Although genetic variants that cause CHD are necessarily held to low frequency, the same does not hold true for noncausative alleles against which there is no strong selection. One example is the family of cytochrome P450 enzymes important in drug metabolism. Unlike the modest effects that common polymorphisms often have on complex trait susceptibility, polymorphic alleles of the cytochrome P450 2D6 enzyme can result in large differences in the ability to metabolize medications [31, 32]. More than 25% of all known clinically utilized drugs are affected, including up to a five-fold change in dosing requirements for common cardiac medications such as metoprolol and procainamide [33, 34]. Another study utilizing this common variant approach focused on polymorphic alleles of the Alzheimer's disease-associated gene Apolipoprotein E (*ApoE*). An association of the *ApoE* ε2 allele was correlated with a reduction in neurodevelopmental testing scores following complex congenital cardiac surgery [35, 36]. Although the effect was modest, it served as an important proof of principle experiment that common polymorphism may impact disease outcome by modifying biologic systems important in treatment strategies for CHD. Other work has demonstrated that genotypes common to CHD populations impacting the renin angiotensin aldosterone system are associated with adverse ventricular remodeling. Following second-stage single-ventricle palliation, infants homozygous for two SNPs, rs833069 in *VEGFA* and rs2758331 in *SOD2* allele, have a combined 16-fold increased risk of death or heart transplant after cardiac surgery [37, 38]. As clinical outcomes for repair of CHD have improved dramatically over the past 20 years, these comorbidities have become as an important a focus as the pathogenesis of the disease itself.

5.4   Rare de novo Variants and Whole Exome Sequencing

By 2010, capacity for whole-exome sequencing technology matured sufficiently to permit cost-effective, genome-wide discovery of rare sequence variants [39]. Unlike prior candidate gene approaches in which potential gene targets need to be identified a priori, all ~20,000 expressed genes that comprise the human exome could be evaluated for sequence variability and tested agnostically for disease association. Prior whole genome approaches such as GWAS did not directly evaluate gene sequence but rather identified variants through polymorphic markers scattered throughout genome, using linkage for association. Although the exome only represents 1% of the human genome, it is critically important. By 2018, sequencing methods for examining the genome outside of the exome, or whole genome sequencing (WGS), have become commonplace [40, 41]. Although there are many advantages, including the identification of important gene regulatory sites, WGS approach has yet to be used extensively for CHD [42].

To account for the benign and noncausative variability between individual genomes, investigators in the NIH/NHLBI-supported Pediatric Cardiac Genomics Consortium (PCGC) sequenced proband trio exomes that consists the affected proband and his or her unaffected parents. De novo mutations found in the affected child but absent from the parental genomes are candidates for disease causality. After further restricting candidate genes to those that were both predicted to be damaging and highly expressed in the developing heart, the PCGC demonstrated a 7.5-fold excess of damaging, rare de novo mutations in CHD cohorts. They further estimated that these rare sequence mutations account for ~10% of CHD, predicting ~400 likely pathogenic CHD genes [8]. Among the newly identified mutations, chromatin remodeling genes related to histone H3K4 were overrepresented. Chromatin remodeling genes regulate the transcription of many developmental genes and are hypothesized to serve as intermediaries between environmental stimuli and gene expression. These candidates were similar to genes found in previous genome-wide studies, previous candidate gene studies, and early animal knockout studies and highlight two general principles of the genetic architecture of CHD [43, 44].

First, CHD mutations commonly impact transcriptional regulatory proteins and cell-signaling pathways [26, 45, 46]. This finding is in contrast to CHD-related conditions such as inherited cardiomyopathies, inherited arrhythmias, and thoracic vascular disease for which structural, rather than regulatory protein, mutations appear to predominate. As signaling cascades are not only dependent upon multiple component proteins but are themselves intertwined with multiple other development systems, it becomes readily understandable why CHD is genetically heterogeneous: disparate gene mutations result in phenotypically similar CHD. The biologic redundancy of these cell-signaling pathways and the shared use of common transcription factors throughout the development of multiple organ systems, such as the heart, the brain, and the gastrointestinal tract, may also help explain the frequent occurrence of extracardiac features in CHD patients [47–49].

Second, mutations in CHD genes usually result in altered gene dosage or haploinsufficiency of the affected gene. Unlike oncologic conditions where mechanisms of uncontrolled growth or gene expression predominate, in CHD, mutations that result in increased gene expression are rare. Instead, CHD mutations are more commonly associated with loss of gene function, analogous to the heterozygous knockdown of a single disease allele in mutant mouse models [30]. Not yet extensively evaluated is the contribution of mosaicism and somatic mutations to the architecture of CHD [50, 51].

The architectural of the genetic variants appear independent of the affected gene, predicting that de novo CHD may occur as a random event. Supporting this hypothesis is the observation that essentially equal numbers of de novo mutations exists between CHD patients and controls, with both groups accumulating slightly more than one exonic mutation per patient per generation. This suggests CHD genomes are not inherently more unstable or prone to mutation than the normal population. There is no evidence that pathogenic genes have an intrinsically higher mutation rate. Genes associated with CHD are not hypermutatable, prone to DNA injury, or found in areas prone to mutation. Last, despite epidemiological evidence suggesting a higher incidence of CHD in Asia, there is no robust evidence of racial or anatomic clustering of identified variants [52]. Taken together, it appears that 1 de novo mutation occurs randomly in the exome every generation and CHD occurs when that mutation occurs in a gene or gene pathway critical to heart development by chance alone.

The contribution of de novo mutations to CHD is significant, yet these mutations represent only a fraction of the observed cases of CHD. Enabled by the technological advances described above, only recently have rare, inherited variants been amenable to study. One recent study used a candidate gene sequencing approach in 610 syndromic and 1281 nonsyndromic CHD patients to determine if inherited rare damaging mutations with presumed incomplete penetrance could account for additional heritability. Despite finding a significantly increased burden of inherited rare damaging variants in nonsyndromic patients, these variants only accounted for disease in just over 1% of nonsyndromic CHD patients [53]. As an extension of their earlier studies, the PCGC recently completed a more comprehensive analysis, examining the effect of recessive inherited variants. In this study of 2871 CHD probands, including 2645 parent-offspring trios, whole exome sequencing identified rare, inherited mutations in 1.8% of patients. As in their prior study, the PCGC found de novo mutations in 8% of cases, including 28% with both neurodevelopmental and extracardiac congenital anomalies. The significant overlap between genes with damaging de novo mutations in probands with CHD and autism suggests shared developmental mechanisms [7].

Syndromes are commonly described as diseases with more than one identifying feature or symptom. For several CHD phenotypes, there are situations where the CHD phenotype may be associated either with a syndrome (e.g., Down syndrome) or simply seen in isolation: this is the difference between what is referred to as syndromic or nonsyndromic CHD. One example is atrioventricular septal defect, that can be seen either in isolation, or in children with Trisomy 21. While several syndromic forms of CHDs have been characterized and causative genes identified, multiple studies screening nonsyndromic CHD patients for mutations in these genes have been unrewarding. It is likely the approach of screening coding regions of syndromic CHD genes for causative mutations in isolated CHDs is fundamentally flawed. One would predict that mutations in the coding region would be present in all cells of the organism and expressed in all tissues where the gene is expressed. More likely, in nonsyndromic patients, mutations would be found in regulatory elements that control expression of these genes in the developing heart. Alterations in regulatory elements of genes involved in syndromic forms of congenital heart defects will likely result in isolated congenital heart defects through cardiac-specific, disruption of expression. For most genes, the region immediately upstream of the minimal promoter contains the most important transcription factor-binding sites. However, for many genes, multiple cis-acting distal elements (enhancers, repressors, and insulators) are required for correct spatiotemporal expression [54, 55]. These regulatory elements may be located upstream, downstream, or within introns and can reside greater than 1 Mb from the target gene [56]. Disruption of these long-range regulatory interactions can result in human disease phenotypes either through global or partial tissue-specific loss or gain of expression. Distal regulatory elements have been shown to be key tissue-specific regulators of cardiac development in animal models, and are often evolutionarily conserved in humans [57, 58].

5.5   Copy Number Variants

De novo, single-nucleotide variants are an important form of sequence variation contributing to human disease. Larger, structural variations resulting from aberrant meiotic recombination are called copy number variants (CNVs). Although commonly defined as exceeding 1000 base pairs in size, most CNVs are substantially larger, with some containing millions of bases coding over 150 genes. CNVs are the dominant form of intermediate structural variation known to significantly impact human disease. Multiple studies demonstrate the association of damaging CNVs with ~10% of CHD cases [59–61]. Notably, CNVs comprise well-recognized deletion syndromes including DiGeorge and Williams syndromes and 7q11.23 deletion syndrome. When 22q11.2 deletion patients are included, up to ~15% of CHD patients are associated with pathogenic CNVs. As with individual genes with SNVs, there are significant differences between the pathogenicity of intermediate structural variants that impact disease inheritance and phenotypic expression. This observation suggests hypotheses regarding the nature CHD independent of specific anatomic classification.

In CHD patients with isolated (without the presence of extracardiac defects) disease, the most frequent detectable genetic abnormality is a copy number variant. Approximately 8% of isolated CHD patients have an identifiable CNV, of which the majority is likely to be de novo; half may have recessive inheritance. Although inherited (recessive) SNVs are mildly enriched in isolated CHD patients, damaging de novo SNVs demonstrate no significant enrichment in nonsyndromic proband cohorts [62]. This data suggests that despite increasingly aggressive clinical genetic screening for CHD, sequencing for isolated CHD is likely to be low yield [63–65].

In CHD patients with extracardiac anomalies and/or neurodevelopmental delay, ~20–30% of syndromic CHD patients harbor a damaging de novo SNV. Inherited (recessive) SNVs are not enriched in syndromic CHD patient cohorts. An additional ~2–3% of CHD patients may harbor a damaging CNV of which half may be inherited. Del 22q11.2 may comprise an additional ~5% of patients of which the majority are overwhelming likely to be de novo. Syndromic CHD patients may benefit from chromosomal microarray testing for the detection of 22q11, and appear to be the target population that would most benefit from clinical whole exome sequencing. Due to the high proportion of de novo mutations, sibling recurrence is likely to be low, although the risk of an affected offspring may be considerable due to the high transmission rate of de novo mutations [66, 67].

The clinical implications of this genetic landscape are now becoming evident. The presence of a pathologic CNV is often associated with serious adverse clinical consequences, including reduced somatic growth, neurodevelopmental delay, and a

2.5-fold increase in death following cardiac surgery [37, 68]. This unanticipated result implies that CNV burden results in shared adverse effects independent of the causative gene sets within the affected segment. Alternatively, the mechanisms that underlie these disparate clinical outcomes may overlap in ways not currently understood. Pathologic CNVs specific to CHD have been found throughout the genome in association with a wide variety of anatomically diverse CHD. For example, it is not intuitive that a CNV resulting in common arterial trunk and a CNV located elsewhere in the genome associated with heterotaxy would share a common adverse impact on survival [42, 69].

5.6   Conclusions

Over the last decade, there has been exponential growth in our ability to decipher the genetic underpinnings of congenital heart disease. Enabled by significant advances in sequencing technology, and the dramatic reduction in sequencing costs, the genetic landscape that underlies CHD is now becoming more clear. However, even with existing technologies, fully 60–70% of CHD cannot be currently explained with a genetic focus. Understanding noncoding variants with combinatorial and recessive gene effects may further elucidate the genetic and epigenetic landscape CHD, with the remainder due to nongenetic etiologies.

In summary, well-described chromosomal aneuploidies such as Trisomy 21 comprise ~10% of CHD. Other genetic forms of CHD result from different classes of genomic alterations and involve a common set of ~400 causative genes that play roles in transcriptional regulation or cell-signaling pathways. The majority of known mutations are rare de novo SNVs (~10%) and CNVs (~10%) that lead to haploinsufficiency or altered gene dosage. An additional small percentage of SNVs are inherited from incompletely penetrant parents, as well as a higher percentage of CNVs. Common sequence polymorphisms or SNPs do not contribute strongly to CHD risk, but likely modify the penetrance of causative rare mutations playing a significant accessory role in CHD pathogenesis and clinical outcomes. Finally, patients with isolated CHD are less likely to harbor a de novo SNV than patients with associated neurodevelopmental delay and extracardiac anomalies than syndromic patients.

References

1. International Human Genome Sequencing, C. Finishing the euchromatic sequence of the human genome. Nature. 2004;431(7011):931–45.
2. Lander ES, et al. Initial sequencing and analysis of the human genome. Nature. 2001;409 (6822):860–921.
3. Venter JC, et al. The sequence of the human genome. Science. 2001;291(5507):1304–51.
4. An Y, et al. Genome-wide copy number variant analysis for congenital ventricular septal defects in Chinese Han population. BMC Med Genet. 2016;9:2.

5. Seidman JG, Seidman C. Transcription factor haploinsufficiency: when half a loaf is not enough. J Clin Invest. 2002;109(4):451–5.
6. Yuan S, Zaidi S, Brueckner M. Congenital heart disease: emerging themes linking genetics and development. Curr Opin Genet Dev. 2013;23(3):352–9.
7. Jin SC, et al. Contribution of rare inherited and de novo variants in 2,871 congenital heart disease probands. Nat Genet. 2017;49(11):1593–601.
8. Zaidi S, et al. De novo mutations in histone-modifying genes in congenital heart disease. Nature. 2013;498(7453):220–3.
9. Bentham J, Bhattacharya S. Genetic mechanisms controlling cardiovascular development. Ann N Y Acad Sci. 2008;1123:10–9.
10. Fahed AC, et al. Genetics of congenital heart disease: the glass half empty. Circ Res. 2013;112 (4):707–20.
11. Wild PS, et al. Large-scale genome-wide analysis identifies genetic variants associated with cardiac structure and function. J Clin Invest. 2017;127(5):1798–812.
12. Baccarelli A, Rienstra M, Benjamin EJ. Cardiovascular epigenetics: basic concepts and results from animal and human studies. Circ Cardiovasc Genet. 2010;3(6):567–73.
13. Chowdhury S, et al. Maternal genome-wide DNA methylation patterns and congenital heart defects. PLoS One. 2011;6(1):e16506.
14. Vallaster M, Vallaster CD, Wu SM. Epigenetic mechanisms in cardiac development and disease. Acta Biochim Biophys Sin Shanghai. 2012;44(1):92–102.
15. Genomes Project C, et al. A global reference for human genetic variation. Nature. 2015;526 (7571):68–74.
16. Sudmant PH, et al. An integrated map of structural variation in 2,504 human genomes. Nature. 2015;526(7571):75–81.
17. International HapMap, C. A haplotype map of the human genome. Nature. 2005;437 (7063):1299–320.
18. Caputo S, et al. Familial recurrence of congenital heart disease in patients with ostium secundum atrial septal defect. Eur Heart J. 2005;26(20):2179–84.
19. Satoda M, et al. Mutations in TFAP2B cause Char syndrome, a familial form of patent ductus arteriosus. Nat Genet. 2000;25(1):42–6.
20. Spencer CC, et al. Designing genome-wide association studies: sample size, power, imputation, and the choice of genotyping chip. PLoS Genet. 2009;5(5):e1000477.
21. Cordell HJ, et al. Genome-wide association study identifies loci on 12q24 and 13q32 associated with tetralogy of Fallot. Hum Mol Genet. 2013;22(7):1473–81.
22. Goodship JA, et al. A common variant in the PTPN11 gene contributes to the risk of tetralogy of Fallot. Circ Cardiovasc Genet. 2012;5(3):287–92.
23. Hu Z, et al. A genome-wide association study identifies two risk loci for congenital heart malformations in Han Chinese populations. Nat Genet. 2013;45(7):818–21.
24. Stevens KN, et al. Common variation in ISL1 confers genetic susceptibility for human congenital heart disease. PLoS One. 2010;5(5):e10855.
25. Winston JB, et al. Heterogeneity of genetic modifiers ensures normal cardiac development. Circulation. 2010;121(11):1313–21.
26. Prendiville T, Jay PY, Pu WT. Insights into the genetic structure of congenital heart disease from human and murine studies on monogenic disorders. Cold Spring Harb Perspect Med. 2014;4(10):a013946.
27. Rogers MS, D'Amato RJ. The effect of genetic diversity on angiogenesis. Exp Cell Res. 2006;312(5):561–74.
28. Abou Hassan OK, et al. NKX2-5 mutations in an inbred consanguineous population: genetic and phenotypic diversity. Sci Rep. 2015;5:8848.
29. McElhinney DB, et al. NKX2.5 mutations in patients with congenital heart disease. J Am Coll Cardiol. 2003;42(9):1650–5.
30. Bruneau BG, et al. A murine model of Holt-Oram syndrome defines roles of the T-box transcription factor Tbx5 in cardiogenesis and disease. Cell. 2001;106(6):709–21.

31. Gao J, et al. From genotype to phenotype: cytochrome P450 2D6-mediated drug clearance in humans. Mol Pharm. 2017;14(3):649–57.
32. Ur Rasheed MS, Mishra AK, Singh MP. Cytochrome P450 2D6 and Parkinson's disease: polymorphism, metabolic role, risk and protection. Neurochem Res. 2017;42(12):3353–61.
33. Lessard E, et al. Role of CYP2D6 in the N-hydroxylation of procainamide. Pharmacogenetics. 1997;7(5):381–90.
34. Mottet F, Vardeny O, de Denus S. Pharmacogenomics of heart failure: a systematic review. Pharmacogenomics. 2016;17(16):1817–58.
35. Gaynor JW, et al. Validation of association of the apolipoprotein E epsilon2 allele with neurodevelopmental dysfunction after cardiac surgery in neonates and infants. J Thorac Cardiovasc Surg. 2014;148(6):2560–6.
36. Gaynor JW, et al. Apolipoprotein E genotype modifies the risk of behavior problems after infant cardiac surgery. Pediatrics. 2009;124(1):241–50.
37. Kim DS, et al. Patient genotypes impact survival after surgery for isolated congenital heart disease. Ann Thorac Surg. 2014;98(1):104–10; discussion 110–1.
38. Mital S, et al. Renin-angiotensin-aldosterone genotype influences ventricular remodeling in infants with single ventricle. Circulation. 2011;123(21):2353–62.
39. Teer JK, Mullikin JC. Exome sequencing: the sweet spot before whole genomes. Hum Mol Genet. 2010;19(R2):R145–51.
40. Cirulli ET, Goldstein DB. Uncovering the roles of rare variants in common disease through whole-genome sequencing. Nat Rev Genet. 2010;11(6):415–25.
41. Veeramah KR, Hammer MF. The impact of whole-genome sequencing on the reconstruction of human population history. Nat Rev Genet. 2014;15(3):149–62.
42. Chung JH, et al. Whole-genome sequencing and integrative genomic analysis approach on two 22q11.2 deletion syndrome family trios for genotype to phenotype correlations. Hum Mutat. 2015;36(8):797–807.
43. Krupp DR, et al. Exonic mosaic mutations contribute risk for autism spectrum disorder. Am J Hum Genet. 2017;101(3):369–90.
44. Menezes J, et al. Exome sequencing reveals novel and recurrent mutations with clinical impact in blastic plasmacytoid dendritic cell neoplasm. Leukemia. 2014;28(4):823–9.
45. Andersen TA, Troelsen Kde L, Larsen LA. Of mice and men: molecular genetics of congenital heart disease. Cell Mol Life Sci. 2014;71(8):1327–52.
46. Stallmeyer B, et al. Mutational spectrum in the cardiac transcription factor gene NKX2.5 (CSX) associated with congenital heart disease. Clin Genet. 2010;78(6):533–40.
47. Dewey FE, et al. Gene coexpression network topology of cardiac development, hypertrophy, and failure. Circ Cardiovasc Genet. 2011;4(1):26–35.
48. Lage K, et al. Genetic and environmental risk factors in congenital heart disease functionally converge in protein networks driving heart development. Proc Natl Acad Sci USA. 2012;109 (35):14035–40.
49. Sperling SR. Systems biology approaches to heart development and congenital heart disease. Cardiovasc Res. 2011;91(2):269–78.
50. Esposito G, et al. Somatic mutations in NKX2-5, GATA4, and HAND1 are not a common cause of tetralogy of Fallot or hypoplastic left heart. Am J Med Genet A. 2011;155A(10):2416–21.
51. Zheng J, et al. Investigation of somatic NKX2-5 mutations in Chinese children with congenital heart disease. Int J Med Sci. 2015;12(7):538–43.
52. van der Linde D, et al. Birth prevalence of congenital heart disease worldwide: a systematic review and meta-analysis. J Am Coll Cardiol. 2011;58(21):2241–7.
53. Sifrim A, et al. Distinct genetic architectures for syndromic and nonsyndromic congenital heart defects identified by exome sequencing. Nat Genet. 2016;48(9):1060–5.
54. Kleinjan DA, van Heyningen V. Long-range control of gene expression: emerging mechanisms and disruption in disease. Am J Hum Genet. 2005;76(1):8–32.
55. West AG, Fraser P. Remote control of gene transcription. Hum Mol Genet. 2005;14(1): R101–11.

56. Velagaleti GV, et al. Position effects due to chromosome breakpoints that map approximately 900 Kb upstream and approximately 1.3 Mb downstream of SOX9 in two patients with campomelic dysplasia. Am J Hum Genet. 2005;76(4):652–62.
57. Saitsu H, Shiota K, Ishibashi M. Analysis of Fibroblast growth factor 15 cis-elements reveals two conserved enhancers which are closely related to cardiac outflow tract development. Mech Dev. 2006;123(9):665–73.
58. Strahle U, Rastegar S. Conserved non-coding sequences and transcriptional regulation. Brain Res Bull. 2008;75(2-4):225–30.
59. Carey AS, et al. Effect of copy number variants on outcomes for infants with single ventricle heart defects. Circ Cardiovasc Genet. 2013;6(5):444–51.
60. Glessner JT, et al. Increased frequency of de novo copy number variants in congenital heart disease by integrative analysis of single nucleotide polymorphism array and exome sequence data. Circ Res. 2014;115(10):884–96.
61. Warburton D, et al. The contribution of de novo and rare inherited copy number changes to congenital heart disease in an unselected sample of children with conotruncal defects or hypoplastic left heart disease. Hum Genet. 2014;133(1):11–27.
62. Gelb BD, Chung WK. Complex genetics and the etiology of human congenital heart disease. Cold Spring Harb Perspect Med. 2014;4(7):a013953.
63. Cowan JR, Ware SM. Genetics and genetic testing in congenital heart disease. Clin Perinatol. 2015;42(2):373–93, ix
64. Geng J, et al. Chromosome microarray testing for patients with congenital heart defects reveals novel disease causing loci and high diagnostic yield. BMC Genomics. 2014;15:1127.
65. Miller DT, et al. Consensus statement: chromosomal microarray is a first-tier clinical diagnostic test for individuals with developmental disabilities or congenital anomalies. Am J Hum Genet. 2010;86(5):749–64.
66. Arndt AK, MacRae CA. Genetic testing in cardiovascular diseases. Curr Opin Cardiol. 2014;29 (3):235–40.
67. Landis BJ, Ware SM. The current landscape of genetic testing in cardiovascular malformations: opportunities and challenges. Front Cardiovasc Med. 2016;3:22.
68. Aiyagari R, et al. Impact of pre-stage II hemodynamics and pulmonary artery anatomy on 12-month outcomes in the Pediatric Heart Network Single Ventricle Reconstruction trial. J Thorac Cardiovasc Surg. 2014;148(4):1467–74.
69. Tomita-Mitchell A, et al. Human gene copy number spectra analysis in congenital heart malformations. Physiol Genomics. 2012;44(9):518–41.

Familial Hypercholesterolemia

6

Ashish Sarraju and Joshua W. Knowles

Contents

6.1 Introduction

Familial Hypercholesterolemia (FH) is one of the most common inherited lipid disorders, with recent studies estimating a prevalence as high as 1 in 200 people [1–3]. Inherited in an autosomal-dominant fashion, it is associated with lifelong, severe elevations in low-density lipoprotein-cholesterol (LDL-c) levels. Individuals

A. Sarraju · J. W. Knowles (✉)
Division of Cardiovascular Medicine and Cardiovascular Institute, Falk Cardiovascular Research Center, Stanford University, Stanford, CA, USA
e-mail: knowlej@stanford.edu

© Springer Nature Switzerland AG 2019
J. Erdmann, A. Moretti (eds.), *Genetic Causes of Cardiac Disease*, Cardiac and Vascular Biology 7, https://doi.org/10.1007/978-3-030-27371-2_6

with FH have a markedly elevated risk of premature ischemic heart disease, 5–20-fold higher than the general population [4–6]. Mutations in the genes for the LDL receptor (*LDLR*), apolipoprotein B-100 (*APOB*), as well as gain of function mutations in the proprotein convertase subtulisin/kexin type 9 protein (*PCSK9*) have all been associated with the pathogenesis of FH [6]. Appropriate management can dramatically improve the life expectancy of those with FH [6]; however, many patients experience delayed diagnosis and inadequate cholesterol lowering [1], underscoring the need for increased awareness, recognition, and timely treatment of this disorder in the general community.

6.2   Inheritance and Family Screening

Familial hypercholesterolemia is generally considered to be autosomal dominant, and thus, those with one pathologic gene variant can have the FH phenotype. Patients can present as either heterozygous FH (HeFH) or homozygous FH (HoFH) [7]. HoFH is a rare (1 in 250,000–one million individuals) but particularly severe form of FH where LDL-c levels usually exceed 400 milligrams per deciliter, mg/dl (in contrast, the average LDL-c in HeFH adults is approximately 200 mg/dl).

For those diagnosed with HeFH, each child (as well as each full sibling) has a 50% chance of inheriting the pathogenic FH variant, and thus, having FH [7]. Family-based screening will usually identify a single affected parent as de novo mutations are rare.

For those with HoFH, each of their parents carry at least one pathogenic variant and therefore will have HeFH. All children of those with HoFH will also have FH, since they will inherit a pathogenic variant from the index patient. Given that parents of HoFH patients are obligate heterozygotes, siblings will have a 50% chance of HeFH and 25% chance of HoFH [7].

Of note, individuals that inherit a large number of common, LDL-c raising alleles (i.e., those with a high genetic risk score for LDL-c variants identified in genome-wide association studies) can present with a phenotype largely indistinguishable from Mendelian forms of FH [8]. In fact, up to 40% of individuals with a phenotypic diagnosis of FH may have this multigenic form of the condition [7]. This has some implications for the use of genetic testing and family screening. However, the attendant CVD risk in such individuals remains high (2–5-fold that of the general population), and so treatment of these individuals is essentially the same as for those with Mendelian forms of FH.

Another clinical mimic of FH is very elevated lipoprotein (a) levels, Lp (a) [9, 10]. Lp(a) particles also carry cholesterol esters, and so individuals with very elevated Lp(a) particle number can also have very elevated LDL-c plasma concentrations. Work from Denmark indicates that in up to 25% of individuals with phenotypic FH, it may be due to Lp(a) [9]. This distinction is important for a number of reasons including family screening. In addition, treatment strategies may also differ as while Lp(a) is a causal risk factor for CVD, there are not yet specific Lp(a)-lowering therapies.

An autosomal-recessive form of severe hypercholesterolemia also deserves mention [11]. Individuals with biallelic mutations in *LDLRAP1* can present with a phenotype that is clinically indistinguishable from FH caused by LDLR mutations including presence of xanthomas, severe LDL-c elevations, and early-onset coronary disease. However, in this case, heterozygote carriers are unaffected.

6.3   Cascade Screening (Family-Based Screening)

After identification of a proband, initiation of cascade screening to assess family members is critical [6]. First-degree family members should be approached promptly with the assistance of the proband and/or the clinic team, and assessed for FH with plasma lipid screening and (potentially) genetic testing, particularly if the index mutation is known. Screening should be extended to second- and third-degree relatives as well. Genetic counseling should be part of the initial treatment plan to help the proband (and family) understand FH, their pedigree (including at-risk relatives), the inheritance of the disease, and health implications. All genetic testing should include pretest genetic counseling that addresses financial considerations of testing, interpretation of genetic tests, and limitations of testing. Counseling may also help in situations involving a proband requesting nondisclosure of test results to family to help understand balancing personal preferences regarding disclosure with potential benefit to family members [6].

6.4   Population Screening

An existing challenge in the management of FH is determining the optimal population screening approach to identify index FH patients, particularly in children. During childhood, the American Academy of Pediatrics recommends universal screening for FH with serum lipid measurements [12]. These recommendations are not universal as the United States Preventive Services Task Force (USPSTF) notes that there is insufficient evidence to assess the balance of risks and harms of lipid screening in children [13]. A recent study from the United Kingdom (UK), in which 10,095 children were screened with lipid levels and gene sequencing during routine immunization visits between 1 and 2 years of age, demonstrated that for every 1000 children screened, 4 adults and 4 children (8 persons total) were diagnosed with FH. The authors estimated a cost of $2900 per person identified as having positive screening results based on their screening protocol, assuming that lipid testing costs $7 and DNA sequencing costs $300 per sample. For adults, universal screening for FH is not current practice. However, the USPSTF does recommend lipid screening in men over 35 and women over 45 and even earlier for those with a family history of coronary heart disease [14].

Table 6.1 Known genetic variants associated with FH [6, 7]

Gene	Mechanism leading to FH phenotype	Estimated proportion of FH attributed to genetic variant
LDLR	Loss-of-function mutations lead to absent or impaired LDLR function, leading to decreased LDL-c uptake by hepatocytes and thus elevated circulating LDL-c levels	60–80%
APOB	Mutations lead to impaired binding between lipoproteins and LDLR, leading to decreased LDL-c uptake by the liver and increased circulating LDL-c levels	1–5%
PCSK9	Gain-of-function mutations cause increased LDLR degradation and, thus, decreased LDLR function and increased circulating LDL-c levels	0–3%

6.5 Genetics and Pathophysiology

The most common pathogenic genetic variants associated with FH are mutations of *LDLR*, which accounts for >80% of FH cases. Mutations of *APOB* and *PCSK9* genes have been described as well, though are less common (Table 6.1). Low-density lipoprotein (LDL) particles are major carriers of cholesterol in the body and contain a single apolipoprotein B molecule which mediates clearance of these LDL particles by binding to LDL receptors (LDLR) on liver cells. Upon binding the ApoB-containing lipoprotein, the LDLR/lipoprotein complex is internalized via endocytosis and directed to the lysosome where the cholesterol is degraded and the LDLR subsequently recycled to the surface of the hepatocyte. Normally circulating PCSK9 (proprotein convertase subtulisin/kexin type 9 protein) binds to the LDLR which ultimately targets the LDLR for degradation, preventing recycling to the surface of the hepatocyte [6]. Thus, mutations of *LDLR* and *APOB* that prevent LDL uptake, along with mutations of *PCSK9* that promote LDLR degradation, cause severely elevated LDL-c levels, creating the FH phenotype.

6.5.1 LDL Receptor

While investigating the genetic basis for FH, Goldstein and Brown identified the LDL receptor (LDLR), work for which they were awarded the Nobel Prize in Physiology or Medicine in 1985 [15]. LDLR is a transmembrane glycoprotein containing 839 amino acids, with six functional domains: signal peptide, ligand-binding domain, epidermal growth factor precursor (EGFP) like, O-linked sugar, transmembrane, and cytoplasmic domain [16]. The *LDLR* gene is located on chromosome 19p13.1–13.3 [6, 16] and contains 18 exons and 17 introns with a length of 45,000 base pairs. The transcription of the *LDLR* gene is controlled through a negative feedback mechanism mediated by intracellular cholesterol content.

LDLR mutations are the most common genetic variants associated with FH [17]. There have been over 1500 characterized mutations of *LDLR* resulting in FH in the University College London and ClinVar databases [18], with the majority being exonic substitutions. Most variants appear to fall within the EGFP-like and ligand-binding domains [7]. Based on their cellular site of impact, LDLR mutations have been previously classified into 6 types: Class 1—absence of synthesis of receptor or precursor protein; Class 2—absent or impaired formation of receptor protein; Class 3—abnormal LDL binding with normal receptor protein synthesis; Class 4—absence of internalization of the receptor complex and clustering in coated pits; Class 5—impaired recycling of receptors, with rapid degradation; and Class 6—impaired targeting of receptors to basolateral membrane [6, 19, 20]. There is increasing evidence that there is a generally graded relationship between the severity of mutation and the severity of the disease presentation (genotype/phenotype correlation) with more severe (i.e., loss of function) *LDLR* mutations conveying higher risk compared to amino acid substitutions [21]. However, because LDL-c levels are also controlled by other genetic and environmental influences, we currently do not specifically tailor therapeutic decisions based solely on mutation status.

6.5.2 Apolipoprotein B-100

Apolipoprotein B-100 (APOB) is the specific ligand responsible for binding to LDLR during LDL-c uptake by liver cells. Mutations of *APOB* that result in impaired binding to LDLR have been associated with autosomal-dominant FH phenotypes similar to those associated with *LDLR* mutations [22, 23]. This has sometimes been referred to as familial defective apolipoprotein B (FDB). *APOB* mutations have been reported to have a milder phenotype compared to those from severe *LDLR* mutations [7].

APOB is a protein with 4560 amino acids with four functional domains. The *APOB* gene spans 42,216 base pairs, with 29 exons and 28 introns. The majority of *APOB* mutations resulting in FH have been described in exon 26 [22], although mutations in other regions have also been reported recently, such as in exon 3 [23]. Exon 26 is responsible for more than half the APOB protein, spanning over 7000 base pairs [7]. Certain founder populations, such as Amish, have been found to have higher prevalence of *APOB* mutations [24]. By far the most common pathologic mutation in *APOB* is a single amino acid substitution (Arg3500Gln) which affects binding to the LDLR [25, 26]. Of note, other mutations of *APOB* can result in very low LDL-c levels, associated with a condition termed familial hypobetalipoproteinemia [27].

6.5.3 Proprotein Convertase Subtulisin/Kexin Type 9

PCSK9 is a circulating peptide that targets the LDLR for lysosomal degradation within liver cells. The PCSK9 protein consists of 692 amino acids and three domains, with a gene spanning 25,378 base pairs and 12 exons [7]. *PCSK9* mutations associated with FH were initially described in 2003 in French families with autosomal-dominant hypercholesterolemia in whom *LDLR* and *APOB* mutations were ruled out [28]. Gain-of-function *PCKS9* mutations have been associated with FH, as increased PCSK9 activity leads to reduced LDLR function, resulting in high circulating LDL-c levels [17]. Compared to *LDLR* mutations, *PCSK9* mutations are thought to be much rarer genetic causes of FH. Of interest, loss-of-function mutations of *PCSK9* can lead to low LDL-c levels and a subsequent reduction in coronary events [29].

6.6 Clinical Presentation

6.6.1 HeFH

Given a lifetime of severely elevated LDL-c levels, HeFH patients are at significant risk for premature coronary disease and mortality compared to the general population. In the prestatin era, the risk of coronary events in those with FH aged 20–39 years was 125-fold higher in women and 48-fold higher in men compared to those with normal lipid levels [4, 30]. Men with untreated FH have a 50% risk of a coronary event by age 50, and women with untreated FH carry a 30% risk of a coronary event by age 60 [7]. Diagnosis in childhood is usually reliant on highly elevated LDL-c measurements (such as those greater than 190 mg/dl) [31]. HeFH can present as angina and myocardial infarctions in adults as early as during the third decade of life [6]. Physical examination findings include stigmata of lifelong severely elevated LDL-c levels, such as arcus cornealis and tendon xanthomas. The presentation of HeFH can be variable, with some patients presenting late in adulthood and others presenting earlier [6]. For reasons that are not clear, the risk of stroke in HeFH patients is not appreciably higher than in the general population.

6.6.2 HoFH

Homozygous FH patients have LDL-c levels usually in excess of 400 mg/dl in early childhood and thus can present as early as the first or second decade of life with coronary heart disease [6]. Presentations range from the presence of striking physical examination findings such as cutaneous xanthomas during childhood, to myocardial infarctions and sudden death during the first decade of life. Interdigital xanthomas, particularly between the thumb and index finger, are considered pathognomonic for HoFH [32]. Clinical diagnosis of HoFH is based on the presence of cutaneous or tendon xanthomas at an early age (<10 years) together with untreated LDL-c greater than 500 mg/dl, treated LDL-c greater than or equal to 300 mg/dl, or a nonhigh density lipoprotein cholesterol level greater than or equal to 300 mg/dl [32]. Because cholesterol can be deposited in multiple areas (including coronary arteries, aortic valve cusps, aortic root, other major arteries, tendons, and cutaneous tissues), aortic

Table 6.2 MEDPED criteria for the diagnosis of FH [33]

Age	General population (mg/dl; mmol/L)	First-degree relative with FH (mg/dl; mmol/L)	Second degree relative with FH (mg/dl; mmol/L)	Third degree relative with FH (mg/dl; mmol/L)
Less than 20	270; 7.0	220; 5.7	230; 5.9	240; 6.2
20–29	290; 7.5	240; 6.2	250; 6.5	260; 6.7
30–39	340; 8.8	270; 7.0	280; 7.2	290; 7.5
40 or greater	360; 9.3	290; 7.5	300; 7.8	310; 8.0

The diagnosis is made if total cholesterol level exceeds the relevant threshold, expressed below in mg/dl and millimoles per liter (mmol/L)

Table 6.3 Simon-Broome registry criteria for diagnosis of FH [30]

Definite FH
Total cholesterol >290 mg/dl or LDL-c > 190 mg/dl in adults, OR
Total cholesterol >260 mg/dl or LDL-c > 155 mg/dl in a child under 16 years of age
PLUS
Tendon xanthomas in patient or first- or second-degree relative, OR
DNA-based evidence of functional LDLR, APOB, or PCSK9 mutation
Possible FH
Total cholesterol >290 mg/dl or LDL-c > 190 mg/dl in an adult, OR
Total cholesterol >260 mg/dl or LDL-c > 155 mg/dl in a child under 16 years of age
PLUS
Family history of MI before 50 years of age in a second-degree relative or 60 years of age in a first-degree relative, OR
Family history of total cholesterol >290 mg/dl in adult first- or second-degree relative, or > 260 mg/dl in a child or sibling under 16 years of age

A diagnosis of either definite FH or possible FH can be made based on the criteria met by the proband

valve disease—including aortic and supravalvular stenosis—is an important consideration in these patients in addition to coronary disease. Early recognition and treatment is critical, as untreated HoFH can quickly lead to clinically significant coronary artery and aortic valve disease [6].

6.7 Diagnosis

There are three widely known country-specific criteria for the diagnosis of FH: The US Make Early Diagnosis to Prevent Early Deaths (MEDPED) criteria (Table 6.2) [33], the UK Simon Broome register criteria (Table 6.3) [30], and the Dutch Lipid Clinic Network criteria (Table 6.4) [34]. Detailed history, physical examination, and appropriate laboratory testing—including genetic testing—are all key components

Table 6.4 Dutch Lipid Clinic Network scoring criteria for diagnosis of FH [34]

	Score (points)
Family history	
First-degree relative with premature coronary and/or vascular disease (men ≤55 years, woman ≤60 years), OR First-degree relative with known LDL-c ≥ 95th percentile for age and sex	1
First-degree relative with tendon xanthoma and/or arcus cornealis, OR Children 18 years of age or younger with LDL-c ≥ 95th percentile for age and sex	2
Clinical history	
Patient with premature coronary artery disease (as defined above)	2
Patient with premature cerebral or peripheral vascular disease	1
Physical examination	
Tendon xanthomas on exam	6
Arcus cornealis at age 45 or younger	4
LDL-c in mg/dl (mmol/L)	
330 (8.5) or greater	8
250–329 (6.5–8.4)	5
190–249 (5.0–6.4)	3
155–189 (4.0–4.9)	1
Genetic testing	
Functional LDLR, APOB, or PCSK9 mutation	8

Based on the total score, the diagnosis of definite FH (8 or more points), probable FH (6–7 points), possible FH (3–5 points), or unlikely FH (less than 3 points) is made

of establishing a clinical diagnosis of FH [35]. The relevance of these criteria may be limited, however, due to improved anticholesterol therapies and recent secular trends in the United States. For example, the prevalence of physical examination findings such as xanthomas appears much lower than before, from the Spanish Familial Hypercholesterolemia Longitudinal Cohort Study or SAFEHEART registry [36–38], and there is concern that accurate family histories are becoming more challenging to obtain based on data from the Cascade Screening for Awareness and Detection of FH or CASCADE FH registry [1, 36].

6.8   Management

The mainstay of FH therapy is cholesterol lowering through lifestyle modification and pharmacotherapy.

6.8.1   HeFH

Statins are the first-line therapy for FH and should be used in all adults with FH and sometimes in children [39]. The addition of nonstatin lipid-lowering therapies such as ezetimibe, bile acid sequestrants, and newer therapeutics including PCSK9 inhibitors is often required. In fact, data from national and international registries (SAFEHEART, CASCADE FH, and The Netherlands) suggest that less than half of FH patients can achieve optimal LDL-c on statins alone [1, 37].

HeFH patients experience a notable reduction in morbidity and mortality with the use of appropriate cholesterol-lowering therapy. Statin-based therapy is the mainstay for treatment and was associated with a 37% reduction in coronary heart disease-related mortality in a prospective cohort study of 3382 HeFH patients [40]. It has also been shown that the risk of MIs in a cohort of statin-treated FH patients (treated relatively early in life) was not significantly different from that of the general population [41]. In addition to pharmacotherapy, dietary modification is recommended as well for HeFH patients, though not likely to achieve adequate cholesterol lowering by itself [6].

In adults with HeFH for primary prevention, the recommended general treatment goal with cholesterol-lowering therapy is a reduction in LDL-c levels by at least 50% [6] though many guidelines suggest targeting an LDL-c of 100 mg/dl. The preferred cholesterol-lowering agents are statins. A recommended approach to pharmacotherapy intensification based on the American Heart Association (AHA) Scientific Statement on FH is to start with a high-intensity statin, such as rosuvastatin or atorvastatin. If adequate lowering is not achieved with monotherapy after 3 months, a second agent, ezetimibe may be added. A third agent is indicated if LDL-c lowering does not reach the target after 3 months of dual therapy, with options including bile acid sequestrants such as colesevelam and PCSK9 inhibitors such as alirocumab or evolocumab which are favored given the much more potent effect on LDL-c and the decrease in CVD events when added to statins in the FOURIER trial. It should be noted that randomized controlled trials examining hard endpoints such as mortality are lacking for adults with FH [42].

With regard to children with HeFH, pravastatin has been approved for use starting at age 8, and other statins have been approved for children as young as 10 years old [6]. Randomized controlled trials have confirmed the lipid-lowering efficacy of statins in children with FH, and statin therapy has also been associated with favorable changes in markers of atherosclerosis such as flow-mediated dilatation of the brachial artery and carotid intima media thickness [43, 44]. However, randomized controlled trials exploring "hard" outcomes such as cardiovascular mortality or addressing long-term safety profiles of statin use in children are lacking. Baseline creatinine, creatine kinase, and hepatic enzyme levels are obtained prior to initiation of treatment [39]. Recommended goal LDL-c levels are <130 mg/dl for those 10 years or older and a 50% reduction of pretreatment LDL-c levels for children 8–10 years of age. Hepatic enzymes should be periodically monitored, and other tests should be performed if there is clinical concern for adverse effects, such as creatine kinase for myopathy, and fasting glucose levels for diabetes.

6.8.2 HoFH

Patients with HoFH should generally be referred to a specialist given the severity of their phenotype. Consensus guidelines have noted LDL-c lowering targets of <100 mg/dl (<135 mg/dl in children), or <70 mg/dl in adults with atherosclerotic cardiovascular disease (ASCVD) [45].

Therapies such as statins or ezetimibe, however, have had relatively modest lipid-lowering impact in HoFH patients especially in those with null mutations in the LDL receptor. Statins are estimated to reduce LDL-c by only 10–25% in most HoFH patients [6], albeit still associated with significant reduction in mortality [46]. The addition of ezetimibe has been estimated to provide only an additional 10–15% LDL-c reduction [6]. PCSK9 inhibitors may be beneficial for HoFH patients with defective (rather than null) LDLR activity given that impaired PCSK9 activity may lead to decreased LDLR degradation and increased LDLR function [4, 47].

Lipoprotein apheresis—removal of lipoproteins from the circulation—is an option for those who are unable to achieve treatment goals with pharmacotherapy [31] (generally reserved for HoFH). Guidelines recommend that apheresis be considered for all patients with HoFH, particularly children and no later than age 8, and often as soon as possible [39]. The AHA Scientific Statement on FH outlines certain indications for initiation of apheresis including LDL-c reduction of less than 50% with other pharmacotherapy and a residual LDL-c greater than 200 mg/dl with cardiovascular disease, or greater than 300 mg/dl [6].

In addition, novel therapeutics, such as PCSK9 inhibitors, mipomersen, and microsomal triglyceride transfer protein (MTTP) inhibitors such as lomitapide, have shown promise as additional LDL-c-lowering agents in HoFH, although their precise role in conjunction with traditional lipid-lowering agents and apheresis is still being elucidated [4, 6]. Lomitapide is an oral MTTP inhibitor approved by the FDA in 2012 for adult HoFH patients requiring additional lipid lowering [4, 48, 49]. Mipomersen is another novel first-in-class medication. It is an injectable synthetic single-strand antisense oligonucleotide analog that binds mRNA encoding for *APOB* [49]. Multiple randomized controlled trials have demonstrated the lipid-lowering efficacy of mipomersen in those with FH [49–51]. Finally, orthotopic liver transplantation may be a consideration for HoFH children and adults with rapidly progressive atherosclerosis, those intolerant of other therapies, or those with inadequate LDL-c-lowering despite other therapies. While there is an absence of trial evidence addressing the role of transplantation, case reports have noted favorable outcomes such as rapid normalization of LDL-c levels [6, 52, 53].

6.9 Summary of Key Points

- Familial hypercholesterolemia is an autosomal-dominant condition resulting in severely elevated levels of total cholesterol and LDL-c associated with premature coronary artery disease, myocardial infarctions, and aortic valve disease manifesting in childhood or early adulthood.

- Pathogenic variants of the gene associated with the LDL receptor (*LDLR*) are the most common genetic mutations associated with the FH phenotype.
- Mutations of the genes encoding for apolipoprotein B-100 (*APOB*) and the proprotein convertase subtulisin/kexin type 9 protein (*PCSK9*) have been associated with FH as well.
- Patients with clinical FH who are negative for pathogenic *LDLR*, *APOB*, and *PCSK9* variants may have multiple mutations in various LDL-c lowering genes, i.e., a polygenic cause. Elevated lipoprotein (a) levels may also mimic the FH phenotype.
- Clinical criteria for the diagnosis of FH include the US MEDPED criteria, the UK Simon Broome registry criteria, and the Dutch Lipid Clinic Network criteria.
- Genetic testing for culprit mutations of *LDLR*, *APOB*, and/or *PCSK9* is an important part of the diagnosis of FH.
- Once a diagnosis is made, cascade screening and genetic counseling should be promptly initiated.
- Heterozygous FH should be treated with early lifestyle modification and cholesterol-lowering therapy starting with statins (pravastatin starting at age 8 and other statins starting at age 10) and often require additional nonstatin lipid-lowering therapies to achieve adequate LDL-c lowering. Lipoprotein apheresis—the removal of lipoproteins from circulation—should be considered if traditional lipid-lowering therapies are not tolerated or are not efficacious or not tolerated.
- Homozygous FH can be recalcitrant to traditional lipid-lowering therapies, and these patients may require early consideration of lipoprotein apheresis, along with novel therapeutics such as mipomersen, lomitapide, and PCSK9 inhibitors.
- Orthotopic liver transplantation has shown favorable results in case reports of patients with FH.

References

1. deGoma EM, et al. Treatment gaps in adults with heterozygous familial hypercholesterolemia in the United States: data from the CASCADE-FH registry. Circ Cardiovasc Genet. 2016;9 (3):240–9.
2. Benn M, et al. Familial hypercholesterolemia in the Danish general population: prevalence, coronary artery disease, and cholesterol-lowering medication. J Clin Endocrinol Metab. 2012;97(11):3956–64.
3. Nordestgaard BG, et al. Familial hypercholesterolaemia is underdiagnosed and undertreated in the general population: guidance for clinicians to prevent coronary heart disease: consensus statement of the European Atherosclerosis Society. Eur Heart J. 2013;34(45):3478–90a.
4. Parizo J, Sarraju A, Knowles JW. Novel therapies for familial hypercholesterolemia. Curr Treat Options Cardiovasc Med. 2016;18(11):64.
5. Ahmad ZS, et al. US physician practices for diagnosing familial hypercholesterolemia: data from the CASCADE-FH registry. J Clin Lipidol. 2016;10(5):1223–9.
6. Gidding SS, et al. The agenda for familial hypercholesterolemia: a scientific statement From the American Heart Association. Circulation. 2015;132(22):2167–92.

7. Youngblom E, Pariani M, Knowles JW. Familial hypercholesterolemia. In: Pagon RA, et al., editors. GeneReviews(R). Seattle, WA: University of Washington; 2016.
8. Talmud PJ, et al. Use of low-density lipoprotein cholesterol gene score to distinguish patients with polygenic and monogenic familial hypercholesterolaemia: a case-control study. Lancet. 2013;381(9874):1293–301.
9. Langsted A, et al. High lipoprotein(a) as a possible cause of clinical familial hypercholesterolaemia: a prospective cohort study. Lancet Diabetes Endocrinol. 2016;4(7):577–87.
10. Nordestgaard BG, et al. Lipoprotein(a) as a cardiovascular risk factor: current status. Eur Heart J. 2010;31(23):2844–53.
11. Soutar AK, Naoumova RP. Mechanisms of disease: genetic causes of familial hypercholesterolemia. Nat Clin Pract Cardiovasc Med. 2007;4(4):214–25.
12. Expert Panel on Integrated Guidelines for Cardiovascular, H, et al. Expert panel on integrated guidelines for cardiovascular health and risk reduction in children and adolescents: summary report. Pediatrics. 2011;128(Suppl 5):S213–56.
13. US Preventive Services Task Force. Screening for lipid disorders in children and adolescents: US preventive services task force recommendation statement. JAMA. 2016;316(6):625–33.
14. Helfand M, Carson S. Screening for lipid disorders in adults: selective update of 2001 US preventive services task force review. Rockville (MD): Agency for Healthcare Research and Quality; 2008.
15. Goldstein JL, Brown MS. The LDL receptor. Arterioscler Thromb Vasc Biol. 2009;29(4):431–8.
16. De Castro-Orós I, Pocoví M, Civeira F. The genetic basis of familial hypercholesterolemia: inheritance, linkage, and mutations. Appl Clin Genet. 2010;3:53–64.
17. Mabuchi H, et al. Genotypic and phenotypic features in homozygous familial hypercholesterolemia caused by proprotein convertase subtilisin/kexin type 9 (PCSK9) gain-of-function mutation. Atherosclerosis. 2014;236(1):54–61.
18. Usifo E, et al. Low-density lipoprotein receptor gene familial hypercholesterolemia variant database: update and pathological assessment. Ann Hum Genet. 2012;76(5):387–401.
19. Hobbs HH, Brown MS, Goldstein JL. Molecular genetics of the LDL receptor gene in familial hypercholesterolemia. Hum Mutat. 1992;1(6):445–66.
20. Koivisto UM, Hubbard AL, Mellman I. A novel cellular phenotype for familial hypercholesterolemia due to a defect in polarized targeting of LDL receptor. Cell. 2001;105(5):575–85.
21. Khera AV, et al. Diagnostic yield and clinical utility of sequencing familial hypercholesterolemia genes in patients with severe hypercholesterolemia. J Am Coll Cardiol. 2016;67(22):2578–89.
22. Borén J, et al. The molecular mechanism for the genetic disorder familial defective apolipoprotein B100. J Biol Chem. 2001;276(12):9214–8.
23. Thomas ER, et al. Identification and biochemical analysis of a novel APOB mutation that causes autosomal dominant hypercholesterolemia. Mol Genet Genomic Med. 2013;1(3):155–61.
24. Shen H, et al. Familial defective apolipoprotein B-100 and increased low-density lipoprotein cholesterol and coronary artery calcification in the old order Amish. Arch Intern Med. 2010;170(20):1850–5.
25. Arnold KS, et al. Isolation of allele-specific, receptor-binding-defective low density lipoproteins from familial defective apolipoprotein B-100 subjects. J Lipid Res. 1994;35(8):1469–76.
26. Soria LF, et al. Association between a specific apolipoprotein B mutation and familial defective apolipoprotein B-100. Proc Natl Acad Sci USA. 1989;86(2):587–91.
27. Musunuru K, et al. Exome sequencing, ANGPTL3 mutations, and familial combined hypolipidemia. N Engl J Med. 2010;363(23):2220–7.
28. Abifadel M, et al. Mutations in PCSK9 cause autosomal dominant hypercholesterolemia. Nat Genet. 2003;34(2):154–6.
29. Cohen JC, et al. Sequence variations in PCSK9, low LDL, and protection against coronary heart disease. N Engl J Med. 2006;354(12):1264–72.

30. Risk of fatal coronary heart disease in familial hypercholesterolaemia. Scientific Steering Committee on behalf of the Simon Broome Register Group. BMJ. 1991;303(6807):893–6.
31. Watts GF, et al. Integrated guidance on the care of familial hypercholesterolemia from the International FH Foundation. J Clin Lipidol. 2014;8(2):148–72.
32. Raal FJ, Santos RD. Homozygous familial hypercholesterolemia: current perspectives on diagnosis and treatment. Atherosclerosis. 2012;223(2):262–8.
33. Williams RR, et al. Diagnosing heterozygous familial hypercholesterolemia using new practical criteria validated by molecular genetics. Am J Cardiol. 1993;72(2):171–6.
34. Civeira F, I.P.o.M.o.F. Hypercholesterolemia. Guidelines for the diagnosis and management of heterozygous familial hypercholesterolemia. Atherosclerosis. 2004;173(1):55–68.
35. Foody JM. Familial hypercholesterolemia: an under-recognized but significant concern in cardiology practice. Clin Cardiol. 2014;37(2):119–25.
36. Kindt I, Mata P, Knowles JW. The role of registries and genetic databases in familial hypercholesterolemia. Curr Opin Lipidol. 2017;28(2):152–60.
37. Mata N, et al. Clinical characteristics and evaluation of LDL-cholesterol treatment of the Spanish Familial Hypercholesterolemia Longitudinal Cohort Study (SAFEHEART). Lipids Health Dis. 2011;10:94.
38. Perez de Isla L, et al. Coronary heart disease, peripheral arterial disease, and stroke in familial hypercholesterolaemia: insights from the SAFEHEART registry (Spanish familial hypercholesterolaemia cohort study). Arterioscler Thromb Vasc Biol. 2016;36(9):2004–10.
39. Wiegman A, et al. Familial hypercholesterolaemia in children and adolescents: gaining decades of life by optimizing detection and treatment. Eur Heart J. 2015;36(36):2425–37.
40. Neil A, et al. Reductions in all-cause, cancer, and coronary mortality in statin-treated patients with heterozygous familial hypercholesterolaemia: a prospective registry study. Eur Heart J. 2008;29(21):2625–33.
41. Versmissen J, et al. Efficacy of statins in familial hypercholesterolaemia: a long term cohort study. BMJ. 2008;337:a2423.
42. Knowles JW. Statins in familial hypercholesterolemia: translating evidence to action. J Am Coll Cardiol. 2016;68(3):261–4.
43. Vuorio A, et al. Statins for children with familial hypercholesterolemia. Cochrane Database Syst Rev. 2014;7:CD006401.
44. Kusters DM, et al. Ten-year follow-up after initiation of statin therapy in children with familial hypercholesterolemia. JAMA. 2014;312(10):1055–7.
45. Cuchel M, et al. Homozygous familial hypercholesterolaemia: new insights and guidance for clinicians to improve detection and clinical management. A position paper from the consensus panel on familial hypercholesterolaemia of the European Atherosclerosis Society. Eur Heart J. 2014;35(32):2146–57.
46. Raal FJ, et al. Reduction in mortality in subjects with homozygous familial hypercholesterolemia associated with advances in lipid-lowering therapy. Circulation. 2011;124(20):2202–7.
47. Raal FJ, et al. Inhibition of PCSK9 with evolocumab in homozygous familial hypercholesterolaemia (TESLA Part B): a randomised, double-blind, placebo-controlled trial. Lancet. 2015;385(9965):341–50.
48. Cuchel M, et al. Efficacy and safety of a microsomal triglyceride transfer protein inhibitor in patients with homozygous familial hypercholesterolaemia: a single-arm, open-label, phase 3 study. Lancet. 2013;381(9860):40–6.
49. Rader DJ, Kastelein JJ. Lomitapide and mipomersen: two first-in-class drugs for reducing low-density lipoprotein cholesterol in patients with homozygous familial hypercholesterolemia. Circulation. 2014;129(9):1022–32.
50. Stein EA, et al. Apolipoprotein B synthesis inhibition with mipomersen in heterozygous familial hypercholesterolemia: results of a randomized, double-blind, placebo-controlled trial to assess efficacy and safety as add-on therapy in patients with coronary artery disease. Circulation. 2012;126(19):2283–92.

51. McGowan MP, et al. Randomized, placebo-controlled trial of mipomersen in patients with severe hypercholesterolemia receiving maximally tolerated lipid-lowering therapy. PLoS One. 2012;7(11):e49006.
52. Maiorana A, et al. Preemptive liver transplantation in a child with familial hypercholesterolemia. Pediatr Transplant. 2011;15(2):E25–9.
53. Moyle M, Tate B. Homozygous familial hypercholesterolaemia presenting with cutaneous xanthomas: response to liver transplantation. Australas J Dermatol. 2004;45(4):226–8.

Long Noncoding RNAs in Cardiovascular Disease

7

Lesca M. Holdt, Alexander Kohlmaier, and Daniel Teupser

Contents

Abstract

Recent advances in high-throughput sequencing of nucleic acids have led to the fascinating insight that the majority of the human genome is transcribed. This includes tens of thousands of RNAs sized larger than 200 nucleotides that are not translated into proteins and are referred to as long noncoding RNAs (lncRNAs).

L. M. Holdt (✉) · A. Kohlmaier · D. Teupser
Institute of Laboratory Medicine, University Hospital, Munich, Germany
e-mail: lesca.holdt@med.uni-muenchen.de

© Springer Nature Switzerland AG 2019
J. Erdmann, A. Moretti (eds.), *Genetic Causes of Cardiac Disease*, Cardiac and Vascular Biology 7, https://doi.org/10.1007/978-3-030-27371-2_7

Until now, only a few of these lncRNAs have been functionally characterized. Here we highlight lncRNAs related to cardiovascular physiology and disease (CVD). We start with an overview of lncRNA classification schemes and of molecular functions of lncRNAs, giving examples of lncRNAs with cardiovascular function in each class. The main focus then is to systematically review 57 lncRNAs implicated in atherosclerosis, myocardial infarction, aortic aneurysm, cardiomyopathy, angiogenesis, arrhythmia, and stroke. We discuss the evidence how these lncRNAs partake in the regulation of cell lineage specification, differentiation potential, cell proliferation, and cell survival in the cardiovascular cell lineages. Specific emphasis is put on recently published lncRNA knockout approaches, and on lncRNAs that have been implicated as important regulators in animal in vivo models of cardiovascular diseases and/or identified in human patient cohorts.

7.1   Introduction

For a long time, proteins have been considered the main functionally active molecules in mammalian cells. Yet, in a few selected cases, functionally important non-protein-coding cellular RNAs had been identified. Most prominently, the long noncoding RNA (lncRNA) *XIST* has been known for decades to be essential for establishing X-chromosome inactivation in female mammals, and studies on *XIST* have become blueprints for approaches in studying lncRNA function (see [1] for review). Similarly, several noncoding RNAs have been identified early on in several important imprinted gene clusters, among them RNAs with relevance for cardiovascular disease, like *H19*, *Meg3*, or *Kcnq1ot1*. Their study has inspired more recent investigations on how lncRNAs interact with chromatin regulators and affect transcription. But only with the advent of high-throughput nucleic acid sequencing analyses in the 2000s, and soon after the detection of the large class of regulatory small interfering RNAs and microRNAs, lncRNAs have entered the focus of the investigation on a genome-wide scale. It is becoming clear by work of the ENCODE or FANTOM consortia that protein-coding genes account for only as little as 1.5% of the genome. In recent years, thousands of lncRNAs have been identified. LncRNAs are transcribed from **thousands of previously unannotated non-protein-coding genes** in our genomes [2–5]. This raises the question if and to what extent the many thousands of lncRNA transcripts are functional. Whether only few or many of these lncRNA transcripts carry cellular functions is a matter of currently ongoing research.

Overall, on a molecular level, it seems that lncRNAs do not have catalytic ribozyme functions. Only ribosomal RNAs (rRNA) and small nuclear RNAs (snRNA) function as catalytic entities in ribosomes and the spliceosome, respectively. Rather, lncRNAs serve as regulators of other molecules by binding to them. In this function, and as will be reviewed in detail in this book chapter, **lncRNAs guide protein complexes to specific DNA sequences, or scaffold multiprotein complexes, or increase or inhibit the activity of enzymes, or affect mRNA stability and translation capacity**. In functional terms, a growing number of lncRNAs are being found to impact the development of cardiovascular cell types and organs in the vertebrate embryo. To name a few, *Braveheart* [6] *Fendrr* [7],

Upperhand [8], and *HoxBlinc* [9] are important recently identified lncRNAs in this class. Other lncRNAs, in some cases first identified in independent systems like in cancer models or in generic cellular screens, were then also found to be misexpressed in diseases of the heart or vasculature and to causally contribute to cardiovascular disease. *ANRIL* [10–12], *MALAT1* [13, 14], *Myheart* [15], *Chaer* [16], *lincRNA-p21* [17], *CARL* [18], *CARMEN* [19], *Rffl-lnc1* [20], and *ROR* [21] are prominent examples of this latter class. Extending the compendium of lncRNAs, recent experiments have documented that besides the thousands of linear lncRNAs, thousands of genes also express circular RNAs (circRNAs), most of which are noncoding [22]. circRNAs emerge from unconventional backsplicing (tail-to-head splicing) of a downstream exon to a more upstream-located exon, resulting in covalent linkage of RNA *in cis* by a covalent $3'$–$5'$ phosphodiester bonding. Compared to linear lncRNAs, an even smaller number of circRNAs has been functionally studied. But from the little existing insight it has become clear that also some circRNAs, though not encoding proteins, can carry regulatory functions and contribute to cardiovascular disease when misregulated, such as circRNAs emerging from the *ANRIL* locus on the well-known cardiovascular risk region on chromosome 9p21 [23, 24].

7.2 General Characteristics and Classification of lncRNAs

Here we describe the molecular characteristics of lncRNAs, linear and circular (Sects. 7.2.1–7.2.2.8), and their well-known molecular functions (Sects. 7.3.1–7.3.10), always by focusing on the description of those lncRNAs that have been implicated in cardiovascular physiology and disease. Their specific roles in physiology and cardiovascular disease are described later (Sects. 7.4.1–7.4.2.7) following the introductory parts on classification and molecular function.

7.2.1 Characteristics of lncRNAs

lncRNAs have been defined rather arbitrarily by a length *>200 nucleotides* (nts), a threshold rooting in cutoffs during biochemical separation from shorter RNAs like microRNAs using a commercial DNA/RNA isolation kit [25, 26]. Earlier functionally annotated noncoding RNAs, such as ribosomal RNA, transfer RNAs, small nuclear RNAs, small nucleolar RNAs, microRNAs, endogenous small interfering RNAs, or Piwi-associated RNAs, even when >200 nts, are not classified as lncRNAs [27] (see Box 7.1 for a list of general characteristics of lncRNAs). The key criterion for being a lncRNA is that, firstly, lncRNAs do not carry prominent open reading frames for protein translation. Secondly, lncRNAs are **unlikely to be translated** into proteins even when carrying open reading frames. In fact, although a significant number (50%) of lncRNAs do associate with ribosomes, this association is not productive, and no translation ensues [28]. Over 27,000 lncRNA genes are predicted to exist in humans, leading to over 100,000 different lncRNA transcripts (https://lncipedia.org) [29]. This number rivals the number of protein-coding genes in our genomes (19,817; www.ensembl.org), feeding the hypothesis that much of our organismic complexity as higher metazoans may be related to the function of noncoding RNAs [30]. A common question is whether lncRNAs are more likely to act in the nucleus or in the cytoplasm. When assessing the transcript abundance of 1339 robustly expressed lncRNAs and of 13,933 mRNAs on a genome-wide scale, 17% of the tested lncRNAs were found to be exclusively localized to the nucleus, a ratio that is slightly but significantly larger than the ratio of exclusively nuclear protein-coding mRNAs among all mRNAs (15%). In contrast, the frequency of exclusively cytoplasmic lncRNAs is small (4%) compared to mRNAs (26%). Still a majority of lncRNAs and mRNAs are present in both nucleus and cytoplasm [31–34]. One exception is the class of circular lncRNAs (circRNAs, see below), which are mainly cytoplasmatic [22, 35]. Overall, the cellular localization patterns do not indicate a preferred subcellular compartment of function for the class of lncRNAs. Instead lncRNA function appears to be highly diverse. Also, on average, lncRNAs are similarly stable compared to coding mRNAs, and only specific classes of lncRNAs, few in the overall lncRNA number, qualify as being specifically unstable, as specified below [36].

Box 7.1 Characteristics of lncRNAs

- Up to 93% of the human genome is transcribed [2–5]
- >200 nucleotides in length [25, 26]
- LncRNAs are a heterogeneous class of tens of thousands of noncoding transcripts [29, 37]
- Similarity to protein-coding mRNAs [37]:
  - RNA polymerase II transcripts
  - Expressed from genes and organized in exons that are defined by chromatin states alike protein-coding mRNAs
  - Carrying 5′cap (depending on class)
  - Spliced and displaying 3′ polyA tails (depending on class)
- Unique features:
  - Unlikely to carry open reading frames >300 nucleotides (see text for details) [38, 39].
  - Shorter than mRNAs on average, with fewer but longer exons [31].
  - Expressed at lower levels [40].
  - Faster primary sequence evolution than mRNAs, but still often with orthologs in other species [41].
  - Less efficient splicing compared to mRNAs [42].
  - 95% of lncRNAs do not productively associate with translating ribosomes [28].
  - Prominent tissue-specificity [31] and potential to regulate gene expression *in cis* and in *trans* [37, 43–46].
  - Sometimes specialized 3′ ends (polyA-independent; triple helices, tRNA/snoRNA-like ends) [47].
  - A large class of lncRNAs can be circular (5′ and 3′ ends are covalently linked; circRNAs do not have 5′ cap or polyA tail) [22].

7.2.2   Classification of lncRNAs

An important approach in trying to grasp the diversity of lncRNA function is to **classify lncRNAs according to the location of their gene bodies in the genome**. This relates to the observation that in many cases the relative positioning of lncRNA genes to other functional elements nearby is important. For example, a common function of lncRNAs is to affect the transcription of nearby genes, either directly or indirectly, by influencing enhancers or the local chromatin state at promoters. However, there is more than one classification scheme, and lncRNAs may also be grouped according to their molecular biogenesis, or their type of functionality, and the group affiliation changes depending on the grouping principle (see Box 7.2 for different classification schemes). In the following paragraphs we present the major classification scheme for lncRNAs that roots in the genomic organization of genes encoding these noncoding RNAs relative to neighboring protein-coding gene elements.

7.2.2.1 Large Intergenic Noncoding RNAs (lincRNAs)

The genetic loci of lincRNAs **do not overlap with protein-coding genes**. lincRNAs are on average 1 kb long, contain exons, and are polyadenylated and spliced. Compared to mRNAs, lincRNAs are not as efficiently co-transcriptionally spliced and not so well polyadenylated [40, 48]. As a consequence, without efficient end-processing and when still carrying introns, such types of lincRNAs can be unstable and become subject to early degradation by the nuclear exosome, the major RNA degradation complex [48]. The degradation process can become active even before lincRNA transcription is terminated. However, a number of lincRNAs, even when lacking a polyA tail can become stabilized by unconventional molecular features, such as structured RNA folds at their 3′ ends. For example, backfolding into RNA triple helices, tRNA-like or snoRNA-like folds, can stabilize certain lincRNAs [47, 49–51]. Several noncoding RNAs have been identified as lincRNAs and have later been implicated in cardiovascular disease, such as *MALAT1* [13, 14, 47], *lincRNA-p21* [17], *Chaer* [16], *ROR* [21], *HOTAIR* [52], *Rncr3* [53], *Gas5* [54], *Mirt1* [55], *UCA1* [55], *linc00305* [56], and *lincRNA-DYNLRB2-2* [57]. Among these, *MALAT1* is an especially well studied case where unconventional 3′ end processing determines the 3′ end: *MALAT1* is not polyadenylated after endonucleolytic cleavage as the vast majority of long RNA polymerase II transcripts, but it is end-processed by RNase P, which cleaves off a tRNA-like RNA-fold from the 3′ end. After that cut, an A/U-rich sequence is exposed that stabilizes *MALAT1* RNA by folding into a triple helical RNA structure that stabilizes *MALAT1* in the absence of a polyA tail [47].

7.2.2.2 Natural Antisense Transcripts (NATs, asRNAs)

A large majority of transcripts in our genomes represent natural antisense transcripts (NATs, asRNAs) relative to neighboring transcripts. NATs are encoded on the strand **opposite to the strand transcribed in the primary locus**. Mostly on the edges of the host gene, NATs can either overlap the promoter or the terminator of a neighboring primary locus. NATs are less frequently spliced or polyadenylated than mRNAs or lincRNAs [58–65]. Examples with relevance to cardiovascular physiology and disease are *ANRIL* [66], *SENCR* [67], *MALAT1* [68] *MANTIS* [69], *Myheart* [15], *HOXC-AS1* [70], *HIF1αAS1* [71], *KCNQ1OT1* [72], and *FosDT* [73].

7.2.2.3 Promoter Upstream Antisense Transcripts (PROMPTS, uaRNAs)

There is a rather diverse class of lncRNA emerging **from regions 5′ to promoters of established genes**, and specifically in antisense orientation to these. These RNAs are termed promoter upstream transcripts (PROMPTS), and are also known as upstream antisense RNAs (uaRNAs) [74–76]. They are expressed in an antisense direction because mammalian RNAP II transcription initiation sites, both at promoters and at enhancers, typically contain oppositely oriented core promoter elements within a single nucleosome-depleted region that defines these transcription start sites. These transcripts are diverse in size and are 5′ capped. Transcription upstream of promoters likely occurs because many gene promoters consist of two separate core promoter elements, which drive divergent transcription events. Often, only the major direction of transcription leads to productive elongation and to stable RNAs. PROMPTS/ uaRNAs are rather unstable due to the absence of transcription start site (TSS)- proximal 5′ splice sites in sequences upstream of promoters. They are also unstable because of the premature occurrence of TSS-proximal polyA sites. As a consequence, they are degraded rapidly by the nuclear exosome [74]. Before the class of uaRNAs was identified and coined in vertebrate genomes, similar unstable upstream RNAs had been found in yeast and named differently: Yeast has two major types of these transcripts. First, there are cryptic unstable transcripts (CUTs). CUTs are expressed in the opposite direction from nonoverlapping bidirectional promoters and their ends can lie in the starts of other genes [61, 77]. The expression of CUTs is limited by rapid degradation [78]. Secondly, hundreds of stable unannotated transcripts (SUTs) exist in yeast [79]. As such, CUTs and SUTs also belong to the previously mentioned class of antisense RNAs [44, 80]. PROMPTs and uaRNAs occur in all gene classes, in principle, and thus have not been specifically studied in the context of cardiovascular disease or cardiovascular genes or lncRNAs. An exception is the lncRNA *Upperhand*, whose expression from a bidirectional promoter, shared with the cardiac transcription factor *Hand2*, is involved in its role in cardiac development [8]. Also *ANRIL* shares a bidirectional promoter with *p14*, but no conclusive insight into the relevance of this positioning has yet been published [66].

7.2.2.4 Enhancer RNAs (eRNA) and Activating Noncoding RNAs with Enhancer Function (ncRNA-a)

As the fourth class of lncRNAs, enhancers of genes have been found to be transcribed. As a consequence, **enhancer elements in the genome express enhancer eRNAs** in 5′ and 3′ direction [81–84]. eRNAs are 50–2000 nts long and do not carry 5′ cap or polyA tail [85]. The classification of eRNAs is at the moment not satisfactorily resolved, as many eRNAs do not appear in lncRNA databases like GENCODE [31, 85]. This could be due to the fact that eRNAs are rather unstable and often present at even lower levels than other lncRNAs [85]. In fact, eRNA analysis requires different, more sensitive and specialized biochemical preparation methods (CAGE

[86, 87], GRO-seq [82], or PRO-seq [88, 89]). Apart from true eRNAs, a number of lncRNAs have been found, called ncRNA-a, that are not classified as eRNAs, but, similar to enhancers, stimulate targets genes in their vicinity [89]. Up to 3000 bona fide lncRNAs from ENCODE databases exist, whose initiation sites overlap predicted enhancers, as determined by the H3K4me1 and H3K27 acetylation during chromatin profiling approaches [90]. With relevance to the cardiovascular system, *Carmen* [19], *Wisper* [91], and *Upperhand* [8] have been proposed to function by enhancing transcription as enhancer RNAs. Also, *HOTTIP* [92], *linc-HOXA1* [93], and *SMILR* [94] activate genes in their immediate neighborhood, but it is not clear whether as eRNA or ncRNA-a or by another mechanism.

7.2.2.5 Long Intronic ncRNAs
The fifth class of lncRNAs is **encoded within introns** of some multi-exon genes and these are called long intronic ncRNAs [4, 95]. Overall, their expression is not a passive bystander to the expression of their parental gene, but they display a differential expression. A subclass of intronic ncRNAs is stable intronic sequences (sisRNAs), which is a broad class of noncoding RNAs that comprises both linear as well as circular RNAs [96, 97].

7.2.2.6 Transcribed Pseudogenes
The sixth class of lncRNAs can be defined as transcripts of a subset (2–20%) of pseudogenes [4]. Pseudogenes have originated in evolutionary history through tandem duplication of a parental gene or retrotransposition of an mRNA of a protein-coding gene. Without the selective pressure of encoding a protein, the rate by which mutations were retained in them increased. As a consequence, most pseudogenes have been inactivated and are not expressed anymore. The best-studied example is the *Xist* RNA, which also has a function in the cardiovascular system, and has emerged through a complex process from a pseudogenized protein-coding gene [98]. On the other hand, when expressed, **transcribed pseudogenes can exert regulatory functions as lncRNAs** and possibly affect their coding host genes [99].

7.2.2.7 Circular RNAs
Based on advances in RNA/DNA sequencing and in bioinformatics tools to map transcriptomes onto genomes, it has been revealed that thousands of genes express circular circRNAs [22]. As much as 20% of all genes expressed at any point in cells produced circRNAs [100–102]. These are produced by spliceosomal action from multiexon host genes: Instead of conventional colinear splicing of exons in their genomic $5' \rightarrow 3'$ order, where introns are excised and later degraded as intronic lariats, an atypical form of splicing (**backsplicing**) occurs during the biogenesis of circRNAs: A downstream exon is ligated to a more upstream exon, causing a **covalent $3' \rightarrow 5'$ linkage of exonic sequences** (see [103] for review). circRNA molecules are rather stable because they are not accessible anymore by cellular RNA degradation machineries, which are primarily exonuclease based (see [104] for review). circRNAs vary in size, but are >500 nucleotides on average [100], and thus classify as lncRNAs, though not usually reviewed together. In contrast to linear

lncRNAs, circRNAs are mostly cytoplasmic [22]. One important function of circRNAs is to control transcriptional initiation and splicing in the nucleus, but also cytoplasmic functions have been documented (see in detail below). So far, only a few circRNAs have been part of studies that explored functions in the cardiovascular system in vivo, and these include *circANRIL* [23, 24], *HRCR* [105], and *MFACR* [106].

7.2.2.8 lncRNAs with Translated Small Open Reading Frames (sORFs)

Though generally not encoding classical protein-coding open reading frames (ORFs) much longer than 300 nts, small ORFs do occur by chance in any sufficiently long RNA transcript, and even in the most classical lncRNA, *Xist* [39]. Only the development of ribosome release profiling technology has since firmly documented that lncRNAs may associate with ribosomes, but that **95% of lncRNAs are indeed not functionally translated** [28]. This means also, however, that hundreds of lncRNAs are potentially protein-coding or at least encoding small peptides. Recent studies revealed that several hundred **micropeptide-encoding short sORFs** are potentially translated from our genomes. Functionally active regulatory and signaling molecules encoded as true micropeptides have already been found in animals [107–112] and plants [113]. Together, a continuum seems to exist between noncoding and coding sequence content in RNA messages [114], but the number of functional translated sORFs in bona fide lncRNAs is considered rather small overall [115]. Some of the best understood micropeptide-generating lncRNAs have indeed a function in the cardiovascular system, namely, in regulating heart muscle contractility [110, 112, 116].

Since different lncRNA classification schemes are used in the literature, key concepts in lncRNA classification are summarized in Box 7.2.

Box 7.2 Classification of lncRNAs
- Classification based on neighborhood to protein-coding genes:
    - Linear intergenic lncRNAs originating from enhancers (bidirectionally transcribed and unstable)
    - Linear lncRNAs originating from host gene promoters (transcribed opposite to host mRNAs, unstable)
    - Linear lncRNAs as stand-alone genes (more stable, longer)
    - Long intronic ncRNAs
    - Circular lncRNAs (stable, produced from internal exonic and intronic sequences of expressed multiexonic genes)
- Classification based on molecular biogenesis (see text for explanation of acronyms):
    - lincRNAs
    - NATs, asRNAs
    - PROMPTS, uaRNAs, CUTs, SUTs
    - eRNAs, ncRNA-a

(continued)

> **Box 7.2** (continued)
> - Classification based on function (majority of lncRNAs are functional; see Fig. 7.1):
>   - lncRNAs with functions as RNA molecules
>   - lncRNAs for which transcriptional act executes functionality
>   - Small minority of lncRNAs that are translated to polypeptides
>   - Transcription noise

7.3   Molecular Functions of lncRNAs

Here we review lncRNA functions that have been studied in the specific context of cardiovascular disease (Fig. 7.1). We start with functions in the nucleus (Fig. 7.1a–e) and then describe functions in the cytoplasm (Fig. 7.1f–j). A majority of lncRNAs can also directly affect transcriptional initiation and elongation by RNA polymerase II at promoters of protein-coding genes by scaffolding, tethering, and regulating the activity of enzyme complexes that modify histone tails or reposition nucleosomes (Fig. 7.1a). Related to this function, lncRNAs that interact with chromatin regulators can regulate large-scale chromatin fiber folding and influence repositioning of fibers relative to heterochromatic subnuclear domains (Fig. 7.1b). Another large fraction of lncRNAs emerge from enhancer regions of protein-coding genes and determine enhancer-dependent activation of the promoter of protein-coding genes (Fig. 7.1c). Many other lncRNAs are positioned in the genome in antisense to protein-coding genes and thereby affect their transcription (Fig. 7.1d). Yet other lncRNAs, but much fewer in number, are known to participate in splicing regulation (Fig. 7.1e). In the cytoplasm, some lncRNAs sequester and inactivate mRNA-regulating microRNAs (Fig. 7.1f). Likely for technical reasons in the experimental assessment, much fewer lncRNAs are known to bind mRNAs and thereby affect mRNA stability (Fig. 7.1g) or translation potential of mRNAs (Fig. 7.1h). lncRNAs can also bind protein complexes in the cytoplasm and affect their function, as shown in the example of *circANRIL* in inhibiting rRNA and ribosome maturation (Fig. 7.1i). Finally, a small minority of linear and circular lncRNAs actually encode small ORFs and are translated to micropeptides or larger parts of proteins, depending on the size of the ORF (Fig. 7.1j). Since no dedicated studies have been performed on the specific cardiovascular role of lncRNAs functioning as regulators of centromeres or of telomeres, or in DNA repair, even if important for general cell functionality, we will not specifically review these latter lncRNAs. In Sects. 7.3.1–7.3.10, we review each functional lncRNA class in detail. We list cardiovascular lncRNAs in each functional class (Fig. 7.1), and refer to these classes also in the following paragraphs on the roles of lncRNAs in cardiovascular physiology and diseases (Sects. 7.4.1–7.4.2.7; Tables 7.1, 7.2, and 7.3).

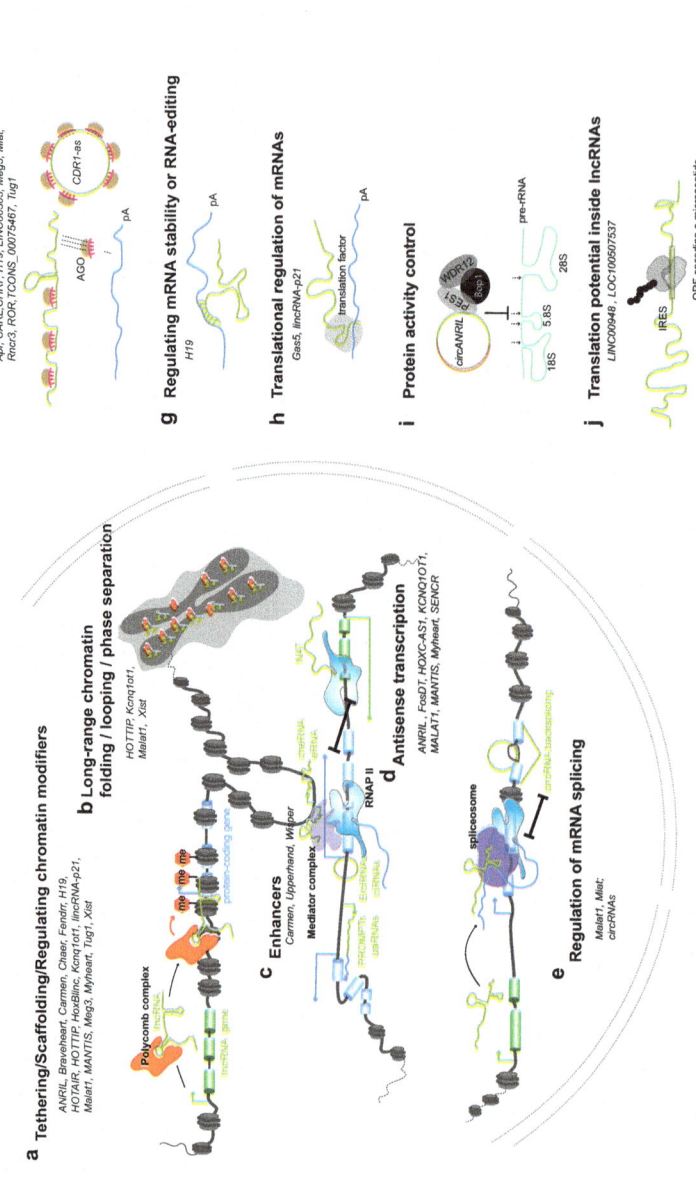

Fig. 7.1 Molecular functions of lncRNAs in eukaryotic cells. Roles of linear and of circular long noncoding RNAs in the nucleus (**a–e**) and in the cytoplasm (**f–j**). Names of lncRNAs with cardiovascular relevance are indicated for each class, as far as known. Chromatinized DNA (dark grey), lncRNAs (light green), and protein-coding genes (light blue). (**a**) Tethering/Scaffolding/Regulating chromatin modifiers. As an example, a lncRNA is shown to tether a repressive

Fig. 7.1 (continued) chromatin regulator (Polycomb complex) via a lncRNA hairpin motif to the transcription start site of a protein-coding target gene to inhibit transcriptional activation. (**b**) Long-range chromatin folding/looping. As an example, *Xist*-dependent chromatin compaction is shown on the female inactive X-chromosome, whose inactivation is promoted by localization to a heterochromatic territory close to the nuclear membrane (light grey). (**c**) Enhancing transcription from gene promoters. Transcription of an enhancer region, which leads to the production of eRNAs, to chromatin fiber looping towards a promoter, and to the activation of the RNAP II preinitiation complex through the Mediator complex (Top). At the promoter, certain classes of intron-containing circRNAs (ElciRNAs, ciRNAs) independently stimulate RNAP II (Bottom). (**d**) Antisense transcription: transcription of lncRNAs in antisense to a protein-coding gene, shown to inhibit sense target mRNA transcription. (**e**) Regulation of mRNA splicing by lncRNAs: linear and circular lncRNAs affect the spliceosome (and associated splicing factors) during alternative splicing. The processes of circRNA formation ("backsplicing") and linear mRNA splicing mutually impair each other. circRNAs can independently affect splicing through R-loop formation and by blocking RNAP II progression. (**f**) Linear lncRNAs (top) and circRNAs (bottom) can harbor multiple microRNA seed sequences, which leads to sequestration and inactivation of microRNA:Argonaute 2 complexes without lncRNA degradation (termed microRNA sponge or competing endogenous RNA effect). Reducing the availability of free microRNAs de-represses mRNAs that are natural targets of these microRNAs. (**g**) Binding of lncRNAs to a stability-regulating motif in a target mRNA, thereby regulating mRNA stability. (**h**) Regulation of mRNA translation: lncRNAs interacting with translation initiation factor (grey) shown as an example. (**i**) Regulation of cytoplasmic proteins: as an example *circANRIL* is shown, which inhibits the PeBoW protein complex in its activity of rRNA processing upstream of ribosome maturation. (**j**) Rare cases where the translation of ORF encoded on linear or circular lncRNAs is observed. A ribosome is depicted to translate a *micropeptide* from a small ORF after associating via an internal ribosome entry site (IRES). *AGO* Argonaute 2-containing RNA-induced silencing complex, *cheRNA* chromatin-enriched RNAs with enhancer function, *ciRNA* intronic circular RNA, *ElciRNA* exon-and-intron-containing circular RNA, *eRNA* enhancer RNA, *IRES* internal ribosome entry site, *me* H3 lysine K9/K27 trimethylation of histone tails, *ORF* open reading frame, *pA* polyA tail of RNAs, *RNAP II* RNA polymerase II transcription apparatus

Table 7.1 In vivo roles of lncRNAs during cardiovascular development

Name (species of KO)	Classification	Cardiovascular phenotype of mutant	Role in cardiovascular cell types	Molecular role
A. *Fendrr* (m.m.)	lncRNA	Cardiogenic mesoderm defect [7]	Cell proliferation and differentiation of lateral plate mesoderm derivatives during formation of myocardium and body wall [7]	Transcription: Scaffold/Tether/Inhibitor of activating and of repressive histone modifier [7]
A. *Braveheart* (m.m.)	lncRNA	Cardiogenic mesoderm defect [6]	Cell fate specification, differentiation of cardiac progenitors, and cardiac cell fate maintenance [6]	Transcription: Scaffold/Tether/Inhibitor of TF [117] and of repressive histone modifier [6]
A. *HoxBlinc* (m.m.)	lincRNA	Cardiogenic mesoderm defect [9]	Cell fate specification in cardiac lineage, also regulation of early hematopoiesis [9]	Transcription: Scaffold/Tether of activating chromatin modifiers, and establishment of activating long-range chromatin interactions at *Hoxb* gene cluster [9]
B. *Terminator* (d.r, m.m. MO)	lncRNA	Essential for gastrulation, survivors with vascular vessel branching defect [118]	Proliferation of pluripotent stem cells, also affecting ECs [118]	n.a.
B. *DEANR1/Alien* (d.r. MO)	lincRNA	Cardiogenic mesodermal defects and vascular vessel branching defect [118]	Differentiation of endoderm from vascular progenitors [119]	Transcription: activation of neighbor gene by tethering TF [119]
B. *Punisher* (d.r. MO)	NAT	Vascular vessel formation and branching defect [118]	Proliferation, branching, tube formation of mature ECs [118]	n.a.
B. *Tie-1AS#* (m.m.)	NAT	Vascular vessel integrity (mouse *Tie-1* KO [120])	Vascular EC cell-cell junction organization [120–122]	Transcription: Antisense inhibition of sense gene *Tie-1* [121]
C. *Upperhand* (m.m)	lncRNA	Heart morphogenesis defective (right ventricular chamber missing) [8]	Stimulation of *Hand2*-dependent cardiac morphogenesis program [8]	Transcription: enhancer for neighboring genes [8]

C. Xist (m.m.)	lncRNA	Essential for survival in females [1] Smaller hearts—perinatal heart growth defect [123], blood cancers in conditional mutants[124]	Hematopoietic stem cell maturation to all lineages, especially myeloid cells [124], potentially affecting cardiomyocyte growth [123]	Transcription: Scaffold/Tether of DNA/chromatin regulators on X-chromosome [1], unclear why specific cardiovascular effects occur [123–125]
D. LINC00948 AK009351 (m.m.)	lncRNA (micropeptide)	Heart physiology: Muscle/Heart excitation-contraction coupling [112]	Peptide-dependent repression of sarcoplasmic reticulum Ca^{2+} pump for myocyte relaxation [112]	Encoding translated micropeptide from lncORF (Myoregulin) [112]
D. LOC100507537/ NONMMUG026737 (m.m.)	lncRNA (micropeptide)	Heart physiology: Muscle/Heart excitation-contraction coupling [116]	Peptide-dependent derepression of sarcoplasmic reticulum Ca^{2+} pump for myocyte contractility [116]	Encoding translated micropeptide from lncORF (DWORF) [116]
E. Malat1 (m.m.)	lincRNA	Nonessential in embryogenesis [126–128]	Proangiogenic, proproliferative in neonatal retina MALAT1 [13, 14] Stress-specific role in cardiovascular cells, see Tables 7.2 and 7.3	Splicing regulator in vitro [217] Transcription: Inhibitor of repressive histone modifiers in vitro [206] Source of small RNAs in vitro [51]
E. Gomafu/MIAT/ Rncr2 (m.m.)	lncRNA	Nonessential in embryogenesis (mild behavioral defects) [219]	Stress-specific role in cardiovascular cells, see Tables 7.2 and 7.3	Splicing regulator [195] ceRNA/microRNA sponge [148]
E. Chaer	lncRNA	Nonessential in embryogenesis [16]	Stress-specific role in cardiovascular cells, see Tables 7.2 and 7.3	Transcription: decoy for repressive histone modifier [16]
E. Rncr3 (m.m.)	lncRNA	Nonessential in embryogenesis, CNS phenotype [205]	Stress-specific role in cardiovascular cells, see Tables 7.2 and 7.3	Parent molecule for microRNA [205]
E. CDR1-as	circRNA	Nonessential in embryogenesis (behavioral defects), no cardiovascular defects in vivo [213]	Stress-specific role in cardiovascular cell, see Tables 7.2 and 7.3	ceRNA/microRNA sponge [214, 215]

(continued)

Table 7.1 (continued)

Name (species of KO)	Classification	Cardiovascular phenotype of mutant	Role in cardiovascular cell types	Molecular role
E. H19 (m.m.)	lncRNA	Embryonic growth restriction, muscle differentiation; unknown role in heart [1]	Stress-specific role in cardiovascular cells, HSC quiescence, tumor suppression; see Tables 7.2 and 7.3	Source of microRNA [220] Transcription: Tether of repressive DNA methylation regulator [200] mRNA binding (decay) [201]

Roles of lncRNAs as determined by knockout of lncRNA locus, or by transgenic expression. lncRNAs involved in specification of the cardiogenic mesoderm (A), followed by lncRNAs involved in vascular vessel formation and branching (B). lncRNAs required for heart morphogenesis (C). lncRNAs encoding heart contractility-regulating micropeptides (D), and lncRNAs whose depletion did not impact heart structure or function but where stress-induced roles became apparent in adults (E). Within each group lncRNAs are ranked by when in embryonic development the first cardiovascular phenotype occurred. Note that the indicated molecular roles (right-most column) have in most cases not been formally shown to be causal for the tissue phenotype (see text for details)

# indicates a natural antisense lncRNA, where the major functional evidence has been extrapolated from the separate knockout analysis of the sense protein-coding host gene

KO knockout, *MO* Morpholino-based depletion in vivo, *TF* transcription factor, *HSC* hematopoietic stem cell

Species: *m.m. mouse musculus*, mouse, *d.r. danio rerio*, zebrafish morpholino knockdown *in toto*, *MO* morpholino treatment in mouse embryo culture *in toto*, *n.a.* not analyzed in publication

Table 7.2 In vivo evidence for lncRNAs in cardiovascular pathology

lncRNA name	Studied orthologs	Classification	CVD entities						
			Atherosclerosis	Myocardial infarction	Aortic aneurysm	Cardiomyopathies and congenital heart disease	Vascularization and angiogenesis	Arrhythmia	Stroke and cerebrovascular aneurysms
A. ANRIL (linear)	h.s.	lncRNA, NAT to p15; bidirectional promoter with p14 [66]	• Up in CAD; SNPs • Detrimental [10–12, 129–134]	• Up in MI • Detrimental [10–12, 129, 131–135]	• Up in AA • Detrimental [136]				• Up in stroke; • SNPs associating with intracranial aneurysms • Detrimental [137–140] [141–143]
A. ANRIL circular (circANRIL)	h.s.	circrna	• Down in CAD; SNPs • Protective [23, 24]						
A. Gas5	m.m. h.s. r.n.	lincrna, snoRNA host	• Up in CAD • n.a. [144]						
A. Gomafu/MIAT/Rncr2	m.m. h.s. r.n.	lncRNA	• Up in CAD; SNPs • (Detrimental?, n.a.) [145]	• Up in MI • Detrimental [146]		• Up in DCM, and in chagas disease • n.a. [147]	• Up in diabetic retina • Detrimental [148]		
A. H19	m.m. h.s. r.n.	lncRNA (imprinted)	• Up in CAD; SNPs • Detrimental [149–151]			• Up in CAVD and DCM • Detrimental [152]			• Up in stroke • Detrimental [153, 154]
A. HAS2-AS1	h.s. m.m.	lncRNA, NAT to HAS2	• Up in CAD. • (Detrimental? n.a.) [155]						
A. HOTTIP	h.s.	lncRNA, bidirectional with HOXA13	• Up in CAD • n.a. [156]				• n.a. (context dep. roles of HOX gene regulation) [156]		
A. HOXC-AS1	h.s.	lncRNA, NAT to HOXC6/9	• Down in CAD • n.a. [70]				• n.a (context dep. roles in HOX gene regulation) [70]		
A. LINC00305	h.s.	lincRNA	• Up in CAD • Detrimental [56]						

(continued)

Table 7.2 (continued)

lncRNA name	Studied orthologs	Classification	CVD entities						
			Atherosclerosis	Myocardial infarction	Aortic aneurysm	Cardiomyopathies and congenital heart disease	Vascularization and angiogenesis	Arrhythmia	Stroke and cerebrovascular aneurysms
A. *lincRNA-DYNLRB2-2*	m.m. h.s.	lincRNA	• n.a. • (Protective via regulating *GPR119*? n.a.) [157]						
A. *lincRNA-p21*	m.m. h.s.	lincRNA, close but opposite to p21	• Down in CAD, SNPs • Protective [158, 159]						
A. *Meg3*	m.m h.s. r.n.	lncRNA (imprinted)	• Down in CAD • n.a. [160]	• Up in TAC • Detrimental [161]					• Up in MCAO • Detrimental [162]
A. *RNCR3 / LINC00599*	m.m. h.s.	lincRNA	• Up in CAD • Protective [53]						
A. *SENCR*	h.s.	lncRNA, NAT to *FLI1*	• Down in CAD, PAD • n.a. [163]				• Down in ischemia • n.a. (proangiogenic) [164]		
A. *SMILR*	h.s.	lncRNA	• Up in CAD • (Detrimental? n.a.) [94]						
A. *Tug1*	m.m. r.n.	lncRNA	• Up in CAD • Detrimental [165]						• Up in MCAO • Detrimental [166]
B. *CHAER*	m.m. h.s.	lincRNA		• Up in TAC • Detrimental [16]					
B. *CHAST*	m.m. h.s.	lncRNA		• Up in TAC and in aortic stenosis patients • Protective [167]					
B. *CHRF*	m.m. h.s.	lncRNA within *Dcc*		• Up in TAC and after • AngII treatment • Detrimental [168]					

B. Hotair	m.m.	lincRNA	• Down in TAC • (Protective?, n.a.) [169, 170]		• Down in ischemic CM • (Protective?, n.a.) [169, 170]
B. KCNQ1OT1	m.m. h.s.	lncRNA, NAT to KCNQ1 (imprinted)	• Up in MI [171] • (Detrimental, n.a.) [172]		
B. Mirt1 / NR_028427	m.m.	lincRNA	• Up in MI (early phase) [55] • Detrimental [173]		
B. Myheart	m.m. h.s.	lncRNA, NAT to Myh7	• Down in TAC • Protective [15]		• Down in hypertrophic/ischemic/idiopathic CM • (Protective? n.a.) [15]
B. ROR /lincRNA-ST8SIA3	m.m. h.s.	lincRNA	• Up in TAC • Detrimental [174]		
B. Wisper	m.m. h.s.	eRNA of Wisp2	• Up in MI • Detrimental [91]		
C. HIF1aAS1	h.s.	lncRNA, NAT to HIF1a		• Up in serum in TAA • n.a. [71]	
D. Apf /AC246817	m.m.	lncRNA			• Up in I/R • Detrimental [175]
D. CARL	m.m	lncRNA			• Down in I/R and anoxia • Protective [18]
D. Carmen	h.s. m.m.	eRNA for miR-143 and miR-145	• Up post-MI • (Protective? n.a.) [19]		• Up in DCM and AOS • (Protective? n.a.) [19]
D. Cdr1-as	m.m	circRNA			• Up in LAD • Detrimental [176]

(continued)

Table 7.2 (continued)

lncRNA name	Studied orthologs	Classification	CVD entities						
			Atherosclerosis	Myocardial infarction	Aortic aneurysm	Cardiomyopathies and congenital heart disease	Vascularization and angiogenesis	Arrhythmia	Stroke and cerebrovascular aneurysms
D. Heartbrake	h.s.	lncRNA				• n.a. link to HAND2/ congenital heart defects [177]			
D. HRCR /mm9-circ-012559	m.m	circRNA				• Down in ISO • Protective [178]			
D. MFACR/mm9-circ-016597	m.m	circRNA				• Up in I/R • Detrimental [106]			
D. UCA1	r.n.	lincRNA				• Up in I/R • n.a. [179]			
E. MALAT1	m.m. r.n.	lincRNA, NAT to TALAM1 lncRNA				• Up in diabetic heart • Detrimental [180, 181]	• Up in hypoxic and diabetic conditions Detrimental/Protective (context-dep.)[13, 14]		• Up in ischemia • Protective [154, 162, 182, 183]
E. MANTIS / AK125871	h.s. m.f.	lncRNA, NAT to Annexin A4					• Down in IPAH • (Context-dep., n.a.) [69]		
E. Myoslid	h.s.	lncRNA, NAT to lncRNAs					• Down in arteries and venes in renal disease • (Detrimental?, n.a.) [184]		
F. AKO55347	h.s. r.n.	lncRNA						• Up in AF [185] • n.a. [185]	
F. Rffl-lnc1	r.n.	lncRNA, in Rffl						• Indel associated with short QT interval and high blood pressure • Protective [20]	

Name	Species	Type			
F. TCONS_00032546 and TCONS_00026102	c.l.	lncRNA	• Down in ANR • Detrimental–Protective [186]		
F. TCONS_00075467	o.c.	lncRNA	• Down in AF • Protective [187]		
G. C2dat1 / AK153573	r.n.	lncRNA, in CaMKIIδ			• Up in MCAO • Detrimental [188]
G. FosDT	r.n.	lncRNA, NAT to Fos			• Up in MCAO • Detrimental [73]
G. NILR	r.n.	lncRNA			• Up in MCAO • Protective [189]
G. SNHG14	m.m.	lncRNA			• Up in MCAO • Detrimental [190]

This table contains 46 lncRNAs that have been implicated in cardiovascular diseases based on genetic knockdown or overexpression in animal disease *models* in vivo, or, in 5 cases, based on decisive evidence from human GWAS studies where disease risk SNPs have been linked to differential lncRNA expression. These cases are labeled "SNPs" in the respective cells of the table. lncRNAs belonging to the following seven CVD entities: atherosclerosis (A), myocardial infarction (B), aortic aneurysm (C), cardiomycopathies and congenital heart disease (D), vascularization and angiogenesis (E), arrhythmia (F), stroke and cerebrovascular aneurysms (G). Within each class, lncRNAs are ranked in alphabetical order. lncRNAs implicated in several disease entities have been assigned to the one single class for which the most compelling evidence has been established. "Up" or "Down" refers to increased or decreased lncRNA abundance in disease conditions compared to controls. "Detrimental" or "Protective" refers to the function of the wildtype lncRNA in normal conditions. *n.a.* not analyzed in publications

AA aortic aneurysm, *AF* atrial fibrillation, *AngII* angiotensin II-induced heart injury, *ANR* abnormal autonomic nervous system remodeling; responsible for altered heart structure and function, *CAD* coronary artery disease, *CAVD* calcific aortic valve disease, *CM* cardiomyopathy, *DCM* dilated cardiomyopathy, *IPAH* idiopathic pulmonary arterial hypertension; lung small vessel disease, *I/R* ischemia-reperfusion-induced heart damage; classified here as model to study cardiomyopathy, *ISO* isoproterenol injection-induced heart injury, *LAD* permanent ligation of the left anterior descending artery, *MCAO* middle cerebral artery occlusion; cerebral infarction model, *MI* myocardial infarction, heart infarction, *PAD* peripheral artery disease, *QT* heart electric cycle interval, *TAA* thoraco-abdominal aneurysm, *TAC* transaortic constriction heart injury model inducing pressure overload; classified here under MI

Species: *h.s.* human, *m.m.* mouse, *r.n.* rat, *c.l.* dog, *m.f.* macaque; primate, *o.c.* rabbit

Table 7.3 Cellular and molecular roles of lncRNAs associated with cardiovascular diseases

lncRNA name	Classification	Role in tissue development	Cellular role	Molecular role
A. ANRIL	lncRNA, (NAT)	n.a	• Proproliferative, antiapoptotic in iPSC-derived monocytes and VSMCs [191] • Proadhesive, proinflammatory in monocytes and endothelial cells [191]	• Transcription: Tether for repressive chromatin modifier [192] • cis-/trans-regulator [191, 192]
A. circANRIL	circRNA	n.a	Antiproliferative, proapoptotic in iPSC-derived monocytes and VSMCs [24]	Protein regulation: Inhibitor of rRNA/ribosome formation [24]
A. Gas5	lncRNA	n.a.	Proapoptotic and antiproliferative in diverse cell types [144]	• Parent molecule for small RNAs[a] [193] • ceRNA/microRNA sponge [54] • Transcription: Decoy for TFs[a] [194]
Gomafu/MIAT/Rncr2	lncRNA	Nonessential	• Profibrotic in cardiac fibroblasts [146] • Pro/antiapoptotic (context-dep.) [148, 181] • Proproliferative, migrative in various cells including ECs [148] • Stem cell maintenances; switching to differentiated state[a] [195, 196]	• Splicing of cell fate determinants[a] [195] • ceRNA/microRNA sponge [148]
H19	lncRNA (imprinted)	Parent-of-origin-dep. effect (see text)	• Proinflammatory in macrophages [151] • Neuroinflammatory, proapoptotic in neurons in ischemia [153] • Regulating differentiation signaling and cardiomyocyte survival in heart [152, 197] • Proliferative invasive in various cells[a]	• Parent molecule for microRNAs[a] [198] • Transcription: Antagonizing microRNAs [199] • Transcription: Tether for repressive DNA/chromatin modifiers[a] [200] • mRNA decay: binding mRNAs[a] [201]
A. HAS2-AS1	lncRNA (NAT)	n.a.	Regulating ECM properties [155]	Transcription: Activating sense transcript via chromatin decompaction [155]

A. *HOTTIP*	lncRNA, bi-directional with *HOXA13*	n.a.	Proproliferative, promigrative in ECs and other cell types [156]	• Transcription: Tethering chromatin modifiers for long-range chromatin activation[a] • Transcription: eRNA-like[92]
A. *HOXC-AS1*	lncRNA (NAT)	n.a.	• Inhibition of cholesterol accumulation in macrophages [70]	n.a.
LINC00305.	lncRNA	n.a.	• Proinflammatory in blood monocytes • Proapoptotic in ECs [56]	• Protein regulation: Scaffolding protein interaction of membrane receptor • ceRNA/microRNA sponge [56]
A. *lincRNADYNLRB2-2*	lncRNA	n.a.	Anti-inflammatory and stimulating cholesterol efflux in foam cells [157]	n.a.
lincRNA-p21	lincRNA	n.a.	• Antiproliferative, proapoptotic in various cells including VSMCs [17]	• Transcription: Tether for repressive hnRNPK, corepressor for p53 [17] • Protein interaction regulation: Inhibiting binding of E3 ligase to target protein • Suppression of mRNA translation: binding to mRNAs[a] [202]
Meg3	lncRNA	n.a.	• Antiproliferative, antimigrative, proapoptotic in many cell types [162, 203] • Profibrotic, prohypertrophic in cardiomyocytes [161]	• Transcription: Tether of repressive chromatin modifier on genes *in trans* [204] • Protein interaction regulation: Interacting with DNA-binding domain of TF [203] • ceRNA/microRNA sponge [160]
A. *RNCR3/ LINC00599*	lncRNA	CNS defect	• Antiapoptotic in ECS; Proproliferative and migrative in VMCS [53]	• Parent molecule of microRNA [205]
SENCR	lncRNA (NAT)	n.a	• Required for VSMC contractility [67] • Affecting EC commitment from hESCs; Promigrative in ECs [164]	• n.a.
A. *SMILR*	lncRNA (eRNA?)	n.a.	• Proproliferative in VSMCs; regulating ECM properties [94]	Transcription: Enhancer of Upstream neighboring gene [94]

(continued)

Table 7.3 (continued)

lncRNA name	Classification	Role in tissue development	Cellular role	Molecular role
A. *Tug1*	lncRNA	n.a.	• Context-dep. Anti-/proproliferative [206] • Proapoptotic in ECs [165] • Antiadhesive in ECs [165] • actin polymerization in VSMCs [207]	• Transcription/Nuclear architecture: Scaffolding loci in nuclear Polycomb bodies [206] • Protein regulation: Scaffolding methyltransferase-non-histone target in cytoplasm [165] • ceRNA/microRNA sponge
B. *CHAER*	lncRNA	n.a.	Hypertrophic growth of cardiomyocytes [16]	Transcription: Competition with other lncRNAs in access to repressive chromatin modifier[a] [16]
B. *CHAST*	lncRNA	n.a.	n.a.	n.a.
B. *CHRF*	lncRNA	n.a.	Proapoptotic; hypertrophy, cardiomyocytes [168]	ceRNA/microRNA sponge [168]
B. *Hotair*	lincRNA	n.a.	Antihypertrophic growth of cardiomyocytes [169]	• Transcription: Tethering repressive chromatin modifiers *in cis*[a] [208] • Transcription: PRC2-independent repression[a] [209] • ceRNA/microRNA sponge [169]
B. *KCNQ1OT1*	lncRNA, NAT to *KCNQ1*	n.a.	Proapoptotic [172]	Transcription: Inhibiting neighboring gene by tethering DNA/chromatin repressors[a] [210, 211]
B. *Mirt1*	lincRNA	n.a.	Proapoptotic, proinflammatory in fibroblasts [173]	n.a.
B. *Myheart*	lncRNA	n.a.	Antihypertrophic, cardiomyocytes [15]	Transcription: Regulator of chromatin remodeling complex [15]
B. *ROR / lincRNA-ST8SIA3*	lincRNA	n.a.	Hypertrophic growth of cardiomyocytes [174]	ceRNA/microRNA sponge[a] [212]

Name	Type		Function	Mechanism
B. Wisper	lncRNA	n.a.	Proliferative, migrative, antiapoptotic in cardiac fibroblasts [91]	n.a
C. HIF1αAS1	lncRNA, NAT to HIF1α	n.a.	Antiapoptotic in ECs and VSMCs [71]	n.a.
D. Apf	lncRNA	n.a	Balance of autophagy activity for cell survival [175]	ceRNA/microRNA sponge [175]
D. CARL	lncRNA	n.a.	Proproliferative; cardiomyocytes and other cells (Balance of mitochondrial fission) [18]	ceRNA/microRNA sponge [18]
D. Carmen	lncRNA	n.a	Myocardial differentiation [19]	• Binding repressive chromatin modifier [19] • eRNA [19]
D. Cdr1-as	circRNA	CNS: sensomotoric gating	• Neurons[a] [213] • Cardiovascular: suggested to be proapoptotic in cardiomyocytes [176]	ceRNA/microRNA sponge [214, 215]
D. Heartbrake	lncRNA	n.a	Antidifferentiative in cardiomyocytes [177]	Transcription: microRNA suppressor [177]
D. HRCR	circRNA	n.a.	Antihypertrophic in cardiomyocytes [105]	ceRNA/microRNA sponge [105]
D. MFACR	circRNA	n.a.	Prosurvival, mitochondrial fission, cardiomyocytes [106]	n.a.
D. UCA1	lincRNA	n.a.	Proproliferative in diverse cell types including cardiomyocytes [179]	n.a.
E. MALAT1	lincRNA		• Proangiogenic, proproliferative in neonatal retina [13, 14] • Proproliferative in ECs, sprouting [13] • Antiapoptotic [182] • Invasive (cancer)[a] [216]	• Transcription: Repressor of local (240 kb) gene neighborhood[a] [126] • Inhibitor of repressive histone modifier in vitro[a] [206] • Splicing regulator in vitro[a] [126, 127, 217, 218] • Parent molecule for small RNAs[a] [51]

(continued)

Table 7.3 (continued)

lncRNA name	Classification	Role in tissue development	Cellular role	Molecular role
E. *MANTIS*	lncRNA (intronic NAT)	n.a	EC migration, sprouting, tube formation [69]	Transcription: Scaffold/regulator of chromatin remodeling complex (activator?) [69]
E. *Myoslid*	lncRNA (NAT)	n.a	VSMCs differentiation, proliferation, contraction, antimigrative [184]	n.a.
F. *AK055347*	lncRNA	n.a.	Prosurvival in cardiomyocytes [185]	n.a.
F. *Rffl-lnc1*	lncRNA in Rffl	n.a.	n.a.	n.a.
F. *TCONS_00032546 and TCONS_00026102*	lncRNA	n.a.	n.a.	n.a.
F. *TCONS_00075467*	lncRNA	n.a.	n.a.	n.a.
G. *C2dat1/AK153573*	lncRNA, in CaMKIIδ	n.a.	Proapoptotic in neurons[a] [188]	Transcription: Stimulating sense gene [188]
G. *FosDT*	lncRNA, NAT to Fos	n.a.	n.a.	Transcription: Tethering corepressor chromatin regulating complex [73]
G. *NILR*	lncRNA	n.a.	n.a.	n.a.
G. *SNHG14*	lncRNA	n.a.	Proinflammatory in microglia[a] [190]	Transcription: Repressing microRNA [190]

Roles of lncRNAs listed as relevant for cardiovascular diseases in Table 7.2

ceRNA competing endogenous RNA, *CNS* central nervous system, *eRNA* enhancer RNA, *hESC* human embryonic stem cells, *iPSC* induced pluripotent stem cell, *rRNA* ribosomal RNA, *TF* transcription factor

[a]Functions that are well established in the literature with a specific lncRNA, but which have so far been determined only in cells other than cardiovascular cell types (and have, thus, not yet been tested in CVD models)

7.3.1 Tethering/Scaffolding/Regulating Chromatin Modifiers

One of the most intensively studied function of lncRNAs is their **interaction with chromatin regulatory proteins** during the control of gene transcription (Fig. 7.1a). This includes interactions with proteins involved in chromatin fiber regulation by DNA methylation systems or histone tail-modifications, in nucleosome-remodeling, and in chromatin fiber folding and long-range looping in the 3D architecture of the nucleus.

The interaction with chromatin regulators is one of the best-understood function of nuclear lncRNAs. Some of the best-known lncRNAs have been shown to **tether chromatin-regulating complexes to specific sites in the genome**, and thereby to promote or repress transcription of protein-coding genes at selected target genes (Fig. 7.1a). Besides tethering, another central task in this interaction is the ability of lncRNAs to **scaffold** (link together) chromatin-regulating factors, for example, to **link several different chromatin readers and writers to specific genomic loci** [221–226]. Recent experiments show, however, that the picture is more complex than that: lncRNAs not only specifically **activate** chromatin regulators, but can also **inhibit** them. The hypothesis has formed that at least some of the long noncoding transcripts in our genome are used to inhibit the activity of chromatin regulators. The aim is to inhibit aberrant, low-affinity misassociation of chromatin regulators on the genome. With such a strategy, many nascent RNA transcripts from many types of genes are employed as noncoding RNAs in making gene expression more accurate [227].

Examples for lncRNAs that bind chromatin-regulating complexes at protein-coding gene promoters, and where relevance to cardiovascular pathophysiology has been established, are: *Fendrr* binding to both the repressive PRC2 and the activating TrxG/MLL complexes [7]; *Braveheart* binding to the repressive PRC2 [6]; *Malat1* binding the activating MLL, LSD1, BAF57, and unmethylated Polycomb 2 proteins in their complexes [206] and indirectly binding to nascent transcripts of active genes [228]; *HoxBlinc* binding to the activating Setd1a/MLL1 [9]; *Xist* to at least 10 repressive chromatin-binding and chromatin-regulating protein complexes including hnRNPK/U, noncanonical repressive PRC1, SHARP/Spen, SAF-A, and LBR [229, 230]; *H19* binding the repressive MBD1 [200]; *ANRIL* binding to activating as well as repressive PRC1 and PRCR2 complexes [191, 192]; *lincRNA p21* binding to repressive hnRNPK [17]; *Meg3* binding to repressive JARID2-PRC2 complexes [204]; *Tug1* binding to repressive PRC2, *RIZ1*, *Sin3A*, and *JARID1A* [206]; *Kcnq1ot1* binding to repressive PRC2 [210]; *MANTIS* binding an activating BRG1-containing SWI/SNF chromatin remodeling complex [69]; *HOTTIP* binding activating WDR5/MLL complexes [92]; *Carmen* binding PRC2 complexes with unclear outcome [19]; *Myheart* binding and inhibiting the BRG1-containing SWI/SNF chromatin remodeling complex [15]; *Chaer* binding and inhibiting PRC2 [16]; and *HOTAIR* binding the repressive PRC2 [208]. By interacting with these chromatin regulators, such lncRNAs can scaffold multiprotein complexes, tether them to specific genomic sites, **serve as decoys during target site association**, and change their activity. In the example of the study of *HOTAIR* it becomes obvious, however, that a careful follow-up analysis is required to confirm the functional relevance of a previously observed physical interaction with a chromatin regulator: While the *HOTAIR* lncRNA does bind the repressive PRC2 complex without doubt, and while there is a correlation between the presence of *HOTAIR* at a genetic locus and the repression of this locus, later decisive experiments challenged the simple picture that *HOTAIR* silenced target genes through PRC2. In fact, it was revealed that PRC2 was not primarily required for *HOTAIR*-mediated target gene repression and that PRC2:*HOTAIR* binding was likely a consequence of later PRC2 recruitment to an already silenced locus [209]. Therefore, care is advised in drawing conclusions on effector mechanisms from the many studies that report a binding of a lncRNA to Polycomb or Trithorax complexes, which are often recruited to loci to maintain but not to induce chromatin states.

What decides if a lncRNAs acts only *in cis* on few neighboring genes, and/or also *in trans* (on loci on other chromosomes on many targets in the genome)? This question has begun to be addressed also by studying lncRNAs with cardiovascular relevance. For example, *MALAT1*, representing one class of lncRNAs, binds with high affinity to hundreds of target genes [228, 231], as opposed to a different class of lncRNAs that bind only to a single locus in the genome [232]. It is thought that **the local concentration of a lncRNA on chromatin is decisive for this differential behavior**: Low-abundance lncRNAs interact rather with few genes in close proximity (like *HOTTIP*) (Fig. 7.1a), while *XIST* with 100 copies/cell can spread further using low-affinity sites, and *MALAT1* with thousands of stable RNA copies can diffuse also *in trans* throughout the nucleus and gain access to hundreds of targets [233]. The presented studies offer blueprints for thinking about how any novel lncRNAs may dynamically interact with chromatin.

7.3.2   lncRNAs Regulate Long-Range Chromosomal Looping in the Nuclear Space

How lncRNAs first bind to selected focal genomic points and then spread over larger domains on the genome to impose long-range gene expression control is currently being investigated (Fig. 7.1b). The *Xist* RNA has been instructive to understand lncRNA spreading over large chromosomal domains, which coincides with establishing a unique chromatin state over the domain, and with affecting the three-dimensional folding of the chromosome fiber. Through these processes, the lncRNA transduces a repressed or active chromatin state over large domains also by repositioning the chromosomal domain to preexisting repressive or activating nuclear subdomains. Intranuclear domains of relevance are different repressive domains (Polycomb bodies, heterochromatic regions close to the nuclear membrane, etc.) and active domains (interchromatin granules, transcription factories, etc.). How lncRNAs mediate chromosome fiber folding is a matter of current research, and this topic is probably best understood for Xist-dependent X-chromosome inactivation. There, unexpectedly, a small number of only 50 *Xist* RNA/PRC2 complexes are at work to compact the nucleosomes over the >150 megabases long X-chromosome. It has been concluded that lncRNA-chromatin interaction is highly dynamic, and a dynamic hit-and-run model has been invoked to explain spreading and coincident compaction of long chromosomal domains [233–235]. For spreading over such long distances, *Xist* RNA is initially locally stabilized by co-transcriptional binding to proteins at a nucleation within its own gene body [236]. Only after that it can start spreading on chromatin *in cis*. *Xist* uses the 3D architecture of the chromosome to spread to noncontiguous loci [237]. In the course of spreading, *Xist* changes chromatin loop structures through evicting cohesin rings which entrap fibers [238] (Fig. 7.1b).

Conceptually similarly, at least three other cardiovascular lncRNAs affect gene activity of target genes by determining looping of chromosomal domains encompassing the target genes. *MALAT1* was found to reposition target genes in the **3D nuclear space**. In particular, *MALAT1* associated with the Polycomb complex member Pc2, inhibited its preferred binding of repressive chromatin marks inside heterochromatic Polycomb bodies, and thereby led to looping of target genes into more active interchromatin granules in the nucleus [206]. *Kcnq1ot1* is involved in regulating chromatin loops, in particular by binding two distinct sequences that are 200 kb apart from each in the *Kcnq1* imprinting locus, an interaction that may contribute to imprinted monoallelic repression of the *Kcnq1* promoter [211]. *HOTTIP* expression and long-range chromosomal looping between the *HOTTIP* locus and the 5' HOXA sites correlate with the recruitment of the activating WDR5/MLL complexes and expression of the 5' HOXA cluster. Thus, lncRNA function can involve aspects that apply to single loci as well as to large chromosomal domains.

Paramount for interacting with chromatin regulatory proteins, for example, during folding chromatin fibers, lncRNAs use short conserved and functional subsegments or **secondary RNA structure folds** [41, 230, 239–241] (Fig. 7.1a). As a second general feature, lncRNAs biophysically have more **flexible joints** than proteins linkers [242, 243]. Thirdly, lncRNAs can often be **highly modular**, exhibiting a number of RNA folds in order to scaffold multiple effectors at once. The cardiovascular *ANRIL* lncRNA is such a case, and it can interact with different repressive Polycomb complexes [192], while using other RNA domains to bind and regulate other effector proteins [24], or to interact with specific genomic regions [191]. In an exemplary case that may be generally instructive for the study of lncRNAs, *Xist* has been found to be able to interact with an unexpectedly large (80–250) number of proteins [238]. In these interactions, *Xist* uses a dozen different RNA segments that separately recruit, for example, histone deacetylases [229], methyltransferases [244], ubiquitin ligase complexes [245], and corepressors [117, 230, 246] and tie the RNA to the nuclear lamina [247], all of which contributes to the heterochromatinization of *Xist* RNA-painted chromosomal regions. In the near future it can be anticipated that some of the recently identified cardiovascular lncRNAs will be biochemically studied in more detail, and one can expect that the number of interacting factors and possible effector mechanisms will significantly increase for each noncoding RNA.

7.3.3 Regulation of the RNAP II Preinitiation Complex at Gene Promoters

The regulation of transcriptional initiation is a multidimensional problem, and involves DNA, RNA, and chromatin-dependent regulation, all of which are mutually influencing each other [248, 249]. A major role of lncRNAs in this context is to regulate the activation of the promoter-bound RNA polymerase II preinitiation complex. In this context, lncRNAs can contribute to RNAP II regulation at several levels (Fig. 7.1c): (1) eRNAs stimulate enhancer function, (2) cheRNAs and ncRNA-a promote enhancer:promoter looping, (3) specialized circular RNAs (ElciRNAs and ciRNAs) activate the RNAP II holocomplex at promoters, and (4) other lncRNAs control transcription factor activity.

7.3.3.1 Activation of Promoters by Noncoding Transcription of Enhancers

Enhancers are known to loop over large distances to allow contact with promoter regions for specifying spatial and temporal or signal-dependent control of gene activation [248]. lncRNAs have taken a central role in how enhancers become active (Fig. 7.1b). The major finding in this respect was that **enhancers are transcribed** and express specific enhancer RNAs (eRNAs) [46, 81–83, 250–255]. Functionally, these eRNAs have been found to function, for example, as decoys for the NELF complex, a known repressor of the core RNAP II complex, thus promoting transcription [251]. Secondly, eRNAs bind and activate the Mediator complex, the central protein connector between enhancers and promoters, which also results in activating the RNAP II preinitiation complex at the promoter [256]. One has to be cautious, however, because whether enhancer transcription is a cause or consequence of promoter activation needs to be carefully tested in each case [257]. Finally, it may be the transcriptional act over an enhancer that activates a nearby gene [250, 258]. To understand this phenomenon, one must consider that during transcription, DNA is partly unwrapped from nucleosomes [259] and such chromatin opening can affect neighboring genes [43, 250]. As such, much remains to be learned about how eRNA-like lncRNAs affect looping between enhancers and promoters. Recent work has revealed how DNA-interacting proteins like CTCF and cohesin contribute to organizing chromosomes into 1–5 megabase-sized domains, so-called topologically associated domains (TAD), within which specific enhancer-promoter interactions can occur. But details of how specificity in the interaction between enhancer and promoter are constrained, and where lncRNAs can become active, are not yet fully understood (see [260] for a recent review on TADs).

7.3.3.2 cheRNAs and ncRNA-a Exert Enhancer-Like Functions

Surprisingly, also *conventional* **genes, either coding or noncoding, can behave as enhancers** [89]. For example, a group of lncRNA genes have been identified by virtue of the encoded lncRNA to reach out to about 3 kb to their neighborhood by acting as **eRNA-like molecule** (called ncRNA-a). In a similar way, but identified in a separate study, a class of chromatin-enriched cheRNAs have been identified as eRNA-like RNAs that associate with the chromatin fraction [261, 262] (Fig. 7.1c). In fact, many well-known lncRNAs with cardiovascular relevance like *HOTTIP* [92], *Kcnq1ot1* [263], or *linc-HOXA1* [93] activate genes in their immediate neighborhood [92, 264, 265], and it is not always clear whether as classical eRNA or as ncRNA-a or as cheRNA.

7.3.3.3 Circular ElciRNAs and ciRNAs Activate the RNAP II Holoxomplex

lncRNAs can also rather directly **stimulate the RNAP II complex at promoters**, and two classes of circular RNAs have been implicated: $3'$-$5'$-linked circular intronic RNAs (ciRNAs) and Exon-Intron-containing circular RNAs (ElciRNAs) [103, 104]. Both co-immunoprecipitate with RNAP II and stimulate its activity [266, 267]. The molecular details of how RNAP II is activated in each case are mostly unclear, but in the case of ElciRNAs the activation depends on the small nuclear U1 snRNA and leads to the activation of TFIIH and P-TEFb within the RNAP II preinitiation complex (Fig. 7.1c).

7.3.3.4 Transcriptional Regulation Through lncRNA:DNA R-Loop Formation

Compared to proteins, RNAs exhibit efficient and easily evolvable modes to both interact with other nucleic acids, as well as be a template for nucleic acid synthesis. Single-stranded RNAs, including nascent **lncRNAs and circRNAs, can hybridize with base-complementary DNA sequences** by threading in and partially opening the DNA helix through complementary base pairing. This can form an R-loop in the form of a rigid A-type-like RNA:DNA double helix [268]. Firstly, R-loops affect transcription: they can decompact nucleosome arrays and stimulate the formation of histone modifications conducive to transcription [268]. Secondly, R-loops cause RNAP II stalling, and associated with this an enhancer looping, or they can affect any type of chromatin-dependent process, such as splicing [269]. In fact, some lncRNAs induce genes *in trans* by establishing R-loops by threading into actively transcribed genes that already offer partially single-stranded DNA regions [231, 270]. Conversely, also the opposite can happen: degradation of R-loop-forming eRNAs on enhancer regions has been found to activate enhancers [252]. The functions of R-loops are still intensively studied, and the role of R-loop formation as a lncRNA effector mechanism is novel. No relevant studies have explored this effector mechanism for cardiovascular lncRNAs yet.

7.3.3.5 Transcriptional Regulation Through Binding Transcription Factors

Finally, different lncRNAs can also directly modulate transcription by binding to certain transcription factors and thereby modulating their binding ability. For example, lncRNA *RMST* was found to interact with the SOX2 transcription factor and to be important for its binding to SOX2 target gene promoters [271]. Similar interactions allow other lncRNAs to co-activate steroid hormone receptor targets or SP1 targets. Not all lncRNAs stimulate transcription though. For example, *Gas5*, a lncRNA with cardiovascular relevance, was found to bind and inhibit SMAD3 during TGFβ signaling [272], and the cardiovascular *Braveheart* lncRNA bound and inhibited the nucleic-acid-binding ZNF9, a factor previously implicated in dilated cardiomyopathy [273].

7.3.4 Antisense Transcription

Genome sequencing of many eukaryotes has shown that as much as 30% of human genes have antisense noncoding transcription partner genes that can overlap in part with the sense gene (Fig. 7.1d). These NATs or asRNAs are a rather diverse class of lncRNAs that function through different effector mechanisms and affect partner genes depending on the relative positioning of their gene bodies.

Transcriptomic analyses indicated that an antisense transcript is often orders of magnitude less abundant than its partner and remains nuclear [274]. Historically, after focusing on selected candidate NATs, a number of studies have begun to address the function of asRNAs on a global scale. Initial studies indicated that repression of the sense coding partner gene was a common function [80] (Fig. 7.1d). Subsequent genome-wide tests suggested, however, that **only a minority (one quarter) of stable asRNAs was functional in competitively silencing their sense gene**, and that many asRNAs may not be functional, at least under standard conditions [44]. Mechanistically, silencing via asRNAs can occur through two principal mechanisms: direct transcriptional interference modes and DNA methylation, or chromatin compaction, or combinations thereof (see [80] for review). Generally, the grade of asRNA-dependent silencing was found to be more pronounced when asRNA transcription reached over the start of the sense gene [61, 80, 208, 275–277]. Most recent experiments suggest that many lncRNAs only weakly affect the magnitude of expression of their sense partners, but that they rather reduce noisy spurious transcription events [44].

7.3.5 Regulation of mRNA Splicing

Several lncRNAs are known to be nuclear, and some of these have been shown to play important regulatory roles in splicing regulation (Fig. 7.1e). As a best understood example with cardiovascular relevance, *MALAT1* was found to localize in nuclear speckles, sites of coordinated transcription and splicing. This **lncRNA binds splicing regulators** of the SR splicing factor family and is essential for their localization in nuclear speckles and for alternative splicing of specific target genes therein [126, 127, 217, 218]. The exact regulatory mechanism has recently been investigated in detail: *MALAT1* interacts with the 3' ends of actively expressed and alternatively spliced pre-mRNAs, likely the RNA-binding SR splicing factors. Other cardiovascular lncRNAS implicated in regulating splicing are *Miat* [195] and, possibly, *Wisper* [91]. By affection splicing indirectly by influencing chromatin structure at gene loci, or as transcriptional regulators of splicing factors, many lncRNAs are expected to be involved in splicing regulation, but only a few mammalian lncRNAs have been identified as clear splicing regulators, and the details of their effector mechanism are still investigated [278–280]. Also, circRNAs have been implicated in regulating splicing. In fact, several independent studies suggest that co-transcriptional generation of circRNAs by the spliceosome, termed backsplicing, competes with linear splicing of mRNAs inside protein-coding genes. The picture emerges that on a genome-wide level, the function of most circRNAs might be to fine-tune the expression of their host mRNA. Underlying this mutual inhibition, several scenarios have been suggested: First, co-transcriptional backsplicing is thought to block the progression of the elongating RNAP II, which affects alternative exon skipping in mRNAs. Secondly, co-transcriptional backsplicing causes an intramolecular covalent linkage in the linear mRNA molecule, an internal junction with a single-strand RNA overhang. This may be the entry site for RNA exonucleases that degrade such mRNA molecules. For a detailed description of the different models that try to explain how backsplicing negatively affects coinciding linear splicing of mRNAs, see a recent review [104].

7.3.6   microRNA Sponging

An intensely studied and highly discussed function of lncRNAs is their capacity to **regulate the stability of coding mRNAs by sequestering microRNAs** that would usually target mRNAs for destruction (Fig. 7.1f). During sequestering, lncRNAs bind microRNAs only over limited stretches of 6–8 nucleotides in length, and therefore lncRNA sponges are spared from degradation [214, 215]. lncRNAs that sponge microRNAs are also called competing endogenous RNAs (ceRNAs). Sponging can occur in all classes of lncRNAs, in linear lncRNAs, in transcribed pseudogenes, and in circRNAs [99, 281–283]. Regarding cardiovascular-relevant lncRNAs, this relates to *H19, LINC00305, Meg3, Tug1, CHRF, ROR, CARL, Rncr*, and *TCONS_00075467* as well as to the circRNAs *HRCR* and *CDR1-as*. Since microRNA-binding regions are short and abundant, theoretically many microRNA sponging events may occur simultaneously on one lncRNA, and this greatly expands gene regulatory complexity [281]. Yet, given the comparatively low abundance of a given single lncRNA, and the relatively high abundances of miRNA and/or target mRNA(s), there is still some controversy about details if and when sponging actually occurs as a sufficiently potent gene-regulator mechanism in vivo [284–288]. A large fraction of lncRNAs, and therefore also of lncRNAs with cardiovascular function, have been implicated in microRNA sponging. We list sponging as a potential effector mechanism in Tables 7.1 and 7.3, but note that experimental evidence on sponging is not equally conclusive in each report. In particular, regarding circRNAs, *CDR1-as* may be one of the very few circRNAs that are actually effective as microRNA sponge in vivo [103].

7.3.7   Regulating mRNA Stability or RNA/DNA Editing

Certain noncoding RNAs pair with target RNA molecules. Thereby, lncRNAs can **promote or impair the stability of mRNAs** in the cytoplasm. This can be due to direct lncRNA:mRNA binding or by influencing the stability of protein:mRNA complexes (Fig. 7.1g). For example, the well-known Staufen 1 protein leads to nonsense-mediated RNA decay when it binds to double-stranded RNA regions in the $3'$ UTR of mRNAs [289]. lncRNAs were found to bind such stability-determining sequences in a target mRNA and protect from decay [289, 290]. Currently, the first mapping approaches by sequencing explore the entirety of all RNA:RNA duplex interactions in the transcriptome, including lncRNA:mRNA interactions [228, 291, 292]. More than 8000 intermolecular RNA interactions have been described therein, >3000 duplex structures are shared between human and mouse, and >100 lncRNA:mRNA interaction exist in single-cell types, indicating a potentially huge regulatory space [292]. As RNAs with cardiovascular relevance, *H19* has been described to bind the RNA-binding protein KSRP, and thereby promote KSRP-dependent target mRNA decay [201].

Secondly, lncRNAs bind to other mRNAs to promote **posttranslational modifications at double-strand RNA regions** (Fig. 7.1g). For example, hydrolytic deamination of adenosine leads to the production of the base inosine, referred to as A-to-I editing. Causal for this editing, ADAR, the responsible enzyme, requires double-stranded RNA folds as substrate. A-to-I editing has roles in many cellular processes, and first insights also showed a role in vascular smooth muscle cells [293] and misregulation in atherosclerosis [294]. Conceptually related, special noncoding RNAs have been identified that serve as guides for posttranscriptional modification of DNA bases by other enzymes in cells of the immune system [295].

7.3.8   Translational Regulation of mRNAs

In the cytoplasm, both linear and circular lncRNAs exhibit a number of regulatory roles relating to the activity of protein complexes as well as to the translation capacity of coding mRNAs (Fig. 7.1h).

A number of regulatory mechanisms have been identified for how lncRNAs impact the protein translation machinery. For example, some lncRNA are thought to positively stimulate the association of protein-coding mRNAs with translating ribosomes under conditions of translation stress [296]. Regarding lncRNAs with cardiovascular relevance, *lincRNA-p21* has been found to bind specific mRNAs and to impair their translation by interacting with the mRNAs and translation factors inside polysomes [202].

7.3.9   Protein Activity Control

Concerning circRNAs with cardiovascular relevance, two circRNAs have been suggested to interact with proteins in the cytosol, and, thereby, to impact central cellular functions (Fig. 7.1i). First, *circANRIL* has been found to bind to the protein PES1 and inhibit its function in a dominant-negative way [24]. PES1 is a member of the evolutionarily conserved PeBoW complex, consisting of Pes1 (Pescadillo), Bop1 (block of proliferation) and WDR12 (WD-repeat protein). This complex is essential from yeast to mammals to instruct the endonucleolytic excision of the *internal* and *external transcribed spacer* elements from the pre-rRNA and the formation of mature 28S and 5.8S rRNAs, failure of which denies the formation of a functional 60S large ribosomal subunit and, consequently, the proper translation of any kind of protein. *circANRIL* stems from the *ANRIL* locus at chromosome 9p21, the most prominent genetic factor of atherosclerotic cardiovascular disease identified by genome-wide association studies [12, 23, 129, 191]. *ANRIL* is transcribed as a linear and a circular lncRNA (*linANRIL* and *circANRIL*, respectively). While *linANRIL* is thought to be a major effector of CVD risk, circANRIL appears to be protective. The function of *linANRIL* will be described in detail separately below, but it is not involved in translational regulation, as far as known. In contrast, *circANRIL*-dependent PeBoW complex regulation impairs cellular protein translation capacity, and this function has been suggested to underlie circANRIL's protection from atherosclerosis [24] (Fig. 7.1i). In a different type of effector mechanism, a second circRNA, *circPABPN1*, is thought to function as a decoy for the HuR protein. By sequestering HuR, circPABPN1 has been suggested to inhibit the translation of a subset of mRNAs that depended on this RNA-binding protein [297]. In fact, HuR is known to promote the stability of many mRNAs and noncoding RNAs [298]. As a direct or indirect consequence of sponging HuR, the protein translation from HuR target mRNAs is affected.

7.3.10   Translation Potential of lncRNAs

Linear lncRNAs do not contain open reading frames (ORFs) that are longer than 300 nucleotides (encoding for more than 100 amino acids), which is the central definition of their character [38, 299]. The distinction between noncoding and coding is not so clear after all in RNAs, and this for several reasons: First, many lncRNAs do contain small ORFs just by chance, and more sensitive mass-spectroscopic analyses and genetic screens revealed that some of these are translated to micropeptides that are as small as 24 amino acids in length on average (Fig. 7.1j). A relatively well-studied class of micropeptides is the family of structurally conserved sarcolamban peptides. These are 34–46 amino acids in length, and are encoded on classically defined lncRNAs. They are functionally important because they bind and regulate the activity of the sarcoplasmic Ca^{2+} pump SERCA and regulate its activity. SERCA is an ATPase important for contraction-relaxation coupling in myocytes by pumping Ca^{2+} from the cytosol into the lumen of the sarcoplamatic reticulum, which is important also in the heart muscle [112]).

On the other hand, circRNAs do stem from protein-coding genes, and thus contain one or several exons. Consequently, circRNAs can contain longer open reading frames. A number of independent ribosome profiling studies have concurred that the large majority of the many thousands of circRNAs are not translated [35, 100, 300]. From the thousands of circRNAs only a few dozen had the potential to encode a polypeptide [301] (Fig. 7.1j), and so far, translation of only two circRNA molecules, *circMbl* [301] and *circZNF609*, have been documented with confidence [302]. Paramount for their translation was that RNA circularization led to the inclusion of the endogenous start codon, as well as of 5′ untranslated regions (UTRs) that folded into specific secondary RNA structures with internal ribosome entry site (IRES)-like properties [301, 302]. Translation from circRNAs can usually not happen, as the circularization that occurs in the internal regions of the gene is likely a consequence of circularization from internal portions of genes and, consequently, the absence of a 5′ Cap, of linear ends, and of the Kozak sequence for ribosome entry and translation initiation in linear 5′ capped mRNAs (see [303] for review). In the rare case of translation from circRNAs, usually the protein produced suffers from premature truncation compared to the native full-length protein of the endogenous linear host mRNA. Whether truncated proteins translated from circRNAs are functional is still unknown.

Summarizing, as heterogeneous as their biogenesis are the functions exerted by lncRNAs. But a common overarching feature is that many lncRNAs participate in the expression control of protein-coding genes. For this, lncRNAs can act *in cis* and *in trans* and affect gene expression on multiple levels, such as by regulating transcription, or affecting pre-mRNA splicing, mRNA stability, and mRNA translational control. As such, many lncRNAs function because of engaging in molecular interactions of the lncRNA with proteins or mRNAs directly, but also the transcriptional act over a lncRNA gene body can per se have functional consequences on neighboring genes, while the lncRNA product formed in this case may be a side product without any function [43, 52, 304]. Lastly, lncRNA genes may even be functional because the length of their gene bodies determines the relative distance between the left and right neighboring genes or regulatory DNA sequences, as learned, for example, by multiple knockout experiments in the complex *Hox* gene cluster [305]. Thus, functional studies of lncRNAs have to be carefully designed to take account of all possible levels of functions of a lncRNA.

7.4   LncRNAs in Cardiovascular Health and Disease

Compared to protein-coding genes, the function of lncRNAs is less well studied. This is especially true for lncRNAs implicated in cardiovascular diseases because many disease-relevant lncRNAs have only been identified within the recent 1–5 years. Consequently, only a few knockout studies on lncRNAs have been reported, and most of these were performed on lncRNAs that had been studied already before in more general cellular functions or in cancer biology.

In the following we first review developmental roles of lncRNAs at the organ level in the context of the embryonic cardiovascular system. These studies used classical knockouts and transgenic overexpression, and assessed the embryonic development of the heart, angiogenesis, or more specifically, lineage specification of the cardiovascular system, for example, from early mesoderm precursor cells. Table 7.1 summarizes the 17 lncRNAs studied by knockout analysis, which were found to have a function in cardiovascular physiology. We then continue by highlighting lncRNAs that have been linked to cardiovascular disease in rodent disease models in vivo (Table 7.2). For a good part of these lncRNAs, evidence exists on differential gene expression of the orthologous human lncRNAs in patients. For only a very few, additional genome-wide association studies (GWAS) in humans have determined single nucleotide polymorphisms (SNPs) in the lncRNA genes, which are associated with disease risk and with differential lncRNA expression (Table 7.2). In Table 7.3, we summarize the evidence on cellular and molecular functions of lncRNAs from Table 7.2.

7.4.1 lncRNAs Regulating Cardiovascular Development in the Embryo

For the heart to form, and be induced by signaling cues, pluripotent embryonic stem cells progressively differentiate into mesodermal and cardiac precursor cells that subsequently terminally differentiate [306]. Transcriptional networks are directed to control lineage commitment and cardiac cell specification [307]. By participating in gene expression regulation, lncRNAs are an intricate part of heart development. A large number of lncRNAs (>200) were found to be differentially expressed in different steps of cardiac commitment during in vitro differentiation from embryonic stem cells [308]. Moreover, hundreds of lncRNAs are known to be differentially abundant at different points in fetal heart development in vivo based on whole-tissue profiling [309]. Also, about 300 lncRNAs are cardiac-specific in the adult heart. Based on coexpression analysis and considering which known protein-coding genes were the closest neighbors of the relevant lncRNAs, some lncRNAs were prioritized as candidates in regulating key developmental determinants of heart development [310].

As a first example of a lncRNA studied by knockout analysis, the mouse lncRNA *Fendrr* has been found to be rather specifically enriched in the lateral plate mesoderm during mid-gestation, and its deletion by insertion of polyA signals, known to cause a transcriptional stop, disrupted the development of ventral structures, including the heart and body wall [7]. As a consequence of myocardial dysfunction, the mutant embryos died. A mere depletion down to 40% of lncRNA levels by RNA interference caused no mutant phenotype in this case, which may be an important consideration for similar developmental studies of other lncRNAs [7]. Mechanistically, *Fendrr* was found to scaffold a repressive Polycomb group complex and, separately, also an activating Trithorax-MLL complex, and to tether it to gene promoters by triple helix formation. This coincided with a long-term effect on target gene expression in the descendants of cells of the cardiac mesoderm [7]. Likely in this function *Fendrr* affected the transcriptional regulators of heart development, such as *Gata6*, and lateral plate mesoderm control genes like *Foxf1*, *Irx3*, and *Pitx2*. Some of the many target genes were regulated *in cis*, some *in trans* [7]. When studied in an independent knockout where internal lncRNA sequence was replaced by a lacZ cassette, a slightly delayed perinatal mutant phenotype was observed, and these *Fendrr* mutants showed lung defects [311]. Together, the discrepancy in mutant phenotypes between different knockouts is not without precedence and may even be expected given that both the RNA transcript and transcription through the lncRNA locus may be relevant [304, 312].

Similar to *Fendrr*, the mouse lncRNA **Braveheart** (*Bvht*) interacts with a chromatin regulator. *Bvht* binds and inhibits the repressive Polycomb complex [6]. Bvht also binds and inhibits the transcription factor ZNF9 [273]. Thereby, *Bvht* regulates an entire cardiac transcription factor network to promote early cardiac cell fate [6]. This network is upstream of master cardiac transcription factors, like *MesP1*, at least in an in vitro model of cardiomyocyte differentiation from embryonic stem cells (ESCs) [6] [273]. MesP1 activity is known to be essential for the specification of all different cardiac cell types, cardiomyocytes, vascular smooth muscle cells (VSMCs), and endothelial cells (ECs). Consistent with the proposed function in cultured cells, in vivo, *Bvht* is essential for early heart development and for maintenance of neonatal cardiac cell fate [6]. No clear *Bvht* ortholog was found in humans, but the possibility exists that different lncRNAs have taken over a conserved role in establishing a cardiogenic transcription factor network also in other vertebrate species.

HoxBlinc is a lincRNA residing in the Hoxb gene locus and has been associated with lineage commitment during cardiovascular development by knockout studies [9]. *HoxBlinc* tethers the activating trithorax Setd1a/MLL1 histone methylating complex to *HoxB*. Loss of function experiments showed that *HoxBlinc* is required to activate *HoxB* during embryogenesis when precursor cells initiate mesoderm formation in the primitive streak. In the mutant, cardiogenic and hemangiogenic mesoderm cell fates are not specified.

Based on high-throughput RNA sequencing of cells differentiating from human ESCs to cardiovascular progenitors and finally to terminally differentiated fetal-like vascular endothelial cells in culture, two independent studies established that several hundred lncRNAs were specific for each investigated stage [118, 119]. From each stage-specific set, one lncRNA was randomly chosen to be functionally analyzed in more detail. Corresponding to their expression during EC differentiation, in vivo expression analysis in both mouse and zebrafish embryos confirmed that **Terminator** was specifically expressed early after fertilization, **DEANR1/Alien** only later in the lateral plate mesoderm and **Punisher** only when the vasculature had formed [118]. Injection of antisense morpholinos binding and degrading these lncRNAs in zebrafish showed stage-specific requirements consistent with the roles inferred from cell culture profiling: *Terminator* was important for gastrulation, with survivors showing defects in the vasculature. *Alien* was important for mesoderm specification, including subsequent vascular patterning, dorsal and intersegmental blood vessel and cardiac chamber formation. And *Punisher* was important for vascular vessel formation, extension, and branching as well as for cardiac development [118]. The molecular effector mechanisms of *Terminator* and *Punisher* remain unknown. *DEANR1/Alien* was also the focus of another study that investigated stage-specific lncRNA expression during definitive endoderm and pancreatic cell specification from hESCs [119]. *DEANR1/Alien* was shown to be encoded close to *FOXA2*, an endoderm marker gene, and to stimulate *FOXA2* expression. It was at the same time also important for endoderm specification. The latter function was at least in part due to *FOXA2* regulation by *DEANR1/Alien*, as confirmed by genetic rescue experiments in cells [119]. While the mechanism of regulation remains to be confirmed, the first experiments suggested that in the forming endoderm, *DEANR1/Alien* may be important to recruit or stabilize SMAD2/3 at the *FOXA2* locus [119].

Tie-1AS is a NAT that binds and downregulates the sense mRNA of its host gene, *Tie-1*. Based on overexpression of *Tie-1AS* and on inhibition of *Tie-1* in zebrafish, the lncRNAs was suggested to have a mild impact on the proper organization of cell-cell junction markers in vivo, including those between vascular endothelial cells [121]. Corroborating these data, the human orthologous lncRNA inhibited human *Tie-1* and was important for tube formation of HUVECs in collagen gels in vitro [121]. These observations are relevant because an earlier mouse knockout for *Tie-1* had documented that this gene was required for vessel integrity [120]. Also, mutations in the related *TIE2/TEK* are known in humans to lead to different defects in venous morphogenesis based on a role of *TIE2* in EC:VSMC (vascular smooth muscle cell) interaction [122].

Upperhand (Uph) is a lncRNA gene that has been found to be important for heart morphogenesis in a mouse knockout study. The mouse *Uph* has a human orthologue, called *HAND2-AS1*, and both mouse and human *Uph/UPH* share a bidirectional promoter with *Hand2/HAND2*, a well-known and important transcription regulator of heart development [8]. Experiments on the mouse *Uph* locus revealed that *Uph* functioned as an enhancer for *Hand2* in embryonic heart tissue and was essential for *Hand2* activation. TALEN-mediated insertion of premature polyA signals that stopped *Uph* transcription led to the embryonic death of knockout mice. *Uph* KO death was ascribed to a failure in forming a right ventricular chamber, a phenotype identical to the *Hand2* KO, and corroborating that Uph functioned through activating *Hand2* in vivo [8]. For its enhancer activity towards *Hand2*, the lncRNA transcribed from the *Uph* locus was not essential. Instead it was the transcriptional act over the heart enhancers in the *Uph* locus that seemed to make these enhancer elements active.

Xist is the central regulator of dosage compensation (X-inactivation), which is inherently essential for female survival, and will not be discussed here (see [1] for a review). A recent conditional *Xist* knockout using *Sox2*-Cre drivers allowed some *Xist* mutant females to survive to adulthood. Surprisingly, the survivors showed rather specific organ defects, and these included defects in heart and spleen maturation [123]. Specifically, perinatal heart growth was delayed and mutants had smaller hearts, as cardiomyocytes did not sufficiently mature by cytoplasmic enlargement [123]. Separately, another study had deleted *Xist* in the fetal hematopoietic lineage and had also found a rather specific phenotype. *Xist* mutant hematopoietic stem cells (HSCs) in females were impaired in differentiating to all lineages [124]. The latter manifested in multilineage blood cell defects, and in adulthood, mutants eventually died from aggressive blood cancers, especially in the myeloid lineage [124]. Together, although the available evidence is skewed by which drivers have been used for *Xist* deletion in the two knockout studies [123, 124], it is surprising that such a general pathway like *Xist*-dependent X-inactivation is somewhat selective for regulating blood stem cells, cardiac, and spleen development. There is a tangible explanation at least for why HSCs are affected: Hematopoietic precursor cells are special among other cell types because they selectively regain the capacity to initiate *de novo* X-inactivation, while other cell types at this advanced stage of embryogenesis lack the silencing factors [125].

Bioinformatics screens had initially identified muscle-specific human *LINC00948* and mouse *AK009351*. These were later found to encode small protein-coding ORFs that were translated **to *myoregulin (MLN)***, a micropeptide that bound and activated the sarcoplasmic Ca^{2+} pump SERCA [112]). SERCA is an ATPase important for contraction-relaxation coupling in myocytes by pumping Ca^{2+} from the cytosol into the lumen of the sarcoplamatic reticulum. These micropeptides belong to the larger family of structurally conserved sarcolamban peptides. LncRNAs of this class have been studied by knockout approaches in vivo. It has been found that micropeptides in this family, like MLN, phospholamban (PLN), or sarcolipin (SLN), can repress the activity of SERCA to terminate muscle contraction [110, 112]. In contrast, the related micropeptide **DWORF**, which is encoded on human *LOC100507537* and mouse *NONMMUG026737* lncRNAs, displaces these SERCA repressors and thereby enhances contraction [116]. SERCA regulation is of relevance also for heart muscle function, due to its regulation of cardiomyocyte contractility.

The ***Metastasis-Associated Lung Adenocarcinoma Transcript 1 (MALAT1)*** is one of the molecularly best-studied lncRNAs. Based on assays in standard cell lines, *MALAT1* has been implicated in regulating alternative splicing [217] and in promoting selective transcriptional activation of target genes. The latter involves binding of *MALAT1* to Pc2, a central member of the Polycomb group silencing complex. Upon binding to *MALAT1*, Pc2 loses its preference for reading heterochromatic histone modifications and gains preference for binding to active chromatin, causing reactivation of Pc2-marked cell-cycle stimulating genes [206]. Beyond that, *MALAT1* interacts with dozens of other proteins [313] and is even the origin of functional small RNAs excised from its 3′ end during maturation (mascRNAs) [51], indicating a possibly wide range of still elusive functions. In the face of all this knowledge, it was surprising that three independent knockouts of the lncRNA *MALAT1* in mouse showed that *MALAT1* was not essential for survival or for organ development and function under normal growth conditions in embryos or adults [126–128]. Neither was global transcription or splicing affected in these mice, which could be due to redundancy with yet unknown different lncRNAs. Only a selective role in regulating genes encoded close to the *MALAT1* locus was found [126, 127]. In contrast to the lack of clear in vivo phenotypes, in vitro *MALAT1* had some functions under stress conditions: *MALAT1* was upregulated under hypoxia, and exerted proangiogenic effects, as will be described in detail in Sect. 7.4.2.3.

Human **MIAT**, also known as **Gomafu/MIAT/Rncr2**, has been linked to cardiovascular disease as will be summarized in the following section on lncRNAs in disease [314]. Separate studies showed that it binds to several splicing regulators and regulates a relatively small number of genes. *MIAT* was implicated before in stem cell, neuronal, and retinal cell differentiation [195]. A recent mouse knockout was established for *Miat* but did not reveal any obvious anatomical defects. Only selective behavioral defects were observed, and only a small number of splicing alterations were documented in primary neuron cultures from these mutant mice [219]. How the lncRNA regulates selective neuronal functions in specific brain parts, and by which molecular mechanisms, is still unknown.

Among 150 lncRNAs deregulated by pressure overload after transaortic constriction in a mouse model, **Chaer, cardiac hypertrophy associated epigenetic regulator**, is a mouse lncRNA conserved also in humans, which shows enriched expression in the heart [16]. It will be described in detail in the context of lncRNAs with roles in cardiovascular diseases in Sect. 7.4.2.1. In the course of studying *Chaer*, a genomic deletion was inserted into the *Chaer* lncRNA locus. The mutant mice did not show obvious morphological organ deficits or functional heart problems in normal conditions. Only in experimental pressure-overload models in the mouse was a function in cardiomyocyte growth control revealed [16].

Rncr3, retinal noncoding RNA 3, also known as **LINC00599**, is a lncRNA with cardiovascular relevance as will be described in detail in the description of lncRNAs involved in atherosclerosis in Sect. 7.4.2.1. A knockout of the *Rncr3* locus exists, but the mutants did not show any morphological or functional abnormalities of the heart or of the vasculature, as far as reported [205]. Instead, the initially viable *Rncr3*-deficient mice become debilitated and later die, likely because of defects observed in multiple types of neurons in the central nervous system.

CDR1-as, cerebellar degeneration-related protein 1 antisense RNA, is a circular lncRNA. It was the first circRNA that was found to have a dedicated function in eukaryotes [214, 215]. It will be described in detail in Sect. 7.4.2.1 because reports exist that suggest this circRNA to be misregulated and misfunctioning in cardiovascular disease. Inconsistent with these reports, loss-of-function studies have been performed in mouse and in zebrafish in vivo, but *CDR1-as* showed no obvious function in hearts of the vasculature. Instead, and consistent with the expression pattern in vivo, a requirement for neuronal functions was revealed. Recently, the *CDR1-as* circRNA was knocked out in mouse, which was technically possible, because of the peculiarity of this locus to only express a circRNA but no linear RNA. The *CDR1-as* circRNA knockout did not show anatomical alterations, and also no heart defects, but defects in central nervous system function [213].

The **H19** lncRNA is located in the *H19/Igf2* imprinted locus, where *H19* and *Igf2* are reciprocally imprinted, such that the two genes are expressed from maternal and paternal chromosomes, respectively, but not from both alleles in a cell. Relevant for the cardiovascular system, *H19* and *Igf2* are known to be expressed in muscle and other mesodermal organs in the embryo. *Igf2* has been studied in the context of embryonic heart development, where it was found to be an epicardial mitogen that was important for ventricular wall proliferation. Heart-specific functions of *H19* remain unknown, but it is peculiar that *H19* is downregulated in adult tissues except in skeletal muscle and heart and that *H19* is reactivated in the cardiovascular system during stress signaling, as will be described in Sect. 7.4.2.1. *H19*'s major function has been found to be limiting for the growth of the placenta and to inhibit cell proliferation, for example, during tumorigenesis. One major effector mechanism is the production of microRNAs from the *H19* sequence. In this function, *H19* was the parent molecule for a microRNA that repressed an *IGF* receptor for maintaining quiescence in long-term quiescent hematopoietic stem cells (HSCs) [198], and for another microRNA that promoted muscle differentiation from myoblasts and muscle regeneration.

7.4.2 Cardiovascular Disease-Associated lncRNAs

There are three principal approaches to how lncRNAs have been implicated in cardiovascular disease. First, lncRNAs may reside in genomic regions, which have been associated with cardiovascular disease (CVD) in genome-wide association studies (GWAS). This approach has so far been reported in only a few cases because genetic variants identified through GWAS are often found in regulatory sequences in some distance to genes, and it is far from trivial to establish causal links in the functionality (expression, splicing, and sequence) of a lncRNA. Secondly, and following a different rationale, a number of lncRNAs have been linked to CVD because they were found to be differentially regulated in disease conditions through genome-wide RNA expression profiling. In a third strategy, lncRNAs were investigated in candidate gene approaches because they regulated disease-relevant processes or resided in or close to protein-coding genes with already established disease relevance. Accordingly, here we will highlight lncRNAs implicated in vivo as causal effectors of the following seven disease entities: **atherosclerosis, myocardial infarction, aortic aneurysm, cardiomyopathies and congenital heart disease, vascularization and angiogenesis, arrhythmia, and stroke and cerebrovascular aneurysms.** Especially for the more established and better-studied lncRNAs, in vivo functions in more than one disease entity have emerged, all of which will be discussed in separate sections. Table 7.2 summarizes all lncRNAs, their change in expression in disease (up/down), and whether their normal function is to protect from disease (protective) or to exacerbate disease (detrimental), as far as determined from genetic knockdown or overexpression in animal disease models. Table 7.3 lists the cellular and molecular functions of all lncRNAs from Table 7.2, as determined from accompanying experiments in relevant cell culture systems.

7.4.2.1 Atherosclerosis and Myocardial Infarction

Briefly, atherosclerosis is characterized by the formation of fibro-fatty lesions (plaques) in arteries, whereby vascular wall cells (SMCs, fibroblasts, ECs) and

diverse immune cells (neutrophils, macrophages, T cells, B cells) aberrantly prolif-
erate, undergo cell death, change cell fate, and trigger nonresolving immune
reactions. The lesions accumulate cells and lipoprotein material and are covered
by a fibrous cap. Upon necrosis in the lesions, and extracellular matrix (ECM)
remodeling, the cap thins, plaques can rupture, and plaque material ends up in the
bloodstream and trigger luminal thrombosis also at distant sites, leading to
myocardial infarction (MI, heart attack) or stroke in brain tissue. In this book chapter
we are reviewing lncRNAs that have been linked by functional analysis to the onset
and severity of atherosclerosis, and to myocardial infarction, which occurs as
acconsequence of atherosclerotic processes. Here we will not review lncRNAs
associated with CAD risk factors such as dyslipidemia, diabetes, obesity, metabolic
syndrome, or systemic blood pressure regulation. So far, 16 lncRNAs have been
implicated in the pathogenesis of atherosclerosis in animal disease models in vivo, of
which 3 were also found to play a role when tested in myocardial infarction models
in vivo. Among CAD-related lncRNAs, as of date, only five lncRNAs emerging
from GWAS have been functionally studied in the context of CAD. There are,
however, dozens of other known CAD risk loci from GWAS datasets, which contain
noncoding RNAs and have not yet been explored in the same way. How lncRNAs
determine risk at GWAS loci will be interesting to explore. In our review, nine
different lncRNAs have been associated only with myocardial infarction in relevant
models, without being implicated in atherosclerosis through in vivo evidence. We
start out with describing the CAD-related lncRNAs: First, a number of tightly
clustering SNPs have been found on the Chr9p21 locus in several independent
genome-wide association studies testing the susceptibility for coronary artery disease (CAD) [130, 131], peripheral artery disease [136], and myocardial infarction
(MI) [135]. A separate haplotype block associated with type II diabetes risk is
encoded nearby [315–317]. Regarding CAD/MI, the dosage of the risk SNPs was
found to be associated with atherosclerosis severity. Also, homozygosity for the risk
SNPs correlated with a twofold increased risk for CAD/MI, which was relevant for
approximately 20% of the population [131, 318]. As explained before, the major
effector of this locus is not a protein-coding gene but a long noncoding RNA, human
ANRIL. *ANRIL* expression levels are increased in more severe, atherosclerosis
phenotypes [10, 12, 129, 131–134]. Its molecular effector mechanism is under
intensive investigation, not least because this lncRNA is tightly linked also to
different diseases, including cancer. Relevant for one proposed effector mechanism,
ANRIL is expressed adjacent to the human INK4-ARF tumor suppressor gene locus
on chromosome 9p21, which encompasses $p16^{INK4a}$, $P14^{ARF}$, and $P15^{INK4b}$,
whereby transcription of *ANRIL* overlaps in antisense with transcription of
$P15^{INK4b}$, hence its name ***Antisense noncoding RNA in the INK4 locus***.

ANRIL has become a paradigm for physiological control of cell proliferation and
survival by a lncRNA. For example, by binding to the PRC1 and PRC2 complex,
ANRIL has been suggested to impose selective repressive histone modifications at
the *INK4-ARF* locus, a central tumor suppressor locus whose repression is known to
result in a cell cycle G0-G1 entry in specific contexts [192, 319, 320]. Such a cell-
cycle stimulating role may hypothetically be relevant for *ANRIL*'s described roles in

activating atherosclerosis [321]. Indeed overexpression of *ANRIL* in several cell types, including PBMCs as well as cells of the vessel wall, was found to be proproliferative [191]. On the other hand, *ANRIL* can also interact with transcriptional coactivators to alter the transcription of target genes *in trans* and, thereby, affect cell adhesion and apoptosis in ways that stimulate atherosclerotic plaque formation [191, 322]. Although the *ANRIL* sequence is primate specific, the structure of the 9p21 locus is conserved in mice, where this locus is also a gene desert with a lncRNA. *AK148321*, the mouse lncRNA in this locus, has a conserved exon structure. Corroborating the importance of the locus for disease, deletion of the entire region led to cardiac-specific misregulation of the *INK4-ARF* orthologs *Cdkn2a/b*, to the misregulated proliferation of mutant aortic VSMCs [323], and to increased vascular aneurysm formation [324]. However, the knockout was not associated with altered atherosclerosis plaque formation, thus raising the question whether the mouse locus is a lncRNA-dependent atherosclerosis risk factor in vivo. In any case, the noncoding region of human Chr9p21, and likely its transcription, is important for CAD susceptibility in humans.

Making the regulation even more complex, the *ANRIL* locus also expresses **circular ANRIL (circANRIL)** RNA isoforms [23, 24]. *circANRIL* abundance was decreased, while linear *ANRIL* levels were increased, in carriers of CAD/MI risk alleles in an association study of 1000–2000 patients with suspected coronary artery disease in PBMCs and whole blood. These findings were recapitulated also in more than 200 human carotid endarterectomy tissue specimens [24]. Moreover, *circANRIL* was decreased in CAD patients also in blood T-lymphocytes [23]. The notion emerged that *circANRIL* protects from CAD [24]. Specifically, *circANRIL* impairs rRNA processing in cells by inhibiting the rRNA-processing PeBoW protein complex [24]. This results in nucleolar stress, p53 activation, and reduced translation capacity, functions that are antiproliferative and, conceptually, would antagonize the proproliferative functions of linear *ANRIL*. Indeed, in cultured cells *circANRIL* functions independently of linear *ANRIL* and impairs proliferation and sensitizes for apoptosis [24], functions that have been associated as to be anti-atherogenic [24]. Together, the *ANRIL* locus highlights a situation, where both the linear and the circular RNA products are functional, and where linear and circular RNA may have antagonistic functional outputs (pro- and antiproliferative, respectively).

GAS5, growth arrest-specific 5, has been known as a noncoding transcript before genome-wide detection of lncRNAs became routine. *GAS5* is conserved in mouse and human and was found to be upregulated during conditions of cell growth arrest, and contribute to cell cycle arrest and apoptosis when induced in lymphocytes and many other cell types. *GAS5* also harbors 10 snRNAs in its introns, some of which give further rise to small piRNAs that induce activating chromatin remodeling at the locus encoding the proapototic *TRAIL* gene. Beyond GAS's role as tumor suppressor in cancers, *GAS5* has also been studied in atherosclerosis. *GAS5* was found to be more abundant in atherosclerotic plaques in patients, and in a rat CAD model [144]. Consistent with its role in other cell types, GAS5 contributed to apoptosis in cultured macrophages and SMCs [325]. In which cells and at which stage of plaque development GAS5 contributes to atherosclerosis will be important to test in the future.

Another lncRNA has been associated with cardiovascular disease in a GWAS case-control study of over 3000 patients: six SNPs in a previously unknown lncRNA on human Chr22q12.1were found to be associated with myocardial infarction, and the lncRNA was, therefore, named *Gomafu, MIAT, myocardial infarction associated transcript*, later also designated as *Rncr2*. Of these SNPs, one was linked to increased *MIAT* transcription [314]. *MIAT* is also known to be upregulated in atherosclerotic plaques [145]. The mouse *MIAT* orthologous lncRNA was found to be aberrantly upregulated in the myocardium shortly after MI in a mouse model for induced infarction [146]. *MIAT* knockdown reduced cardiac fibrosis and improved cardiac function, which indicated that overactivation of *MIAT* was detrimental and contributed to heart fibrosis [146]. The underlying molecular mechanisms remained unknown in this study. Molecular effector functions of mouse *Miat* have, however, been delineated in a number of unrelated mouse models, which give an idea of how *Miat* might function in organs in general: For example, mouse *Miat* was found to be one of the few highly conserved noncoding targets of Oct4, and it was required for mouse embryonic stem cell maintenance. Conversely, increased levels promoted mesodermal and ectodermal lineage specification [196]. This may be of relevance also for the heart and progenitor cells therein. Also in other cell lineages, this proposed role in switching from progenitor to differentiated cell state was corroborated: It could be shown that *Miat* was involved in prosurvival signaling of differentiated neuronal cell progeny in mouse [195]. Apart from binding and potentially sponging microRNAs, in all these studies its only other described molecular function was in splicing regulation, possibly in guiding splicing of cell fate determinants [195]. How this function in progenitor cell differentiation relates to the function of progenitor cells in damaged tissues and organs remains to be tested.

H19, an imprinted maternally expressed transcript, is a lncRNA known as a model for studying the regulation of an imprinted locus, and for how imprinted growth-regulating loci are implicated in tumorigenesis. Two SNPs in *H19* on human Chr11p15.5 have been associated with CAD in a GWAS of 700 individuals [149]. Increased *H19* levels have been found in the plasma of atherosclerosis patients [150, 151]. It is also known already for long that *H19* is low expressed but that it can be reinduced in the aorta or VSMCs upon stress signaling or in atherosclerotic plaques. Also, when tested in cultured macrophages upon treatment with oxidized LDL, *H19* contributed to the inflammatory transcription response [151]. How *H19* contributes to cardiovascular diseases on a molecular level is still to be determined.

HAS2-AS1, HAS2 antisense RNA 1, is located at the *HAS2* gene locus [155]. This locus is interesting, because it is neighboring another cardiovascular lncRNA, *SMILR (smooth muscle–induced lncRNA enhances replication*, also known as *RP11-94a24.1). SMILR* and *HAS2* were transcribed from the same strand and in the same direction, while *HAS2-AS1* is transcribed as antisense RNA. *HAS2-AS1* expression has been found to be stimulated by inflammatory signaling and to activate *HAS2* expression. *HAS2* shows increased levels in human plaque tissue as compared to healthy controls. Thereby, *HAS2*, a hyaluronan synthetase, is thought to promote atherosclerotic neointima formation. On the other hand, also *SMILR* has been found to be increased in human plaque tissue. Additionally, *HAS2, but not HAS2-AS1* levels dropped after *SMILR* siRNA-mediated knockdown. Together, by a still elusive molecular mechanism, *SMILR* is thought to be required for the full transcriptional stimulation of *HAS2* by opening the chromatin compaction state of the *HAS2* locus [155]. Thus two lncRNAs may independently relay the detrimental overactivation of *HAS2*.

Some members of the evolutionarily conserved HOX gene family have been specifically associated with regulatory roles in the cardiovascular system. For example, several selected HOX genes are involved in cardiac progenitor proliferation and differentiation; others must be repressed for angiogenesis to proceed. Two human HOX gene-regulating lncRNAs have been studied: *HOTTIP* and *HOXC-AS1*. **HOTTIP, HOXA distal transcript antisense RNA**, is a well-studied lncRNA that had originally been found as a *cis*-regulator of the *HOXA* locus from where it is transcribed, particularly as an activator that recruited stimulating histone-modifying WDR5/MLL complexes for long-range chromatin activation. *HOTTIP* has recently also been found to be increased in arterial tissues samples of CAD patients [156]. Overexpression and knockdown experiments showed that *HOTTIP* levels regulated EC proliferative activity. *HOTTIP* also affected migration capacity, as determined by experiments in cell culture. Whether and how *HOTTIP* contributed to atherosclerotic plaque formation, however, remained unclear. A more general role of *HOTTIP* is likely, since overactivation of *HOTTIP* has also been found in cancers, and was implicated in prostate cancer hormone-independent growth, likely in large parts via *HOX* gene activation.

The second relevant lncRNA is **HOXC-AS1, HOXC cluster antisense RNA**. Microarray RNA profiling of carotid atherosclerosis revealed an antisense lncRNA in the human HOXC cluster, **HOXC-AS1, HOXC cluster antisense RNA**, as downregulated compared to normal arterial intimal tissue [70]. Investigating how this downregulation might occur, the authors turned to cultured monocytes and found that administration of oxidized LDL to cultures decreased *HOXC-AS1* expression. In these conditions, overexpression of *HOXC-AS1* abolished the accumulation of cholesterol in monocytes as well as the inhibitory effect of ox-LDL on the expression of the nearby *HOXC6* gene. The authors suggested that the *HOXC-AS1* antisense lncRNA might somehow regulate cholesterol metabolism, and that the possibility existed that this occurred via transcriptionally affecting HOX genes in the HOXC cluster [70]. This finding is of particular importance because independent work on *HOXC9* has revealed that this HOX gene can promote EC quiescence and vascular vessel morphogenesis [70]. Therefore, HOX-regulating lncRNAs have been studied with a particular focus on angiogenesis, as described in Sect. 7.4.2.4.

By searching the GWASdb database for SNPs already previously linked to atherosclerosis by GWAS but which are less significant ($P < 1.0 \times 10^{-3}$) and were manually curated from the literature [326], one group focused on one such SNP that resided in the intron of a lncRNA on Chr18q22.1: This lncRNA, **LINC00305**, had so far not been characterized and its expression was found to be enriched in atherosclerotic plaque samples as well as in blood monocytes of atherosclerosis patients, showing that the SNP associated with lncRNA expression [56]. To study the function of *LINC00305*, gene expression profiling was performed upon lncRNAs overexpression in THP-1 monocyte cell lines, as well as in human aortic smooth muscle cells cocultured with these. *LINC00305* promoted the expression of inflammatory genes by activating NF-κB signaling in monocytes, which translated to shifting SMCs from a contractile to a pathological synthetic state [56]. Using immunoprecipitation experiments and mass spectrometry to screen for lncRNA-interacting proteins, the authors found that *LINC00305* bound to membrane-resident lipocalin-interacting membrane receptor and increased its interaction with the aryl hydrocarbon receptor repressor (AHRR). This interaction altered downstream AHRR signaling, such that more nuclear AHRR and increased NF-kB signaling were observed after *LINC00305* overexpression. Whether and how AHRR regulated NF-kB signaling components is not yet clear [56].

A recent RNA expression analysis of foam cell formation upon oxLDL administration to cultured human macrophages revealed induction of **lincRNA-DYNLRB2-2** [157]. This lncRNA was further studied because it was found that it induced expression of the evolutionarily conserved *GPR119*. This regulation potentially is of relevance for CAD because GPR119 is a G protein-coupled receptor that is known from earlier work to be involved in metabolic homeostasis by suppressing food intake and reducing body weight gain in rat models [327]. In the study on lincRNA-DYNLRB2-2, it was found that both lincRNA and GPR119 also promoted cholesterol efflux from human foam cells. The efflux driven by lincRNA-DYNLRB2-2 depended on GPR119 [157]. Conversely, lincRNA-DYNLRB2-2 repressed *TLR2*, a well-known upstream Nf-κB signaling stimulator, and this repression was necessary for cholesterol efflux [57]. These findings were substantiated by the finding that *GPR119* overexpression in $apoE^{-/-}$ mice conferred antiatherosclerotic effects, while *TLR2* overexpression conferred proatherosclerotic effects [157]. *lincRNA-DYNLRB2-2* was not tested in this in vivo context.

Another lncRNA is **lincRNA-p21,** located on mouse Chr17, 15 kb upstream of the tumor suppressor *p21 (cyclin-dependent kinase inhibitor 1A)*, but transcribed from the opposite strand and in the opposite direction. *lincRNA-p21* was not primarily identified by GWAS studies in humans, but by a genomic study of mouse lincRNAs that were direct transcriptional targets of p53 and that were induced by p53 [17]. Only later studies focused on lincRNA-p21's role in atherosclerosis. Four SNPs were found in human *lincRNA-21* on Chr9p21.2, locating in a single haplotype block and correlating with reduced risk for CAD and MI in over 600 patients in a case-control study [158]. The effect of risk alleles on *lincRNA-p21* expression was not reported in this study, and the molecular function of *lincRNA-p21* not explored. In another study, *lincRNA-p21* was also found to be downregulated in atherosclerotic plaques in arteries in an $apoE^{-/-}$ mouse model for CAD, as well as in arterial tissue and in PBMCs of human CAD patients [159]. Knocking-down *lincRNA-p21* locally in injured mouse carotid arteries with siRNA technology caused hyperplasia of the neointima in the lesion, coinciding with increased proliferation and reduced apoptosis levels in cells of the vessel [159]. How does this relate to p53? Through decades of earlier research on p53, it has become clear that p53 affects many cellular facets, most notably the switch between cell survival and apoptosis, cell proliferation, and DNA repair. The majority of studies have focused on p53's role in tumorigenesis, but some studies on its role in atherosclerosis exist: In atherosclerotic tissue, *p53* is not mutated, but in mouse models of atherosclerosis, *p53* knockouts showed increased disease severity, and VSMCs and macrophages were thought to be affected [328]. This shows that *lincRNA-p21* phenocopies *p53*'s roles, consistent with the notion that this lincRNA was a p53 effector also in the process of atherogenesis [17, 159]. In fact, earlier functional tests had shown that mouse *lincRNA-p21* served a function as a corepressor for p53 in its second known function as a transcriptional repressor of a set of target genes, including *p21* [17]. It could, thus, be that normally *lincRNA-p21* repressed antiproliferation genes like *p21* to induce apoptosis and reduce proliferation in an atheroprotective function.

Meg3, maternally expressed gene 3, has been well studied. It is known to inhibit proliferation and migration, in part via inhibiting TGFβ signaling, and promote apoptosis in the context of cancer studies. In a cardiovascular context, a study reported that human *MEG3* was downregulated in tissues of CAD patients and that overexpression of *MEG3* in endothelial cells suppressed EC proliferation in vitro [160]. It is important to mention in this context that *MEG3* is expressed from the conserved *DLK1-MEG3* imprinted domain. Studies of this imprinted locus in mouse showed that *Meg3* function as a repressor by tethering the Polycomb PRC2 complex at target genes *in trans* [204]. This is relevant as mouse *Dlk1* has been independently linked to atherosclerosis: The suppression of *Dlk1* protected from atherosclerosis through promoting regenerative EC division for EC turnover in injured arteries, while disturbed blood flow inhibited this pathway [329]. Thus, MEG3/*Meg3* may have several cell-type-specific roles in the cardiovascular system, and EC-related functions may be particularly interesting to study in the future.

RNCR3, retinal noncoding RNA 3, also known as **LINC00599**, is a lincRNA that is a source of the microRNA *miR-124a*. *RNCR3* was found to be induced in mouse and human aorta atherosclerotic lesions, and specifically also by ox-LDL treatment of ECs and VSMCs [53]. In a mouse model of atherosclerosis, systemic injection of shRNA for *RNCR3* increased lesion size. This proceeded with EC apoptosis, reduced VSCM proliferation and migration, and with increased cholesterol levels and higher levels of TNF-α and IL-6 in the circulation [53]. Though the exact underlying mechanism is still unknown, these data suggest that *RNCR3* is an atheroprotective factor. Possibly *RNCR3* exerts a role in protecting from apoptosis because an earlier independent knockout of *RNCR3* had shown a phenotype in the central nervous system, with defects in brain development, axonal morphogenesis of dentate gyrus granule cells, and retinal cone cell death.

SENCR, smooth muscle and endothelial cell-enriched migration/differentiation-associated long noncoding RNA, is a lncRNA that is encoded as NAT of the *FLI1* gene. *FLI1* is a transcription factor known to regulate EC and blood cell formation. *SENCR* has described pleiotropic functions in ECs and VSMCs, making it difficult to propose how it specifically functions in CAD [67]. Directed tests revealed that *SENCR* was downregulated in ECs from vessels of patients with premature CAD, as well as in patients suffering from occlusive peripheral artery disease [163]. Whether any of the cellular roles of *SENCR* contributed to disease was not further functionally investigated. Although being a NAT, *SENCR* functions independent of *FLI1* regulation, at least while regulating VSMC contractility. How *SENCR* functions is unclear, but it has been shown that *SENCR* is cytoplasmic [67].

SMILR, smooth muscle–induced lncRNA enhances replication, also known as **RP11-94a24.1**, has already been mentioned above in the context of *HAS2* locus regulation. SMILR has, however, also been studied on its own: While profiling RNA expression following stimulation of human VSMCs with the inflammatory interleukin-1α and the mitogen and chemoattractant PDGF, 200–300 noncoding transcripts were found to be differentially regulated [94]. *SMILR* was among the top differentially abundant lncRNAs. *SMILR* also became the focus of interest for this study because IL-1α and PDGF treatment did not change its abundance in other vascular cell types such as ECs [94]. *SMILR* was shown to be more abundant in unstable human atherosclerotic plaques, as compared to healthy adjacent vessel tissue and prospectively defined by positron emission tomography imaging. *SMILR* knockdown in cultured VSMCs reduced VSMC proliferation, while overexpression enhanced it [94]. When investigating how *SMILR* functioned on a molecular level, the authors found that genes in the genomic neighborhood of *SMILR* were coordinately upregulated together with *SMILR* in their cell culture model.

Tug1, taurine upregulated gene 1, is another well-studied lncRNA. It had originally been found in studies of the developing retina, but subsequent studies revealed a multitude of functions in diverse cell types: Overarching some functions, it has been proposed that *Tug1* repressed growth control genes though scaffolding and promoting their intranuclear association with repressive Polycomb group complexes in heterochromatic Polycomb bodies in the nucleus [206]. Since *Tug1* is also expressed in ECs, and the role of PcG members is of general importance for many cell types, *Tug1*'s function also in the cardiovascular system was investigated. Here, it was found in cell culture studies that *Tug1* activated the enzymatic activity of the PcG member Ezh2 in methylating α-actin as a nonhistone target in the cytoplasm of VSMCs. This type of regulation has been found in earlier independent studies and correlates with increased actin polymerization. Also, *Tug1* was shown to regulate EC tight junctions. Lastly, *Tug1* was found to be upregulated in an $ApoE^{-/-}$ mouse model of atherosclerosis, and to be less upregulated when these mice were fed the anti-atherosclerotic and antianginal compound Tanshinol, a carbocyclic catechol compound in herbal extracts of sage in traditional Chinese medicine [165]. Whether the human *TUG1* orthologue was also differentially expressed in human atherosclerotic lesions was not tested. Nevertheless, Tanshinol was found to repress the induction of mouse *Tug1* and of human *TUG1* by ox-LDL in mouse and human ECs, in cell culture, respectively. Conversely, *TUG1* overexpression inhibited the antiapoptotic effect of Tanshinol in ox-LDL-treated human ECs [165]. This suggested that human *TUG1* may be a therapeutically interesting lncRNA in human CAD.

In the following section we highlight lncRNAs that have been implicated in myocardial infarction through the analysis of relevant animal models of MI, but which have not been specifically studied in atherosclerosis models. Data mainly derive from mouse MI models, in many cases experimentally triggered by transverse aortic constriction (TAC). In some cases, differential lncRNA expression data have been obtained from orthologous lncRNAs in humans myocardial infarction.

Among 150 lncRNAs deregulated by pressure overload after transaortic constriction in a mouse model, *Chaer, cardiac hypertrophy associated epigenetic regulator*, showed enriched expression in hearts and was studied carefully [16]. A human *CHAER* ortholog exists that is also expressed in the human heart. There is at least a trend for *Chaer* to be induced in hearts from patients with dilated cardiomyopathy. The *Chaer* locus was inactivated by deleting exon 2, but the knockout mice did not show obvious morphological organ deficits or heart problems. Yet, under pressure overload, cardiac hypertrophy and pathological fibrosis were reduced in the absence of *Chaer* [16]. The knockouts also showed overall increased heart function in this model. Molecularly, *Chaer* was found to interact with Ezh2, the central histone methyltransferase of the PRC2 polycomb complex. Further analyses suggested that *Chaer* competed with other lncRNAs such as *Hotair* or *Fendrr* in binding to PRC2, while leaving the enzymatic activity of Ezh2 intact. This interaction of *Chaer and the PRC2* complex seemed to negatively impact the association of PRC2 with selected target genes and thus their repression [16]. The human *CHAER* ortholog shows similar interactions with the human PRC2 complex and stimulated the expression of hypertrophic genes in human cardiomyocytes. Why *Chaer/CHAER* gained access to PRC2 over *Hotair* during a hypertrophic growth phase in cardiomyocytes is not yet fully clear, but a modification of the PRC2 complex by the growth-promoting mTOR complex was indicated in the first analysis. Targeting this switch is of specific interest also from a therapeutic aspect.

While screening for lncRNAs differentially expressed in conditions of cardiomyocyte hypertrophy upon transverse aortic constriction in mice, *Chast, cardiac hypertrophy-associated transcript*, was identified [167]. This lncRNAs was also upregulated in human hypertrophic hearts when analyzing heart tissue of aortic stenosis patients. Systemic overexpression of *Chast* ameliorated heart pathology in TAC-operated mice, indicating that it might be exploited as a protective factor in the future.

CHRF, cardiac hypertrophy related factor, was identified in a study on the microRNA-dependent regulation of cardiac hypertrophy. Among other microRNAs, the conserved *miR-489* was recently shown to be downregulated in pressure-overload models following transverse aortic constriction (TAC) in mice, as well as in human heart failure samples, and in cultured cardiomyocytes upon treatment with angiotensin II [168]. Ang-II is a vasorestrictive peptide binding to its membrane receptors, known for a long time to promote cardiomyocyte hypertrophy also in vitro. Apart from many other important physiological functions in the kidney and other organs, Ang-II is known to be released upon hemodynamic overload and can stimulate cardiac hypertrophy. Ang-II treatment of cardiomyocytes is an in vitro model to study aspects of the downstream effects. Interestingly, *miR-489* depletion by RNAi caused hypertrophy, and further promoted hypertrophy upon Ang-II treatment in vitro [168]. Conversely, the cardiac-specific overexpression of *miR-*

489 in conditional transgenic mice did not show anatomical aberrations, but displayed a reduced hypertrophic response in the heart after in vivo Ang-II treatment. This effect was ascribed, at least in part, to micro-RNA-dependent inhibition of translation of Myd88, an adaptor protein for Nf-κB signaling that was already previously involved in regulation of hypertrophy. *CHRF*, on the other hand, acted upstream, was induced by Ang-II and in TAC mouse models, sequestered *miR-489*, and when overexpressed impaired miR-489. Together, *CHRF* was suggested to promote cardiac hypertrophy and caused apoptosis [168].

Hotair, Hox transcript antisense intergenic RNA, is another famous lncRNA studied in the context of embryonic development as well as in cancer. In a recent study, Hotair was found to be decreased after TAC surgery in mouse hearts [169, 170]. There is also a human *HOTAIR* lncRNA, and it is known to be downregulated in left heart ventricle biopsies as well as in PBMCS in dilated ischemic cardiomyopathy patients [170]. Overexpression of mouse *Hotair* in cultured cardiomyocytes reduced angiotensin II-triggered hypertrophic growth in vitro [169]. The authors suggested that for this function *Hotair* sequestered *miR-19* to derepress *PTEN* [169], a pair of factors implicated directly and indirectly before in other cells in growth-dependent DNA replication control [330]. Whether this is the effector mechanism in vivo is open, also because *Hotair* has been previously implicated in *Hox* gene regulation. A number of studies put forward evidence that *Hotair* repressed transcription of posterior Hox (*Hoxd*) genes *in trans*, through tethering several independent repressive chromatin-modifying complexes. Most recently, dedicated knockout analyses of the entire *Hotair* locus in vivo challenged the *trans*-regulation model that based on near-complete or partial exonic knockouts of Hotair. The knockouts showed that *Hotair* only had a very subtle role, and this was in *cis*-regulation of nearby *Hoxc* genes [52].

KCNQ1OT1, KCNQ1 opposite strand/antisense transcript 1, is known as a classical example of a lncRNA expressed at an imprinted growth regulating locus. In a recent study, human *KCNQ1OT1* levels were also found to be increased in neutrophils and monocytes in whole blood preparations in patients with myocardial infarction (MI), where it was associated with hypertension, and not so much with inflammation markers [171]. This study showed that *KCNQ1OT1* together with *ANRIL* might be useful in prediction of left ventricular dysfunction after MI. Extending this study, a recent report tested the role of *KCNQ1OT1* in infarcted tissue. Silencing of *KCNQ1OT1* antagonized cell death in myocardial cells upon oxygen/glucose deprivation in vitro [172]. Thus, *KCNQ1OT1* has heart-specific functions, and may not only be a marker of MI, but also actively promote injury in ischemic hearts when overacted [172]. The effector mechanisms of *KCNQ1OT1* during MI are still unknown. *KCNQ1OT1*'s molecular roles are only known from unrelated studies in other organ systems. Knockout analyses showed that dysregulation of mouse *Kcnq1ot1* correlated with embryonic overgrowth phenotypes and mental disabilities but also with some cancers. In these studies it became clear that *Kcnq1ot1* is paternally expressed and recruits the repressive PRC2 and the G9a histone methyltransferase for silencing *Kcnq1* and other protein-coding genes in the locus *in cis*. In the heart, *Kcnq1ot1* represses *Kcnq1* during cardiac development, but whether this is the causal determinant for becoming detrimental during MI remains open.

Mirt1, myocardial infarction-associated transcript 1 or *NR_028427,*) and **Mirt2** (*ENSMUST00000100512*) were identified in a group of 30 lncRNAs that were differentially expressed in a mouse model of myocardial infarction based on coronary ligation [55]. Within the 24 h after permanent occlusion of the anterior interventricular artery, *Mirt1* and *Mirt2* were upregulated 10–20-fold. *Mirt1* levels then decreased to baseline levels by 48 h. *Mirt1* located in fibroblasts more than in cardiomyocytes. When *Mirt1* was depleted by injecting shRNA-expressing constructs into the myocardium before MI, cardiac functions were ameliorated after MI and the infarction size slightly reduced, suggesting that *Mirt1* upregulation contributed to pathological changes in the infarcted heart [173]. Similarly, in vitro, hypoxia-induced *Mirt1* expressed in cultured mouse cardiac fibroblasts, and *Mirt1* contributed to their apoptosis, to Nf-κB activation, and to the expression of inflammatory cytokines [173].

Myheart (Mhrt), is a NAT conserved in mouse and humans, and is expressed antisense to the Myosin Heavy Chain 7 RNA transcript, a molecular motor protein allowing heart muscles to contract. *Mhrt* is a myocardial-specific lncRNA and expressed with several isoforms in nuclei of cardiomyocytes [15]. Mouse *Mhrt* lncRNAs were found to be downregulated in a mouse TAC-induced heart pressure overload model, which was used as a trigger for cardiac hypertrophy and fibrosis and impairing hearts functionally. Likewise, human *MHRT* levels were downregulated in heart tissue of patients suffering from left ventricular hypertrophy, or from ischemic cardiomyopathy or from idiopathic dilated cardiomyopathy. Conditional forced transgenic expression of mouse *Mhrt* in cardiomyocytes of adult mouse hearts in vivo reduced stress-induced cardiac hypertrophy and fibrosis, as well as left ventricular dilation and fractional shortening [15]. This offered evidence for the notion that Mhrt is a lncRNA that protects from hypertrophy-induced heart injury. Mechanistically, mouse *Mhrt* was shown to bind SWI/SNF chromatin remodeling complex, which is well known to regulate a number of target genes by modulating nucleosome positioning and occupancy at genetic loci. The binding of *Mhrt* to Brg1 was found to repress Brg1 activity, since *Mhrt* was shown to compete with binding of Brg1 to chromatinized DNA. Since Brg1 was additionally found to repress *Mhrt1* lncRNA expression, a feedback loop was suggested that may fine-tune *Mhrt* RNA levels and Brg1 activity levels to protect hearts from hypertrophy of cardiomyocytes [15]. Given its pronounced therapeutic effects in the mouse model, *Myheart* may become one of the first lncRNAs that might be used for therapeutic purposes in humans in the future.

ROR, regulator of reprogramming, or previously known as *lincRNA-ST8SIA3*, has been identified in unrelated work as lincRNA that is upregulated during derivation of human-induced pluripotent stem cells from fibroblasts and contributed to gaining an undifferentiated proliferative cellular phenotype in pluripotent cells [21]. At least part of this role was ascribed to functioning as a sponge for *miR-145*, *miR-181*, and *miR-99* [212]. This allowed the core pluripotency transcription factors OCT4, SOX2, and NANOG, which were otherwise targeted by these three microRNAs, to become more stably expressed, especially to withstand low-level aberrant differentiation signals. Since these transcription factors also stimulated *ROR* expression, a feed-forward regulatory loop had been proposed that stabilized stem cell self-renewal, but also made differentiation progress more robust since *ROR* levels decreased during programmed progenitor cell differentiation [212]. These findings were noticeable because they placed, for the first time, a lncRNA in the core pluripotency transcription factor network. *ROR* has more recently also been found to be upregulated in a mouse heart injury model following transverse aortic constriction, as well as in models of cardiomyocytes hypertrophy in cell culture [174]. siRNA-mediated depletion of *ROR* impaired hypertrophic growth in vitro. Since ROR could be depleted by *miR-133*, a known antihypertrophic microRNA, the authors suggested a pathological role of *ROR* in promoting hypertrophic growth downstream of this microRNA [174]. How the stimulation of hypertrophy in cardiomyocytes of the diseased heart relate to the hypothetical reactivation of pluripotency-related genes, as suggested from the previous unrelated studies, remains to be addressed.

The authors of a recent study focused on a set of lncRNAs that were heart-enriched, conserved also in human and differentially induced after myocardial infarction in mouse and human [331]. Five percent of these lncRNAs could be mapped to heart-selective super-enhancers [91]. Super-enhancers, also known as stretch-enhancers, represent a subgroup of enhancers that have been identified by a body of recently published work, particularly influential enhancer elements in the genome: Super-enhancers have a high multiplicity and density of enhancer elements, and locate close to genes that encode for important cell-fate-determining factors, master microRNAs, and for central signaling pathway components with relevance for development and disease (see a recent publication for characteristics of super-enhancers [332]). Among the top induced lncRNAs in the class of heart-enriched lncRNAs, a recent study found the mouse **Wisper** lncRNA, named after the location close to the protein-coding gene *Wisp2*, a nonstructural signaling protein in the ECM [91]. *Wisper* was shown to be present in cardiac fibroblasts and to peak in expression two weeks after MI in mice in an LAD mouse MI Model, as well as in the proliferative fibrotic phase in hearts in a model of cardiac fibrosis induced by hypertension due to left renal artery clipping. Depletion of *Wisper* upregulated *Wisp2*, decreased proliferation and migration, impaired the transdifferentiation of cardiac fibroblasts to myofibroblasts, at least based on marker gene expression profiling in vitro, and induced proapoptotic genes. Since *Wisp2* depletion alone did not trigger all these effects, [91] must have additional effectors. Among many possible mechanisms, the authors found four RNA-processing factors and splicing regulators, TIAR, PTB3, DIS3L2, and CELF2, to specifically bind the lncRNA *Wisper* [91]. The relevance of these interactions is not yet clear, but using GapmeRs to deplete *Wisper* in vivo reduced the extent of cardiac fibrosis and the infarction size. A human ortholog of *Wisper* exists, and shows fibrosis-associated induction in AOS (Adams Oliver syndrome) patients, who characteristically show myocardial fibrosis and LV dysfunction. These data suggest that *Wisper/WISPER* has a conserved detrimental function in MI [91]. It is interesting to note that *Wisper* is not the only example of a lncRNA residing at a super-enhancer locus. Also *Carmen* and *Upperhand* (see above) may be super-enhancer RNAs, and have been investigated in relation to the cardiovascular system.

7.4.2.2 Aortic Aneurysms

Aortic aneurysms are focal asymmetrical dilations of the vessel. They occur most commonly in the infrarenal abdominal aorta, but can also be found elsewhere, for example in the chest. The biomechanical integrity of the vessel is in large parts determined by fibers, such as collagen and elastin fibers, in the ECM, and the ECM of the vessel depends in large parts on the VSMCs. An SNP in *ANRIL* has been linked not only to abdominal aortic aneurysms, but also to sporadic and familial intracranial aneurysm [136, 333, 334]. A nearby SNP in *ANRIL* that associated with diabetes did not associate with aneurysms in this context. How *ANRIL* molecular contributes to changes in the extracellular matrix or in the contractility of VSMCS has not been tested in the context of aneurysm formation. Yet, in studies of cancerous cells, *ANRIL* was implicated in promoting metastasis, which may point to an underlying role in promoting ECM remodeling, a hypothesis that remains to be tested in more detail [335–338].

Also *HIF1aAS1, HIF1A antisense RNA 1*, became the focus of interest because it was found to be differentially abundant in the serum of patients with thoracoabdominal aortic aneurysms (TAA) [71]. *HIF1aAS* was more abundant in TAA. The authors tested its role in affecting the survival of vascular cell types in culture. Knockdown of *HIF1aAS1* partially protected from experimentally induced apoptosis in VSCMs and ECs in vitro. Whether and how this function in VSMCS and ECs related to aneurysm formation was not further tested [71].

7.4.2.3 Cardiomyopathies and Congenital Heart Disease

So far, knockout analyses have shown that 10 lncRNAs have roles in embryonic heart development, and that a further 6 lncRNAs do not show any obvious morphological or functional abnormalities in the embryo but develop dysfunction under stress conditions in the adult (Table 7.1). These latter lncRNAs are part of the group that will be described in the following: In this section we review 12 lncRNAs linked to cardiomyopathies and heart diseases, some of which are genetically inherited (congenital). LncRNAs in this group have often been experimentally studied in rodent ischemia-reperfusion (I/R) heart injury models, and specifically in the context of hypertrophic overgrow of cardiomyocytes, which is triggered thereby. When cardiomyocyte hypertrophy is sustained it may eventually increase the risk of heart failure. So, the functional separation of this class from the class of MI-related lncRNAs (see above) is not always clear-cut. Beyond the lncRNAs that have been functionally studied and are reported here, many more lncRNAs have been found to be differentially expressed in human heart diseases but have so far not been functionally explored. For these we refer to the relevant studies that profiled lncRNA expression in dilated cardiomyopathy [170, 339], ischemic heart failure, inherited hypertrophic cardiomyopathy [340], and congenital heart defects that range from ventricular septal defects over atrial septal defects to tetralogy of Fallot

[341, 342]. The following lncRNAs have, however, been studied in functional terms and we present them in alphabetical order:

Apf, the lncRNA *autophagy promoting factor*, was found to be induced in infarcted hearts in ischemia-reperfusion (I/R)-mediated mouse models of heart injury [175]. *Apf* was found to overactivate autophagy by sequestering *miR-188-3p* and, thus, inhibiting *miR-188-3p*-dependent repression of *ATG7* [175]. A balanced repression of *ATG7* seems, however, important, as ATG7 is one of the central known components in forming the autophagosome vesicle. Consistent with this idea, *Apf* RNAi resulted in a reduced size of infarcted heart tissue in the I/R model and *ATG7* overexpression rescued this effect. Therefore, the model was proposed that *Apf* induction led to a detrimental overactivation of autophagy that mediated cell death in the infarcted heart, possibly in cardiomyocytes [175]. Autophagy is normally operating in physiology for the homeostasis of cells' molecular building. During metabolic stresses like starvation or hypoxia, autophagy is especially important, also to remove dysfunctional molecules and organelles. Both overactivation and blockage of autophagy have been linked by multiple earlier studies to many diseases and contribute to cancer, aging, degenerative diseases, and cardiovascular diseases (see [343] for review). Although *Apf*-dependent autophagy was detrimental in the current study in the I/R model [175], the proposed mechanism should not be generalized to other cardiovascular diseases, where autophagy also plays a role. For example, autophagosome number was found to be increased in macrophages of atherosclerotic plaques, but there autophagy is thought to curb plaque necrosis and to stabilize plaques by reducing macrophage apoptosis [344].

Carl, cardiac apoptosis-related lncRNA, is a lncRNA that has come to attention in the context of studying the regulation of mitochondrial energy production in energetically active cardiomyocytes. *Carl* is a mouse lncRNA that is expressed in the heart, and is repressed by hypoxia in cultured mouse cardiomyocytes. The authors showed that *Carl* sequestered *miR-539*, while *miR-539* was found to represses PHB2 [18]. PHB2 is a mitochondrial protein belonging to the class of prohibitins and its function is to limit mitochondrial fission in anoxic conditions. Limiting fission is a proproliferative adaptation necessary to survive anoxia. Since *Carl* expression is, however, downregulated in hypoxia, and *miR-539*-dependent PHB2 inhibition begins to prevail, the end result is myocardial death. In vivo, overexpressing mouse *Carl* in the aortic root in a mouse ischemia/reperfusion model reduced myocardial death and limited heart infarction size [18]. *Carl*, thus, seems to ensure sufficiently high levels of PHB2, which has protective consequences in the post-infarcted heart and is due to ensuring a correctly balanced level of mitochondrial energy production [345]. How the situation is in the human heart is not known. Conceptually, the regulation of the mitochondrial fission-fusion cycle for balancing energy production is involved in a number of diseases, and in particular also in cardiovascular diseases, since the heart is the most metabolically active organ, carries the highest content of mitochondria, and since mitochondrial capacity is impaired, for example, in heart failure (see [346] for review).

Carmen, cardiac mesoderm enhancer-associated noncoding RNA, is a lncRNA that is required for the specification of the cardiac lineage [19]. It was found upon profiling lncRNAs that were induced upon differentiation of fetal human cardiac precursor cells in culture. *Carmen*'s gene body overlapped with a previously characterized fetal and adult human heart enhancer that had been determined by chromatin-IP experiments. RNAi experiments showed a requirement for *Carmen* in the specification to cardiac SMCs and differentiation to cardiomyocytes in a differentiation model of mouse embryonic stem cells in culture [19]. The authors suggested that *Carmen* functioned as an enhancer RNA to potentiate enhanced expression of genes close to the lncRNA locus *in cis*. But genetic rescue experiments showed that additional effectors had to exist elsewhere, and that *Carmen*, maybe, also had a *trans*-acting role on still unknown target genes. Since *Carmen* bound the PRC2 complex, its *trans*-repressive function may be tethering this repressive chromatin regulator to target loci, but whether this was indeed the case and how this related to its eRNA function remains to be shown [19]. How *Carmen* impacts human heart diseases is less clear. Specific isoforms of *Carmen* were found to be induced in hearts after myocardial infarction, in human hearts of dilated cardiomyopathy (DCM), and aortic stenosis patients [19]. Since RNAi-mediated depletion of *Carmen* led to a reduction of structural heart proteins and of heart transcription factors Gata4 and Nkx2–5 in neonatal murine cardiomyocytes, a role in maintaining the differentiated fate of cardiomyocytes has been proposed [19].

The evidence that misregulated levels of these two circRNAs contribute to heart disease is rather solid, at least in the experimental mouse models. The effector mechanism is, however, disputed. microRNA sponging by circRNA is not considered to be a physiological function for the vast majority of circRNAs. Only a few circRNAs are expressed at such high levels and with such a large amount of microRNA binding seed regions that microRNA binding could possibly occur at relevant rates. ***CDR1-as, cerebellar degeneration-related protein 1 antisense RNA***, is a classic example of a microRNA-sponging circRNA, but likely also the exception. It contains 74 microRNA seed regions for the miR-7 microRNA. Since *CDR1-as* lacks full sequence complementarity with this microRNA, *CDR1-as* degradation is thought to be avoided [214, 215]. Indeed, *CDR1-as* does form stabilized microRNA:AGO2 endonuclease complexes, as would be expected for a microRNA sponge [215]. That sponging was indeed a biological function for *CDR1-as* was concluded from genetic experiments [213, 214]. Due to its expression domain, and as evidenced by the loss of function studies in mouse and zebrafish in vivo, *CDR1-as* had a neuronal function in the CNS. A knockout exists in mouse, but it does not show anatomical alterations, and also no heart defects [213]. Despite this evidence, *CDR1-as* has also been studied in the cardiovascular context. *CDR1-as* circRNA was found to be upregulated in infarcted heart tissue in a mouse model for myocardial infarction. Following the previously proposed model that *CDR1-as* served as a sponge for *miR-7a*, they suggested that in this function *CDR1-as* inhibited the antiapoptotic effects of *miR-7a* in cardiomyocytes in vitro [176]. Further experiments have to substantiate the proposed model in the disease model in vivo.

The well-known ***H19*** lncRNA was found to be induced after chemical induction of dilated cardiomyopathy in rats. Depletion of *H19* by intracoronary injection of shRNA-expressing constructs improved heart function and reduced cardiomyocyte apoptosis [197]. Induction of *H19* seems to be also involved in calcific aortic valve disease (CAVD) in humans [152]. This disease is characterized by the abnormal mineralization of the aortic valve, which leads to its thickening, impaired leaflet motion and stenosis. In this case, *H19* was pathologically upregulated and this induction promoted cultured human valve interstitial cells to start expressing genes, like osteocalcin, *BGLAP*, *BMP2*, or *RUNX2*, which are normally only expressed during osteogenesis [152]. *H19*-depended repression of *NOTCH1* expression was implicated in the upstream regulation of these genes because *H19* was found to inhibit *NOTCH1*, and NOTCH1 was known from earlier studies to repress the osteogenic genes. The authors offered further evidence for a SNP in the *H19* promoter of CAVD patients, which did not cause loss of imprinting as in other *H19*-dependent diseases, but which correlated with reduced DNA methylation and, thus, increased *H19* expression [152].

The ***Heartbrake lncRNA 1 (HBL1)*** is a special case of a human-specific cardiac regulator that is highly conserved in primates but not in other vertebrates. Human *HBL1* became the focus of interest because *it* is a cardiovascular lineage-specific lncRNA, as determined by expression profiling in differentiating human pluripotent stem cells in culture [177]. *HBL1* was shown to become deactivated after cells exited from pluripotency during the onset of cardiomyocyte differentiation. *HBL1* was found to sequester *miR-1*, a microRNA known from before to be important for cardiomyocyte development. The authors of the current study showed that *HBL1* restrained *miR-1*'s ability in promoting ventricular cardiomyocyte expansion. This is likely of relevance also for humans because one target of mouse *miR-1* is the cardiogenic transcription factor *Hand2* [347], and a heterozygous missense mutation in human *HAND2* has been linked to tetralogy of Fallot syndrome, a disease showing ventricular heart septal defects and ventricular heart hypertrophy [348]. This connection indirectly implicates *HBL1* as a mediator of congenital heart defects in humans, possibly in a function in undifferentiated cardiomyocyte progenitors and at the onset of cardiomyocyte cell fate differentiation [177].

Also, some circRNAs have been implicated in hypertrophic growth control in the heart. *mm9-circ-012559* is a 5'-3'-linked circRNA in the mouse. Based on its role it has later been renamed ***HRCR, heart-related circular RNA*** [105]. *HRCR* was identified through a circRNA profiling upon TAC-induced cardiac hypertrophy and heart failure, and in parallel also in a heart injury model based on the infusion of isoproterenol. In these models *HRCR* abundance was reduced compared to controls. Since seed regions for the microRNA *miR-223* were found on *HRCR*, experiments were performed to test HCRC's role in sequestering this microRNA. *miR-223* is relevant for heart physiology because it serves a role as an inhibitor of the known *apoptosis repressor with Card domain (ARC)*, which is known to protect from hypertrophic effects in injured hearts. Since overexpression of *HRCR* reduced the degree of cardiac hypertrophy depending on ARC, the authors concluded that reducing the levels of *HRCR* contributed to heart failure because *miR-223* became active to deplete ARC, and ARC could, thus, no longer protect against heart hypertrophy [105].

The function of *Malat1, metastasis-associated lung adenocarcinoma transcript 1*, in cardiomyopathy was recently tested in a rat model. *Malat1* was found to be induced in cardiac tissue in diseased hearts of diabetic rats. Depleting *Malat1* by intracoronary injected shRNA-expressing constructs improved heart function as measured in left ventricular function [349].

Specific isoforms of the mouse *Meg3* lncRNA were among the most abundant lncRNAs specific to cardiac fibroblasts. *Meg3* was further enriched after TAC-induced heart injury. Silencing *Meg3* systemically by injecting inhibiting oligonucleotides in mice ameliorated *cardiac* fibrosis and diastolic dysfunction, and reduced cardiomyocyte hypertrophy in the TAC-injury model [161]. Thus, *Meg3* is cardioprotective. There is a human MEG3 ortholog, and we have described its functions in CAD. The role and regulation of human *MEG3* in cardiomyopathies, and independent of CAD, have so far not been studied.

MFACR, mitochondrial fission and apoptosis-related circRNA, is another 5'-3'-linked circRNA in the mouse. Its levels were found to be upregulated in a heart ischemia/reperfusion injury model in mouse [106]. In this study it was suggested that *miR-652-3p* was sequestered by *MFACR*. *miR-652-3p* had the deleterious capacity to repress MTP18, a factor important for cellular viability by facilitating mitochondrial fission. In the injury model in vivo, and following systemic manipulation of *MFACR* levels, the authors showed that *MFACR* activation derepressed MTP18, mitochondrial fission, and that the unbalanced overshooting of mitochondrial apoptosis led to heart dysfunction [106].

MIAT has been described in the sections on CAD-related lncRNAs and will be described also for problems in angiogenesis. *MIAT* has also been linked to cardiac hypertrophy, and the underlying roles are not necessarily the same. First, *MIAT* one was found to be induced in heart ventricle samples of patients suffering from noninflammatory dilated cardiomyopathy (DCM), and even more so in Chagas disease patients [147]. This disease typically is triggered by infection with Trypanosomes and develops a form of aggressive inflammatory (DCM) with myocarditis, hypertrophy, and fibrosis. *MIAT* upregulation was corroborated in a mouse model of Chagas disease [147], as well as in Angiotensin II-triggered cardiac hypertrophy models in mice [180] and in a rat diabetic cardiomyopathy model [181]. In the latter model, depleting *MIAT* by shRNAs in vivo ameliorated heart function in diabetic mouse cardiomyopathy and antagonized high glucose-mediated apoptosis in cultured cardiomyocytes. How *MIAT* upregulation affected cardiac hypertrophy in each case is unclear, and whether the different diseases share a common function of *MIAT* has so far not been addressed.

UCA1, urothelial cancer associated 1, is a lncRNA that is the top induced noncoding RNA in the heart in a rat model of partial cardiac ischemia/reperfusion. Overexpression increased proliferation and depletion decreased viability in cultured cardiomyocytes [179]. Thus, *UCA1* has antiapoptotic, proliferative functions. The relevance for humans has not been tested.

7.4.2.4 Vascularization and Angiogenesis

Also in this section on lncRNAs linked to vascularization and angiogenesis we will only review lncRNAs that have been implicated through evidence from animal models in vivo. Depending on lncRNA, additional evidence may exist that a lncRNAs is differentially expressed in diseased human tissue under hypoxic conditions or in animal cell culture models of vascularization. On a cellular level, these lncRNAs are involved in aspects of endothelial cell biology, to vessel formation and branching in normal conditions, or in ischemic conditions such as during atherosclerosis or heart infarction. LncRNAs implicated in cerebrovascular stroke, where vasculariziation plays a role, will be presented in Sect. 7.4.2.6. Relevant lncRNAs in section encompass *HOTTIP*, *Malat1*, *MANTIS*, and *SENCR*. Secondly, and although we do not consider lncRNAs linked to CAD risk factors like diabetes or high blood pressure per se, in this section we also consider reports on lncRNAs related to studies of ECs during vascularization in diabetic retinopathy (*Miat*, *Malat1*) or in end-stage renal disease (*Myoslid*). The lncRNAs are discussed in alphabetical order.

Given the role of some HOX genes in EC biology, **HOTTIP** has been investigated in the context of angiogenesis. Overexpression of *HOTTIP* was found to promote EC proliferation and migration in vitro, while siRNA-mediated knockdown had the opposite effects [156]. This proliferative and migrative function resembled the role of *HOTTIP* in other previously tested cells. With relevance to the cardiovascular system, so far it is also known that *Hoxa13* is important for endothelial cell specification in the vasculature connecting embryo and placenta. Thus, whether *HOTTIP* would be also functioning in ECs in the adult vascular wall, or be misregulated in disease, remains to be seen.

Malat1 has a number of molecular functions as far as studied in cultured cells, but no obvious anatomical defect when the gene was knocked out in mice in vivo (see the previous section). Human *MALAT1* was, however, found to be specifically upregulated in stress conditions, such as in cultured human umbilical vein endothelial cells (HUVECs) exposed to hypoxia, which is a pro-angiogenic condition. *Malat1*, *MEG3*, *TUG1*, and *SNHG5* were among the top induced lncRNAs. In this condition *Malat1* knockdown by siRNAs or GapmeRs decreased hypoxic EC proliferation, and impaired EC migration and sprouting in spheroid and scratch assay in vitro [13]. Based on these results, the mouse *Malat1* knockouts were reinvestigated and a specific deficit in vascularization was found also in vivo, and specifically in the neonatal retina [13]. Stress-specific roles may also underlie *Malat1*'s roles in cancer cells, where *Malat1* is upregulated in tumors, while knockout impaired cancer cell migration and metastasis, as well as alternative splicing in mammary tumor models [216]. The relevance of *MALAT1/Malat1* has since been studied also in ischemic disease models in vivo. When GapmeRs directed against *Malat1* were intraperitoneally injected into mice during experimentally triggered hindlimb ischemia, blood flow recovery was found to be reduced [13]. Thus, *Malat1* was suggested to be required for vascular growth also in this context, and its role in stimulating EC proliferation and sprouting, as determined in cultured cells, may underlie vascularization. In contrast to its beneficial proangiogenic role in CAD models, during diabetic retinopathy, retinal vessels growth instructed by *Malat1* can also be detrimental when uncontrolled: *Malat1* was found to be induced in hypoxic and in diabetic conditions, as shown, for example, in rat and mouse models. In the context of diabetic retinal pathology, *Malat1* contributed to pathological vascular retinal pathology [14]. *MALAT1* is also induced in ischemic conditions in cerebrovascular stroke models and will be treated in Sect. 7.4.2.6 in more detail.

A lncRNA encoded in antisense within an intron of *Annexin A4* was recently investigated in some more detail and named *MANTIS, n342419* [69]. So far, *MANTIS* expression has been observed in endothelial cells, and was tested in several cardiovascular diseases. *MANTIS* was found to be downregulated in lung tissue of idiopathic pulmonary arterial hypertension in humans, a small-vessel disease characterized by endothelial apoptosis and proliferation. Secondly, *MANTIS* was found to be upregulated in vessel tissue in atherosclerosis regression in a monkey CAD disease model, which could be due to a hypothetical role of *MANTIS* in promoting vascular regeneration [69]. Though not specifically expressed only in endothelial cells (ECs), a recent study focused on ECs and studied functions of human *MANTIS* based on a CRISPR/Cas9-mediated knockout in human umbilical vein endothelial cells (HUVECs) or depletion by siRNA in other endothelial cell lines [69]. Molecularly, *MANTIS* localized to the nucleus, where it bound BRG1, known as the catalytic subunit of the SWI/SNF ATP-dependent chromatin remodeling complex [69]. *MANTIS* was suggested to contribute to BRG1 helicase activity and, thus, to the chromatin-remodeling activity of BRG1-containing complexes. For this function it was suggested that *MANTIS* stabilized the interaction of BRG1 with selected members of the SWI/SNF chromatin remodeling complex. In this function, *MANTIS* might activate BRG1-dependent target genes, but this model remains to be experimentally corroborated. In ECs, human *MANTIS* was shown to be pleiotropically important for proper vascular tube formation in matrigel assays, growth factor-dependent spheroid outgrowth, and cell migration of ECs in vitro [69]. Together, MANTIS is a *trans*-acting lncRNA that activates angiogenic genes like *SOX18* or *SMAD6* by nucleosome remodeling.

MIAT, **myocardial infarction-associated transcript**, is a lncRNA that is involved in pathological angiogenesis. *MIAT* is upregulated in retinas and in the diseased fibrovascular membranes of type 2 diabetes patients. This induction was confirmed in cell culture in a number of different endothelial cell types, including in HUVECs under glucose stress. In a hyperglycemic rat model in vivo, *Miat* knockdown ameliorated retinal function in the disease state, for example, by reducing apoptosis of retinal cells [148]. On a cellular level, *Miat* was implicated in promoting cell death triggered by hyperglycemic stress and in stimulating proliferation and migration of ECs, and on a tissue level in advancing vessel leakage. It is likely that multiple molecular roles underlie these diverse functions. As tested in macaque retinal endothelial cells, one attractive model is that *Miat* de-represses VEGF by microRNA sponging, which may account for *Miat*'s role in ECs, and, therein, in angiogenesis and neovascularization [148].

SENCR which has already been described during in vivo studies in the context of CAD (see above) was also found to be relevant for angiogenesis. In limb muscle samples of human critical limb ischemia patients, and in vessel wall ECs of patients with premature CAD, *SENCR* levels were shown to be reduced [164]. In this study it was also found that human *SENCR* promoted endothelial commitment and differentiation and angiogenesis-related cell functions, like proliferation migration and tube formation, in a differentiation model of human embryonic stem cells in vitro. This suggested that *SENCR*'s functions are more complex and potentially cell type specific [164].

The differentiation of vascular smooth muscle cells is known to require the transcription factor serum response factor (SRF) and its coactivator *Myocardin*. In a sequencing approach, dozens of lncRNAs were found to be induced after overexpression of Myocardin in MYOCD in human coronary artery SMCs [184]. Among them, **Myoslid, myocardin-induced smooth muscle lncRNA,** was identified, a cytoplasmically enriched lncRNA. *Myoslid* is VSMC-specific, expressed in blood vessels and bladder, is encoded in overlapping patterns with three other uncharacterized lncRNAs, and is a NAT of at least one of them. Knockdown experiments suggested a role in promoting the contractile, antiproliferative, and antimigrative phenotype of VSMCs, which is consistent with a role in promoting Myocardin-dependent differentiation [184]. Myoslid was subsequently found to be downregulated in arteriovenous tissue samples from patients with end-stage renal disease. Therein, a hyperplastic and stenotic response has been associated with disease. This suggests that reduction of Myoslid in vascular disease may contribute to the disease-promoting de-differentiation of VSMCs.

7.4.2.5 Arrhythmia

In the fifth functional class we summarize lncRNAs linked to heart electrophysiology, heartbeat, and rhythm. Compared to other cardiovascular disease entities, few lncRNAs have been associated with this disease type and the depth of functional insight is limited. Cardiac arrhythmias can be, but do not need to be, associated with structural heart diseases, with defects involving aspects of the conductions system. Heart structures relevant for rhythm include the sinus node, the specialized atrial tracts, the atrioventricular node, and the bundle of His. Characteristically shaped electrical currents reach the myocytes of the heart via ion channels, specialized proteins and gap junctions. Most SNPs associated with arrhythmias through GWAS affect one of the multiple ion channel subunits. Also structural heart defects can lead to heart arrhythmias. In this case it is most often cardiac fibrosis or hypertrophy that impairs proper conduction in the heart. In the following paragraphs, we only focus on those lncRNAs that have been functionally investigated in relation to heart rhythm by in vivo models.

One study investigated lncRNAs differentially expressed in patients with atrial fibrillation when comparing two left atrial regions, one being the region of the pulmonary vein and the surrounding left atrial area (LA-PV), and the other being the area of the left atrial appendage (LAA) [185]. The reason for comparing these two regions lies in the earlier observations that the LA-PV junction is implicated as most important for arrhythmogenicity based on catheter ablation experiments, while the LAA is involved in other functions, it is, for example, a site of thrombus formation. Among more than 90 such lncRNAs, *AK055347* was found to be enriched in the LA-PV, and was the most significantly altered lncRNA there. In cultured rat cardiomyocyte cell lines, *AK055347* knockdown decreased viability [185]. Whether and how the proproliferative function of *AK055347* related to arrhythmia remained, however, open.

A recent study performed quantitative trait locus (QTL) mapping in two rat strains that differed in how their blood pressure was affected by salt levels with the aim to identify genes regulating blood pressure. The researchers identified a QTL responsible for high blood pressure, which was, coincidently, also responsible for shortening the cardiac QT-interval [20]. It was mapped to the Rffl locus, and specifically the noncoding RNA therein, *Rffl-lnc1*. Interestingly, these two phenotypes have also been co-associated within the synthetic genomic region in humans, as shown by data from human GWAS studies [350]. A pathologically shortened Q-wave/T-wave interval, as recorded in the electrocardiogram, often associates with mutations in cardiac ion channels, and even without involving structural changes in the heart tissue is known to correlate with an increased risk for atrial and ventricular arrhythmia in humans. In the QTL study in the rat, the researchers found that 19 bp indel in a previously undescribed lncRNA, *Rffl-lnc1*, was responsible for the mutant phenotype in the rat. *Rffl-lnc1* locates in the 5′ UTR of the single protein-coding gene in this region, and *Rffl* is a **ring finger and FYVE-like domain containing E3 ubiquitin protein ligase** [20]. Deletion of the *Rffl-lnc1* and, independently, CRISPR/Cas9-mediated reintroduction of the missing 19 nucleotides into the mutant rat strain showed that *Rffl-lnc1* indeed was the causative gene [20]. Together, how *Rffl-lnc1* polymorphisms partake in the control of cardiac features will be interesting to investigate. At the moment it is thought that *Rffl-lnc1* may either directly regulate the cardiac conduction system or the neural autonomic pathways in the heart, or it may indirectly affect heart rhythm through the regulation of cardiomyocyte hypertrophy in the context of increased blood pressure [20].

Another study performed lncRNA expression profiling at sites of enervation on the surface of the heart, where collections of autonomous nerves form ganglia plexus, mainly in island-like fat pad structures or in the adipose tissue below the epicardial membrane [186]. Autonomic nervous system remodeling (ANR) at such sites is known to contribute to arrhythmia. Canine models have been traditionally used to study this aspect. After receiving heart tachypacing via electrodes, dogs developed ANR, and **TCONS_00032546** and **TCONS_00026102** were found to be downregulated [186]. Injecting knockdown constructs for the two lincRNAs into the relevant fat pads in dogs was sufficient to slightly shorten, or prolong, respectively, the atrial effective refractory period and increase, or inhibit, respectively, the onset of tachycardia and atrial fibrillation [186]. How these two lncRNAs differentially impact neuronal remodeling in the onset of arrhythmia remained unknown, as well as whether human orthologs existed that carried conserved functions.

In conceptually similar experiments in a rabbit model for studying conduction remodeling in atrial fibrillation, lncRNA expression profiling was performed in right atrial tissue samples [187]. The authors then used coexpression network analysis to focus on pathologically potentially interesting lncRNAs, such as the downregulated *TCONS_00075467*. A human sequence with 70% conservation exists, opening the possibility that the lncRNA was conserved. Depleting *TCONS_00075467* in the right atrium in the rabbit model shortened the atrial effective refractory period and increased the frequency of fibrillation onset in the animal model [187]. The authors suggested that this lncRNA functioned as a protector by inducing the expression of the calcium voltage-gated *CACNA1C* ion channel as a microRNA sponge. Thereby, this work related to earlier published insight on the role of *CACNA1C* downregulation by microRNAs in human patients with fibrillation [351] and to the earlier identification of a loss-of-function mutation in *CACNA1C* in Brugada syndrome patients, a disease with characteristic ST-segment elevation [352].

7.4.2.6 Stroke and Cerebrovascular Aneurysms

In the sixth functional class, lncRNAs related to brain stroke are reviewed. This class is not treated separately to indicate a separation in terms of atherosclerosis pathways leading to thrombus and ischemia formation, but because brain neurons are the cell type hit most by ischemia, and because specialized pathogenic mechanisms occur in the brain involving distinct sets of lncRNAs. The occlusion of the middle cerebral artery is the most common route to cerebral stroke, and lack of oxygen and glucose rapidly inflicts neurons during hypoxia-ischemia. Depolarization and impaired neurotransmitter uptake subsequently cause toxic amounts of extracellular glutamate to accumulate, which further contributes to mitochondrial dysfunction and oxidative stress. Finally, ischemia also triggers inflammatory signaling, which can further increase brain tissue damage. To date, nine lncRNAs have been associated with stroke. For two of them there is some evidence that the relevant lncRNAs were also important in human stroke patients. Of the nine lncRNAs, two had protective functions in stroke-induced brain damage, while seven aggravated neuronal problems after brain stroke.

Chr9p21 is well known as a major CAD risk locus. Different risk genotypes in the human Chr9p21 locus have been found to associate with different cardiovascular disease entities. Initial analyses offered conflicting evidence whether Chr9p21 genotypes were also associated with stroke, with some studies reporting an association [137–139], and some studies showing no significant association [136, 353, 354]. A meta-analysis of relevant GWAS explored the association and did document a robust but relatively weak association [140]. The investigation also suggested a subtype-specific association of Chr9p21 with ischemic stroke, namely, with large artery stroke [140]. Also, a more recent study corroborated this notion by showing that specific SNPs in *ANRIL* lncRNA associated with susceptibility and recurrence rate of atherothrombotic and hemorrhagic cerebrovascular stroke even in those patients that did not have a history of heart disease [141]. Most recently, another study performed an eQTL analysis of five previously identified risk SNPs in *ANRIL* [12], and identified genes whose expressions associated with at least one of the five risk SNPs. Among 87 of such potential eQTL hits, *CARD8* was selected as a potential *ANRIL* effector, which was further substantiated as *CARD8* expression also dropped when *ANRIL* was knocked down in cultured cells [142]. Since an inactivating mutation in *CARD8* was associated with ischemic stroke, and since *CARD8* is known from before as negative regulator of NF-κB signaling, the possibility exists that *ANRIL* impacts ischemic stroke also via *CARD8*-dependent NF-kB regulation [142]. While this hypothetical pathway remains to be rigorously tested, an implication of *ANRIL* in stroke is rather well documented by multiple studies. Studies investigating a direct role of *ANRIL* in stroke are so far limited: In a single study, the role of *ANRIL* was studied in a model for type 2 diabetes in rats under a high-fat diet and after middle cerebral artery occlusion to trigger stroke [143]. *ANRIL* was found to be increased in brain tissue in this context. Overexpression and knockdown of *ANRIL* systemically in the rat model showed that *ANRIL* augmented endothelial microvessel density, which correlated with increased VEGF and NF-κB expression levels. Since neuronal functionality post stroke was not tested after *ANRIL* overexpression or knockdown, one can yet only speculate whether the observed *ANRIL*-promoted angiogenesis aggravated complications of a stroke in this disease model.

AK153573 is a predominantly nuclear lncRNA that partially overlaps in sense the *CaMKIIδ* gene. Renamed ***C2dat1, CaMKIIδ-associated transcript 1***, this lncRNA was found to be induced in neurons surrounding ischemic regions in a stroke model of middle cerebral artery occlusion in mouse and rat in vivo [188]. *C2dat1* was also upregulated after experimental oxygen-glucose-deprivation/reoxygenation (OGD/R) in neuronal cells in vitro. CaMKIIδ levels did decrease in the ischemic core, but were induced, similar to *C2dat1*, in the periphery of the lesion. Knocking down C2dat1 in cultured neurons reduced *CaMKIIδ* expression and stress-dependent induction of NF-κB signaling and, while neuronal survival was increased. These findings are consistent with the large body of earlier evidence that acute inhibition of CaMKII is neuroprotective [188]. C2dat1 may, thus, likely function via acute CaMKIIδ locus stimulation after stroke, and this pathway may exacerbate apoptosis during stroke.

FosDT, Fos downstream transcript, has been identified as lncRNA that was induced in the rat cerebral cortex after transient middle cerebral artery occlusion [73]. Further, *FosDT* bound Sin3 and coREST, which are members of the repressive REST chromatin regulating complex, whereby REST is known from independent work to become active and to be necessary for neuronal apoptosis during stroke. In the present work, it was found that *FosDT* contributed to the repressive function of REST, and that silencing *FosDT* reduced brain infarction size and ameliorated neurological functions after infarction, as determined after intracerebral injection of siRNAs [73]. One possibility, thus, is that *FosDT* scaffolds REST to repress selected target genes that are otherwise necessary for their survival.

Beyond the involvement in CAD and in cardiomyopathies, ***H19*** also plays a role in cerebral stroke. The first insight into this association was obtained in rodent models. When focal cerebral ischemia was induced by middle cerebral artery occlusion/reperfusion in mice or rats, *H19* levels rose in blood plasma, and also in infarcted brain tissue [153, 154]. Corroborating this finding, *H19* was also upregulated in human patients within hours after a stroke [153]. Thereby, *H19* expression was higher in plasma, neutrophils, and lymphocytes. Since an intraventricular injection of siRNAs targeting *H19* reduced brain infarct volume and ameliorated neurological functions in the mouse stroke model, it can be deduced that *H19* induction usually exerts a detrimental function during stroke. How *H19* is detrimental is less clear. Current insights into this question base on work with cultured cells: For example, *H19* was found to promote neuroinflammatory TNFα and interleukin expression. Secondly, *H19* was shown to impair the formation of neuroprotective M2 microglia. Microglial cells are the major cell type in mounting stroke-induced inflammation [153]. And third, albeit only tested in a human neuroblastoma cell line, and not necessarily related, *H19* contributed to apoptosis in an OGD/reperfusion model in vitro [154]. *H19* has not appeared in GWAS studies on stroke, but selected SNPs in *H19* were still found associated to some degree with stroke [154].

Consistent with its antiproliferative functions in many cell types, **Meg3** was recently found to be proapoptotic in the context of cerebral stroke in mouse [162]. While mouse *Meg3* is known to be downregulated in CAD-affected vessels, it was, however, upregulated in ischemic brain tissue. *Meg3* promoted neuronal cell death, and the authors suggested that *Meg3* bound to p53's DNA binding domain. Whether and how such an interaction would activate p53 in a proapoptotic pathway is of interest not only for the cardiovascular field [162]. However, whether *Meg3* indeed functioned in neurons in vivo, or also affected angiogenesis in the infarcted brain, as shown in before in CAD model, or both, is still not clear.

N1LR, also known as **MRAK051854**, was identified by microarray-based transcriptional profiling of ischemic brain tissue after intraluminal middle cerebral artery occlusion to induce a stroke in the rat model system. Based on coexpression network and pathway analysis, *N1LR* became the focus of interest as a highly connected upregulated lncRNA [189]. Also, this lncRNA was upregulated in a mouse stroke model. Overexpressing *N1LR* after injection into the mouse cortex reduced the infarct volume and neuronal cell death, while siRNA against *N1LR* enhanced the lesions, suggestive of protective functions of this lncRNA. The molecular effector mechanism still remains to be interrogated.

In another approach, lncRNAs were bioinformatically screened for harboring seeds for *miR-145-5p*, a microRNA whose induction had been previously associated with ischemia in the heart. Specifically, **SNHG14,** also known as **small nucleolar RNA host gene 1** or **UBE3A-ATS**, was such a lncRNA. It is expressed in antisense to the ubiquitin protein ligase E3A gene, and was found to be induced during the first days after stroke in a middle cerebral artery occlusion model in mice, while *miR-145-5p* levels correspondingly dropped [190]. In cultured microglial cell lines, *SNHG14* stimulated expression of inflammatory factors, while it reduced *miR-145-5p* levels. The inflammatory phenotype was rescued by adding *miR-145-5p* mimics, consistent with the interpretation that *SNHG14* impaired the protective function of *miR-145-5p* by keeping its levels low [190].

A role in promoting neuronal cell death was ascribed to the lncRNA **Tug1** during stroke [166]. *Tug1* levels increased in a middle cerebral artery occlusion model in the rat in vivo, as well as in cultured neurons after oxygen/glucose deprivation. *Tug1* knockdown reduced apoptosis under OGD conditions in vitro, and this effect correlated with upregulation of *miR-9* levels, suggesting a detrimental role in stroke [166]. *Tug1* has been linked to apoptosis, but also to other roles in other cellular contexts before, particularly in tumorigenesis. Given the intense research on *miR-9* in neurogenesis, it will be interesting to dissect how *Tug1* connects to the rather well-understood functions of *miR-9* in the CNS, and whether the regulation of apoptosis is indeed the central function of interest. For example, *miR-9* is known from earlier work to be expressed in self-renewing neural progenitors and is required for neurogenesis by promoting the timely cell cycle exit of differentiating neurons. At the same time *miR-9* has also previously been found to promote angiogenesis to proceed in concordance with neurogenesis by curbing premature neuronal VEGF expression [355]. Thus, the region- and time-specific coordination of neurogenesis and angiogenesis may be how *Tug1* determines the manifestation of stroke.

Together, a number of studies have described RNA expression profiles in brain tissue after stroke, and lists of relevant lncRNAs residing close to known stroke risk loci have been compiled as well [356]. Compared to CAD and MI models, the level of mechanistic in vivo insight is, however, still limited. Questions, like in which cell types and via which effector mechanisms lncRNAs function during stroke are still open. Nevertheless, there is a rich literature on transcriptional regulation of neuronal cell types, and fast advances may be possible if relevant lncRNAs can be convincingly linked to already known and well-studied transcriptional effectors.

7.4.2.7 lncRNAs in Basic Cellular Processes that Overarch CVD Entities

Finally, an important set of lncRNAs has been implicated in the regulation of inflammatory signaling, and derailing immune signaling is a central factor in all entities of cardiovascular diseases. We are not reviewing this inflammation-related set of lncRNAs in depth though, as we are not reviewing lncRNAs implicated in generic proliferation control either. Excellent recent reviews have been published on lncRNAs in immune control [357] and in cell cycle and growth control [358]. Also, as of yet, most of the lncRNAs in these two broad classes have not been specifically studied in the context of cardiovascular disease. This has, in part, also to do with the fact that there is limited knowledge about the exact cell lineage transitions contributing to cell fate changes that underlie atherosclerosis.

Box 7.3 Reading Highlights

1. **Reports on linear cardiovascular lncRNAs:**
   (a) Klattenhoff, C. A. et al., (2013) Braveheart, a long noncoding RNA required for cardiovascular lineage commitment, *Cell.* 152, 570–83.
   (b) Han, P. et al., (2014) A long noncoding RNA protects the heart from pathological hypertrophy, *Nature.* 514, 102–6.
   (c) Anderson, K. M., et al., (2016) Transcription of the non-coding RNA *Upperhand* controls Hand2 expression and heart development, *Nature.* 539, 433–436.

(continued)

Box 7.3 (continued)

(d) Wang, Z et al., (2016) The long noncoding RNA *Chaer* defines an epigenetic checkpoint in cardiac hypertrophy, *Nat Med.* 22, 1131–1139.

(e) Liu, J. et al., (2017) HBL1 Is a Human Long Noncoding RNA that Modulates Cardiomyocyte Development from Pluripotent Stem Cells by Counteracting MIR1, *Dev Cell.* 42, 333–348 e5.

(f) Hon, C. C. et al., (2017) CRISPRi-based genome-scale identification of functional long noncoding RNA loci in human cells, *Science.* 355.

2. **Reports on cardiovascular circRNAs:**

(a) Holdt, L. M. et al., (2016) Circular non-coding RNA ANRIL modulates ribosomal RNA maturation and atherosclerosis in humans, *Nat Commun.* 7, 12,429.

(b) Wang et al., (2016) A circular RNA protects the heart from pathological hypertrophy and heart failure by targeting miR-223, *Eur Heart J.* 37, 2602–11.

(c) review: Barrett, S. P. & Salzman, J. (2016) Circular RNAs: analysis, expression and potential functions, *Development.* 143, 1838–47.

3. **Reports addressing generally important concepts in lncRNA biology:**

(a) Wide-spread role of lncRNAs as enhancers of transcription: Engreitz, J. M. et al., (2016) Local regulation of gene expression by lncRNA promoters, transcription and splicing, *Nature.* 539, 452–455.

(b) The relevance of RNA:PRC2 interaction is under scrutiny: Kaneko, S et al., (2014) Nascent RNA interaction keeps PRC2 activity poised and in check, *Genes Dev.* 28, 1983–8.

(c) LncRNAs can be functional despite lack of sequence conservation Ulitsky, I., Shkumatava, A., Jan, C. H., Sive, H. & Bartel, D. P. (2011) Conserved function of lincRNAs in vertebrate embryonic development despite rapid sequence evolution, *Cell.* 147, 1537–50.

(d) Widespread role antisense transcription of lncRNA genes: Huber, F. et al., (2016) Protein Abundance Control by Non-coding Antisense Transcription, *Cell Rep.* 15, 2625–36.

(e) Translation of micropeptides from lncRNAs: Micropeptides encoded on lncRNAs: Anderson, D. M. et al., (2015) A micropeptide encoded by a putative long noncoding RNA regulates muscle performance, *Cell.* 160, 595–606.

(f) microRNA sponging may be less common than thought: Denzler, R., et al., (2016) Impact of MicroRNA Levels, Target-Site Complementarity, and Cooperativity on Competing Endogenous RNA-Regulated Gene Expression, *Mol Cell.* 64, 565–579.

7.5   Summary and Outlook

So far 57 lncRNAs have been functionally investigated in the context of cardiovascular physiology and disease. Many more lncRNAs are differentially expressed, but have not yet been functionally assessed. Therefore, and given the number of different cell types and processes implicated in development, homeostasis and regulation of the heart, vasculature, and vascularized tissue, it is expected that many more lncRNAs will be linked to some aspect of coronary artery physiology and disease in the future. As can be seen already from the work presented in this book chapter, a single lncRNA can exhibit different functions in different contexts, and in extreme cases, can engage with dozens of different binding partners through specialized interaction domains. Therefore, although lncRNAs sometimes show low copy numbers, a limited stability, and a low level of evolutionary selection in sequence, thousands of lncRNAs are expressed with high confidence, with a high degree of cell type specificity, and with functional consequences.

The question of how many of the thousands of lncRNAs are indeed functional genetic elements has until recently not been possible to address. Yet, with the emergence of CRISPR/Cas9-based genetic loss of function experiments, genome-wide genetic screens have begun to explore this question experimentally. When testing 16,000 lncRNAs for effects on cell proliferation capacity in cultured cells, nearly 500 individual lncRNA transcripts (or 3% of all lncRNAs) were found to be important for cell division [359]. This number is unexpectedly large when compared to equivalent genetic tests on the overall functionality of protein-coding genes, of which 10–12% are essential for optimal cell proliferation [360, 361]. Since only a few types of cells and only a single functional readout were assayed in this screen, one can project that a large percentage of lncRNAs will be annotated as functional, when carefully tested [29, 362]. Since every single lncRNA may further produce several spliced isoforms, and since ribonucleotides can additionally be modified posttranscriptionally, for example, by methylation [363], there is a huge regulatory space how lncRNAs can affect physiology and disease.

In a study with genome-wide relevance, the evolutionarily conserved coexpression of lncRNAs was measured relative to other transcripts, and colocalization with eQTL-associated SNPs was taken into consideration. Also, this type of approach suggested that more lncRNAs than previously thought (up to 40%) appear to be potentially functional because they associate with at least one known trait [29]. Future experimental approaches will have to take into account lncRNAs and their time- and location-specific transcription as an integral part of gene-regulatory networks.

Based on the already known important roles of lncRNAs in the cardiovascular system as well as in other physiological processes and diseases, linear and circular noncoding RNAs become a medically interesting target molecule in different respects: First, to lncRNAs and circRNAs as biomarkers for pathophysiological states, both in tissue samples, as well in cell-free form in the circulating blood. In the blood, cell-free noncoding RNAs are found inside exosomes or other membrane-contained vesicles, or in association with proteins, and reporting on cellular states in remote internal organs. To date, most technologies to sequence cell-free nucleic acids in preclinical settings focus on DNA and on finding mutations therein. Combined DNA and RNA sequencing from blood may show increased power in monitoring early disease onset, or therapy success and recurrence of disease after therapy. Second, lncRNAs may be directly used to alter cell physiology in situ: Delivering synthetically produced topically applied lncRNAs to diseased organs may be just one option. Especially if technologies are advanced that stabilize lncRNA half-lives, for example, exchanging the phosphodiester linkage by phosphorothioate bonds in the RNA backbone, or that allow to more reliably target diseased cells in a body, RNA therapeutic agents may become a reasonable option for treatment of some conditions [364]. Editing lncRNA transcripts with modified CRISPR-Cas variants directly, instead of editing the genomic template, is another novel therapeutic approach that will gain importance [364, 365]. In an alternative approach, although it is early days in our understanding of noncoding RNAs, the elaborate secondary structures of a lncRNA may allow us to find or design specific inhibitory drugs that enhance or inhibit the function of a specific lncRNA in its interaction with a protein complex.

In any case, already today, the study of lncRNAs in CAD entities has proven highly valuable, especially because it has yielded a major shift in our understanding of genetic networks that regulate cell fate, and because it has helped to functionally annotate disease-associated genetic loci from GWAS, which often reside in the noncoding DNA sequence space. To conclude, the picture emerges that transcription of noncoding RNAs, linear or circular, is in most cases not a failure in our gene expression program, but an integrated regulatory feature that affects all levels of cell function.

Acknowledgments This work was funded by the University Hospital Munich and by the German Research Foundation (DFG) as part of the Collaborative Research Center CRC1123 "Atherosclerosis—Mechanisms and Networks of Novel Therapeutic Targets" (project B1). Part of this project was also funded by the Deutsche Forschungsgemeinschaft (DFG, German Research Foundation, # 394237736).

Compliance with Ethical Standards Not applicable.

Disclosures None.

Conflicts of Interest There are no competing interests.

References

1. Lee JT, Bartolomei MS. X-inactivation, imprinting, and long noncoding RNAs in health and disease. Cell. 2013;152:1308–23.
2. Bertone P, Stolc V, Royce TE, Rozowsky JS, Urban AE, Zhu X, Rinn JL, Tongprasit W, Samanta M, Weissman S, Gerstein M, Snyder M. Global identification of human transcribed sequences with genome tiling arrays. Science. 2004;306:2242–6.

3. Carninci P, Kasukawa T, Katayama S, Gough J, Frith M, Maeda C, Oyama N, Ravasi R, Lenhard T, Wells B, Kodzius C, Shimokawa R, Bajic K, Brenner VB, Batalov SE, Forrest S, Zavolan AR, Davis M, Wilming MJ, Aidinis LG, Allen V, Ambesi-Impiombato JE, Apweiler A, Aturaliya R, Bailey RN, Bansal TL, Baxter M, Beisel L, Bersano KW, Bono T, Chalk H, Chiu AM, Choudhary KP, Christoffels V, Clutterbuck A, Crowe DR, Dalla ML, Dalrymple E, de Bono BP, Gatta BD, di Bernardo G, Down D, Engstrom T, Fagiolini P, Faulkner M, Fletcher G, Fukushima CF, Furuno T, Futaki M, Gariboldi S, Georgii-Hemming M, Gingeras P, Gojobori TR, Green T, Gustincich RE, Harbers S, Hayashi M, Hensch Y, Hirokawa TK, Hill N, Huminiecki D, Iacono L, Ikeo M, Iwama K, Ishikawa A, Jakt T, Kanapin M, Katoh A, Kawasawa M, Kelso Y, Kitamura J, Kitano H, Kollias H, Krishnan G, Kruger SP, Kummerfeld A, Kurochkin SK, Lareau IV, Lazarevic LF, Lipovich D, Liu L, Liuni J, McWilliam S, Babu SM, Madera M, Marchionni M, Matsuda L, Matsuzawa H, Miki S, Mignone H, Miyake F, Morris S, Mottagui-Tabar K, Mulder S, Nakano N, Nakauchi N, Ng H, Nilsson P, Nishiguchi R, Nishikawa S, et al. The transcriptional landscape of the mammalian genome. Science. 2005;309:1559–63.
4. Consortium EP, Birney E, Stamatoyannopoulos JA, Dutta A, Guigo R, Gingeras TR, Margulies EH, Weng Z, Snyder M, Dermitzakis ET, Thurman RE, Kuehn MS, Taylor CM, Neph S, Koch CM, Asthana S, Malhotra A, Adzhubei I, Greenbaum JA, Andrews RM, Flicek P, Boyle PJ, Cao H, Carter NP, Clelland GK, Davis S, Day N, Dhami P, Dillon SC, Dorschner MO, Fiegler H, Giresi PG, Goldy J, Hawrylycz M, Haydock A, Humbert R, James KD, Johnson BE, Johnson EM, Frum TT, Rosenzweig ER, Karnani N, Lee K, Lefebvre GC, Navas PA, Neri F, Parker SC, Sabo PJ, Sandstrom R, Shafer A, Vetrie D, Weaver M, Wilcox S, Yu M, Collins FS, Dekker J, Lieb JD, Tullius TD, Crawford GE, Sunyaev S, Noble WS, Dunham I, Denoeud F, Reymond A, Kapranov P, Rozowsky J, Zheng D, Castelo R, Frankish A, Harrow J, Ghosh S, Sandelin A, Hofacker IL, Baertsch R, Keefe D, Dike S, Cheng J, Hirsch HA, Sekinger EA, Lagarde J, Abril JF, Shahab A, Flamm C, Fried C, Hackermuller J, Hertel J, Lindemeyer M, Missal K, Tanzer A, Washietl S, Korbel J, Emanuelsson O, Pedersen JS, Holroyd N, Taylor R, Swarbreck D, Matthews N, Dickson MC, Thomas DJ, Weirauch MT, et al. Identification and analysis of functional elements in 1% of the human genome by the ENCODE pilot project. Nature. 2007;447:799–816.
5. Djebali S, Davis CA, Merkel A, Dobin A, Lassmann T, Mortazavi A, Tanzer A, Lagarde J, Lin W, Schlesinger F, Xue C, Marinov GK, Khatun J, Williams BA, Zaleski C, Rozowsky J, Roder M, Kokocinski F, Abdelhamid RF, Alioto T, Antoshechkin I, Baer MT, Bar NS, Batut P, Bell K, Bell I, Chakrabortty S, Chen X, Chrast J, Curado J, Derrien T, Drenkow J, Dumais E, Dumais J, Duttagupta R, Falconnet E, Fastuca M, Fejes-Toth K, Ferreira P, Foissac S, Fullwood MJ, Gao H, Gonzalez D, Gordon A, Gunawardena H, Howald C, Jha S, Johnson R, Kapranov P, King B, Kingswood C, Luo OJ, Park E, Persaud K, Preall JB, Ribeca P, Risk B, Robyr D, Sammeth M, Schaffer L, See LH, Shahab A, Skancke J, Suzuki AM, Takahashi H, Tilgner H, Trout D, Walters N, Wang H, Wrobel J, Yu Y, Ruan X, Hayashizaki Y, Harrow J, Gerstein M, Hubbard T, Reymond A, Antonarakis SE, Hannon G, Giddings MC, Ruan Y, Wold B, Carninci P, Guigo R, Gingeras TR. Landscape of transcription in human cells. Nature. 2012;489:101–8.
6. Klattenhoff CA, Scheuermann JC, Surface LE, Bradley RK, Fields PA, Steinhauser ML, Ding H, Butty VL, Torrey L, Haas S, Abo R, Tabebordbar M, Lee RT, Burge CB, Boyer LA. Braveheart, a long noncoding RNA required for cardiovascular lineage commitment. Cell. 2013;152:570–83.
7. Grote P, Wittler L, Hendrix D, Koch F, Wahrisch S, Beisaw A, Macura K, Blass G, Kellis M, Werber M, Herrmann BG. The tissue-specific lncRNA Fendrr is an essential regulator of heart and body wall development in the mouse. Dev Cell. 2013;24:206–14.
8. Anderson KM, Anderson DM, McAnally JR, Shelton JM, Bassel-Duby R, Olson EN. Transcription of the non-coding RNA upperhand controls Hand2 expression and heart development. Nature. 2016;539:433–6.

9. Deng C, Li Y, Zhou L, Cho J, Patel B, Terada N, Li Y, Bungert J, Qiu Y, Huang S. HoxBlinc RNA recruits Set1/MLL complexes to activate Hox gene expression patterns and mesoderm lineage development. Cell Rep. 2016;14:103–14.
10. Broadbent HM, Peden JF, Lorkowski S, Goel A, Ongen H, Green F, Clarke R, Collins R, Franzosi MG, Tognoni G, Seedorf U, Rust S, Eriksson P, Hamsten A, Farrall M, Watkins H, Consortium P. Susceptibility to coronary artery disease and diabetes is encoded by distinct, tightly linked SNPs in the ANRIL locus on chromosome 9p. Hum Mol Genet. 2008;17:806–14.
11. Schaefer AS, Richter GM, Groessner-Schreiber B, Noack B, Nothnagel M, El Mokhtari NE, Loos BG, Jepsen S, Schreiber S. Identification of a shared genetic susceptibility locus for coronary heart disease and periodontitis. PLoS Genet. 2009;5:e1000378.
12. Cunnington MS, Santibanez Koref M, Mayosi BM, Burn J, Keavney B. Chromosome 9p21 SNPs associated with multiple disease phenotypes correlate with ANRIL expression. PLoS Genet. 2010;6:e1000899.
13. Michalik KM, You X, Manavski Y, Doddaballapur A, Zornig M, Braun T, John D, Ponomareva Y, Chen W, Uchida S, Boon RA, Dimmeler S. Long noncoding RNA MALAT1 regulates endothelial cell function and vessel growth. Circ Res. 2014;114:1389–97.
14. Liu JY, Yao J, Li XM, Song YC, Wang XQ, Li YJ, Yan B, Jiang Q. Pathogenic role of lncRNA-MALAT1 in endothelial cell dysfunction in diabetes mellitus. Cell Death Dis. 2014;5:e1506.
15. Han P, Li W, Lin CH, Yang J, Shang C, Nurnberg ST, Jin KK, Xu W, Lin CY, Lin CJ, Xiong Y, Chien HC, Zhou B, Ashley E, Bernstein D, Chen PS, Chen HS, Quertermous T, Chang CP. A long noncoding RNA protects the heart from pathological hypertrophy. Nature. 2014;514:102–6.
16. Wang Z, Zhang XJ, Ji YX, Zhang P, Deng KQ, Gong J, Ren S, Wang X, Chen I, Wang H, Gao C, Yokota T, Ang YS, Li S, Cass A, Vondriska TM, Li G, Deb A, Srivastava D, Yang HT, Xiao X, Li H, Wang Y. The long noncoding RNA Chaer defines an epigenetic checkpoint in cardiac hypertrophy. Nat Med. 2016;22:1131–9.
17. Huarte M, Guttman M, Feldser D, Garber M, Koziol MJ, Kenzelmann-Broz D, Khalil AM, Zuk O, Amit I, Rabani M, Attardi LD, Regev A, Lander ES, Jacks T, Rinn JL. A large intergenic noncoding RNA induced by p53 mediates global gene repression in the p53 response. Cell. 2010;142:409–19.
18. Wang K, Long B, Zhou LY, Liu F, Zhou QY, Liu CY, Fan YY, Li PF. CARL lncRNA inhibits anoxia-induced mitochondrial fission and apoptosis in cardiomyocytes by impairing miR-539-dependent PHB2 downregulation. Nat Commun. 2014;5:3596.
19. Ounzain S, Micheletti R, Arnan C, Plaisance I, Cecchi D, Schroen B, Reverter F, Alexanian M, Gonzales C, Ng SY, Bussotti G, Pezzuto I, Notredame C, Heymans S, Guigo R, Johnson R, Pedrazzini T. CARMEN, a human super enhancer-associated long noncoding RNA controlling cardiac specification, differentiation and homeostasis. J Mol Cell Cardiol. 2015;89:98–112.
20. Cheng X, Waghulde H, Mell B, Morgan EE, Pruett-Miller SM, Joe B. Positional cloning of quantitative trait nucleotides for blood pressure and cardiac QT-interval by targeted CRISPR/Cas9 editing of a novel long non-coding RNA. PLoS Genet. 2017;13:e1006961.
21. Loewer S, Cabili MN, Guttman M, Loh YH, Thomas K, Park IH, Garber M, Curran M, Onder T, Agarwal S, Manos PD, Datta S, Lander ES, Schlaeger TM, Daley GQ, Rinn JL. Large intergenic non-coding RNA-RoR modulates reprogramming of human induced pluripotent stem cells. Nat Genet. 2010;42:1113–7.
22. Salzman J, Gawad C, Wang PL, Lacayo N, Brown PO. Circular RNAs are the predominant transcript isoform from hundreds of human genes in diverse cell types. PLoS One. 2012;7: e30733.
23. Burd CE, Jeck WR, Liu Y, Sanoff HK, Wang Z, Sharpless NE. Expression of linear and novel circular forms of an INK4/ARF-associated non-coding RNA correlates with atherosclerosis risk. PLoS Genet. 2010;6:e1001233.

24. Holdt LM, Stahringer A, Sass K, Pichler G, Kulak NA, Wilfert W, Kohlmaier A, Herbst A, Northoff BH, Nicolaou A, Gabel G, Beutner F, Scholz M, Thiery J, Musunuru K, Krohn K, Mann M, Teupser D. Circular non-coding RNA ANRIL modulates ribosomal RNA maturation and atherosclerosis in humans. Nat Commun. 2016;7:12429.
25. Kapranov P, Cheng J, Dike S, Nix DA, Duttagupta R, Willingham AT, Stadler PF, Hertel J, Hackermuller J, Hofacker IL, Bell I, Cheung E, Drenkow J, Dumais E, Patel S, Helt G, Ganesh M, Ghosh S, Piccolboni A, Sementchenko V, Tammana H, Gingeras TR. RNA maps reveal new RNA classes and a possible function for pervasive transcription. Science. 2007;316:1484–8.
26. Guttman M, Garber M, Levin JZ, Donaghey J, Robinson J, Adiconis X, Fan L, Koziol MJ, Gnirke A, Nusbaum C, Rinn JL, Lander ES, Regev A. Ab initio reconstruction of cell type-specific transcriptomes in mouse reveals the conserved multi-exonic structure of lincRNAs. Nat Biotechnol. 2010;28:503–10.
27. Cech TR, Steitz JA. The noncoding RNA revolution-trashing old rules to forge new ones. Cell. 2014;157:77–94.
28. Guttman M, Russell P, Ingolia NT, Weissman JS, Lander ES. Ribosome profiling provides evidence that large noncoding RNAs do not encode proteins. Cell. 2013;154:240–51.
29. Hon CC, Ramilowski JA, Harshbarger J, Bertin N, Rackham OJ, Gough J, Denisenko E, Schmeier S, Poulsen TM, Severin J, Lizio M, Kawaji H, Kasukawa T, Itoh M, Burroughs AM, Noma S, Djebali S, Alam T, Medvedeva YA, Testa AC, Lipovich L, Yip CW, Abugessaisa I, Mendez M, Hasegawa A, Tang D, Lassmann T, Heutink P, Babina M, Wells CA, Kojima S, Nakamura Y, Suzuki H, Daub CO, de Hoon MJ, Arner E, Hayashizaki Y, Carninci P, Forrest AR. An atlas of human long non-coding RNAs with accurate 5′ ends. Nature. 2017;543:199–204.
30. Ulitsky I. Evolution to the rescue: using comparative genomics to understand long non-coding RNAs. Nat Rev Genet. 2016;17:601–14.
31. Derrien T, Johnson R, Bussotti G, Tanzer A, Djebali S, Tilgner H, Guernec G, Martin D, Merkel A, Knowles DG, Lagarde J, Veeravalli L, Ruan X, Ruan Y, Lassmann T, Carninci P, Brown JB, Lipovich L, Gonzalez JM, Thomas M, Davis CA, Shiekhattar R, Gingeras TR, Hubbard TJ, Notredame C, Harrow J, Guigo R. The GENCODE v7 catalog of human long noncoding RNAs: analysis of their gene structure, evolution, and expression. Genome Res. 2012;22:1775–89.
32. Khalil AM, Guttman M, Huarte M, Garber M, Raj A, Rivea Morales D, Thomas K, Presser A, Bernstein BE, van Oudenaarden A, Regev A, Lander ES, Rinn JL. Many human large intergenic noncoding RNAs associate with chromatin-modifying complexes and affect gene expression. Proc Natl Acad Sci U S A. 2009;106:11667–72.
33. van Heesch S, van Iterson M, Jacobi J, Boymans S, Essers PB, de Bruijn E, Hao W, MacInnes AW, Cuppen E, Simonis M. Extensive localization of long noncoding RNAs to the cytosol and mono- and polyribosomal complexes. Genome Biol. 2014;15:R6.
34. Cabili MN, Dunagin MC, McClanahan PD, Biaesch A, Padovan-Merhar O, Regev A, Rinn JL, Raj A. Localization and abundance analysis of human lncRNAs at single-cell and single-molecule resolution. Genome Biol. 2015;16:20.
35. Jeck WR, Sorrentino JA, Wang K, Slevin MK, Burd CE, Liu J, Marzluff WF, Sharpless NE. Circular RNAs are abundant, conserved, and associated with ALU repeats. RNA. 2013;19:141–57.
36. Clark MB, Johnston RL, Inostroza-Ponta M, Fox AH, Fortini E, Moscato P, Dinger ME, Mattick JS. Genome-wide analysis of long noncoding RNA stability. Genome Res. 2012;22:885–98.
37. Guttman M, Amit I, Garber M, French C, Lin MF, Feldser D, Huarte M, Zuk O, Carey BW, Cassady JP, Cabili MN, Jaenisch R, Mikkelsen TS, Jacks T, Hacohen N, Bernstein BE, Kellis M, Regev A, Rinn JL, Lander ES. Chromatin signature reveals over a thousand highly conserved large non-coding RNAs in mammals. Nature. 2009;458:223–7.

38. Clamp M, Fry B, Kamal M, Xie X, Cuff J, Lin MF, Kellis M, Lindblad-Toh K, Lander ES. Distinguishing protein-coding and noncoding genes in the human genome. Proc Natl Acad Sci U S A. 2007;104:19428–33.
39. Dinger ME, Pang KC, Mercer TR, Mattick JS. Differentiating protein-coding and noncoding RNA: challenges and ambiguities. PLoS Comput Biol. 2008;4:e1000176.
40. Cabili MN, Trapnell C, Goff L, Koziol M, Tazon-Vega B, Regev A, Rinn JL. Integrative annotation of human large intergenic noncoding RNAs reveals global properties and specific subclasses. Genes Dev. 2011;25:1915–27.
41. Ulitsky I, Shkumatava A, Jan CH, Sive H, Bartel DP. Conserved function of lincRNAs in vertebrate embryonic development despite rapid sequence evolution. Cell. 2011;147:1537–50.
42. Tilgner H, Knowles DG, Johnson R, Davis CA, Chakrabortty S, Djebali S, Curado J, Snyder M, Gingeras TR, Guigo R. Deep sequencing of subcellular RNA fractions shows splicing to be predominantly co-transcriptional in the human genome but inefficient for lncRNAs. Genome Res. 2012;22:1616–25.
43. Engreitz JM, Haines JE, Perez EM, Munson G, Chen J, Kane M, McDonel PE, Guttman M, Lander ES. Local regulation of gene expression by lncRNA promoters, transcription and splicing. Nature. 2016;539:452–5.
44. Huber F, Bunina D, Gupta I, Khmelinskii A, Meurer M, Theer P, Steinmetz LM, Knop M. Protein abundance control by non-coding antisense transcription. Cell Rep. 2016;15:2625–36.
45. Joung J, Engreitz JM, Konermann S, Abudayyeh OO, Verdine VK, Aguet F, Gootenberg JS, Sanjana NE, Wright JB, Fulco CP, Tseng YY, Yoon CH, Boehm JS, Lander ES, Zhang F. Genome-scale activation screen identifies a lncRNA locus regulating a gene neighbourhood. Nature. 2017;548(7667):343–6.
46. Lai F, Gardini A, Zhang A, Shiekhattar R. Integrator mediates the biogenesis of enhancer RNAs. Nature. 2015;525:399–403.
47. Wilusz JE, JnBaptiste CK, Lu LY, Kuhn CD, Joshua-Tor L, Sharp PA. A triple helix stabilizes the 3′ ends of long noncoding RNAs that lack poly(A) tails. Genes Dev. 2012;26:2392–407.
48. Schlackow M, Nojima T, Gomes T, Dhir A, Carmo-Fonseca M, Proudfoot NJ. Distinctive patterns of transcription and RNA processing for human lincRNAs. Mol Cell. 2017;65:25–38.
49. Wilusz JE. Long noncoding RNAs: re-writing dogmas of RNA processing and stability. Biochim Biophys Acta. 2016;1859:128–38.
50. Yin QF, Yang L, Zhang Y, Xiang JF, Wu YW, Carmichael GG, Chen LL. Long noncoding RNAs with snoRNA ends. Mol Cell. 2012;48:219–30.
51. Zhang B, Mao YS, Diermeier SD, Novikova IV, Nawrocki EP, Jones TA, Lazar Z, Tung CS, Luo W, Eddy SR, Sanbonmatsu KY, Spector DL. Identification and characterization of a class of MALAT1-like genomic loci. Cell Rep. 2017;19:1723–38.
52. Amandio AR, Necsulea A, Joye E, Mascrez B, Duboule D. Hotair is dispensible for mouse development. PLoS Genet. 2016;12:e1006232.
53. Shan K, Jiang Q, Wang XQ, Wang YN, Yang H, Yao MD, Liu C, Li XM, Yao J, Liu B, Zhang YY, J Y, Yan B. Role of long non-coding RNA-RNCR3 in atherosclerosis-related vascular dysfunction. Cell Death Dis. 2016;7:e2248.
54. Xu C, Zhang Y, Wang Q, Xu Z, Jiang J, Gao Y, Gao M, Kang J, Wu M, Xiong J, Ji K, Yuan W, Wang Y, Liu H. Long non-coding RNA GAS5 controls human embryonic stem cell self-renewal by maintaining NODAL signalling. Nat Commun. 2016;7:13287.
55. Zangrando J, Zhang L, Vausort M, Maskali F, Marie PY, Wagner DR, Devaux Y. Identification of candidate long non-coding RNAs in response to myocardial infarction. BMC Genomics. 2014;15:460.
56. Zhang DD, Wang WT, Xiong J, Xie XM, Cui SS, Zhao ZG, Li MJ, Zhang ZQ, Hao DL, Zhao X, Li YJ, Wang J, Chen HZ, Lv X, Liu DP. Long noncoding RNA LINC00305 promotes inflammation by activating the AHRR-NF-kappaB pathway in human monocytes. Sci Rep. 2017;7:46204.

57. Li Y, Shen S, Ding S, Wang L. LincRNA DYN-LRB2-2 upregulates cholesterol efflux by decreasing TLR2 expression in macrophages. J Cell Biochem. 2018;119:1911–21.
58. Berretta J, Morillon A. Pervasive transcription constitutes a new level of eukaryotic genome regulation. EMBO Rep. 2009;10:973–82.
59. Kramer C, Loros JJ, Dunlap JC, Crosthwaite SK. Role for antisense RNA in regulating circadian clock function in Neurospora crassa. Nature. 2003;421:948–52.
60. Nagano T, Mitchell JA, Sanz LA, Pauler FM, Ferguson-Smith AC, Feil R, Fraser P. The Air noncoding RNA epigenetically silences transcription by targeting G9a to chromatin. Science. 2008;322:1717–20.
61. Camblong J, Iglesias N, Fickentscher C, Dieppois G, Stutz F. Antisense RNA stabilization induces transcriptional gene silencing via histone deacetylation in S. cerevisiae. Cell. 2007;131:706–17.
62. Prescott EM, Proudfoot NJ. Transcriptional collision between convergent genes in budding yeast. Proc Natl Acad Sci U S A. 2002;99:8796–801.
63. Hobson DJ, Wei W, Steinmetz LM, Svejstrup JQ. RNA polymerase II collision interrupts convergent transcription. Mol Cell. 2012;48:365–74.
64. Lee JT, Lu N. Targeted mutagenesis of Tsix leads to nonrandom X inactivation. Cell. 1999;99:47–57.
65. Sado T, Hoki Y, Sasaki H. Tsix silences Xist through modification of chromatin structure. Dev Cell. 2005;9:159–65.
66. Pasmant E, Laurendeau I, Heron D, Vidaud M, Vidaud D, Bieche I. Characterization of a germ-line deletion, including the entire INK4/ARF locus, in a melanoma-neural system tumor family: identification of ANRIL, an antisense noncoding RNA whose expression coclusters with ARF. Cancer Res. 2007;67:3963–9.
67. Bell RD, Long X, Lin M, Bergmann JH, Nanda V, Cowan SL, Zhou Q, Han Y, Spector DL, Zheng D, Miano JM. Identification and initial functional characterization of a human vascular cell-enriched long noncoding RNA. Arterioscler Thromb Vasc Biol. 2014;34:1249–59.
68. Zong X, Nakagawa S, Freier SM, Fei J, Ha T, Prasanth SG, Prasanth KV. Natural antisense RNA promotes 3' end processing and maturation of MALAT1 lncRNA. Nucleic Acids Res. 2016;44:2898–908.
69. Leisegang MS, Fork C, Josipovic I, Richter F, Preussner J, Hu J, Miller MJ, Epah JN, Hofmann P, Gunther S, Moll F, Valasarajan C, Heidler J, Ponomareva Y, Freiman TM, Maegdefessel L, Plate KH, Mittelbronn M, Uchida S, Kunne C, Stellos K, Schermuly RT, Weissmann N, Devraj K, Wittig I, Boon RA, Dimmeler S, Pullamsetti SS, Looso M, Miller FJ, Brandes RP. Long noncoding RNA MANTIS facilitates endothelial angiogenic function. Circulation. 2017;136(1):65–79.
70. Huang C, Hu YW, Zhao JJ, Ma X, Zhang Y, Guo FX, Kang CM, Lu JB, Xiu JC, Sha YH, Gao JJ, Wang YC, Li P, Xu BM, Zheng L, Wang Q. Long noncoding RNA HOXC-AS1 suppresses Ox-LDL-induced cholesterol accumulation through promoting HOXC6 expression in THP-1 macrophages. DNA Cell Biol. 2016;35:722–9.
71. Zhao Y, Feng G, Wang Y, Yue Y, Zhao W. Regulation of apoptosis by long non-coding RNA HIF1A-AS1 in VSMCs: implications for TAA pathogenesis. Int J Clin Exp Pathol. 2014;7:7643–52.
72. Lee MP, DeBaun MR, Mitsuya K, Galonek HL, Brandenburg S, Oshimura M, Feinberg AP. Loss of imprinting of a paternally expressed transcript, with antisense orientation to KVLQT1, occurs frequently in Beckwith-Wiedemann syndrome and is independent of insulin-like growth factor II imprinting. Proc Natl Acad Sci U S A. 1999;96:5203–8.
73. Mehta SL, Kim T, Vemuganti R. Long noncoding RNA FosDT promotes ischemic brain injury by interacting with REST-associated chromatin-modifying proteins. J Neurosci. 2015;35:16443–9.
74. Preker P, Nielsen J, Kammler S, Lykke-Andersen S, Christensen MS, Mapendano CK, Schierup MH, Jensen TH. RNA exosome depletion reveals transcription upstream of active human promoters. Science. 2008;322:1851–4.

75. Ntini E, Jarvelin AI, Bornholdt J, Chen Y, Boyd M, Jorgensen M, Andersson R, Hoof I, Schein A, Andersen PR, Andersen PK, Preker P, Valen E, Zhao X, Pelechano V, Steinmetz LM, Sandelin A, Jensen TH. Polyadenylation site-induced decay of upstream transcripts enforces promoter directionality. Nat Struct Mol Biol. 2013;20:923–8.
76. Almada AE, Wu X, Kriz AJ, Burge CB, Sharp PA. Promoter directionality is controlled by U1 snRNP and polyadenylation signals. Nature. 2013;499:360–3.
77. Neil H, Malabat C, d'Aubenton-Carafa Y, Xu Z, Steinmetz LM, Jacquier A. Widespread bidirectional promoters are the major source of cryptic transcripts in yeast. Nature. 2009;457:1038–42.
78. Schulz D, Schwalb B, Kiesel A, Baejen C, Torkler P, Gagneur J, Soeding J, Cramer P. Transcriptome surveillance by selective termination of noncoding RNA synthesis. Cell. 2013;155:1075–87.
79. Xu Z, Wei W, Gagneur J, Perocchi F, Clauder-Munster S, Camblong J, Guffanti E, Stutz F, Huber W, Steinmetz LM. Bidirectional promoters generate pervasive transcription in yeast. Nature. 2009;457:1033–7.
80. Pelechano V, Steinmetz LM. Gene regulation by antisense transcription. Nat Rev Genet. 2013;14:880–93.
81. Kim TK, Hemberg M, Gray JM, Costa AM, Bear DM, Wu J, Harmin DA, Laptewicz M, Barbara-Haley K, Kuersten S, Markenscoff-Papadimitriou E, Kuhl D, Bito H, Worley PF, Kreiman G, Greenberg ME. Widespread transcription at neuronal activity-regulated enhancers. Nature. 2010;465:182–7.
82. Wang D, Garcia-Bassets I, Benner C, Li W, Su X, Zhou Y, Qiu J, Liu W, Kaikkonen MU, Ohgi KA, Glass CK, Rosenfeld MG, Fu XD. Reprogramming transcription by distinct classes of enhancers functionally defined by eRNA. Nature. 2011;474:390–4.
83. Li W, Notani D, Ma Q, Tanasa B, Nunez E, Chen AY, Merkurjev D, Zhang J, Ohgi K, Song X, Oh S, Kim HS, Glass CK, Rosenfeld MG. Functional roles of enhancer RNAs for oestrogen-dependent transcriptional activation. Nature. 2013;498:516–20.
84. Puc J, Kozbial P, Li W, Tan Y, Liu Z, Suter T, Ohgi KA, Zhang J, Aggarwal AK, Rosenfeld MG. Ligand-dependent enhancer activation regulated by topoisomerase-I activity. Cell. 2015;160:367–80.
85. Li W, Notani D, Rosenfeld MG. Enhancers as non-coding RNA transcription units: recent insights and future perspectives. Nat Rev Genet. 2016;17:207–23.
86. Shiraki T, Kondo S, Katayama S, Waki K, Kasukawa T, Kawaji H, Kodzius R, Watahiki A, Nakamura M, Arakawa T, Fukuda S, Sasaki D, Podhajska A, Harbers M, Kawai J, Carninci P, Hayashizaki Y. Cap analysis gene expression for high-throughput analysis of transcriptional starting point and identification of promoter usage. Proc Natl Acad Sci U S A. 2003;100:15776–81.
87. Murakawa Y, Yoshihara M, Kawaji H, Nishikawa M, Zayed H, Suzuki H, Fantom C, Hayashizaki Y. Enhanced Identification of Transcriptional Enhancers Provides Mechanistic Insights into Diseases. Trends Genet. 2016;32:76–88.
88. Kwak H, Fuda NJ, Core LJ, Lis JT. Precise maps of RNA polymerase reveal how promoters direct initiation and pausing. Science. 2013;339:950–3.
89. Orom UA, Derrien T, Beringer M, Gumireddy K, Gardini A, Bussotti G, Lai F, Zytnicki M, Notredame C, Huang Q, Guigo R, Shiekhattar R. Long noncoding RNAs with enhancer-like function in human cells. Cell. 2010;143:46–58.
90. Vucicevic D, Corradin O, Ntini E, Scacheri PC, Orom UA. Long ncRNA expression associates with tissue-specific enhancers. Cell Cycle. 2015;14:253–60.
91. Micheletti R, Plaisance I, Abraham BJ, Sarre A, Ting CC, Alexanian M, Maric D, Maison D, Nemir M, Young RA, Schroen B, Gonzalez A, Ounzain S, Pedrazzini T. The long noncoding RNA Wisper controls cardiac fibrosis and remodeling. Sci Transl Med. 2017;9
92. Wang KC, Yang YW, Liu B, Sanyal A, Corces-Zimmerman R, Chen Y, Lajoie BR, Protacio A, Flynn RA, Gupta RA, Wysocka J, Lei M, Dekker J, Helms JA, Chang HY. A

long noncoding RNA maintains active chromatin to coordinate homeotic gene expression. Nature. 2011;472:120–4.

93. Maamar H, Cabili MN, Rinn J, Raj A. linc-HOXA1 is a noncoding RNA that represses Hoxa1 transcription in cis. Genes Dev. 2013;27:1260–71.

94. Ballantyne MD, Pinel K, Dakin R, Vesey AT, Diver L, Mackenzie R, Garcia R, Welsh P, Sattar N, Hamilton G, Joshi N, Dweck MR, Miano JM, McBride MW, Newby DE, McDonald RA, Baker AH. Smooth muscle enriched long noncoding RNA (SMILR) regulates cell proliferation. Circulation. 2016;133:2050–65.

95. Nakaya HI, Amaral PP, Louro R, Lopes A, Fachel AA, Moreira YB, El-Jundi TA, da Silva AM, Reis EM, Verjovski-Almeida S. Genome mapping and expression analyses of human intronic noncoding RNAs reveal tissue-specific patterns and enrichment in genes related to regulation of transcription. Genome Biol. 2007;8:R43.

96. Gardner EJ, Nizami ZF, Talbot CC Jr, Gall JG. Stable intronic sequence RNA (sisRNA), a new class of noncoding RNA from the oocyte nucleus of Xenopus tropicalis. Genes Dev. 2012;26:2550–9.

97. Pek JW, Osman I, Tay ML, Zheng RT. Stable intronic sequence RNAs have possible regulatory roles in Drosophila melanogaster. J Cell Biol. 2015;211:243–51.

98. Duret L, Chureau C, Samain S, Weissenbach J, Avner P. The Xist RNA gene evolved in eutherians by pseudogenization of a protein-coding gene. Science. 2006;312:1653–5.

99. Poliseno L, Salmena L, Zhang J, Carver B, Haveman WJ, Pandolfi PP. A coding-independent function of gene and pseudogene mRNAs regulates tumour biology. Nature. 2010;465:1033–8.

100. Guo JU, Agarwal V, Guo H, Bartel DP. Expanded identification and characterization of mammalian circular RNAs. Genome Biol. 2014;15:409.

101. Salzman J, Chen RE, Olsen MN, Wang PL, Brown PO. Cell-type specific features of circular RNA expression. PLoS Genet. 2013;9:e1003777.

102. Rybak-Wolf A, Stottmeister C, Glazar P, Jens M, Pino N, Giusti S, Hanan M, Behm M, Bartok O, Ashwal-Fluss R, Herzog M, Schreyer L, Papavasileiou P, Ivanov A, Ohman M, Refojo D, Kadener S, Rajewsky N. Circular RNAs in the mammalian brain are highly abundant. Conserved, and Dynamically Expressed, Mol Cell. 2015;58:870–85.

103. Barrett SP, Salzman J. Circular RNAs: analysis, expression and potential functions. Development. 2016;143:1838–47.

104. Holdt LM, Kohlmaier A, Teupser D. Molecular roles and function of circular RNAs in eukaryotic cells. Cell Mol Life Sci. 2018;75(6):1071–98.

105. Wang K, Long B, Liu F, Wang JX, Liu CY, Zhao B, Zhou LY, Sun T, Wang M, Yu T, Gong Y, Liu J, Dong YH, Li N, Li PF. A circular RNA protects the heart from pathological hypertrophy and heart failure by targeting miR-223. Eur Heart J. 2016;37:2602–11.

106. Wang K, Gan TY, Li N, Liu CY, Zhou LY, Gao JN, Chen C, Yan KW, Ponnusamy M, Zhang YH, Li PF. Circular RNA mediates cardiomyocyte death via miRNA-dependent upregulation of MTP18 expression. Cell Death Differ. 2017;24:1111–20.

107. Kondo T, Hashimoto Y, Kato K, Inagaki S, Hayashi S, Kageyama Y. Small peptide regulators of actin-based cell morphogenesis encoded by a polycistronic mRNA. Nat Cell Biol. 2007;9:660–5.

108. Hanyu-Nakamura K, Sonobe-Nojima H, Tanigawa A, Lasko P, Nakamura A. Drosophila Pgc protein inhibits P-TEFb recruitment to chromatin in primordial germ cells. Nature. 2008;451:730–3.

109. Kondo T, Plaza S, Zanet J, Benrabah E, Valenti P, Hashimoto Y, Kobayashi S, Payre F, Kageyama Y. Small peptides switch the transcriptional activity of Shavenbaby during Drosophila embryogenesis. Science. 2010;329:336–9.

110. Magny EG, Pueyo JI, Pearl FM, Cespedes MA, Niven JE, Bishop SA, Couso JP. Conserved regulation of cardiac calcium uptake by peptides encoded in small open reading frames. Science. 2013;341:1116–20.

111. Pauli A, Norris ML, Valen E, Chew GL, Gagnon JA, Zimmerman S, Mitchell A, Ma J, Dubrulle J, Reyon D, Tsai SQ, Joung JK, Saghatelian A, Schier AF. Toddler: an embryonic signal that promotes cell movement via Apelin receptors. Science. 2014;343:1248636.
112. Anderson DM, Anderson KM, Chang CL, Makarewich CA, Nelson BR, McAnally JR, Kasaragod P, Shelton JM, Liou J, Bassel-Duby R, Olson EN. A micropeptide encoded by a putative long noncoding RNA regulates muscle performance. Cell. 2015;160:595–606.
113. Rohrig H, Schmidt J, Miklashevichs E, Schell J, John M. Soybean ENOD40 encodes two peptides that bind to sucrose synthase. Proc Natl Acad Sci U S A. 2002;99:1915–20.
114. Pauli A, Valen E, Schier AF. Identifying (non-)coding RNAs and small peptides: challenges and opportunities. BioEssays. 2015;37:103–12.
115. Slavoff SA, Mitchell AJ, Schwaid AG, Cabili MN, Ma J, Levin JZ, Karger AD, Budnik BA, Rinn JL, Saghatelian A. Peptidomic discovery of short open reading frame-encoded peptides in human cells. Nat Chem Biol. 2013;9:59–64.
116. Nelson BR, Makarewich CA, Anderson DM, Winders BR, Troupes CD, Wu F, Reese AL, McAnally JR, Chen X, Kavalali ET, Cannon SC, Houser SR, Bassel-Duby R, Olson EN. A peptide encoded by a transcript annotated as long noncoding RNA enhances SERCA activity in muscle. Science. 2016;351:271–5.
117. Monfort A, Di Minin G, Postlmayr A, Freimann R, Arieti F, Thore S, Wutz A. Identification of Spen as a crucial factor for Xist function through forward genetic screening in haploid embryonic stem cells. Cell Rep. 2015;12:554–61.
118. Kurian L, Aguirre A, Sancho-Martinez I, Benner C, Hishida T, Nguyen TB, Reddy P, Nivet E, Krause MN, Nelles DA, Esteban CR, Campistol JM, Yeo GW, Belmonte JCI. Identification of novel long noncoding RNAs underlying vertebrate cardiovascular development. Circulation. 2015;131:1278–90.
119. Jiang W, Liu Y, Liu R, Zhang K, Zhang Y. The lncRNA DEANR1 facilitates human endoderm differentiation by activating FOXA2 expression. Cell Rep. 2015;11:137–48.
120. Puri MC, Rossant J, Alitalo K, Bernstein A, Partanen J. The receptor tyrosine kinase TIE is required for integrity and survival of vascular endothelial cells. EMBO J. 1995;14:5884–91.
121. Li K, Blum Y, Verma A, Liu Z, Pramanik K, Leigh NR, Chun CZ, Samant GV, Zhao B, Garnaas MK, Horswill MA, Stanhope SA, North PE, Miao RQ, Wilkinson GA, Affolter M, Ramchandran R. A noncoding antisense RNA in tie-1 locus regulates tie-1 function in vivo. Blood. 2010;115:133–9.
122. Limaye N, Wouters V, Uebelhoer M, Tuominen M, Wirkkala R, Mulliken JB, Eklund L, Boon LM, Vikkula M. Somatic mutations in angiopoietin receptor gene TEK cause solitary and multiple sporadic venous malformations. Nat Genet. 2009;41:118–24.
123. Yang L, Kirby JE, Sunwoo H, Lee JT. Female mice lacking Xist RNA show partial dosage compensation and survive to term. Genes Dev. 2016;30:1747–60.
124. Yildirim E, Kirby JE, Brown DE, Mercier FE, Sadreyev RI, Scadden DT, Lee JT. Xist RNA is a potent suppressor of hematologic cancer in mice. Cell. 2013;152:727–42.
125. Savarese F, Flahndorfer K, Jaenisch R, Busslinger M, Wutz A. Hematopoietic precursor cells transiently reestablish permissiveness for X inactivation. Mol Cell Biol. 2006;26:7167–77.
126. Zhang B, Arun G, Mao YS, Lazar Z, Hung G, Bhattacharjee G, Xiao X, Booth CJ, Wu J, Zhang C, Spector DL. The lncRNA Malat1 is dispensable for mouse development but its transcription plays a cis-regulatory role in the adult. Cell Rep. 2012;2:111–23.
127. Nakagawa S, Ip JY, Shioi G, Tripathi V, Zong X, Hirose T, Prasanth KV. Malat1 is not an essential component of nuclear speckles in mice. RNA. 2012;18:1487–99.
128. Eissmann M, Gutschner T, Hammerle M, Gunther S, Caudron-Herger M, Gross M, Schirmacher P, Rippe K, Braun T, Zornig M, Diederichs S. Loss of the abundant nuclear non-coding RNA MALAT1 is compatible with life and development. RNA Biol. 2012;9:1076–87.
129. Holdt LM, Beutner F, Scholz M, Gielen S, Gabel G, Bergert H, Schuler G, Thiery J, Teupser D. ANRIL expression is associated with atherosclerosis risk at chromosome 9p21. Arterioscler Thromb Vasc Biol. 2010;30:620–7.

130. Cooper JD, Walker NM, Smyth DJ, Downes K, Healy BC, Todd JA, Type IDGC. Follow-up of 1715 SNPs from the Wellcome Trust Case Control Consortium genome-wide association study in type I diabetes families. Genes Immun. 2009;10(Suppl 1):S85–94.

131. McPherson R, Pertsemlidis A, Kavaslar N, Stewart A, Roberts R, Cox DR, Hinds DA, Pennacchio LA, Tybjaerg-Hansen A, Folsom AR, Boerwinkle E, Hobbs HH, Cohen JC. A common allele on chromosome 9 associated with coronary heart disease. Science. 2007;316:1488–91.

132. Liu Y, Sanoff HK, Cho H, Burd CE, Torrice C, Mohlke KL, Ibrahim JG, Thomas NE, Sharpless NE. INK4/ARF transcript expression is associated with chromosome 9p21 variants linked to atherosclerosis. PLoS One. 2009;4:e5027.

133. Jarinova O, Stewart AF, Roberts R, Wells G, Lau P, Naing T, Buerki C, McLean BW, Cook RC, Parker JS, McPherson R. Functional analysis of the chromosome 9p21.3 coronary artery disease risk locus. Arterioscler Thromb Vasc Biol. 2009;29:1671–7.

134. Holdt LM, Teupser D. Recent studies of the human chromosome 9p21 locus, which is associated with atherosclerosis in human populations. Arterioscler Thromb Vasc Biol. 2012;32:196–206.

135. Helgadottir A, Thorleifsson G, Manolescu A, Gretarsdottir S, Blondal T, Jonasdottir A, Jonasdottir A, Sigurdsson A, Baker A, Palsson A, Masson G, Gudbjartsson DF, Magnusson KP, Andersen K, Levey AI, Backman VM, Matthiasdottir S, Jonsdottir T, Palsson S, Einarsdottir H, Gunnarsdottir S, Gylfason A, Vaccarino V, Hooper WC, Reilly MP, Granger CB, Austin H, Rader DJ, Shah SH, Quyyumi AA, Gulcher JR, Thorgeirsson G, Thorsteinsdottir U, Kong A, Stefansson K. A common variant on chromosome 9p21 affects the risk of myocardial infarction. Science. 2007;316:1491–3.

136. Helgadottir A, Thorleifsson G, Magnusson KP, Gretarsdottir S, Steinthorsdottir V, Manolescu A, Jones GT, Rinkel GJ, Blankensteijn JD, Ronkainen A, Jaaskelainen JE, Kyo Y, Lenk GM, Sakalihasan N, Kostulas K, Gottsater A, Flex A, Stefansson H, Hansen T, Andersen G, Weinsheimer S, Borch-Johnsen K, Jorgensen T, Shah SH, Quyyumi AA, Granger CB, Reilly MP, Austin H, Levey AI, Vaccarino V, Palsdottir E, Walters GB, Jonsdottir T, Snorradottir S, Magnusdottir D, Gudmundsson G, Ferrell RE, Sveinbjornsdottir S, Hernesniemi J, Niemela M, Limet R, Andersen K, Sigurdsson G, Benediktsson R, Verhoeven EL, Teijink JA, Grobbee DE, Rader DJ, Collier DA, Pedersen O, Pola R, Hillert J, Lindblad B, Valdimarsson EM, Magnadottir HB, Wijmenga C, Tromp G, Baas AF, Ruigrok YM, van Rij AM, Kuivaniemi H, Powell JT, Matthiasson SE, Gulcher JR, Thorgeirsson G, Kong A, Thorsteinsdottir U, Stefansson K. The same sequence variant on 9p21 associates with myocardial infarction, abdominal aortic aneurysm and intracranial aneurysm. Nat Genet. 2008;40:217–24.

137. Matarin M, Brown WM, Singleton A, Hardy JA, Meschia JF, ISGS Investigators. Whole genome analyses suggest ischemic stroke and heart disease share an association with polymorphisms on chromosome 9p21. Stroke. 2008;39:1586–9.

138. Smith JG, Melander O, Lovkvist H, Hedblad B, Engstrom G, Nilsson P, Carlson J, Berglund G, Norrving B, Lindgren A. Common genetic variants on chromosome 9p21 confers risk of ischemic stroke: a large-scale genetic association study. Circ Cardiovasc Genet. 2009;2:159–64.

139. Wahlstrand B, Orho-Melander M, Delling L, Kjeldsen S, Narkiewicz K, Almgren P, Hedner T, Melander O. The myocardial infarction associated CDKN2A/CDKN2B locus on chromosome 9p21 is associated with stroke independently of coronary events in patients with hypertension. J Hypertens. 2009;27:769–73.

140. Anderson CD, Biffi A, Rost NS, Cortellini L, Furie KL, Rosand J. Chromosome 9p21 in ischemic stroke: population structure and meta-analysis. Stroke. 2010;41:1123–31.

141. Zhang W, Chen Y, Liu P, Chen J, Song L, Tang Y, Wang Y, Liu J, Hu FB, Hui R. Variants on chromosome 9p21.3 correlated with ANRIL expression contribute to stroke risk and recurrence in a large prospective stroke population. Stroke. 2012;43:14–21.

142. Bai Y, Nie S, Jiang G, Zhou Y, Zhou M, Zhao Y, Li S, Wang F, Lv Q, Huang Y, Yang Q, Li Q, Li Y, Xia Y, Liu Y, Liu J, Qian J, Li B, Wu G, Wu Y, Wang B, Cheng X, Yang Y, Ke T, Li H, Ren X, Ma X, Liao Y, Xu C, Tu X, Wang QK. Regulation of CARD8 expression by ANRIL and association of CARD8 single nucleotide polymorphism rs2043211 (p.C10X) with ischemic stroke. Stroke. 2014;45:383–8.

143. Zhang B, Wang D, Ji TF, Shi L, Yu JL. Overexpression of lncRNA ANRIL up-regulates VEGF expression and promotes angiogenesis of diabetes mellitus combined with cerebral infarction by activating NF-kappaB signaling pathway in a rat model. Oncotarget. 2017;8:17347–59.

144. Chen L, Yao H, Hui JY, Ding SH, Fan YL, Pan YH, Chen KH, Wan JQ, Jiang JY. Global transcriptomic study of atherosclerosis development in rats. Gene. 2016;592:43–8.

145. Arslan S, Berkan O, Lalem T, Ozbilum N, Goksel S, Korkmaz O, Cetin N, Devaux Y, Cardiolinc Network. Long non-coding RNAs in the atherosclerotic plaque. Atherosclerosis. 2017;266:176–81.

146. Qu X, Du Y, Shu Y, Gao M, Sun F, Luo S, Yang T, Zhan L, Yuan Y, Chu W, Pan Z, Wang Z, Yang B, Lu Y. MIAT Is a Pro-fibrotic Long Non-coding RNA Governing Cardiac Fibrosis in Post-infarct Myocardium. Sci Rep. 2017;7:42657.

147. Frade AF, Laugier L, Ferreira LR, Baron MA, Benvenuti LA, Teixeira PC, Navarro IC, Cabantous S, Ferreira FM, da Silva Candido D, Gaiotto FA, Bacal F, Pomerantzeff P, Santos RH, Kalil J, Cunha-Neto E, Chevillard C. Myocardial infarction-associated transcript, a long noncoding RNA. Is Overexpressed During Dilated Cardiomyopathy Due to Chronic Chagas Disease, J Infect Dis. 2016;214:161–5.

148. Yan B, Yao J, Liu JY, Li XM, Wang XQ, Li YJ, Tao ZF, Song YC, Chen Q, Jiang Q. lncRNA-MIAT regulates microvascular dysfunction by functioning as a competing endogenous RNA. Circ Res. 2015;116:1143–56.

149. Gao W, Zhu M, Wang H, Zhao S, Zhao D, Yang Y, Wang ZM, Wang F, Yang ZJ, Lu X, Wang LS. Association of polymorphisms in long non-coding RNA H19 with coronary artery disease risk in a Chinese population. Mutat Res. 2015;772:15–22.

150. Zhang Z, Gao W, Long QQ, Zhang J, Li YF, Liu DC, Yan JJ, Yang ZJ, Wang LS. Increased plasma levels of lncRNA H19 and LIPCAR are associated with increased risk of coronary artery disease in a Chinese population. Sci Rep. 2017;7:7491.

151. Han Y, Ma J, Wang J, Wang L. Silencing of H19 inhibits the adipogenesis and inflammation response in ox-LDL-treated Raw264.7 cells by up-regulating miR-130b. Mol Immunol. 2017;93:107–14.

152. Hadji F, Boulanger MC, Guay SP, Gaudreault N, Amellah S, Mkannez G, Bouchareb R, Marchand JT, Nsaibia MJ, Guauque-Olarte S, Pibarot P, Bouchard L, Bosse Y, Mathieu P. Altered DNA methylation of long noncoding RNA H19 in calcific aortic valve disease promotes mineralization by silencing NOTCH1. Circulation. 2016;134:1848–62.

153. Wang J, Zhao H, Fan Z, Li G, Ma Q, Tao Z, Wang R, Feng J, Luo Y. Long noncoding RNA H19 promotes neuroinflammation in ischemic stroke by driving histone deacetylase 1-dependent M1 microglial polarization. Stroke. 2017;48:2211–21.

154. Wang J, Cao B, Han D, Sun M, Feng J. Long non-coding RNA H19 induces cerebral ischemia reperfusion injury via activation of autophagy. Aging Dis. 2017;8:71–84.

155. Vigetti D, Deleonibus S, Moretto P, Bowen T, Fischer JW, Grandoch M, Oberhuber A, Love DC, Hanover JA, Cinquetti R, Karousou E, Viola M, D'Angelo ML, Hascall VC, De Luca G, Passi A. Natural antisense transcript for hyaluronan synthase 2 (HAS2-AS1) induces transcription of HAS2 via protein O-GlcNAcylation. J Biol Chem. 2014;289:28816–26.

156. Liao B, Chen R, Lin F, Mai A, Chen J, Li H, Xu Z, Dong S. Long noncoding RNA HOTTIP promotes endothelial cell proliferation and migration via activation of the Wnt/beta-catenin pathway. J Cell Biochem. 2018;119(3):2797–805.

157. Hu YW, Yang JY, Ma X, Chen ZP, Hu YR, Zhao JY, Li SF, Qiu YR, Lu JB, Wang YC, Gao JJ, Sha YH, Zheng L, Wang Q. A lincRNA-DYNLRB2-2/GPR119/GLP-1R/ABCA1-

dependent signal transduction pathway is essential for the regulation of cholesterol homeostasis. J Lipid Res. 2014;55:681–97.

158. Tang SS, Cheng J, Cai MY, Yang XL, Liu XG, Zheng BY, Xiong XD. Association of lincRNA-p21 haplotype with coronary artery disease in a Chinese han population. Dis Markers. 2016;2016:9109743.

159. Wu G, Cai J, Han Y, Chen J, Huang ZP, Chen C, Cai Y, Huang H, Yang Y, Liu Y, Xu Z, He D, Zhang X, Hu X, Pinello L, Zhong D, He F, Yuan GC, Wang DZ, Zeng C. LincRNA-p21 regulates neointima formation, vascular smooth muscle cell proliferation, apoptosis, and atherosclerosis by enhancing p53 activity. Circulation. 2014;130:1452–65.

160. Wu Z, He Y, Li D, Fang X, Shang T, Zhang H, Zheng X. Long noncoding RNA MEG3 suppressed endothelial cell proliferation and migration through regulating miR-21. Am J Transl Res. 2017;9:3326–35.

161. Piccoli MT, Gupta SK, Viereck J, Foinquinos A, Samolovac S, Kramer FL, Garg A, Remke J, Zimmer K, Batkai S, Thum T. Inhibition of the cardiac fibroblast-enriched lncRNA Meg3 prevents cardiac fibrosis and diastolic dysfunction. Circ Res. 2017;121:575–83.

162. Yan H, Yuan J, Gao L, Rao J, Hu J. Long noncoding RNA MEG3 activation of p53 mediates ischemic neuronal death in stroke. Neuroscience. 2016;337:191–9.

163. Shahmoradi N, Nasiri M, Kamfiroozi H, Kheiry MA. Association of the rs555172 polymorphism in SENCR long non-coding RNA and atherosclerotic coronary artery disease. J Cardiovasc Thorac Res. 2017;9:170–4.

164. Boulberdaa M, Scott E, Ballantyne M, Garcia R, Descamps B, Angelini GD, Brittan M, Hunter A, McBride M, McClure J, Miano JM, Emanueli C, Mills NL, Mountford JC, Baker AH. A role for the long noncoding RNA SENCR in commitment and function of endothelial cells. Mol Ther. 2016;24:978–90.

165. Chen C, Cheng G, Yang X, Li C, Shi R, Zhao N. Tanshinol suppresses endothelial cells apoptosis in mice with atherosclerosis via lncRNA TUG1 up-regulating the expression of miR-26a. Am J Transl Res. 2016;8:2981–91.

166. Chen S, Wang M, Yang H, Mao L, He Q, Jin H, Ye ZM, Luo XY, Xia YP, Hu B. LncRNA TUG1 sponges microRNA-9 to promote neurons apoptosis by up-regulated Bcl2l11 under ischemia. Biochem Biophys Res Commun. 2017;485:167–73.

167. Viereck J, Kumarswamy R, Foinquinos A, Xiao K, Avramopoulos P, Kunz M, Dittrich M, Maetzig T, Zimmer K, Remke J, Just A, Fendrich J, Scherf K, Bolesani E, Schambach A, Weidemann F, Zweigerdt R, de Windt LJ, Engelhardt S, Dandekar T, Batkai S, Thum T. Long noncoding RNA Chast promotes cardiac remodeling. Sci Transl Med. 2016;8:326ra22.

168. Wang K, Liu F, Zhou LY, Long B, Yuan SM, Wang Y, Liu CY, Sun T, Zhang XJ, Li PF. The long noncoding RNA CHRF regulates cardiac hypertrophy by targeting miR-489. Circ Res. 2014;114:1377–88.

169. Lai Y, He S, Ma L, Lin H, Ren B, Ma J, Zhu X, Zhuang S. HOTAIR functions as a competing endogenous RNA to regulate PTEN expression by inhibiting miR-19 in cardiac hypertrophy. Mol Cell Biochem. 2017;432:179–87.

170. Greco S, Zaccagnini G, Perfetti A, Fuschi P, Valaperta R, Voellenkle C, Castelvecchio S, Gaetano C, Finato N, Beltrami AP, Menicanti L, Martelli F. Long noncoding RNA dysregulation in ischemic heart failure. J Transl Med. 2016;14:183.

171. Vausort M, Wagner DR, Devaux Y. Long noncoding RNAs in patients with acute myocardial infarction. Circ Res. 2014;115:668–77.

172. Li X, Dai Y, Yan S, Shi Y, Han B, Li J, Cha L, Mu J. Down-regulation of lncRNA KCNQ1OT1 protects against myocardial ischemia/reperfusion injury following acute myocardial infarction. Biochem Biophys Res Commun. 2017;491:1026–33.

173. Li X, Zhou J, Huang K. Inhibition of the lncRNA Mirt1 attenuates acute myocardial infarction by suppressing NF-kappaB activation. Cell Physiol Biochem. 2017;42:1153–64.

174. Jiang F, Zhou X, Huang J. Long non-coding RNA-ROR mediates the reprogramming in cardiac hypertrophy. PLoS One. 2016;11:e0152767.

175. Bravo-San Pedro JM, Kroemer G, Galluzzi L. Autophagy and mitophagy in cardiovascular disease. Circ Res. 2017;120:1812–24.
176. Geng HH, Li R, Su YM, Xiao J, Pan M, Cai XX, Ji XP. The circular RNA Cdr1as promotes myocardial infarction by mediating the regulation of miR-7a on its target genes expression. PLoS One. 2016;11:e0151753.
177. Liu J, Li Y, Lin B, Sheng Y, Yang L. HBL1 is a human long noncoding RNA that modulates cardiomyocyte development from pluripotent stem cells by counteracting MIR1. Dev Cell. 2017;42:333–348 e5.
178. Lin X, Zhan JK, Wang YJ, Tan P, Chen YY, Deng HQ, Liu YS. Function, role, and clinical application of microRNAs in vascular aging. Biomed Res Int. 2016;2016:6021394.
179. Liu Y, Zhou D, Li G, Ming X, Tu Y, Tian J, Lu H, Yu B. Long non coding RNA-UCA1 contributes to cardiomyocyte apoptosis by suppression of p27 expression. Cell Physiol Biochem. 2015;35:1986–98.
180. Zhu XH, Yuan YX, Rao SL, Wang P. LncRNA MIAT enhances cardiac hypertrophy partly through sponging miR-150. Eur Rev Med Pharmacol Sci. 2016;20:3653–60.
181. Zhou X, Zhang W, Jin M, Chen J, Xu W, Kong X. lncRNA MIAT functions as a competing endogenous RNA to upregulate DAPK2 by sponging miR-22-3p in diabetic cardiomyopathy. Cell Death Dis. 2017;8:e2929.
182. Zhang X, Tang X, Liu K, Hamblin MH, Yin KJ. Long noncoding RNA Malat1 regulates cerebrovascular pathologies in ischemic stroke. J Neurosci. 2017;37:1797–806.
183. Xin JW, Jiang YG. Long noncoding RNA MALAT1 inhibits apoptosis induced by oxygen-glucose deprivation and reoxygenation in human brain microvascular endothelial cells. Exp Ther Med. 2017;13:1225–34.
184. Zhao J, Zhang W, Lin M, Wu W, Jiang P, Tou E, Xue M, Richards A, Jourd'heuil D, Asif A, Zheng D, Singer HA, Miano JM, Long X. MYOSLID is a novel serum response factor-dependent long noncoding RNA that amplifies the vascular smooth muscle differentiation program. Arterioscler Thromb Vasc Biol. 2016;36:2088–99.
185. Chen G, Guo H, Song Y, Chang H, Wang S, Zhang M, Liu C. Long noncoding RNA AK055347 is upregulated in patients with atrial fibrillation and regulates mitochondrial energy production in myocardiocytes. Mol Med Rep. 2016;14:5311–7.
186. Wang W, Wang X, Zhang Y, Li Z, Xie X, Wang J, Gao M, Zhang S, Hou Y. Transcriptome analysis of canine cardiac fat pads: involvement of two novel long non-coding RNAs in atrial fibrillation neural remodeling. J Cell Biochem. 2015;116:809–21.
187. Li Z, Wang X, Wang W, Du J, Wei J, Zhang Y, Wang J, Hou Y. Altered long non-coding RNA expression profile in rabbit atria with atrial fibrillation: TCONS_00075467 modulates atrial electrical remodeling by sponging miR-328 to regulate CACNA1C. J Mol Cell Cardiol. 2017;108:73–85.
188. Xu Q, Deng F, Xing Z, Wu Z, Cen B, Xu S, Zhao Z, Nepomuceno R, Bhuiyan MI, Sun D, Wang QJ, Ji A. Long non-coding RNA C2dat1 regulates CaMKIIdelta expression to promote neuronal survival through the NF-kappaB signaling pathway following cerebral ischemia. Cell Death Dis. 2016;7:e2173.
189. Wu Z, Wu P, Zuo X, Yu N, Qin Y, Xu Q, He S, Cen B, Liao W, Ji A. LncRNA-N1LR enhances neuroprotection against ischemic stroke probably by inhibiting p53 phosphorylation. Mol Neurobiol. 2017;54:7670–85.
190. Qi X, Shao M, Sun H, Shen Y, Meng D, Huo W. Long non-coding RNA SNHG14 promotes microglia activation by regulating miR-145-5p/PLA2G4A in cerebral infarction. Neuroscience. 2017;348:98–106.
191. Holdt LM, Hoffmann S, Sass K, Langenberger D, Scholz M, Krohn K, Finstermeier K, Stahringer A, Wilfert W, Beutner F, Gielen S, Schuler G, Gabel G, Bergert H, Bechmann I, Stadler PF, Thiery J, Teupser D. Alu elements in ANRIL non-coding RNA at chromosome 9p21 modulate atherogenic cell functions through trans-regulation of gene networks. PLoS Genet. 2013;9:e1003588.

192. Yap KL, Li S, Munoz-Cabello AM, Raguz S, Zeng L, Mujtaba S, Gil J, Walsh MJ, Zhou MM. Molecular interplay of the noncoding RNA ANRIL and methylated histone H3 lysine 27 by polycomb CBX7 in transcriptional silencing of INK4a. Mol Cell. 2010;38:662–74.
193. Smith CM, Steitz JA. Classification of gas5 as a multi-small-nucleolar-RNA (snoRNA) host gene and a member of the 5′-terminal oligopyrimidine gene family reveals common features of snoRNA host genes. Mol Cell Biol. 1998;18:6897–909.
194. Kino T, Hurt DE, Ichijo T, Nader N, Chrousos GP. Noncoding RNA gas5 is a growth arrest- and starvation-associated repressor of the glucocorticoid receptor. Sci Signal. 2010;3:ra8.
195. Aprea J, Prenninger S, Dori M, Ghosh T, Monasor LS, Wessendorf E, Zocher S, Massalini S, Alexopoulou D, Lesche M, Dahl A, Groszer M, Hiller M, Calegari F. Transcriptome sequencing during mouse brain development identifies long non-coding RNAs functionally involved in neurogenic commitment. EMBO J. 2013;32:3145–60.
196. Sheik Mohamed J, Gaughwin PM, Lim B, Robson P, Lipovich L. Conserved long noncoding RNAs transcriptionally regulated by Oct4 and Nanog modulate pluripotency in mouse embryonic stem cells. RNA. 2010;16:324–37.
197. Zhang Y, Zhang M, Xu W, Chen J, Zhou X. The long non-coding RNA H19 promotes cardiomyocyte apoptosis in dilated cardiomyopathy. Oncotarget. 2017;8:28588–94.
198. Venkatraman A, He XC, Thorvaldsen JL, Sugimura R, Perry JM, Tao F, Zhao M, Christenson MK, Sanchez R, Yu JY, Peng L, Haug JS, Paulson A, Li H, Zhong XB, Clemens TL, Bartolomei MS, Li L. Maternal imprinting at the H19-Igf2 locus maintains adult haematopoietic stem cell quiescence. Nature. 2013;500:345–9.
199. Kallen AN, Zhou XB, Xu J, Qiao C, Ma J, Yan L, Lu L, Liu C, Yi JS, Zhang H, Min W, Bennett AM, Gregory RI, Ding Y, Huang Y. The imprinted H19 lncRNA antagonizes let-7 microRNAs. Mol Cell. 2013;52:101–12.
200. Monnier P, Martinet C, Pontis J, Stancheva I, Ait-Si-Ali S, Dandolo L. H19 lncRNA controls gene expression of the Imprinted Gene Network by recruiting MBD1. Proc Natl Acad Sci U S A. 2013;110:20693–8.
201. Giovarelli M, Bucci G, Ramos A, Bordo D, Wilusz CJ, Chen CY, Puppo M, Briata P, Gherzi R. H19 long noncoding RNA controls the mRNA decay promoting function of KSRP. Proc Natl Acad Sci U S A. 2014;111:E5023–8.
202. Yoon JH, Abdelmohsen K, Srikantan S, Yang X, Martindale JL, De S, Huarte M, Zhan M, Becker KG, Gorospe M. LincRNA-p21 suppresses target mRNA translation. Mol Cell. 2012;47:648–55.
203. Zhou Y, Zhong Y, Wang Y, Zhang X, Batista DL, Gejman R, Ansell PJ, Zhao J, Weng C, Klibanski A. Activation of p53 by MEG3 non-coding RNA. J Biol Chem. 2007;282:24731–42.
204. Kaneko S, Bonasio R, Saldana-Meyer R, Yoshida T, Son J, Nishino K, Umezawa A, Reinberg D. Interactions between JARID2 and noncoding RNAs regulate PRC2 recruitment to chromatin. Mol Cell. 2014;53:290–300.
205. Sanuki R, Onishi A, Koike C, Muramatsu R, Watanabe S, Muranishi Y, Irie S, Uneo S, Koyasu T, Matsui R, Cherasse Y, Urade Y, Watanabe D, Kondo M, Yamashita T, Furukawa T. miR-124a is required for hippocampal axogenesis and retinal cone survival through Lhx2 suppression. Nat Neurosci. 2011;14:1125–34.
206. Yang L, Lin C, Liu W, Zhang J, Ohgi KA, Grinstein JD, Dorrestein PC, Rosenfeld MG. ncRNA- and Pc2 methylation-dependent gene relocation between nuclear structures mediates gene activation programs. Cell. 2011;147:773–88.
207. Chen R, Kong P, Zhang F, Shu YN, Nie X, Dong LH, Lin YL, Xie XL, Zhao LL, Zhang XJ, Han M. EZH2-mediated alpha-actin methylation needs lncRNA TUG1, and promotes the cortex cytoskeleton formation in VSMCs. Gene. 2017;616:52–7.
208. Rinn JL, Kertesz M, Wang JK, Squazzo SL, Xu X, Brugmann SA, Goodnough LH, Helms JA, Farnham PJ, Segal E, Chang HY. Functional demarcation of active and silent chromatin domains in human HOX loci by noncoding RNAs. Cell. 2007;129:1311–23.

209. Portoso M, Ragazzini R, Brencic Z, Moiani A, Michaud A, Vassilev I, Wassef M, Servant N, Sargueil B, Margueron R. PRC2 is dispensable for HOTAIR-mediated transcriptional repression. EMBO J. 2017;36:981–94.
210. Terranova R, Yokobayashi S, Stadler MB, Otte AP, van Lohuizen M, Orkin SH, Peters AH. Polycomb group proteins Ezh2 and Rnf2 direct genomic contraction and imprinted repression in early mouse embryos. Dev Cell. 2008;15:668–79.
211. Zhang H, Zeitz MJ, Wang H, Niu B, Ge S, Li W, Cui J, Wang G, Qian G, Higgins MJ, Fan X, Hoffman AR, Hu JF. Long noncoding RNA-mediated intrachromosomal interactions promote imprinting at the Kcnq1 locus. J Cell Biol. 2014;204:61–75.
212. Wang Y, Xu Z, Jiang J, Xu C, Kang J, Xiao L, Wu M, Xiong J, Guo X, Liu H. Endogenous miRNA sponge lincRNA-RoR regulates Oct4, Nanog, and Sox2 in human embryonic stem cell self-renewal. Dev Cell. 2013;25:69–80.
213. Piwecka M, Glazar P, Hernandez-Miranda LR, Memczak S, Wolf SA, Rybak-Wolf A, Filipchyk A, Klironomos F, Cerda Jara CA, Fenske P, Trimbuch T, Zywitza V, Plass M, Schreyer L, Ayoub S, Kocks C, Kuhn R, Rosenmund C, Birchmeier C, Rajewsky N. Loss of a mammalian circular RNA locus causes miRNA deregulation and affects brain function. Science. 2017;357
214. Memczak S, Jens M, Elefsinioti A, Torti F, Krueger J, Rybak A, Maier L, Mackowiak SD, Gregersen LH, Munschauer M, Loewer A, Ziebold U, Landthaler M, Kocks C, le Noble F, Rajewsky N. Circular RNAs are a large class of animal RNAs with regulatory potency. Nature. 2013;495:333–8.
215. Hansen TB, Jensen TI, Clausen BH, Bramsen JB, Finsen B, Damgaard CK, Kjems J. Natural RNA circles function as efficient microRNA sponges. Nature. 2013;495:384–8.
216. Arun G, Diermeier S, Akerman M, Chang KC, Wilkinson JE, Hearn S, Kim Y, MacLeod AR, Krainer AR, Norton L, Brogi E, Egeblad M, Spector DL. Differentiation of mammary tumors and reduction in metastasis upon Malat1 lncRNA loss. Genes Dev. 2016;30:34–51.
217. Tripathi V, Ellis JD, Shen Z, Song DY, Pan Q, Watt AT, Freier SM, Bennett CF, Sharma A, Bubulya PA, Blencowe BJ, Prasanth SG, Prasanth KV. The nuclear-retained noncoding RNA MALAT1 regulates alternative splicing by modulating SR splicing factor phosphorylation. Mol Cell. 2010;39:925–38.
218. Bernard D, Prasanth KV, Tripathi V, Colasse S, Nakamura T, Xuan Z, Zhang MQ, Sedel F, Jourdren L, Coulpier F, Triller A, Spector DL, Bessis A. A long nuclear-retained non-coding RNA regulates synaptogenesis by modulating gene expression. EMBO J. 2010;29:3082–93.
219. Ip JY, Sone M, Nashiki C, Pan Q, Kitaichi K, Yanaka K, Abe T, Takao K, Miyakawa T, Blencowe BJ, Nakagawa S. Gomafu lncRNA knockout mice exhibit mild hyperactivity with enhanced responsiveness to the psychostimulant methamphetamine. Sci Rep. 2016;6:27204.
220. Cai X, Cullen BR. The imprinted H19 noncoding RNA is a primary microRNA precursor. RNA. 2007;13:313–6.
221. Wang KC, Chang HY. Molecular mechanisms of long noncoding RNAs. Mol Cell. 2011;43:904–14.
222. Ulitsky I, Bartel DP. lincRNAs: genomics, evolution, and mechanisms. Cell. 2013;154:26–46.
223. Margueron R, Reinberg D. The Polycomb complex PRC2 and its mark in life. Nature. 2011;469:343–9.
224. Cajigas I, Leib DE, Cochrane J, Luo H, Swyter KR, Chen S, Clark BS, Thompson J, Yates JR 3rd, Kingston RE, Kohtz JD. Evf2 lncRNA/BRG1/DLX1 interactions reveal RNA-dependent inhibition of chromatin remodeling. Development. 2015;142:2641–52.
225. Kaneko S, Son J, Bonasio R, Shen SS, Reinberg D. Nascent RNA interaction keeps PRC2 activity poised and in check. Genes Dev. 2014;28:1983–8.
226. Di Ruscio A, Ebralidze AK, Benoukraf T, Amabile G, Goff LA, Terragni J, Figueroa ME, De Figueiredo Pontes LL, Alberich-Jorda M, Zhang P, Wu M, D'Alo F, Melnick A, Leone G, Ebralidze KK, Pradhan S, Rinn JL, Tenen DG. DNMT1-interacting RNAs block gene-specific DNA methylation. Nature. 2013;503:371–6.

227. Portoso M, Ragazzinni R, Brenčič Ž, Moiani A, Michaud A, Vassilev I, Wassef M, Servant N, Sargueil B, Margueron R. PRC2 is dispensable for HOTAIR-mediated transcriptional repression. EMBO J. 2017;36:981–94.

228. Engreitz JM, Sirokman K, McDonel P, Shishkin AA, Surka C, Russell P, Grossman SR, Chow AY, Guttman M, Lander ES. RNA-RNA interactions enable specific targeting of noncoding RNAs to nascent Pre-mRNAs and chromatin sites. Cell. 2014;159:188–99.

229. McHugh CA, Chen CK, Chow A, Surka CF, Tran C, McDonel P, Pandya-Jones A, Blanco M, Burghard C, Moradian A, Sweredoski MJ, Shishkin AA, Su J, Lander ES, Hess S, Plath K, Guttman M. The Xist lncRNA interacts directly with SHARP to silence transcription through HDAC3. Nature. 2015;521:232–6.

230. Chu C, Zhang QC, da Rocha ST, Flynn RA, Bharadwaj M, Calabrese JM, Magnuson T, Heard E, Chang HY. Systematic discovery of Xist RNA binding proteins. Cell. 2015;161:404–16.

231. West JA, Davis CP, Sunwoo H, Simon MD, Sadreyev RI, Wang PI, Tolstorukov MY, Kingston RE. The long noncoding RNAs NEAT1 and MALAT1 bind active chromatin sites. Mol Cell. 2014;55:791–802.

232. Ramirez F, Lingg T, Toscano S, Lam KC, Georgiev P, Chung HR, Lajoie BR, de Wit E, Zhan Y, de Laat W, Dekker J, Manke T, Akhtar A. High-affinity sites form an interaction network to facilitate spreading of the MSL complex across the X chromosome in drosophila. Mol Cell. 2015;60:146–62.

233. Engreitz JM, Ollikainen N, Guttman M. Long non-coding RNAs: spatial amplifiers that control nuclear structure and gene expression. Nat Rev Mol Cell Biol. 2016;17:756–70.

234. Sunwoo H, Wu JY, Lee JT. The Xist RNA-PRC2 complex at 20-nm resolution reveals a low Xist stoichiometry and suggests a hit-and-run mechanism in mouse cells. Proc Natl Acad Sci U S A. 2015;112:E4216–25.

235. Simon MD, Pinter SF, Fang R, Sarma K, Rutenberg-Schoenberg M, Bowman SK, Kesner BA, Maier VK, Kingston RE, Lee JT. High-resolution Xist binding maps reveal two-step spreading during X-chromosome inactivation. Nature. 2013;504:465–9.

236. Jeon Y, Lee JT. YY1 tethers Xist RNA to the inactive X nucleation center. Cell. 2011;146:119–33.

237. Engreitz JM, Pandya-Jones A, McDonel P, Shishkin A, Sirokman K, Surka C, Kadri S, Xing J, Goren A, Lander ES, Plath K, Guttman M. The Xist lncRNA exploits three-dimensional genome architecture to spread across the X chromosome. Science. 2013;341:1237973.

238. Minajigi A, Froberg J, Wei C, Sunwoo H, Kesner B, Colognori D, Lessing D, Payer B, Boukhali M, Haas W, Lee JT. Chromosomes. A comprehensive Xist interactome reveals cohesin repulsion and an RNA-directed chromosome conformation. Science. 2015;349

239. Ilik IA, Quinn JJ, Georgiev P, Tavares-Cadete F, Maticzka D, Toscano S, Wan Y, Spitale RC, Luscombe N, Backofen R, Chang HY, Akhtar A. Tandem stem-loops in roX RNAs act together to mediate X chromosome dosage compensation in Drosophila. Mol Cell. 2013;51:156–73.

240. Somarowthu S, Legiewicz M, Chillon I, Marcia M, Liu F, Pyle AM. HOTAIR forms an intricate and modular secondary structure. Mol Cell. 2015;58:353–61.

241. Spitale RC, Flynn RA, Zhang QC, Crisalli P, Lee B, Jung JW, Kuchelmeister HY, Batista PJ, Torre EA, Kool ET, Chang HY. Structural imprints in vivo decode RNA regulatory mechanisms. Nature. 2015;519:486–90.

242. Martin WJ, Reiter NJ. Structural roles of noncoding RNAs in the heart of enzymatic complexes. Biochemistry. 2017;56:3–13.

243. Blythe AJ, Fox AH, Bond CS. The ins and outs of lncRNA structure: How, why and what comes next? Biochim Biophys Acta. 2016;1859:46–58.

244. da Rocha ST, Boeva V, Escamilla-Del-Arenal M, Ancelin K, Granier C, Matias NR, Sanulli S, Chow J, Schulz E, Picard C, Kaneko S, Helin K, Reinberg D, Stewart AF, Wutz A, Margueron R, Heard E. Jarid2 is implicated in the initial xist-induced targeting of PRC2 to the inactive X chromosome. Mol Cell. 2014;53:301–16.

245. Almeida M, Pintacuda G, Masui O, Koseki Y, Gdula M, Cerase A, Brown D, Mould A, Innocent C, Nakayama M, Schermelleh L, Nesterova TB, Koseki H, Brockdorff N. PCGF3/5-PRC1 initiates Polycomb recruitment in X chromosome inactivation. Science. 2017;356:1081–4.

246. Wutz A, Rasmussen TP, Jaenisch R. Chromosomal silencing and localization are mediated by different domains of Xist RNA. Nat Genet. 2002;30:167–74.

247. Chen CK, Blanco M, Jackson C, Aznauryan E, Ollikainen N, Surka C, Chow A, Cerase A, McDonel P, Guttman M. Xist recruits the X chromosome to the nuclear lamina to enable chromosome-wide silencing. Science. 2016;354:468–72.

248. Hnisz D, Day DS, Young RA. Insulated neighborhoods: structural and functional units of mammalian gene control. Cell. 2016;167:1188–200.

249. Bonev B, Cavalli G. Organization and function of the 3D genome. Nat Rev Genet. 2016;17:661–78.

250. Kaikkonen MU, Spann NJ, Heinz S, Romanoski CE, Allison KA, Stender JD, Chun HB, Tough DF, Prinjha RK, Benner C, Glass CK. Remodeling of the enhancer landscape during macrophage activation is coupled to enhancer transcription. Mol Cell. 2013;51:310–25.

251. Schaukowitch K, Joo JY, Liu X, Watts JK, Martinez C, Kim TK. Enhancer RNA facilitates NELF release from immediate early genes. Mol Cell. 2014;56:29–42.

252. Pefanis E, Wang J, Rothschild G, Lim J, Kazadi D, Sun J, Federation A, Chao J, Elliott O, Liu ZP, Economides AN, Bradner JE, Rabadan R, Basu U. RNA exosome-regulated long non-coding RNA transcription controls super-enhancer activity. Cell. 2015;161:774–89.

253. Ounzain S, Pedrazzini T. Super-enhancer lncs to cardiovascular development and disease. Biochim Biophys Acta. 2016;1863:1953–60.

254. Mousavi K, Zare H, Dell'orso S, Grontved L, Gutierrez-Cruz G, Derfoul A, Hager GL, Sartorelli V. eRNAs promote transcription by establishing chromatin accessibility at defined genomic loci. Mol Cell. 2013;51:606–17.

255. Lam MT, Cho H, Lesch HP, Gosselin D, Heinz S, Tanaka-Oishi Y, Benner C, Kaikkonen MU, Kim AS, Kosaka M, Lee CY, Watt A, Grossman TR, Rosenfeld MG, Evans RM, Glass CK. Rev-Erbs repress macrophage gene expression by inhibiting enhancer-directed transcription. Nature. 2013;498:511–5.

256. Lai F, Orom UA, Cesaroni M, Beringer M, Taatjes DJ, Blobel GA, Shiekhattar R. Activating RNAs associate with Mediator to enhance chromatin architecture and transcription. Nature. 2013;494:497–501.

257. Natoli G, Andrau JC. Noncoding transcription at enhancers: general principles and functional models. Annu Rev Genet. 2012;46:1–19.

258. Gribnau J, Diderich K, Pruzina S, Calzolari R, Fraser P. Intergenic transcription and developmental remodeling of chromatin subdomains in the human beta-globin locus. Mol Cell. 2000;5:377–86.

259. Venkatesh S, Workman JL. Histone exchange, chromatin structure and the regulation of transcription. Nat Rev Mol Cell Biol. 2015;16:178–89.

260. Dekker J, Mirny L. The 3D genome as moderator of chromosomal communication. Cell. 2016;164:1110–21.

261. Werner MS, Ruthenburg AJ. Nuclear fractionation reveals thousands of chromatin-tethered noncoding RNAs adjacent to active genes. Cell Rep. 2015;12:1089–98.

262. Werner MS, Sullivan MA, Shah RN, Nadadur RD, Grzybowski AT, Galat V, Moskowitz IP, Ruthenburg AJ. Chromatin-enriched lncRNAs can act as cell-type specific activators of proximal gene transcription. Nat Struct Mol Biol. 2017;24:596–603.

263. Pandey RR, Mondal T, Mohammad F, Enroth S, Redrup L, Komorowski J, Nagano T, Mancini-Dinardo D, Kanduri C. Kcnq1ot1 antisense noncoding RNA mediates lineage-specific transcriptional silencing through chromatin-level regulation. Mol Cell. 2008;32:232–46.

264. Mohammad F, Pandey RR, Nagano T, Chakalova L, Mondal T, Fraser P, Kanduri C. Kcnq1ot1/Lit1 noncoding RNA mediates transcriptional silencing by targeting to the perinucleolar region. Mol Cell Biol. 2008;28:3713–28.
265. Shechner DM, Hacisuleyman E, Younger ST, Rinn JL. Multiplexable, locus-specific targeting of long RNAs with CRISPR-Display. Nat Methods. 2015;12:664–70.
266. Zhang Y, Zhang XO, Chen T, Xiang JF, Yin QF, Xing YH, Zhu S, Yang L, Chen LL. Circular intronic long noncoding RNAs. Mol Cell. 2013;51:792–806.
267. Li Z, Huang C, Bao C, Chen L, Lin M, Wang X, Zhong G, Yu B, Hu W, Dai L, Zhu P, Chang Z, Wu Q, Zhao Y, Jia Y, Xu P, Liu H, Shan G. Exon-intron circular RNAs regulate transcription in the nucleus. Nat Struct Mol Biol. 2015;22:256–64.
268. Chedin F. Nascent connections: R-loops and chromatin patterning. Trends Genet. 2016;32:828–38.
269. Sanz LA, Hartono SR, Lim YW, Steyaert S, Rajpurkar A, Ginno PA, Xu X, Chedin F. Prevalent. Dynamic, and Conserved R-Loop Structures Associate with Specific Epigenomic Signatures in Mammals, Mol Cell. 2016;63:167–78.
270. Cloutier SC, Wang S, Ma WK, Al Husini N, Dhoondia Z, Ansari A, Pascuzzi PE, Tran EJ. Regulated formation of lncRNA-DNA hybrids enables faster transcriptional induction and environmental adaptation. Mol Cell. 2016;62:148.
271. Ng SY, Bogu GK, Soh BS, Stanton LW. The long noncoding RNA RMST interacts with SOX2 to regulate neurogenesis. Mol Cell. 2013;51:349–59.
272. Tang R, Zhang G, Wang YC, Mei X, Chen SY. The long non-coding RNA GAS5 regulates transforming growth factor beta (TGF-beta)-induced smooth muscle cell differentiation via RNA Smad-binding elements. J Biol Chem. 2017;292:14270–8.
273. Xue Z, Hennelly S, Doyle B, Gulati AA, Novikova IV, Sanbonmatsu KY, Boyer LA. A G-rich motif in the lncRNA braveheart interacts with a zinc-finger transcription factor to specify the cardiovascular lineage. Mol Cell. 2016;64:37–50.
274. Ozsolak F, Kapranov P, Foissac S, Kim SW, Fishilevich E, Monaghan AP, John B, Milos PM. Comprehensive polyadenylation site maps in yeast and human reveal pervasive alternative polyadenylation. Cell. 2010;143:1018–29.
275. Chu C, Qu K, Zhong FL, Artandi SE, Chang HY. Genomic maps of long noncoding RNA occupancy reveal principles of RNA-chromatin interactions. Mol Cell. 2011;44:667–78.
276. Zhao J, Ohsumi TK, Kung JT, Ogawa Y, Grau DJ, Sarma K, Song JJ, Kingston RE, Borowsky M, Lee JT. Genome-wide identification of polycomb-associated RNAs by RIP-seq. Mol Cell. 2010;40:939–53.
277. Hongay CF, Grisafi PL, Galitski T, Fink GR. Antisense transcription controls cell fate in Saccharomyces cerevisiae. Cell. 2006;127:735–45.
278. Beltran M, Puig I, Pena C, Garcia JM, Alvarez AB, Pena R, Bonilla F, de Herreros AG. A natural antisense transcript regulates Zeb2/Sip1 gene expression during Snail1-induced epithelial-mesenchymal transition. Genes Dev. 2008;22:756–69.
279. Munroe SH, Lazar MA. Inhibition of c-erbA mRNA splicing by a naturally occurring antisense RNA. J Biol Chem. 1991;266:22083–6.
280. Krystal GW, Armstrong BC, Battey JF. N-myc mRNA forms an RNA-RNA duplex with endogenous antisense transcripts. Mol Cell Biol. 1990;10:4180–91.
281. Salmena L, Poliseno L, Tay Y, Kats L, Pandolfi PP. A ceRNA hypothesis: the Rosetta Stone of a hidden RNA language? Cell. 2011;146:353–8.
282. Tay Y, Kats L, Salmena L, Weiss D, Tan SM, Ala U, Karreth F, Poliseno L, Provero P, Di Cunto F, Lieberman J, Rigoutsos I, Pandolfi PP. Coding-independent regulation of the tumor suppressor PTEN by competing endogenous mRNAs. Cell. 2011;147:344–57.
283. Cesana M, Cacchiarelli D, Legnini I, Santini T, Sthandier O, Chinappi M, Tramontano A, Bozzoni I. A long noncoding RNA controls muscle differentiation by functioning as a competing endogenous RNA. Cell. 2011;147:358–69.
284. Thomson DW, Dinger ME. Endogenous microRNA sponges: evidence and controversy. Nat Rev Genet. 2016;17:272–83.

285. Denzler R, Agarwal V, Stefano J, Bartel DP, Stoffel M. Assessing the ceRNA hypothesis with quantitative measurements of miRNA and target abundance. Mol Cell. 2014;54:766–76.
286. Bosson AD, Zamudio JR, Sharp PA. Endogenous miRNA and target concentrations determine susceptibility to potential ceRNA competition. Mol Cell. 2014;56:347–59.
287. Karreth FA, Reschke M, Ruocco A, Ng C, Chapuy B, Leopold V, Sjoberg M, Keane TM, Verma A, Ala U, Tay Y, Wu D, Seitzer N, Velasco-Herrera Mdel C, Bothmer A, Fung J, Langellotto F, Rodig SJ, Elemento O, Shipp MA, Adams DJ, Chiarle R, Pandolfi PP. The BRAF pseudogene functions as a competitive endogenous RNA and induces lymphoma in vivo. Cell. 2015;161:319–32.
288. Denzler R, McGeary SE, Title AC, Agarwal V, Bartel DP, Stoffel M. Impact of MicroRNA Levels. Target-Site Complementarity, and Cooperativity on Competing Endogenous RNA-Regulated Gene Expression, Mol Cell. 2016;64:565–79.
289. Gong C, Maquat LE. lncRNAs transactivate STAU1-mediated mRNA decay by duplexing with 3′ UTRs via Alu elements. Nature. 2011;470:284–8.
290. Damas ND, Marcatti M, Come C, Christensen LL, Nielsen MM, Baumgartner R, Gylling HM, Maglieri G, Rundsten CF, Seemann SE, Rapin N, Thezenas S, Vang S, Orntoft T, Andersen CL, Pedersen JS, Lund AH. SNHG5 promotes colorectal cancer cell survival by counteracting STAU1-mediated mRNA destabilization. Nat Commun. 2016;7:13875.
291. Lu Z, Zhang QC, Lee B, Flynn RA, Smith MA, Robinson JT, Davidovich C, Gooding AR, Goodrich KJ, Mattick JS, Mesirov JP, Cech TR, Chang HY. RNA duplex map in living cells reveals higher-order transcriptome structure. Cell. 2016;165:1267–79.
292. Aw JG, Shen Y, Wilm A, Sun M, Lim XN, Boon KL, Tapsin S, Chan YS, Tan CP, Sim AY, Zhang T, Susanto TT, Fu Z, Nagarajan N, Wan Y. In vivo mapping of eukaryotic RNA interactomes reveals principles of higher-order organization and regulation. Mol Cell. 2016;62:603–17.
293. Fei J, Cui XB, Wang JN, Dong K, Chen SY. ADAR1-mediated RNA editing. A Novel Mechanism Controlling Phenotypic Modulation of Vascular Smooth Muscle Cells, Circ Res. 2016;119:463–9.
294. Stellos K, Gatsiou A, Stamatelopoulos K, Perisic Matic L, John D, Lunella FF, Jae N, Rossbach O, Amrhein C, Sigala F, Boon RA, Furtig B, Manavski Y, You X, Uchida S, Keller T, Boeckel JN, Franco-Cereceda A, Maegdefessel L, Chen W, Schwalbe H, Bindereif A, Eriksson P, Hedin U, Zeiher AM, Dimmeler S. Adenosine-to-inosine RNA editing controls cathepsin S expression in atherosclerosis by enabling HuR-mediated post-transcriptional regulation. Nat Med. 2016;22:1140–50.
295. Zheng S, Vuong BQ, Vaidyanathan B, Lin JY, Huang FT, Chaudhuri J. Non-coding RNA generated following lariat debranching mediates targeting of AID to DNA. Cell. 2015;161:762–73.
296. Carrieri C, Cimatti L, Biagioli M, Beugnet A, Zucchelli S, Fedele S, Pesce E, Ferrer I, Collavin L, Santoro C, Forrest AR, Carninci P, Biffo S, Stupka E, Gustincich S. Long non-coding antisense RNA controls Uchl1 translation through an embedded SINEB2 repeat. Nature. 2012;491:454–7.
297. Abdelmohsen K, Panda AC, Munk R, Grammatikakis I, Dudekula DB, De S, Kim J, Noh JH, Kim KM, Martindale JL, Gorospe M. Identification of HuR target circular RNAs uncovers suppression of PABPN1 translation by CircPABPN1. RNA Biol. 2017:1–9.
298. Lebedeva S, Jens M, Theil K, Schwanhausser B, Selbach M, Landthaler M, Rajewsky N. Transcriptome-wide analysis of regulatory interactions of the RNA-binding protein HuR. Mol Cell. 2011;43:340–52.
299. Frith MC, Bailey TL, Kasukawa T, Mignone F, Kummerfeld SK, Madera M, Sunkara S, Furuno M, Bult CJ, Quackenbush J, Kai C, Kawai J, Carninci P, Hayashizaki Y, Pesole G, Mattick JS. Discrimination of non-protein-coding transcripts from protein-coding mRNA. RNA Biol. 2006;3:40–8.
300. You X, Vlatkovic I, Babic A, Will T, Epstein I, Tushev G, Akbalik G, Wang M, Glock C, Quedenau C, Wang X, Hou J, Liu H, Sun W, Sambandan S, Chen T, Schuman EM, Chen

W. Neural circular RNAs are derived from synaptic genes and regulated by development and plasticity. Nat Neurosci. 2015;18:603–10.

301. Pamudurti NR, Bartok O, Jens M, Ashwal-Fluss R, Stottmeister C, Ruhe L, Hanan M, Wyler E, Perez-Hernandez D, Ramberger E, Shenzis S, Samson M, Dittmar G, Landthaler M, Chekulaeva M, Rajewsky N, Kadener S. Translation of CircRNAs. Mol Cell. 2017;66(1):9–21.e7.
302. Legnini I, Di Timoteo G, Rossi F, Morlando M, Briganti F, Sthandier O, Fatica A, Santini T, Andronache A, Wade M, Laneve P, Rajewsky N, Bozzoni I. Circ-ZNF609 is a circular RNA that can be translated and functions in myogenesis. Mol Cell. 2017;66(1):22–37.e9.
303. Sonenberg N, Hinnebusch AG. Regulation of translation initiation in eukaryotes: mechanisms and biological targets. Cell. 2009;136:731–45.
304. Bassett AR, Akhtar A, Barlow DP, Bird AP, Brockdorff N, Duboule D, Ephrussi A, Ferguson-Smith AC, Gingeras TR, Haerty W, Higgs DR, Miska EA, Ponting CP. Considerations when investigating lncRNA function in vivo. elife. 2014;3:e03058.
305. Fabre PJ, Leleu M, Mormann BH, Lopez-Delisle L, Noordermeer D, Beccari L, Duboule D. Large scale genomic reorganization of topological domains at the HoxD locus. Genome Biol. 2017;18:149.
306. Srivastava D. Making or breaking the heart: from lineage determination to morphogenesis. Cell. 2006;126:1037–48.
307. Bruneau BG. Signaling and transcriptional networks in heart development and regeneration. Cold Spring Harb Perspect Biol. 2013;5:a008292.
308. Wamstad JA, Alexander JM, Truty RM, Shrikumar A, Li F, Eilertson KE, Ding H, Wylie JN, Pico AR, Capra JA, Erwin G, Kattman SJ, Keller GM, Srivastava D, Levine SS, Pollard KS, Holloway AK, Boyer LA, Bruneau BG. Dynamic and coordinated epigenetic regulation of developmental transitions in the cardiac lineage. Cell. 2012;151:206–20.
309. Zhu JG, Shen YH, Liu HL, Liu M, Shen YQ, Kong XQ, Song GX, Qian LM. Long noncoding RNAs expression profile of the developing mouse heart. J Cell Biochem. 2014;115:910–8.
310. Matkovich SJ, Edwards JR, Grossenheider TC, de Guzman Strong C, Dorn GW 2nd. Epigenetic coordination of embryonic heart transcription by dynamically regulated long noncoding RNAs. Proc Natl Acad Sci U S A. 2014;111:12264–9.
311. Sauvageau M, Goff LA, Lodato S, Bonev B, Groff AF, Gerhardinger C, Sanchez-Gomez DB, Hacisuleyman E, Li E, Spence M, Liapis SC, Mallard W, Morse M, Swerdel MR, D'Ecclessis MF, Moore JC, Lai V, Gong G, Yancopoulos GD, Frendewey D, Kellis M, Hart RP, Valenzuela DM, Arlotta P, Rinn JL. Multiple knockout mouse models reveal lincRNAs are required for life and brain development. elife. 2013;2:e01749.
312. Delas MJ, Hannon GJ. lncRNAs in development and disease: from functions to mechanisms. Open Biol. 2017;7
313. Chen R, Liu Y, Zhuang H, Yang B, Hei K, Xiao M, Hou C, Gao H, Zhang X, Jia C, Li L, Li Y, Zhang N. Quantitative proteomics reveals that long non-coding RNA MALAT1 interacts with DBC1 to regulate p53 acetylation. Nucleic Acids Res. 2017;45:9947–59.
314. Ishii N, Ozaki K, Sato H, Mizuno H, Saito S, Takahashi A, Miyamoto Y, Ikegawa S, Kamatani N, Hori M, Saito S, Nakamura Y, Tanaka T. Identification of a novel non-coding RNA. MIAT, that confers risk of myocardial infarction, J Hum Genet. 2006;51:1087–99.
315. Diabetes Genetics Initiative of Broad Institute of, H., Mit, L. U., Novartis Institutes of BioMedical, R., Saxena, R., Voight, B. F., Lyssenko, V., Burtt, N. P., de Bakker, P. I., Chen, H., Roix, J. J., Kathiresan, S., Hirschhorn, J. N., Daly, M. J., Hughes, T. E., Groop, L., Altshuler, D., Almgren, P., Florez, J. C., Meyer, J., Ardlie, K., Bengtsson Bostrom, K., Isomaa, B., Lettre, G., Lindblad, U., Lyon, H. N., Melander, O., Newton-Cheh, C., Nilsson, P., Orho-Melander, M., Rastam, L., Speliotes, E. K., Taskinen, M. R., Tuomi, T., Guiducci, C., Berglund, A., Carlson, J., Gianniny, L., Hackett, R., Hall, L., Holmkvist, J., Laurila, E., Sjogren, M., Sterner, M., Surti, A., Svensson, M., Svensson, M., Tewhey, R., Blumenstiel, B., Parkin, M., Defelice, M., Barry, R., Brodeur, W., Camarata, J., Chia, N., Fava, M., Gibbons, J., Handsaker, B., Healy, C., Nguyen, K., Gates, C., Sougnez, C., Gage, D., Nizzari,

M., Gabriel, S. B., Chirn, G. W., Ma, Q., Parikh, H., Richardson, D., Ricke, D. & Purcell, S. Genome-wide association analysis identifies loci for type 2 diabetes and triglyceride levels. Science. 2007;316:1331–6.

316. Scott LJ, Mohlke KL, Bonnycastle LL, Willer CJ, Li Y, Duren WL, Erdos MR, Stringham HM, Chines PS, Jackson AU, Prokunina-Olsson L, Ding CJ, Swift AJ, Narisu N, Hu T, Pruim R, Xiao R, Li XY, Conneely KN, Riebow NL, Sprau AG, Tong M, White PP, Hetrick KN, Barnhart MW, Bark CW, Goldstein JL, Watkins L, Xiang F, Saramies J, Buchanan TA, Watanabe RM, Valle TT, Kinnunen L, Abecasis GR, Pugh EW, Doheny KF, Bergman RN, Tuomilehto J, Collins FS, Boehnke M. A genome-wide association study of type 2 diabetes in Finns detects multiple susceptibility variants. Science. 2007;316:1341–5.

317. Zeggini E, Weedon MN, Lindgren CM, Frayling TM, Elliott KS, Lango H, Timpson NJ, Perry JR, Rayner NW, Freathy RM, Barrett JC, Shields B, Morris AP, Ellard S, Groves CJ, Harries LW, Marchini JL, Owen KR, Knight B, Cardon LR, Walker M, Hitman GA, Morris AD, Doney AS, Wellcome Trust Case Control C, McCarthy MI, Hattersley AT. Replication of genome-wide association signals in UK samples reveals risk loci for type 2 diabetes. Science. 2007;316:1336–41.

318. Dandona S, Stewart AF, Chen L, Williams K, So D, O'Brien E, Glover C, Lemay M, Assogba O, Vo L, Wang YQ, Labinaz M, Wells GA, McPherson R, Roberts R. Gene dosage of the common variant 9p21 predicts severity of coronary artery disease. J Am Coll Cardiol. 2010;56:479–86.

319. Yu W, Gius D, Onyango P, Muldoon-Jacobs K, Karp J, Feinberg AP, Cui H. Epigenetic silencing of tumour suppressor gene p15 by its antisense RNA. Nature. 2008;451:202–6.

320. Kotake Y, Nakagawa T, Kitagawa K, Suzuki S, Liu N, Kitagawa M, Xiong Y. Long non-coding RNA ANRIL is required for the PRC2 recruitment to and silencing of p15 (INK4B) tumor suppressor gene. Oncogene. 2011;30:1956–62.

321. Kuo CL, Murphy AJ, Sayers S, Li R, Yvan-Charvet L, Davis JZ, Krishnamurthy J, Liu Y, Puig O, Sharpless NE, Tall AR, Welch CL. Cdkn2a is an atherosclerosis modifier locus that regulates monocyte/macrophage proliferation. Arterioscler Thromb Vasc Biol. 2011;31:2483–92.

322. Zhou X, Han X, Wittfeldt A, Sun J, Liu C, Wang X, Gan LM, Cao H, Liang Z. Long non-coding RNA ANRIL regulates inflammatory responses as a novel component of NF-kappaB pathway. RNA Biol. 2016;13:98–108.

323. Visel A, Zhu Y, May D, Afzal V, Gong E, Attanasio C, Blow MJ, Cohen JC, Rubin EM, Pennacchio LA. Targeted deletion of the 9p21 non-coding coronary artery disease risk interval in mice. Nature. 2010;464:409–12.

324. Meijer-Spierings AH. Is there a relationship between social status and interpersonal behavior? Tijdschr Voor Ziekenverpl. 1966;19:208–12.

325. Chen L, Yang W, Guo Y, Chen W, Zheng P, Zeng J, Tong W. Exosomal lncRNA GAS5 regulates the apoptosis of macrophages and vascular endothelial cells in atherosclerosis. PLoS One. 2017;12:e0185406.

326. Li MJ, Wang P, Liu X, Lim EL, Wang Z, Yeager M, Wong MP, Sham PC, Chanock SJ, Wang J. GWASdb: a database for human genetic variants identified by genome-wide association studies. Nucleic Acids Res. 2012;40:D1047–54.

327. Overton HA, Babbs AJ, Doel SM, Fyfe MC, Gardner LS, Griffin G, Jackson HC, Procter MJ, Rasamison CM, Tang-Christensen M, Widdowson PS, Williams GM, Reynet C. Deorphanization of a G protein-coupled receptor for oleoylethanolamide and its use in the discovery of small-molecule hypophagic agents. Cell Metab. 2006;3:167–75.

328. Guevara NV, Kim HS, Antonova EI, Chan L. The absence of p53 accelerates atherosclerosis by increasing cell proliferation in vivo. Nat Med. 1999;5:335–9.

329. Schober A, Nazari-Jahantigh M, Wei Y, Bidzhekov K, Gremse F, Grommes J, Megens RT, Heyll K, Noels H, Hristov M, Wang S, Kiessling F, Olson EN, Weber C. MicroRNA-126-5p promotes endothelial proliferation and limits atherosclerosis by suppressing Dlk1. Nat Med. 2014;20:368–76.

330. Feng J, Liang J, Li J, Li Y, Liang H, Zhao X, McNutt MA, Yin Y. PTEN controls the DNA replication process through MCM2 in response to replicative stress. Cell Rep. 2015;13:1295–303.

331. Ounzain S, Micheletti R, Beckmann T, Schroen B, Alexanian M, Pezzuto I, Crippa S, Nemir M, Sarre A, Johnson R, Dauvillier J, Burdet F, Ibberson M, Guigo R, Xenarios I, Heymans S, Pedrazzini T. Genome-wide profiling of the cardiac transcriptome after myocardial infarction identifies novel heart-specific long non-coding RNAs. Eur Heart J. 2015;36:353–68a.

332. Suzuki HI, Young RA, Sharp PA. Super-enhancer-mediated RNA processing revealed by integrative MicroRNA network analysis. Cell. 2017;168:1000–1014 e15.

333. Foroud T, Koller DL, Lai D, Sauerbeck L, Anderson C, Ko N, Deka R, Mosley TH, Fornage M, Woo D, Moomaw CJ, Hornung R, Huston J, Meissner I, Bailey-Wilson JE, Langefeld C, Rouleau G, Connolly ES, Worrall BB, Kleindorfer D, Flaherty ML, Martini S, Mackey J, De Los Rios La Rosa F, Brown RD Jr, Broderick JP, FIA Study Investigators. Genome-wide association study of intracranial aneurysms confirms role of Anril and SOX17 in disease risk. Stroke. 2012;43:2846–52.

334. Hashikata H, Liu W, Inoue K, Mineharu Y, Yamada S, Nanayakkara S, Matsuura N, Hitomi T, Takagi Y, Hashimoto N, Miyamoto S, Koizumi A. Confirmation of an association of single-nucleotide polymorphism rs1333040 on 9p21 with familial and sporadic intracranial aneurysms in Japanese patients. Stroke. 2010;41:1138–44.

335. Lin L, Gu ZT, Chen WH, Cao KJ. Increased expression of the long non-coding RNA ANRIL promotes lung cancer cell metastasis and correlates with poor prognosis. Diagn Pathol. 2015;10:14.

336. Qiu JJ, Lin YY, Ding JX, Feng WW, Jin HY, Hua KQ. Long non-coding RNA ANRIL predicts poor prognosis and promotes invasion/metastasis in serous ovarian cancer. Int J Oncol. 2015;46:2497–505.

337. Sun Z, Ou C, Ren W, Xie X, Li X, Li G. Downregulation of long non-coding RNA ANRIL suppresses lymphangiogenesis and lymphatic metastasis in colorectal cancer. Oncotarget. 2016;7:47536–55.

338. Zhao JJ, Hao S, Wang LL, Hu CY, Zhang S, Guo LJ, Zhang G, Gao B, Jiang Y, Tian WG, Luo DL. Long non-coding RNA ANRIL promotes the invasion and metastasis of thyroid cancer cells through TGF-beta/Smad signaling pathway. Oncotarget. 2016;7:57903–18.

339. Haas J, Mester S, Lai A, Frese KS, Sedaghat-Hamedani F, Kayvanpour E, Rausch T, Nietsch R, Boeckel JN, Carstensen A, Volkers M, Dietrich C, Pils D, Amr A, Holzer DB, Martins Bordalo D, Oehler D, Weis T, Mereles D, Buss S, Riechert E, Wirsz E, Wuerstle M, Korbel JO, Keller A, Katus HA, Posch AE, Meder B. Genomic structural variations lead to dysregulation of important coding and non-coding RNA species in dilated cardiomyopathy. EMBO Mol: Med; 2017.

340. Yang W, Li Y, He F, Wu H. Microarray profiling of long non-coding RNA (lncRNA) associated with hypertrophic cardiomyopathy. BMC Cardiovasc Disord. 2015;15:62.

341. Song G, Shen Y, Zhu J, Liu H, Liu M, Shen YQ, Zhu S, Kong X, Yu Z, Qian L. Integrated analysis of dysregulated lncRNA expression in fetal cardiac tissues with ventricular septal defect. PLoS One. 2013;8:e77492.

342. Gu M, Zheng A, Tu W, Zhao J, Li L, Li M, Han S, Hu X, Zhu J, Pan Y, Xu J, Yu Z. Circulating LncRNAs as Novel. Non-Invasive Biomarkers for Prenatal Detection of Fetal Congenital Heart Defects, Cell Physiol Biochem. 2016;38:1459–71.

343. Levine B, Kroemer G. Autophagy in the pathogenesis of disease. Cell. 2008;132:27–42.

344. Liao X, Sluimer JC, Wang Y, Subramanian M, Brown K, Pattison JS, Robbins J, Martinez J, Tabas I. Macrophage autophagy plays a protective role in advanced atherosclerosis. Cell Metab. 2012;15:545–53.

345. Liesa M, Shirihai OS. Mitochondrial dynamics in the regulation of nutrient utilization and energy expenditure. Cell Metab. 2013;17:491–506.

346. Brown DA, Perry JB, Allen ME, Sabbah HN, Stauffer BL, Shaikh SR, Cleland JG, Colucci WS, Butler J, Voors AA, Anker SD, Pitt B, Pieske B, Filippatos G, Greene SJ, Gheorghiade M. Expert consensus document: mitochondrial function as a therapeutic target in heart failure. Nat Rev Cardiol. 2017;14:238–50.

347. Zhao Y, Samal E, Srivastava D. Serum response factor regulates a muscle-specific microRNA that targets Hand2 during cardiogenesis. Nature. 2005;436:214–20.

348. Lu CX, Gong HR, Liu XY, Wang J, Zhao CM, Huang RT, Xue S, Yang YQ. A novel HAND2 loss-of-function mutation responsible for tetralogy of Fallot. Int J Mol Med. 2016;37:445–51.

349. Zhang M, Gu H, Xu W, Zhou X. Down-regulation of lncRNA MALAT1 reduces cardiomyocyte apoptosis and improves left ventricular function in diabetic rats. Int J Cardiol. 2016;203:214–6.

350. Newton-Cheh C, Eijgelsheim M, Rice KM, de Bakker PI, Yin X, Estrada K, Bis JC, Marciante K, Rivadeneira F, Noseworthy PA, Sotoodehnia N, Smith NL, Rotter JI, Kors JA, Witteman JC, Hofman A, Heckbert SR, O'Donnell CJ, Uitterlinden AG, Psaty BM, Lumley T, Larson MG, Stricker BH. Common variants at ten loci influence QT interval duration in the QTGEN Study. Nat Genet. 2009;41:399–406.

351. Lu Y, Zhang Y, Wang N, Pan Z, Gao X, Zhang F, Zhang Y, Shan H, Luo X, Bai Y, Sun L, Song W, Xu C, Wang Z, Yang B. MicroRNA-328 contributes to adverse electrical remodeling in atrial fibrillation. Circulation. 2010;122:2378–87.

352. Antzelevitch C, Pollevick GD, Cordeiro JM, Casis O, Sanguinetti MC, Aizawa Y, Guerchicoff A, Pfeiffer R, Oliva A, Wollnik B, Gelber P, Bonaros EP Jr, Burashnikov E, Wu Y, Sargent JD, Schickel S, Oberheiden R, Bhatia A, Hsu LF, Haissaguerre M, Schimpf R, Borggrefe M, Wolpert C. Loss-of-function mutations in the cardiac calcium channel underlie a new clinical entity characterized by ST-segment elevation, short QT intervals, and sudden cardiac death. Circulation. 2007;115:442–9.

353. Ikram MA, Seshadri S, Bis JC, Fornage M, DeStefano AL, Aulchenko YS, Debette S, Lumley T, Folsom AR, van den Herik EG, Bos MJ, Beiser A, Cushman M, Launer LJ, Shahar E, Struchalin M, Du Y, Glazer NL, Rosamond WD, Rivadeneira F, Kelly-Hayes M, Lopez OL, Coresh J, Hofman A, DeCarli C, Heckbert SR, Koudstaal PJ, Yang Q, Smith NL, Kase CS, Rice K, Haritunians T, Roks G, de Kort PL, Taylor KD, de Lau LM, Oostra BA, Uitterlinden AG, Rotter JI, Boerwinkle E, Psaty BM, Mosley TH, van Duijn CM, Breteler MM, Longstreth WT Jr, Wolf PA. Genomewide association studies of stroke. N Engl J Med. 2009;360:1718–28.

354. Lemmens R, Abboud S, Robberecht W, Vanhees L, Pandolfo M, Thijs V, Goris A. Variant on 9p21 strongly associates with coronary heart disease, but lacks association with common stroke. Eur J Hum Genet. 2009;17:1287–93.

355. Madelaine R, Sloan SA, Huber N, Notwell JH, Leung LC, Skariah G, Halluin C, Pasca SP, Bejerano G, Krasnow MA, Barres BA, Mourrain P. MicroRNA-9 couples brain neurogenesis and angiogenesis. Cell Rep. 2017;20:1533–42.

356. Dykstra-Aiello C, Jickling GC, Ander BP, Shroff N, Zhan X, Liu D, Hull H, Orantia M, Stamova BS, Sharp FR. Altered expression of long noncoding RNAs in blood after ischemic stroke and proximity to putative stroke risk loci. Stroke. 2016;47:2896–903.

357. Chen YG, Satpathy AT, Chang HY. Gene regulation in the immune system by long noncoding RNAs. Nat Immunol. 2017;18:962–72.

358. Schmitt AM, Chang HY. Long noncoding RNAs in cancer pathways. Cancer Cell. 2016;29:452–63.

359. Liu SJ, Horlbeck MA, Cho SW, Birk HS, Malatesta M, He D, Attenello FJ, Villalta JE, Cho MY, Chen Y, Mandegar MA, Olvera MP, Gilbert LA, Conklin BR, Chang HY, Weissman JS, Lim DA. CRISPRi-based genome-scale identification of functional long noncoding RNA loci in human cells. Science. 2017;355

360. Wang T, Birsoy K, Hughes NW, Krupczak KM, Post Y, Wei JJ, Lander ES, Sabatini DM. Identification and characterization of essential genes in the human genome. Science. 2015;350:1096–101.

361. Hart T, Chandrashekhar M, Aregger M, Steinhart Z, Brown KR, MacLeod G, Mis M, Zimmermann M, Fradet-Turcotte A, Sun S, Mero P, Dirks P, Sidhu S, Roth FP, Rissland OS, Durocher D, Angers S, Moffat J. High-resolution CRISPR screens reveal fitness genes and genotype-specific cancer liabilities. Cell. 2015;163:1515–26.
362. Chi KR. The dark side of the human genome. Nature. 2016;538:275–7.
363. Patil DP, Chen CK, Pickering BF, Chow A, Jackson C, Guttman M, Jaffrey SR. m(6)A RNA methylation promotes XIST-mediated transcriptional repression. Nature. 2016;537:369–73.
364. Dowdy SF. Overcoming cellular barriers for RNA therapeutics. Nat Biotechnol. 2017;35:222–9.
365. Cox DBT, Gootenberg JS, Abudayyeh OO, Franklin B, Kellner MJ, Joung J, Zhang F. RNA editing with CRISPR-Cas13. Science. 2017;358:1019–27.

Mouse Models to Study Inherited Cardiomyopathy

8

Mohammad Bakhtiar Hossain, Mohammad Bohlooly-Y, and Ralph Knöll

Contents

M. B. Hossain · R. Knöll (✉)
Integrated Cardio Metabolic Center (ICMC), Myocardial Genetics, Karolinska Institutet, University Hospital, Heart and Vascular Theme, Huddinge, Sweden

Bioscience Research & Early Development, Cardiovascular, Renal and Metabolic Diseases (CVRM), Biopharmaceuticals R&D, AstraZeneca, Mölndal, Sweden
e-mail: Ralph.Knoell@KI.se

M. Bohlooly-Y
Discovery Sciences, BioPharmaceuticals R&D, AstraZeneca, Gothenburg, Sweden

© Springer Nature Switzerland AG 2019
J. Erdmann, A. Moretti (eds.), *Genetic Causes of Cardiac Disease*, Cardiac and Vascular Biology 7, https://doi.org/10.1007/978-3-030-27371-2_8

Abstract

Cardiovascular diseases, including cardiomyopathy and associated heart failure, are the number one cause of death worldwide, but our ability to interfere with these devastating diseases is limited. Cardiomyopathies, which are mainly due to genetic causes, make a significant portion of heart failure. Current pharmacological treatment of cardiovascular diseases focuses on symptoms rather than the underlying cellular mechanisms, and the gaps in our understanding of cellular mechanisms of the disease are profound. Elucidation of these mechanisms is a central issue in cardiovascular biology and important for designing new treatment for cardiovascular diseases. While significant progress has been made in using in vitro systems in deciphering the mechanisms and finding innovative solutions for cardiovascular disease treatment, including the use of induced pluripotent stem cell (iPSC) derivatives, suitable in vivo models are more difficult to develop. Among several different species, mouse models are rather inexpensive, easily manipulatable, reproducible, physiologically representative of human disease, and ethically acceptable. This chapter will provide a brief overview of genetics of heart failure, largely focusing on genetically altered mouse models and experimental approaches applicable to cardiovascular research.

List of Abbreviations

AD	Autosomal dominant
AR	Autosomal recessive
ARVC	Arrhythmogenic right ventricular cardiomyopathy
BNP	Brain natriuretic peptide
cKO	Conditional knockout
CPVT	Catecholaminergic polymorphic ventricular tachycardia
CRISPR	Clustered regularly interspaced short palindromic repeats
CSRP3/Csrp3 (or MLP/Mlp)	Cysteine-rich protein 3
CVD	Cardiovascular disease
DCM	Dilated cardiomyopathy
ECM	Extracellular matrix
HCM	Hypertrophic cardiomyopathy
KI	Knock in
KO	Knockout
LV	Left ventricle
LVNC	Left ventricular noncompaction
PGE	Precision genetic engineering
PKA	Protein kinase A
PLN/Pln	Phospholamban
RCM	Restrictive cardiomyopathy

SR   Sarcoplasmic reticulum
SRF   Serum response factor
TALEN   Transcription activator-like effector nucleases
ZFN   Zinc finger nuclease

8.1   Introduction

According to the World Health Organization (WHO), cardiovascular disease (CVD) is considered as the major cause of morbidity and mortality [1]. During the course of the last decades, impressive achievements have been made in diagnosis and treatment of cardiovascular disease, relying to a large extent on the use of experimental animal models. Despite this, no permanent cures exist for the overwhelming majority of cardiovascular diseases. As such, heart failure can only be treated by heart transplantation, but only a very limited number of donor hearts are available and its therapeutic potential is limited by the complications of long-term allograft vasculopathy. Innovative therapies can be developed by tissue engineering, autologous and allogenic cell therapies, gene therapy and genome editing—but their development depends to a large extent on extensive animal experimentation.

Human heart samples, either non-transplantable or end-stage failing, usually obtained at the time of transplantation, are valuable for identifying underlying disease causing molecular pathways, but their number is limited and they exhibit significant variability due to differences in genetics, epigenetics, environment, and different therapeutic regimens. Therefore, we need to develop preclinical testing for relevant CVD models, including heart failure. The closer the heart or body weight of the animal is in comparison to humans, the more similar are the hearts. Therefore, large animal models, like nonhuman primates and dogs, are much more relevant, but experimentations with these animals require extensive ethical considerations and are expensive to perform, which limits their widespread use.

Rodent models, especially mice, are often used in CVD research since they are easy to handle, have a short gestation time, are genetically manipulable, and have relatively low maintenance costs—making them much more suitable for high-throughput screening. Despite phylogenetically quite distant from humans and some pathophysiological features of disease, including electrophysiology, and their response to pharmacological interventions being significantly different, mouse models have become the method of choice in biomedical research during the last decades.

8.2   Inherited Cardiomyopathies

Cardiomyopathies are a heterogenous group of myocardial diseases associated with mechanical or electrical dysfunction that usually exhibit inappropriate ventricular hypertrophy or dilation [2]. Primary cardiomyopathies are predominantly confined to the myocardium, while secondary cardiomyopathies result from systemic diseases associated with heart failure. Major types of cardiomyopathy includes hypertrophic cardiomyopathy (HCM), dilated cardiomyopathy (DCM), arrhythmogenic right ventricular cardiomyopathy (ARVC), and restrictive cardiomyopathy (RCM). Genetic mutations have been reported to account for a significant percentage of all cardiomyopathies, and it is likely that a genetic component exists for all (figure).

- Hypertrophic cardiomyopathy (HCM) is characterized morphologically and defined by a hypertrophied, nondilated left ventricle (LV) in the absence of other systemic or cardiac diseases that are capable of producing the magnitude of wall thickening evident (e.g., systemic hypertension, aortic valve stenosis). About 60% of all HCM cases are due to mutations in genes encoding sarcomeric proteins. The MYH7:p.R403Q mutation in the cardiac β-myosin heavy chain was the first mutation shown to cause cardiomyopathy [3, 4].
- Dilated forms of cardiomyopathy (DCM) are characterized by ventricular chamber enlargement and systolic dysfunction with normal LV wall thickness. About 35% of all DCM patients carry mutations in genes encoding sarcomeric or cytoskeletal components.
- Arrhythmogenic right ventricular cardiomyopathy (ARVC) involves the right ventricle predominantly with progressive loss of myocytes and fatty or fibrofatty tissue replacement, resulting in regional (segmental) or global abnormalities. About 50% of all ARVC cases can be traced back to mutations, especially in genes encoding desmosomal components.
- Primary restrictive cardiomyopathy (RCM) is a rare form of heart disease and a cause of heart failure that is characterized by normal or decreased volume of both ventricles associated with biatrial enlargement, normal LV wall thickness and atrioventricular (AV) valves, impaired ventricular filling with restrictive physiology, and normal (or near normal) systolic function.
- Left ventricular noncompaction (LVNC) is a recently recognized congenital cardiomyopathy characterized by adistinctive ("spongy") morphological appearance of the LV myocardium. Noncompaction involves predominantly the distal (apical) portion of the LV chamber with deep intertrabecular recesses (sinusoids) in communication with the ventricular cavity, resulting from an arrest in the normal embryogenesis.

However, while the overwhelming majority of cardiomyopathies are autosomal dominant, more complex modes of inheritance, such as homozygous, double heterozygous, and compound heterozygous mutations, are existent which underlie recessive, X-linked, or mitochondrial types of inheritance (Table 8.1). While a lot of valuable information were gathered from careful clinical and genetic examination of human subjects, significant molecular insights were revealed from the generation and analysis of genetically altered mouse models.

Table 8.1 Most important genetic cardiomyopathies, prevalences, mode of inheritance, and related genes

Disease	Prevalence	Inheritance	Genes/structure
HCM	1:500	Mainly AD	Sarcomere
DCM	1:500–1:2000	Mainly AD, AR, mitochondrial, X-linked	Sarcomere, cytoskeleton, signal transduction ECM
ARVC	1:2000–1:5000		Desmosome
RCM	Rare	AD, AR, mitochondrial, X-linked	Sarcomere (troponin)
LVNC	Rare	AD, AR, mitochondrial, X-linked	DCM related

AD autosomal dominant, *AR* autosomal recessive, *HCM* hypertrophic cardiomyopathy, *DCM* dilated cardiomyopathy, *ECM* extracellular matrix, *ARVC* arrhythmogenic right ventricular cardiomyopathy, *RCM* restrictive cardiomyopathy, *LVNC* left ventricular noncompaction

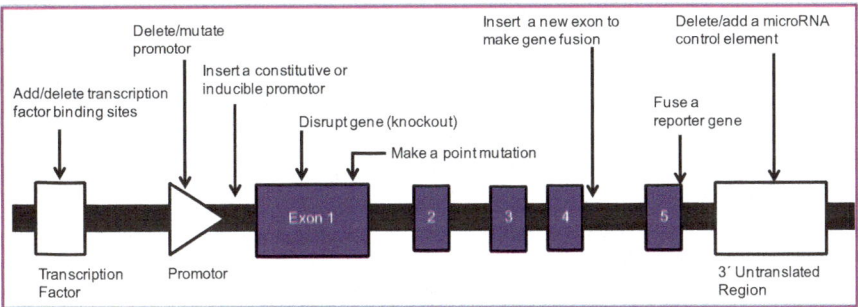

Fig. 8.1 Examples of various genetic changes that can be introduced into mice

8.3   Overview of Genetic Interventions

Genetic engineering can be applied to create model animals that mimic human conditions and gene therapy. Genetically modified mice are the most common genetically engineered animal model and used to study heart disease, including cardiomyopathies. Homologous recombination using a targeting vector that consists of a modified version of the endogenous gene can change the gene or interest. Furthermore, we can introduce different functional changes of a specific gene into a mouse locus through gene targeting (Fig. 8.1).

8.3.1   Random Transgenesis

A transgenic mouse model can be generated by random insertion of a foreign DNA sequence/fragment into the genome by random transgenesis. The insertion of the foreign DNA usually results in a gain of function (expression of a new gene) or overexpression of endogenous genes. The classical method used for the generation of transgenic mice is via pronuclear injection where a transgene is injected into a fertilized mouse egg with subsequent random integration of the gene within the mouse genome. These models are the simplest with very little molecular biology work and require a relatively short development time. However, the biggest disadvantages of this model come from the unpredictability of the location of genomic integration and the number of inserted gene copies. Random insertion of genetic elements into the mouse genome can lead to the misregulated expression of essential genes. The inserted transgene may also be subjected to variegation and positional effects, such as gene silencing, which may lead to altered phenotypic outcomes.

8.3.2   Knockout (KO)

We can determine the function of a specific gene in an organism by creating a KO rendering that specific gene nonfunctional and by inferring the differences between the knockout organism and normal or wild-type littermate controls. The original gene knockout technique was developed by Martin Evans, Oliver Smithies, and Mario Capecchi [5–7]. It is necessary that we know the sequence of the gene to create a KO organism. The traditional process of creating a KO organism starts with constructing an appropriate plasmid where the mutant version of the gene of interest is incorporated. The plasmid construct is engineered with an aim that the faulty version of the gene will recombine with the target gene. Embryonic stem cells are transfected with the plasmid, along with a "transgene" whose overexpression indicates plasmid transfection, and screened by antibiotic selection (Fig. 8.2). Recombination occurs in the region of that sequence within the gene, resulting in the insertion of a faulty sequence to disrupt the gene. The targeted embryonic stem

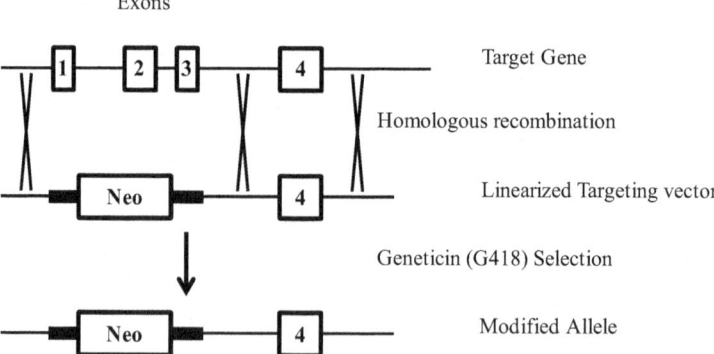

Fig. 8.2 The process of homologous recombination employed in mouse model development. A gene KO model is generated where the neomycin selection cassette replaces part of the gene coding sequence, rendering the gene nonfunctional (Adapted from Gama Sosa et al. [8])

cells are inserted into early embryos, and chimeric organism is generated. Chimeric animals are selectively bred to produce heterozygous animals which are needed to obtain homozygous mutation carriers.

8.3.3 Conditional Knockout

In contrast, a conditional KO allows gene deletion in a tissue (e.g., liver, heart)- or time-specific (a specific stage in development) manner. In tissue-specific conditional KO, a gene is inactivated in target tissue(s) only; in all other tissues, the gene is fully functional. A target gene can also be inactivated at particular time(s) of interest, e.g., mimicking adult-onset condition with more physiological responses and disease relevance. Conditional KO animals are useful to study the role of individual genes in living organisms; as in traditional gene knockout, embryonic death can occur from a gene mutation. Conditional KO organism is most commonly created by introducing short sequences called lox (locus of recombination) sites around the gene of interest and introduced into the germline via the same mechanism as KO (Fig. 8.3). This germline can then be crossed to another line containing Cre recombinase, which is a viral enzyme that can recognize these sequences. This system is inducible by tetracycline or by other means that activate transcription of the Cre recombinase gene or by tamoxifen that activates transport of the Cre recombinase protein to the nucleus. Activation of the system leads to either deletion or inversion of the genes between the two lox sites, depending on their orientation.

8.3.4 Knock In (KI)

Gene KI is a targeted insertion involving one-for-one substitution of DNA sequence information in a genetic locus. This technique usually relies on homologous recombination for gene replacement. Embryonic stem cells with specific gene modification are then implanted into a viable blastocyst, which will grow into a mature chimeric mouse with some cells having the modifications introduced to the embryonic stem cells. Subsequent offspring of the chimeric mouse will then have the gene KI [9]. The first KI mouse model in heart failure research was described by Geisterfer Lowrance et al., where HCM was generated by introducing $Arg^{403} \rightarrow Gln$ mutation into the α cardiac myosin heavy chain (MHC) mouse locus [10]. This elegant study was a "proof-of-concept" showing that a specific mutation is the underlying cause of a devastating human disease. Later on, this mouse model was used to study the role of sarcomere mutations in the development of HCM and successfully used in drug discovery where the beneficial effects of MYK-461 is discovered [11]. We can design KI models where the transgene is introduced into the permissive ROSA26 gene locus, which is well suited for gene overexpression and speedy KI model development [12]. Reporter genes or any other genes of interest can be introduced into ROSA26 gene locus.

A point mutation can be introduced in a single defined base pair location of a gene of interest. The resulting model expresses a mutant protein instead of the wild type if

Mouse expressing Cre recombinase gene controlled by a tissue-specific promoter X

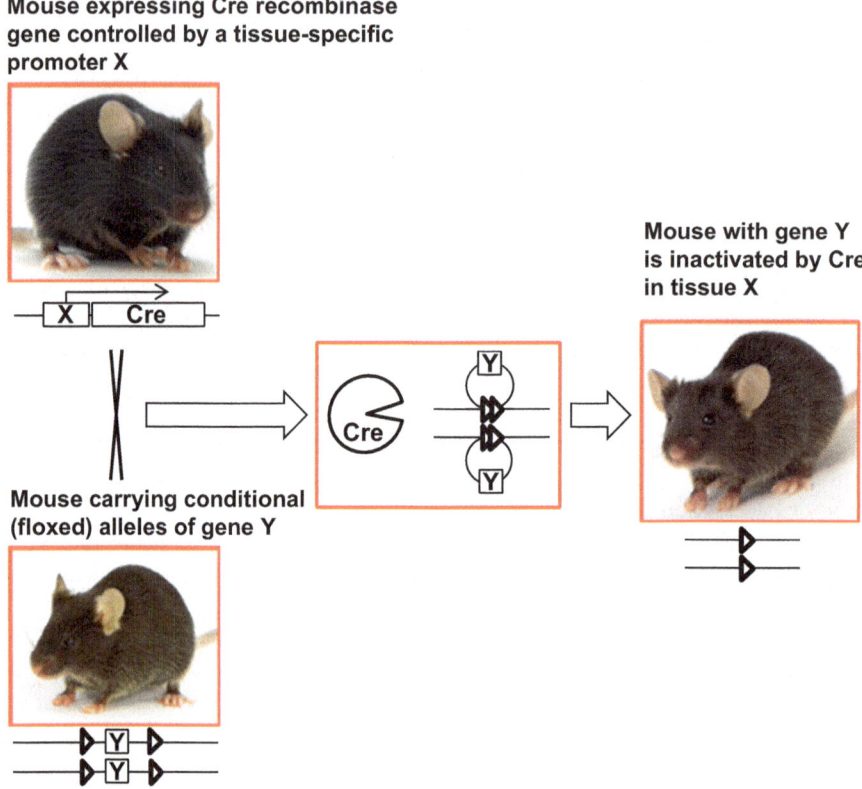

Mouse with gene Y is inactivated by Cre in tissue X

Mouse carrying conditional (floxed) alleles of gene Y

Fig. 8.3 The Cre-loxP system is used extensively in mouse models for cell type- and tissue-specific as well as temporally regulated genetic alteration. Mice that express Cre recombinase under the control of a tissue-specific promoter are crossed with mice that constitutively express a "floxed" genetic region, meaning that the region is flanked by loxP sites. Cre-mediated recombination excises any region of DNA in between, leaving behind a single loxP site

the point mutation is in the coding sequence, but the expression pattern is maintained. Point-mutant models have several potential applications including the study of the causal role of mutations in human pathological conditions, in vivo functionality of an enzyme active site, or protein binding site. Point mutations may also be used as a complementary strategy or an alternative strategy to the more common KO mouse model.

8.3.5   Precision Genetic Engineering (PGE)

PGE focuses on the development of site-directed modification methods in specific DNA sequences to introduce new traits in the models. Zinc finger domains, a zinc ion containing protein domain that can recognize specific DNA sequence, can be engineered to fuse with DNA cleavage domain to form zinc finger nucleases (ZFNs). These engineered DNA-binding proteins that facilitate targeted editing of the genome by creating double-strand breaks in DNA at specific sites [13]. Transcription activator-like effector nucleases (TALEN) are other sets of chimeric nucleases composed of programmable, sequence-specific DNA-binding modules fused with nonspecific DNA cleavage domain, which can cut DNA at specific locations [14]. ZFNs and TALENs enable homology-directed repair at specific genomic locations after inducing DNA double-strand breaks. Recent development of CRISPR/Cas9 system also allows us accurate and successful gene insertions with ease [15]. The CRISPR/Cas system is a prokaryotic immune system that confers resistance to foreign genetic elements [16]. We can manipulate cellular genome specifically at a desired location—thus allowing existing genes to be removed—by delivering the Cas9 nuclease complexed with a synthetic guide RNA (gRNA) into a cell [17], and we can generate complete gene KO mice in a single step [18].

8.3.6   Tet-Off/Tet-On

Different antibiotics, e.g., tetracycline or its derivatives, can induce gene expression. This method was first described by Professors Hermann Bujard and Manfred Gossen at the University of Heidelberg in 1992 [19]. In this system, transcription is turned on/off in the presence of an antibiotic in a reversible way, providing an advantage over KO and KI where the gene inactivation is irreversible. A tetracycline response element (TRE) is 7 repeats of a 19-nucleotide sequence, and this sequence is recognized by tetracycline repressor (tetR). A tetracycline-controlled transactivator (tTA) was developed by Gossen and Bujard by fusing tetR with the C-terminal domain of VP16, an essential transcriptional activation domain from herpes simplex virus. In this system, tet-inducible promoter is repressed by the presence of tetracycline and known as Tet-Off. In a subsequent report, Gossen et al. [20] used random mutagenesis to identify which amino acids of tetR are important for tetracycline-dependent repression. Mutation of these residues created a reverse Tet repressor (rTetR) which relies on the presence of tetracycline for transcription induction (Tet-On).

8.3.7   Humanized Mouse Model

A humanized mouse is a biological model in which a mouse gene is replaced by either a part of or the entire equivalent human gene. The human protein is expressed in mouse in the same cells and tissues instead of the mouse protein. Humanized mouse models are important tools for studying pathological conditions, preclinical research, and compound efficacy in vivo to ensure translatability between various human- and species-specific in vitro and in vivo models. The human protein is expressed typically using the mouse promoter and regulatory regions to ensure natural expression and maintain functionality of the protein. Humanized models provide a far better predictive tool to model human disease and efficacy testing of therapeutic compounds over wild-type mice [21]. By replacing key murine genes with their human counterparts, these models offer greater predictive ability that enables us to increase the impact of research on human diseases, speed up drug discovery, and reduce the time-to-market of new therapeutics.

8.3.8 Reporter Mouse Model

A reporter mouse model is a model in which a target gene is modified for the purpose of monitoring its promoter activity. This is done through the replacement of coding sequence of the gene with the coding sequence of a marker, such as green fluorescent protein (GFP). Reporter mouse models can be used to monitor a target protein's expression, localization, and trafficking and are powerful tools to monitor transcriptional activity of a promoter in vivo. We can also generate reporter models where we avoid knocking out the endogenous gene and use polycistronic technology to couple the expression of the reporter gene with the endogenous promoter, thus allowing co-expression of both target and reporter genes [22].

8.4 Examples: Mouse Models of Cardiomyopathies

To describe a complete overview, listing the myriads of genetically altered mouse models used in heart failure research, is beyond the scope of this review. Therefore, we chose to highlight some of the models we think are probably most important in identifying the roles of key proteins in cardiomyopathies (Fig. 8.4) (we apologize to many authors who contributed to this exciting field but not being mentioned here).

8.4.1 Phospholamban (Pln)

One of the first genetically altered mouse models to study the importance of individual proteins in myocardial contractility was the phospholamban knockout model [23]. Phospholamban is an important regulator of Ca^{2+}-ATPase in the cardiac sarcoplasmic reticulum. This protein was first discovered by Arnold Martin Katz and coworkers in 1974 [24] and is an integral membrane protein of 52 amino acids encoded by *PLN* gene. Phospholamban is a substrate for the cAMP-dependent protein kinase (PKA) in cardiac muscle. Although it functions as an inhibitor of Ca^{2+}-ATPase, phosphorylation by PKA relieves the inhibition. The mouse phospholamban gene was cloned from 129/SvJ mouse. An internal *Hind*III fragment was replaced by *Xho* I-*Sal* I fragment of the polyadenylation signal-deficient *neo* gene from pMCl*neo*. Embryonic stem cells were electroporated with this targeting fragment. Phospholamban-targeted ES cells were injected into C57Bl/6 blastocysts and transferred to pseudopregnant females. Heterozygous mutant mice were mated to generate homozygous mutant. Phospholamban ablation in mice showed no gross

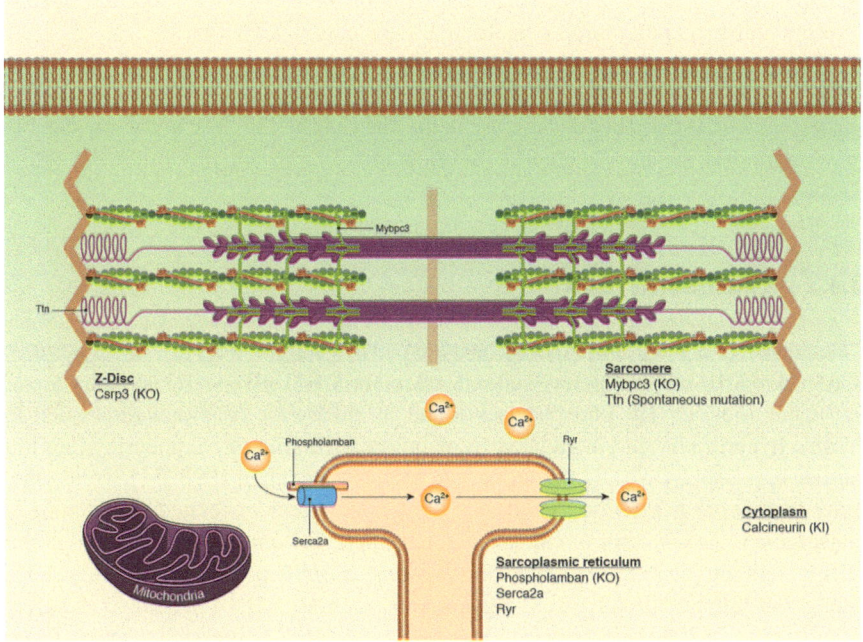

Fig. 8.4 Different cardiomyocyte proteins that were genetically altered in mouse models described in this report. The corresponding mouse models generated to decipher the function and significance are indicated

developmental abnormalities, but mice exhibited enhanced myocardial performance. The time to peak pressure and the time to half-relaxation were significantly shorter in absence of phospholamban. These mice provided an attractive system for elucidation of the regulatory effects of phospholamban on sarcoplasmic reticulum Ca^{2+}-ATPase.

Phospholamban KO mice were crossbred with a well-characterized mouse model of DCM, which harbors a deficiency in muscle-specific LIM protein (MLP; please see also CSRP3 section). Phospholamban ablation led to a dramatic rescue of the cardiomyopathy phenotype of MLP KO mice [25]. Later on, phospholamban KO mice were also crossbred with TNF1.6 mice, which develop heart failure as a consequence of cardiac-specific overexpression of tumor necrosis factor alpha (TNF-α). The resulting offsprings with phospholamban ablation showed improved calcium transients but not cardiac functions [26]. Phospholamban KO mice were also crossbred with tropomyosin-mutant (α-TmE180G) mice, which mimic mutation and features from human hypertrophic cardiomyopathy. The resulting progeny (PLN KO/Tm180) displayed a rescued hypertrophic phenotype with improved morphology and cardiac function [27]. More recent work showed that phospholamban ablation in ryanodine receptor (RyR2)-mutant mice breaks spontaneous Ca^{2+} waves, which is a major cause of Ca^{2+}-mediated arrhythmias [28]. A marked increase in sarcoplasmic reticulum Ca^{2+} leaking was also observed. Despite increased Ca^{2+} leakage, phospholamban ablation in these mice protected against catecholaminergic polymorphic ventricular tachycardia (CPVT) [28].

The PLN:p.R14del, highly prevalent in the Netherlands, is associated with severe DCM and ARVC. Recently, iPS-derived cardiomyocytes from patients have been successfully used to model this disease in the "petri dish" [29, 30]; however, suitable in vivo models are also needed to develop novel cures for this otherwise lethal disease.

8.4.2   Calcineurin (CaN)

Calcineurin is a heterodimer of a 58 to 64 kDa catalytic subunit, calcineurin A (CnA), and a 19 kDa regulatory subunit, calcineurin B (CnB), well known as one of the major four Ser/Thr phosphatases found in eukaryotic cells and implicated in cardiac hypertrophy and associated heart failure. The role of calcineurin (CaN) in cardiac hypertrophy was first described by Molkentin et al. in 1998 [31]. Transgenic mice that express activated forms of calcineurin were generated by cloning a constitutively active form of calcineurin A catalytic subunit with $5'Sal$ I and $3'Hin$ DIII linkers into an expression vector containing α-*MHC* promoter. Similarly, mice expressing human NF-AT3 were generated by cloning a DNA sequence encoding amino acids 317–902 of human NF-AT3 into α-*MHC* expression vector. These sequences were injected into fertilized oocytes, and resulting oocytes were transferred into the oviducts of pseudopregnant mice. Cardiac hypertrophy and heart failure mimicking human heart disease were observed in the transgenic mice expressing calcineurin and NF-AT3. More importantly, the authors showed that pharmacologic agents that inhibit calcineurin activity block in vivo and in vitro hypertrophy. Calcineurin A β-deficient animals (CnAβ−/−) have been generated; the only phenotype reported is smaller hearts in comparison to their wild-type littermate controls [32]. However, calcineurin also appears to be cardioprotective, at least under conditions of ischemia reperfusion [33].

8.4.3   Titin (TTN)

Titin is important in the contraction of striated muscle tissue and connects Z line to M line in sarcomere (for a brief review, please see Tabish et al. [34]). It functions as a molecular spring and responsible for the passive elasticity of muscle. It is the largest known protein with 244 individually folded domains [35]—and these domains unfold when the protein is stretched and refold when the tension is removed [36]. Gerull et al. first reported that titin mutations cause familial DCM in autosomal-dominant fashion in 2002 [37]. Later on, it was shown in a large cohort of subjects that truncating mutations in titin account up to 25% of all causes of dilated cardiomyopathy [38]. These data have largely been confirmed by other authors (for review Tabish et al. [34] and Marston et al. [39]). The first titin mutation reported in mice arose spontaneously on C57BL/6 J background, causing muscular dystrophy with myositis [40]. A KI mouse model was generated with a previously identified 2-bp insertion mutation. Homozygous mutation caused sarcomere formation defects in embryonic heart and was shown to be embryonically lethal [41]. Heterozygous mice were viable and showed normal cardiac morphology and function. However, when these heterozygous mice were chronically exposed to angiotensin II, they developed marked left ventricular dilatation with impaired fractional shortening and diffused myocardial fibrosis—mimicking human DCM [41].

Titin mutation was recently (2014) added to human familial clinical genetic testing of cardiomyopathies, which resulted in an additional 10% genetic diagnosis of DCM [42]. Frameshift mutations in the titin gene are a major cause of DCM. A proof-of-concept was established recently showing that disruption in the *titin* reading frame due to truncating mutations can be restored by exon skipping. This RNA-based strategy was shown to work in both patient cardiomyocytes and in mouse heart with DCM, providing us with a potential treatment option for DCM [43].

8.4.4 Myosin Binding Protein C 3 (Mybpc3)

Myosin binding protein C (MYBPC) is a crucial component of the sarcomere and an important regulator of muscle function. While mutations in the cardiac isoform of MYBPC (MYBPC3) are well-known causes of cardiomyopathies, such as HCM and DCM, the underlying molecular mechanisms are not well understood. A variety of MYBPC3 mutations have been studied in great detail and several corresponding mouse models have been generated. Most MYBPC3 mutations may cause haploinsufficiency, and with it, they may cause a primary increase in calcium sensitivity that could explain major clinical features of HCM patients such as the hypercontractile phenotype and the well-known secondary effects such as myofibrillar disarray, fibrosis, myocardial hypertrophy, and remodeling including arrhythmogenesis. However, the presence of poison peptides, mutant sarcomeric proteins that incorporate into myofibrils and act as dominant negative proteins, in some cases cannot be fully excluded, and most probably other mechanisms are also at play. Here we discuss various MYBPC3 mouse models and their implications in cardiomyopathy.

MYBPC3 mutations causing human HCM were first reported in 1995 [44, 45]. Indeed HCM is a frequent disease, affecting 1:500 individuals [2], and depending on the population analyzed, MYBPC3 mutations are found in up to 40–50% of the genotyped HCM patients [46]. To date more than 350 different MYBPC3 mutations have been reported in patients with HCM[47]. In general, MYBPC3 mutations are associated with a slightly lower penetrance (those with disease-causing mutations that have clinical manifestations), with later onset of disease, and with milder forms of disease progression in comparison to other HCM-causing mutations located, for example, in the beta myosin heavy chain (MYH7) gene [48, 49]. However this general statement may not necessarily be true for all MYBPC3 mutations, and indeed some MYBPC3 mutations are associated with a poor prognosis.

Other MYBPC3 mutations primarily found in DCM patients have also been reported, for example, the Asn948Thr missense mutation [50]. In addition, the MYBPC3 Arg502Trp mutation with a frequency of about 2.4% is the most common HCM-causing mutation among individuals of European descent in the USA [51]. Other mutations may be prevalent in different European populations, such as in the Netherlands, where three different founder mutations are present: (1) the c.2373_2374insG MYBPC3 which is present in the great majority of HCM patients (up to 8 to 25%) and where (2) the c.2864_2865delCT and (3) the c.2827C>T mutations occur in about 5% of HCM patients each [52, 53]. Although MYBPC3 mutations are established as well-known causes of cardiomyopathy, the underlying molecular mechanisms are not well defined. Genetically altered mouse models provided significant new insights, which will be discussed in the next paragraphs.

To gain more insight into the underlying molecular mechanisms, Mybpc3 has been deleted in genetically altered mouse models by two independent groups. Loss of this protein is not associated with any embryonic lethality, and MYBPC3 also is not essential for sarcomere formation, but its absence results in profound eccentric hypertrophy in the homozygous animals [54, 55]. Hemodynamic analysis of the mice where exons 1 and 2 have been deleted was performed by the Carrier group, revealing the presence of normal contractility but severe diastolic defects. In addition, heterozygous animals develop septal hypertrophy, a hallmark of HCM [54]. However, these animals were engineered such that they harbor a complete ablation of the gene and therefore are useful to identify basic mechanisms, but the overwhelming majority of human mutation carriers express mutant mRNAs and probably proteins, which makes it difficult to relate these data directly to humans.

Therefore, in addition to the pure knockout models, wild-type and truncated MYBPC3, which mimic a certain type of human mutations leading to the loss of carboxyterminal domain including titin and myosin binding sites, were overexpressed in mouse models. Overexpression of the truncated form, but not the wild-type protein, led to the appearance of HCM features, including hypertrophy, and an increase in calcium sensitivity [56]. Whereas abovementioned transgenes were well expressed at the mRNA and protein levels, overexpression of a Mybpc3 mutant lacking only the myosin binding domain resulted in the expression of only very modest levels of mutant protein (i.e., about 5%) which led to a mild hypertrophy and heart failure phenotype [57].

Two additional KI mouse models have been engineered to carry Mybpc3 mutations found in patients which affect titin and myosin binding. Interestingly, animals homozygous for these mutations develop a DCM-like phenotype with depressed contractility and hypertrophy [58]. Another KI mouse model was generated such that the mutant protein did not contain the amino terminal myosin binding domain. The mutant protein was readily integrated into the sarcomeres of heterozygous and homozygous animals and was protein kinase A (PKA) phosphorylatable, and no major structural defects of sarcomere were detected. However, this mutation was associated with a significant increase in calcium sensitivity [59]. An additional mouse KI model, based on a G>A transition located on the last nucleotide of exon 6 and which was found in a patient with HCM, has been generated by Lucie Carrier's group. Interestingly this mutation gives rise to three different mRNAs: (1) missense mutation; (2) nonsense due to exon skipping, frameshift, and premature stop codon; and (3) deletion/insertion as nonsense but with additional partial retention of a downstream intron which restores the reading frame and leads to an almost full-length protein. Homozygous animals develop hypertrophy, interstitial fibrosis, and decreased myocardial function, whereas heterozygous animals do not have any obvious phenotype [60]. Additional genetically altered mouse models have been generated to study MYBPC3 phosphorylation, which will be discussed in the next part.

8.4.4.1 MYBPC3 Phosphorylation

Human MYBPC3 contains at least four phosphorylation sites that are localized inside the myosin binding motif (serines 275, 284, 304, with an additional phosphorylation site not unequivocally identified) [61]); the mouse myosin binding motif contains three to four sites (serines 273, 282, 302, (305)) [62–64]. These phosphorylation sites have been studied in various transgenic mouse models; for example, lines have been established where the phosphorylation sites (Ser273, Ser282, and Ser302), along with two adjacent sites that could be potentially phosphorylated (Thr272, Thr281), were converted to alanines. While overexpression of wild-type MYBPC3 was able to rescue the MYBPC3 null phenotype, the non-phosphorylatable MYBPC3 was not [65]. Transgenic mice with complete deletion of phosphorylation motif "LAGAGRRTS" show an increase in cardiac contractility and relaxation and an increase in the phosphorylation of the remaining MYBPC3, troponin I, and phospholamban [66]. On the other hand, overexpression of a phosphomimetic MYBPC3 in a Mybpc3 null background showed subtle changes in sarcomeric ultrastructure characterized by increased distances between the thick filaments, indicating that phosphomimetic MYBPC3 affects thick-thin filament relationship. The loss of MYBPC3 phosphorylation has been observed in failing human hearts, and strategies to increase its phosphorylation may have cardioprotective effects [61, 67].

Although compelling evidences suggest that MYBPC3 phosphorylation modulates contractility by controlling the proximity of the myosin heads to actin, the precise molecular mechanism remains unclear (reviewed in [68]). MYBPC3 phosphorylation can abolish its ability to interact with the S2 region of the myosin heavy chain in vitro, but it may enhance MYBPC3 interactions with the thin filament. Alternatively, dephosphorylation results in strong binding of Mybpc3 to the myosin head, probably preventing its force-generating interaction with actin [67, 69]. However, recent data indicate that MYBPC3 may act synergistically with the myosin regulatory light chain to enhance cross-bridge formation by altering the interaction of the myosin head with actin. It may well be that this interaction depends on phosphorylation of either Mybpc3 or regulatory light chain and may well be able to provide another mechanism how phosphorylation of sarcomeric proteins may affect protein-protein interactions and kinetics of force development [70]. Although it is well known that the distance between thick and thin filaments is a major determinant of calcium sensitivity and the loss of MYBPC3 is associated with increased calcium sensitivity, the precise molecular mechanism of how MYBPC3 phosphorylation affects this system is unknown. Part of this problem is that PKA phosphorylation decreases calcium sensitivity via troponin I phosphorylation, which leads to a decrease in troponin C calcium sensitivity. However Mybpc3 phosphorylation, which appears to decrease calcium sensitivity as well, has no effect on interfilament spacing [70]. Further studies, based on various transgenic animals, will certainly help to answer this question.

8.4.5   Cysteine-Rich Protein 3 (Csrp3/Mlp)

Cysteine-rich protein 3 (CSRP3) or muscle LIM protein (MLP) is a myogenic factor and was first described in 1994 [71]. MLP is a LIM-only protein consisting of 194 amino acids that constitute two LIM domains, each followed by a glycine-rich domain. Soon after its initial description, the protein became important in cardiovascular research when it was demonstrated in 1997 [72] that Csrp3-deficient mice developed hypertrophy followed by DCM later in life—at that time the first genetically altered organism reproducing this devastating disease. Perhaps even more important was that human CSRP3 mutations were able to cause human forms of cardiomyopathy [73–77]. The precise mechanisms by which CSRP3 mutations cause these diseases are not fully elucidated yet. As mentioned before, additional ablation of phospholamban (Pln) in Csrp3$^{-/-}$ mice exhibiting normal or enhanced SR-calcium ATPase function [78] rescues the heart failure phenotype (i.e., no heart failure in Pln and Csrp3 double KO mice; please see PLN section) [25]. It was also shown that Csrp3$^{-/-}$ cardiomyocytes are defective in producing brain natriuretic peptide (BNP), a sensitive cardiac mechanical load marker, when exposed to mechanical stress but not to endothelin (Gq-coupling receptor agonist) or phenylephrine (α-adrenergic agonist), suggesting that the primary defect is not located in the downstream pathway but in the initial sensing of the stretch stimulus [76]. Thus, CSRP3 was suggested to be part of a macromolecular, cardiac mechanical stretch sensor.

CSRP3 was found to be present in different cellular compartments, including the sarcomeric Z-disc/I-band of different species such as drosophila [79], mouse [72, 80–82], and humans [76]. It was also reported to be present in the costamere (where β1-spectrin interacts with MLP) [83, 84], in the intercalated disc (where N-RAP interacts with MLP) [85], and in the cytoplasm (actin) [71], as well as in the nucleus [80]. However, Geier and coworkers found MLP mostly cytoplasmic and not at the Z-disc [86]. Boateng and colleagues demonstrated quite elegantly that CSRP3 in cardiac myocytes is present in two different molecular forms: the oligomeric form, which is present at the sarcolemma and the cytoskeleton, and the monomeric form, which is exclusively located in the nucleus [80]. Interestingly, after myocardial infarction and pressure overload, in both animal model and failing human hearts, nuclear CSRP3 levels in the myocardium increased at the expense of nonnuclear form. In failing human hearts, almost no CSRP3 was detectable outside the nucleus. Additional work by Boateng and coworkers provided stronger evidence for nucleocytoplasmic shuttling of CSRP3 based on putative nuclear localization signal (NLS) [87]. The possible role of NLS was first suggested by Fung et al. when a clone with sequence homology of rat muscle LIM protein was isolated and characterized from a human fetal heart cDNA library [88]. Significant insights on the functional properties of nuclear CSRP3 were provided by using cell-penetrating synthetic peptide containing the putative nuclear localization signal (RKYGPK) of CSRP3—which inhibited any shuttling of endogenously synthesized CSRP3 to the nucleus, thus documenting functionality of this NLS [87]. Inhibition of nuclear translocation prevented the increased protein accumulation, usually seen with phenylephrine treatment, in isolated cardiomyocytes, thus identifying an important role of CSRP3 in controlling agonist-stimulated hypertrophy. Interestingly, cyclic strain of myocytes with prior NLS treatment resulted in disarrayed sarcomeres—an observation that has also been made in cardiomyocytes of Csrp3 KO mice [71]. Inhibition of nuclear shuttling during cyclic-mechanical strain prevents increased protein synthesis and BNP expression, suggesting that CSRP3 is required for remodeling of myofilaments and altered gene expression. A direct link between BNP gene expression and Csrp3 has been shown earlier [76], and research from other groups with heterozygous Csrp3 KO mice with myocardial infarction suggest similar association [82]. Although the functional significance of nuclear Csrp3, like myocyte hypertrophy following biomechanical stress, gene expression, and sarcomere assembly, has been demonstrated in this study, more questions are arising. One of the most important questions is how a small protein-adaptor molecule like MLP can exert its actions in different cellular compartments including the nucleus. The prohypertrophic phosphatase calcineurin has been previously identified as a cytoskeletal target/interacting partner of Csrp3 [82, 89, 90], and it was demonstrated that CSRP3 at the Z-disc is necessary for calcineurin activation in cardiac myocytes [82, 89, 90].

What are the *nuclear* targets of CSRP3 in cardiac myocytes? As CSRP3 does not bind DNA directly, one could speculate that it interacts with DNA by stress-responsive cardiac transcription factors like GATA4, GATA6, AP1, or serum response factor (SRF). In fact, it was previously shown that CSRP2, which is closely related to CSRP3, can bind GATAs and SRF and thereby enhance gene expression in smooth muscle cells [91]. By linking different transcription factors, as shown by the complex formation via protein-protein interaction between two LIM-domain containing proteins [92], CSRP3 may integrate the activities of multiple nuclear regulatory proteins in order to coordinate gene expression. It is also possible that CSRP3 facilitates nucleocytoplasmic shuttling of its nonnuclear target calcineurin, which has been shown to translocate to the nucleus in hypertrophied cardiac myocytes [93]. Research on CSRP3 was predominantly focused on its cytoskeletal function. Although the interaction of CSRP3 with different transcription factors, such as MyoD, MRF4, and myogenin, was shown in skeletal myocytes [94], and CSRP3 nuclear localization in right ventricular murine hypertrophy following increased biomechanical stress [95], the functional significance of this NLS remained elusive until the report by Boateng and coworkers [87]. In this context, we have previously shown that CSRP3 also interacts with the transcription factor ZBTB17, which is important for survival signaling [96].

It should be noted that CSRP3 becomes acetylated within its NLS on position K69 [81], a modification thought to affect calcium sensitivity, which may affect CSRP3's nucleocytoplasmic shuttling. It will be interesting to analyze the effects of human CSRP3 mutations on its nuclear localization and analyze their ability to affect specific genes, particularly of K69R mutation, found in an individual affected by DCM and endocardial fibroelastosis [77]. Besides acetylation, other posttranslational modifications such as phosphorylation, sumoylation, and/or polyubiquitinylation may also affect CSRP3 function [97].

8.5   Conclusion

Due to many advantages over large animal models, mouse models are commonly used in CV research, both in academia and big pharma. They a have a short life span, which allows scientists to follow the natural course of disease at an accelerated pace. More importantly, mice can be genetically modified, allowing for rapid establishment of proof-of-principle that can later be extended to larger animal models. Nevertheless, mice have disadvantages, such as being phylogenetically distant from humans, and some physiological features and their response to pharmacological treatments may not be comparable to humans. Therefore, translational aspects and the value of genetically altered mice must be interpreted with caution. However, genetically altered mouse models provided us with fundamental insights in cardiac Ca^{2+} homeostasis/signaling (Pln), hypertrophism and survival signaling (calcineurin), patho-mechanisms of general heart failure (Csrp3, Mlp), phosphorylation (Mybpc3), and potential routes to therapy (Ttn). Although we made significant progress, more efforts are necessary to better understand the underlying molecular mechanisms of heart failure and to exploit the knowledge for innovative therapies to benefit the mankind.

Acknowledgments Ralph Knöll is supported by funds provided by the Integrated Cardio Metabolic Centre at the Karolinska Institute (ICMC/KI), Leducq Foundation, German Research Foundation (Kn 448/10-2), and Hjärt och Lungfonden in Sweden.

We apologize to the many authors having contributed to this exiting field but have not been mentioned here.

References

1. Stevens G, Mascarenhas M, Mathers C. Global health risks: progress and challenges. Bull World Health Organ. 2009;87(9):646.
2. Maron BJ, Towbin JA, Thiene G, Antzelevitch C, Corrado D, Arnett D, Moss AJ, Seidman CE, Young JB. Contemporary definitions and classification of the cardiomyopathies: an American Heart Association Scientific Statement from the Council on Clinical Cardiology, Heart Failure

and Transplantation Committee; Quality of Care and Outcomes Research and Functional Genomics and Translational Biology Interdisciplinary Working Groups; and Council on Epidemiology and Prevention. Circulation. 2006;113(14):1807–16.

3. Geisterfer-Lowrance AA, Kass S, Tanigawa G, Vosberg HP, McKenna W, Seidman CE, Seidman JG. A molecular basis for familial hypertrophic cardiomyopathy: a beta cardiac myosin heavy chain gene missense mutation. Cell. 1990;62(5):999–1006.

4. Watkins H, Rosenzweig A, Hwang DS, Levi T, McKenna W, Seidman CE, Seidman JG. Characteristics and prognostic implications of myosin missense mutations in familial hypertrophic cardiomyopathy. N Engl J Med. 1992;326(17):1108–14.

5. Capecchi MR. Generating mice with targeted mutations. Nat Med. 2001;7(10):1086–90.

6. Evans MJ. The cultural mouse. Nat Med. 2001;7(10):1081–3.

7. Smithies O. Forty years with homologous recombination. Nat Med. 2001;7(10):1083–6.

8. Gama Sosa MA, De Gasperi R, Elder GA. Animal transgenesis: an overview. Brain Struct Funct. 2010;214(2–3):91–109.

9. Manis JP. Knock out, knock in, knock down--genetically manipulated mice and the Nobel Prize. N Engl J Med. 2007;357(24):2426–9.

10. Geisterfer-Lowrance AA, Christe M, Conner DA, Ingwall JS, Schoen FJ, Seidman CE, Seidman JG. A mouse model of familial hypertrophic cardiomyopathy. Science. 1996;272 (5262):731–4.

11. Green EM, Wakimoto H, Anderson RL, Evanchik MJ, Gorham JM, Harrison BC, Henze M, Kawas R, Oslob JD, Rodriguez HM, Song Y, Wan W, Leinwand LA, Spudich JA, McDowell RS, Seidman JG, Seidman CE. A small-molecule inhibitor of sarcomere contractility suppresses hypertrophic cardiomyopathy in mice. Science. 2016;351(6273):617–21.

12. Soriano P. Generalized lacZ expression with the ROSA26 Cre reporter strain. Nat Genet. 1999;21(1):70–1.

13. Geurts AM, Cost GJ, Freyvert Y, Zeitler B, Miller JC, Choi VM, Jenkins SS, Wood A, Cui X, Meng X, Vincent A, Lam S, Michalkiewicz M, Schilling R, Foeckler J, Kalloway S, Weiler H, Menoret S, Anegon I, Davis GD, Zhang L, Rebar EJ, Gregory PD, Urnov FD, Jacob HJ, Buelow R. Knockout rats via embryo microinjection of zinc-finger nucleases. Science. 2009;325(5939):433.

14. Boch J. TALEs of genome targeting. Nat Biotechnol. 2011;29(2):135–6.

15. Ruan J, Li H, Xu K, Wu T, Wei J, Zhou R, Liu Z, Mu Y, Yang S, Ouyang H, Chen-Tsai RY, Li K. Highly efficient CRISPR/Cas9-mediated transgene knockin at the H11 locus in pigs. Sci Rep. 2015;5:14253.

16. Joung J, Konermann S, Gootenberg JS, Abudayyeh OO, Platt RJ, Brigham MD, Sanjana NE, Zhang F. Genome-scale CRISPR-Cas9 knockout and transcriptional activation screening. Nat Protoc. 2017;12(4):828–63.

17. Jinek M, Chylinski K, Fonfara I, Hauer M, Doudna JA, Charpentier E. A programmable dual-RNA-guided DNA endonuclease in adaptive bacterial immunity. Science. 2012;337 (6096):816–21.

18. Zuo E, Cai YJ, Li K, Wei Y, Wang BA, Sun Y, Liu Z, Liu J, Hu X, Wei W, Huo X, Shi L, Tang C, Liang D, Wang Y, Nie YH, Zhang CC, Yao X, Wang X, Zhou C, Ying W, Wang Q, Chen RC, Shen Q, Xu GL, Li J, Sun Q, Xiong ZQ, Yang H. One-step generation of complete gene knockout mice and monkeys by CRISPR/Cas9-mediated gene editing with multiple sgRNAs. Cell Res. 2017;27(7):933–45.

19. Gossen M, Bujard H. Tight control of gene expression in mammalian cells by tetracycline-responsive promoters. Proc Natl Acad Sci U S A. 1992;89(12):5547–51.

20. Gossen M, Freundlieb S, Bender G, Muller G, Hillen W, Bujard H. Transcriptional activation by tetracyclines in mammalian cells. Science. 1995;268(5218):1766–9.

21. Garcia S, Freitas AA. Humanized mice: current states and perspectives. Immunol Lett. 2012;146(1–2):1–7.

22. Tsukiyama T, Kato-Itoh M, Nakauchi H, Ohinata Y. A comprehensive system for generation and evaluation of induced pluripotent stem cells using piggyBac transposition. PLoS One. 2014;9(3):e92973.

23. Luo W, Grupp IL, Harrer J, Ponniah S, Grupp G, Duffy JJ, Doetschman T, Kranias EG. Targeted ablation of the phospholamban gene is associated with markedly enhanced myocardial contractility and loss of beta-agonist stimulation. Circ Res. 1994;75(3):401–9.

24. Tada M, Kirchberger MA, Repke DI, Katz AM. The stimulation of calcium transport in cardiac sarcoplasmic reticulum by adenosine 3':5'-monophosphate-dependent protein kinase. J Biol Chem. 1974;249(19):6174–80.

25. Minamisawa S, Hoshijima M, Chu G, Ward CA, Frank K, Gu Y, Martone ME, Wang Y, Ross J Jr, Kranias EG, Giles WR, Chien KR. Chronic phospholamban-sarcoplasmic reticulum calcium ATPase interaction is the critical calcium cycling defect in dilated cardiomyopathy. Cell. 1999;99(3):313–22.

26. Janczewski AM, Zahid M, Lemster BH, Frye CS, Gibson G, Higuchi Y, Kranias EG, Feldman AM, McTiernan CF. Phospholamban gene ablation improves calcium transients but not cardiac function in a heart failure model. Cardiovasc Res. 2004;62(3):468–80.

27. Gaffin RD, Pena JR, Alves MS, Dias FA, Chowdhury SA, Heinrich LS, Goldspink PH, Kranias EG, Wieczorek DF, Wolska BM. Long-term rescue of a familial hypertrophic cardiomyopathy caused by a mutation in the thin filament protein, tropomyosin, via modulation of a calcium cycling protein. J Mol Cell Cardiol. 2011;51(5):812–20.

28. Bai Y, Jones PP, Guo J, Zhong X, Clark RB, Zhou Q, Wang R, Vallmitjana A, Benitez R, Hove-Madsen L, Semeniuk L, Guo A, Song LS, Duff HJ, Chen SR. Phospholamban knockout breaks arrhythmogenic Ca(2)(+) waves and suppresses catecholaminergic polymorphic ventricular tachycardia in mice. Circ Res. 2013;113(5):517–26.

29. Karakikes I, Stillitano F, Nonnenmacher M, Tzimas C, Sanoudou D, Termglinchan V, Kong CW, Rushing S, Hansen J, Ceholski D, Kolokathis F, Kremastinos D, Katoulis A, Ren L, Cohen N, Gho JM, Tsiapras D, Vink A, Wu JC, Asselbergs FW, Li RA, Hulot JS, Kranias EG, Hajjar RJ. Correction of human phospholamban R14del mutation associated with cardiomyopathy using targeted nucleases and combination therapy. Nat Commun. 2015;6:6955.

30. Stillitano F, Turnbull IC, Karakikes I, Nonnenmacher M, Backeris P, Hulot JS, Kranias EG, Hajjar RJ, Costa KD. Genomic correction of familial cardiomyopathy in human engineered cardiac tissues. Eur Heart J. 2016;37(43):3282–4.

31. Molkentin JD, Lu JR, Antos CL, Markham B, Richardson J, Robbins J, Grant SR, Olson EN. A calcineurin-dependent transcriptional pathway for cardiac hypertrophy. Cell. 1998;93 (2):215–28.

32. Bueno OF, Wilkins BJ, Tymitz KM, Glascock BJ, Kimball TF, Lorenz JN, Molkentin JD. Impaired cardiac hypertrophic response in Calcineurin Abeta -deficient mice. Proc Natl Acad Sci U S A. 2002;99(7):4586–91.

33. Bueno OF, Lips DJ, Kaiser RA, Wilkins BJ, Dai YS, Glascock BJ, Klevitsky R, Hewett TE, Kimball TR, Aronow BJ, Doevendans PA, Molkentin JD. Calcineurin Abeta gene targeting predisposes the myocardium to acute ischemia-induced apoptosis and dysfunction. Circ Res. 2004;94(1):91–9.

34. Tabish AM, Azzimato V, Alexiadis A, Buyandelger B, Knoll R. Genetic epidemiology of titin-truncating variants in the etiology of dilated cardiomyopathy. Biophys Rev. 2017;9:207–23.

35. Labeit S, Kolmerer B. Titins: giant proteins in charge of muscle ultrastructure and elasticity. Science. 1995;270(5234):293–6.

36. Minajeva A, Kulke M, Fernandez JM, Linke WA. Unfolding of titin domains explains the viscoelastic behavior of skeletal myofibrils. Biophys J. 2001;80(3):1442–51.

37. Gerull B, Gramlich M, Atherton J, McNabb M, Trombitas K, Sasse-Klaassen S, Seidman JG, Seidman C, Granzier H, Labeit S, Frenneaux M, Thierfelder L. Mutations of TTN, encoding the giant muscle filament titin, cause familial dilated cardiomyopathy. Nat Genet. 2002;30 (2):201–4.

38. Herman DS, Lam L, Taylor MR, Wang L, Teekakirikul P, Christodoulou D, Conner L, DePalma SR, McDonough B, Sparks E, Teodorescu DL, Cirino AL, Banner NR, Pennell DJ, Graw S, Merlo M, Di Lenarda A, Sinagra G, Bos JM, Ackerman MJ, Mitchell RN, Murry CE, Lakdawala NK, Ho CY, Barton PJ, Cook SA, Mestroni L, Seidman JG, Seidman CE. Truncations of titin causing dilated cardiomyopathy. N Engl J Med. 2012;366(7):619–28.
39. Marston S, Montgiraud C, Munster AB, Copeland O, Choi O, Dos Remedios C, Messer AE, Ehler E, Knoll R. OBSCN Mutations Associated with Dilated Cardiomyopathy and Haploinsufficiency. PLoS One. 2015;10(9):e0138568.
40. Garvey SM, Rajan C, Lerner AP, Frankel WN, Cox GA. The muscular dystrophy with myositis (mdm) mouse mutation disrupts a skeletal muscle-specific domain of titin. Genomics. 2002;79 (2):146–9.
41. Gramlich M, Michely B, Krohne C, Heuser A, Erdmann B, Klaassen S, Hudson B, Magarin M, Kirchner F, Todiras M, Granzier H, Labeit S, Thierfelder L, Gerull B. Stress-induced dilated cardiomyopathy in a knock-in mouse model mimicking human titin-based disease. J Mol Cell Cardiol. 2009;47(3):352–8.
42. Weintraub RG, Semsarian C, Macdonald P. Dilated cardiomyopathy. Lancet. 2017;390 (10092):400–14.
43. Gramlich M, Pane LS, Zhou Q, Chen Z, Murgia M, Schotterl S, Goedel A, Metzger K, Brade T, Parrotta E, Schaller M, Gerull B, Thierfelder L, Aartsma-Rus A, Labeit S, Atherton JJ, McGaughran J, Harvey RP, Sinnecker D, Mann M, Laugwitz KL, Gawaz MP, Moretti A. Antisense-mediated exon skipping: a therapeutic strategy for titin-based dilated cardiomyopathy. EMBO Mol Med. 2015;7(5):562–76.
44. Bonne G, Carrier L, Bercovici J, Cruaud C, Richard P, Hainque B, Gautel M, Labeit S, James M, Beckmann J, Weissenbach J, Vosberg HP, Fiszman M, Komajda M, Schwartz K. Cardiac myosin binding protein-C gene splice acceptor site mutation is associated with familial hypertrophic cardiomyopathy. Nat Genet. 1995;11(4):438–40.
45. Watkins H, Conner D, Thierfelder L, Jarcho JA, MacRae C, McKenna WJ, Maron BJ, Seidman JG, Seidman CE. Mutations in the cardiac myosin binding protein-C gene on chromosome 11 cause familial hypertrophic cardiomyopathy. Nat Genet. 1995;11(4):434–7.
46. Richard P, Charron P, Carrier L, Ledeuil C, Cheav T, Pichereau C, Benaiche A, Isnard R, Dubourg O, Burban M, Gueffet JP, Millaire A, Desnos M, Schwartz K, Hainque B, Komajda M. Hypertrophic cardiomyopathy: distribution of disease genes, spectrum of mutations, and implications for a molecular diagnosis strategy. Circulation. 2003;107(17):2227–32.
47. Carrier L, Mearini G, Stathopoulou K, Cuello F. Cardiac myosin-binding protein C (MYBPC3) in cardiac pathophysiology. Gene. 2015;573(2):188–97.
48. Niimura H, Bachinski LL, Sangwatanaroj S, Watkins H, Chudley AE, McKenna W, Kristinsson A, Roberts R, Sole M, Maron BJ, Seidman JG, Seidman CE. Mutations in the gene for cardiac myosin-binding protein C and late-onset familial hypertrophic cardiomyopathy. N Engl J Med. 1998;338(18):1248–57.
49. Watkins H, Ashrafian H, Redwood C. Inherited cardiomyopathies. N Engl J Med. 2011;364 (17):1643–56.
50. Dachmlow S, Erdmann J, Knueppel T, Gille C, Froemmel C, Hummel M, Hetzer R, Regitz-Zagrosek V. Novel mutations in sarcomeric protein genes in dilated cardiomyopathy. Biochem Biophys Res Commun. 2002;298(1):116–20.
51. Saltzman AJ, Mancini-DiNardo D, Li C, Chung WK, Ho CY, Hurst S, Wynn J, Care M, Hamilton RM, Seidman GW, Gorham J, McDonough B, Sparks E, Seidman JG, Seidman CE, Rehm HL. Short communication: the cardiac myosin binding protein C Arg502Trp mutation: a common cause of hypertrophic cardiomyopathy. Circ Res. 2010;106(9):1549–52.
52. Alders M, Jongbloed R, Deelen W, van den Wijngaard A, Doevendans P, Ten Cate F, Regitz-Zagrosek V, Vosberg HP, van Langen I, Wilde A, Dooijes D, Mannens M. The 2373insG mutation in the MYBPC3 gene is a founder mutation, which accounts for nearly one-fourth of the HCM cases in the Netherlands. Eur Heart J. 2003;24(20):1848–53.

53. Christiaans I, Nannenberg EA, Dooijes D, Jongbloed RJ, Michels M, Postema PG, Majoor-Krakauer D, van den Wijngaard A, Mannens MM, van Tintelen JP, van Langen IM, Wilde AA. Founder mutations in hypertrophic cardiomyopathy patients in the Netherlands. Netherlands heart journal: monthly journal of the Netherlands Society of Cardiology and the Netherlands Heart Foundation. 2010;18(5):248–54.
54. Carrier L, Knoll R, Vignier N, Keller DI, Bausero P, Prudhon B, Isnard R, Ambroisine ML, Fiszman M, Ross J Jr, Schwartz K, Chien KR. Asymmetric septal hypertrophy in heterozygous cMyBP-C null mice. Cardiovasc Res. 2004;63(2):293–304.
55. Harris SP, Bartley CR, Hacker TA, McDonald KS, Douglas PS, Greaser ML, Powers PA, Moss RL. Hypertrophic cardiomyopathy in cardiac myosin binding protein-C knockout mice. Circ Res. 2002;90(5):594–601.
56. Yang Q, Sanbe A, Osinska H, Hewett TE, Klevitsky R, Robbins J. A mouse model of myosin binding protein C human familial hypertrophic cardiomyopathy. J Clin Invest. 1998;102 (7):1292–300.
57. Yang Q, Sanbe A, Osinska H, Hewett TE, Klevitsky R, Robbins J. In vivo modeling of myosin binding protein C familial hypertrophic cardiomyopathy. Circ Res. 1999;85(9):841–7.
58. McConnell BK, Jones KA, Fatkin D, Arroyo LH, Lee RT, Aristizabal O, Turnbull DH, Georgakopoulos D, Kass D, Bond M, Niimura H, Schoen FJ, Conner D, Fischman DA, Seidman CE, Seidman JG. Dilated cardiomyopathy in homozygous myosin-binding protein-C mutant mice. J Clin Invest. 1999;104(12):1771.
59. Witt CC, Gerull B, Davies MJ, Centner T, Linke WA, Thierfelder L. Hypercontractile properties of cardiac muscle fibers in a knock-in mouse model of cardiac myosin-binding protein-C. J Biol Chem. 2001;276(7):5353–9.
60. Vignier N, Schlossarek S, Fraysse B, Mearini G, Kramer E, Pointu H, Mougenot N, Guiard J, Reimer R, Hohenberg H, Schwartz K, Vernet M, Eschenhagen T, Carrier L. Nonsense-mediated mRNA decay and ubiquitin-proteasome system regulate cardiac myosin-binding protein C mutant levels in cardiomyopathic mice. Circ Res. 2009;105(3):239–48.
61. Copeland O, Sadayappan S, Messer AE, Steinen GJ, van der Velden J, Marston SB. Analysis of cardiac myosin binding protein-C phosphorylation in human heart muscle. J Mol Cell Cardiol. 2010;49(6):1003–11.
62. Jia W, Shaffer JF, Harris SP, Leary JA. Identification of novel protein kinase A phosphorylation sites in the M-domain of human and murine cardiac myosin binding protein-C using mass spectrometry analysis. J Proteome Res. 2010;9(4):1843–53.
63. Yuan C, Guo Y, Ravi R, Przyklenk K, Shilkofski N, Diez R, Cole RN, Murphy AM. Myosin binding protein C is differentially phosphorylated upon myocardial stunning in canine and rat hearts-- evidence for novel phosphorylation sites. Proteomics. 2006;6(14):4176–86.
64. Yuan C, Sheng Q, Tang H, Li Y, Zeng R, Solaro RJ. Quantitative comparison of sarcomeric phosphoproteomes of neonatal and adult rat hearts. Am J Physiol Heart Circ Physiol. 2008;295 (2):H647–56.
65. Sadayappan S, Gulick J, Osinska H, Martin LA, Hahn HS, Dorn GW 2nd, Klevitsky R, Seidman CE, Seidman JG, Robbins J. Cardiac myosin-binding protein-C phosphorylation and cardiac function. Circ Res. 2005;97(11):1156–63.
66. Yang Q, Hewett TE, Klevitsky R, Sanbe A, Wang X, Robbins J. PKA-dependent phosphorylation of cardiac myosin binding protein C in transgenic mice. Cardiovasc Res. 2001;51(1):80–8.
67. Sadayappan S, Osinska H, Klevitsky R, Lorenz JN, Sargent M, Molkentin JD, Seidman CE, Seidman JG, Robbins J. Cardiac myosin binding protein C phosphorylation is cardioprotective. Proc Natl Acad Sci U S A. 2006;103(45):16918–23.
68. Schlossarek S, Mearini G, Carrier L. Cardiac myosin-binding protein C in hypertrophic cardiomyopathy: mechanisms and therapeutic opportunities. J Mol Cell Cardiol. 2011;50 (4):613–20.
69. Kulikovskaya I, McClellan G, Flavigny J, Carrier L, Winegrad S. Effect of MyBP-C binding to actin on contractility in heart muscle. J Gen Physiol. 2003;122(6):761–74.

70. Colson BA, Locher MR, Bekyarova T, Patel JR, Fitzsimons DP, Irving TC, Moss RL. Differential roles of regulatory light chain and myosin binding protein-C phosphorylations in the modulation of cardiac force development. J Physiol. 2010;588(Pt 6):981–93.
71. Arber S, Halder G, Caroni P. Muscle LIM protein, a novel essential regulator of myogenesis, promotes myogenic differentiation. Cell. 1994;79(2):221–31.
72. Arber S, Hunter JJ, Ross J Jr, Hongo M, Sansig G, Borg J, Perriard JC, Chien KR, Caroni P. MLP-deficient mice exhibit a disruption of cardiac cytoarchitectural organization, dilated cardiomyopathy, and heart failure. Cell. 1997;88(3):393–403.
73. Bos JM, Poley RN, Ny M, Tester DJ, Xu X, Vatta M, Towbin JA, Gersh BJ, Ommen SR, Ackerman MJ. Genotype-phenotype relationships involving hypertrophic cardiomyopathy-associated mutations in titin, muscle LIM protein, and telethonin. Mol Genet Metab. 2006;88 (1):78–85.
74. Geier C, Perrot A, Ozcelik C, Binner P, Counsell D, Hoffmann K, Pilz B, Martiniak Y, Gehmlich K, van der Ven PF, Furst DO, Vornwald A, von Hodenberg E, Nurnberg P, Scheffold T, Dietz R, Osterziel KJ. Mutations in the human muscle LIM protein gene in families with hypertrophic cardiomyopathy. Circulation. 2003;107(10):1390–5.
75. Hershberger RE, Parks SB, Kushner JD, Li D, Ludwigsen S, Jakobs P, Nauman D, Burgess D, Partain J, Litt M. Coding sequence mutations identified in MYH7, TNNT2, SCN5A, CSRP3, LBD3, and TCAP from 313 patients with familial or idiopathic dilated cardiomyopathy. Clin Transl Sci. 2008;1(1):21–6.
76. Knöll R, Hoshijima M, Hoffman HM, Person V, Lorenzen-Schmidt I, Bang ML, Hayashi T, Shiga N, Yasukawa H, Schaper W, McKenna W, Yokoyama M, Schork NJ, Omens JH, McCulloch AD, Kimura A, Gregorio CC, Poller W, Schaper J, Schultheiss HP, Chien KR. The cardiac mechanical stretch sensor machinery involves a Z disc complex that is defective in a subset of human dilated cardiomyopathy. Cell. 2002;111(7):943–55.
77. Mohapatra B, Jimenez S, Lin JH, Bowles KR, Coveler KJ, Marx JG, Chrisco MA, Murphy RT, Lurie PR, Schwartz RJ, Elliott PM, Vatta M, McKenna W, Towbin JA, Bowles NE. Mutations in the muscle LIM protein and alpha-actinin-2 genes in dilated cardiomyopathy and endocardial fibroelastosis. Mol Genet Metab. 2003;80(1–2):207–15.
78. Su Z, Yao A, Zubair I, Sugishita K, Ritter M, Li F, Hunter JJ, Chien KR, Barry WH. Effects of deletion of muscle LIM protein on myocyte function. Am J Physiol Heart Circ Physiol. 2001;280(6):H2665–73.
79. Mery A, Taghli-Lamallem O, Clark KA, Beckerle MC, Wu X, Ocorr K, Bodmer R. The Drosophila muscle LIM protein, Mlp84B, is essential for cardiac function. J Exp Biol. 2008;211(Pt 1):15–23.
80. Boateng SY, Belin RJ, Geenen DL, Margulies KB, Martin JL, Hoshijima M, de Tombe PP, Russell B. Cardiac dysfunction and heart failure are associated with abnormalities in the subcellular distribution and amounts of oligomeric muscle LIM protein. Am J Physiol Heart Circ Physiol. 2007;292(1):H259–69.
81. Gupta MP, Samant SA, Smith SH, Shroff SG. HDAC4 and PCAF bind to cardiac sarcomeres and play a role in regulating myofilament contractile activity. J Biol Chem. 2008;283 (15):10135–46.
82. Heineke J, Ruetten H, Willenbockel C, Gross SC, Naguib M, Schaefer A, Kempf T, Hilfiker-Kleiner D, Caroni P, Kraft T, Kaiser RA, Molkentin JD, Drexler H, Wollert KC. Attenuation of cardiac remodeling after myocardial infarction by muscle LIM protein-calcineurin signaling at the sarcomeric Z-disc. Proc Natl Acad Sci U S A. 2005;102(5):1655–60.
83. Flick MJ, Konieczny SF. The muscle regulatory and structural protein MLP is a cytoskeletal binding partner of betaI-spectrin. J Cell Sci. 2000;113.(Pt 9:1553–64.
84. Lange S, Gehmlich K, Lun AS, Blondelle J, Hooper C, Dalton ND, Alvarez EA, Zhang X, Bang ML, Abassi YA, Dos Remedios CG, Peterson KL, Chen J, Ehler E. MLP and CARP are linked to chronic PKCalpha signalling in dilated cardiomyopathy. Nat Commun. 2016;7:12120.

85. Ehler E, Horowits R, Zuppinger C, Price RL, Perriard E, Leu M, Caroni P, Sussman M, Eppenberger HM, Perriard JC. Alterations at the intercalated disk associated with the absence of muscle lim protein. J Cell Biol. 2001;153(4):763–72.
86. Geier C, Gehmlich K, Ehler E, Hassfeld S, Perrot A, Hayess K, Cardim N, Wenzel K, Erdmann B, Krackhardt F, Posch MG, Osterziel KJ, Bublak A, Nagele H, Scheffold T, Dietz R, Chien KR, Spuler S, Furst DO, Nurnberg P, Ozcelik C. Beyond the sarcomere: CSRP3 mutations cause hypertrophic cardiomyopathy. Hum Mol Genet. 2008;17(18):2753–65.
87. Boateng SY, Senyo SE, Qi L, Goldspink PH, Russell B. Myocyte remodelling in response to hypertrophic stimuli requires nucleocytoplasmic shuttling of muscle LIM protein. J Mol Cell Cardiol. 2009;47(4):426–35.
88. Fung YW, Wang R, Liew CC. Characterization of a human cardiac gene which encodes for a LIM domain protein and is developmentally expressed in myocardial development. J Mol Cell Cardiol. 1996;28(6):1203–10.
89. Heineke J, Molkentin JD. Regulation of cardiac hypertrophy by intracellular signalling pathways. Nat Rev Mol Cell Biol. 2006;7(8):589–600.
90. Jeong D, Kim JM, Cha H, Oh JG, Park J, Yun SH, Ju ES, Jeon ES, Hajjar RJ, Park WJ. PICOT attenuates cardiac hypertrophy by disrupting calcineurin-NFAT signaling. Circ Res. 2008;102 (6):711–9.
91. Chang DF, Belaguli NS, Iyer D, Roberts WB, Wu SP, Dong XR, Marx JG, Moore MS, Beckerle MC, Majesky MW, Schwartz RJ. Cysteine-rich LIM-only proteins CRP1 and CRP2 are potent smooth muscle differentiation cofactors. Dev Cell. 2003;4(1):107–18.
92. Schmeichel KL, Beckerle MC. The LIM domain is a modular protein-binding interface. Cell. 1994;79(2):211–9.
93. Hallhuber M, Burkard N, Wu R, Buch MH, Engelhardt S, Hein L, Neyses L, Schuh K, Ritter O. Inhibition of nuclear import of calcineurin prevents myocardial hypertrophy. Circ Res. 2006;99(6):626–35.
94. Kong Y, Flick MJ, Kudla AJ, Konieczny SF. Muscle LIM protein promotes myogenesis by enhancing the activity of MyoD. Mol Cell Biol. 1997;17(8):4750–60.
95. Ecarnot-Laubriet A, De Luca K, Vandroux D, Moisant M, Bernard C, Assem M, Rochette L, Teyssier JR. Downregulation and nuclear relocation of MLP during the progression of right ventricular hypertrophy induced by chronic pressure overload. J Mol Cell Cardiol. 2000;32 (12):2385–95.
96. Buyandelger B, Mansfield C, Kostin S, Choi O, Roberts AM, Ware JS, Mazzarotto F, Pesce F, Buchan R, Isaacson RL, Vouffo J, Gunkel S, Knoll G, McSweeney SJ, Wei H, Perrot A, Pfeiffer C, Toliat MR, Ilieva K, Krysztofinska E, Lopez-Olaneta MM, Gomez-Salinero JM, Schmidt A, Ng KE, Teucher N, Chen J, Teichmann M, Eilers M, Haverkamp W, Regitz-Zagrosek V, Hasenfuss G, Braun T, Pennell DJ, Gould I, Barton PJ, Lara-Pezzi E, Schafer S, Hubner N, Felkin LE, O'Regan DP, Brand T, Milting H, Nurnberg P, Schneider MD, Prasad S, Petretto E, Knoll R. ZBTB17 (MIZ1) Is Important for the Cardiac Stress Response and a Novel Candidate Gene for Cardiomyopathy and Heart Failure. Circ Cardiovasc Genet. 2015;8 (5):643–52.
97. Wende AR. Post-translational modifications of the cardiac proteome in diabetes and heart failure. Proteomics Clin Appl. 2016;10(1):25–38.

Interrogating Cardiovascular Genetics in Zebrafish

9

Jiandong Liu, Marc Renz, and David Hassel

Contents

Emerged as a highly versatile and applicable vertebrate animal model to study embryonic development, zebrafish over the last two decades became a valuable human disease model for cardiovascular research. The unprecedented in vivo imaging capabilities allowing live analysis of organ formation and organ function combined with the ease of precise genetic interrogation established zebrafish as a recognized cardiovascular disease model. This chapter provides an overview of

J. Liu
McAllister Heart Institute, University of North Carolina at Chapel Hill, Chapel Hill, NC, USA

Department of Pathology and Laboratory Medicine, University of North Carolina at Chapel Hill, Chapel Hill, NC, USA

M. Renz · D. Hassel (✉)
Department of Medicine III, Cardiology, Angiology, Pneumology, University Hospital Heidelberg, Heidelberg, Germany

© Springer Nature Switzerland AG 2019
J. Erdmann, A. Moretti (eds.), *Genetic Causes of Cardiac Disease*, Cardiac and Vascular Biology 7, https://doi.org/10.1007/978-3-030-27371-2_9

zebrafish's history in biomedical and particularly cardiovascular research, delivers an outline of available methodologies and resources when using zebrafish, and gives examples of the versatility of zebrafish in cardiovascular research and how it progressed our understanding of a given disease.

9.1   From a Developmental Model to a Disease Model

9.1.1   The Beginnings and the Rise in Popularity

In the 1980s, Dr. George Streisinger and his colleagues at the University of Oregon laid the groundwork for using the zebrafish as a genetic model. He focused initially his pioneering work on developmental genetics based on several practical advantages and genetic amenities. The idea of having a vertebrate model with which genome-wide forward genetic screens are feasible as performed before in worms, flies, and plants to uncover novel gene functions ultimately materialized in 1989 was carried out by a colleague of Streisinger at the University of Oregon [1]. Inspired by the success and the potential of this first forward genetic screen, scientists from two groups in Boston and Tübingen conducted the first large-scale forward genetic screen in a vertebrate model. This screen recovered over 800 mutants with mutations in 372 genes uniquely involved in the development of the central nervous system, intestines, blood, bone, vasculature, and the heart to only name a few [2, 3]. Even over 20 years after conducting this screen, zebrafish mutants from it are still being used and analyzed in labs worldwide, illustrating the huge impact and the value this screen had and still has in the community.

Zebrafish are approximately 3–5-cm-long fish belonging to the *teleost* family with inexpensive housing and modest space requirements. When mated, a single female can produce between 100 and 300 eggs per week. Sexual maturity is reached with the age of 3 months. Particularly tempting as a developmental model is its extrauterine development and its fast ontogeny. Its small larval size of 1–2 mm allows analysis throughout the entire development, while macroscopic cell experiments like cell transplantations remain feasible and efficient. Gastrulation begins at around 5 h postfertilization (hpf) and segmentation, and with that somitogenesis, by 10 hpf (Table 9.1) [4]. Embryonic vasculogenesis is completed by 18 hpf, and angiogenesis can be observed before 20 hpf, and the first primitive heart tube initiates first contractions within the first day of development. Already 3 days postfertilization (dpf), the cardiovascular system of the zebrafish is comparable to that of a newborn mammal and reaches full maturity by 5 dpf, in a by-then organism exhibiting most of the internal organs found in humans [5]. In contrast to mammalian models, zebrafish does not rely on a functioning cardiovascular system during the first 5 days of life, since oxygen and nutrient supply is achieved exclusively through passive diffusion. Even substantial cardiovascular deficiencies can thus be tolerated and fish, even without any observable heart activity or blood flow, develop overly normal without systemic compromising impact due to secondary adverse effects within the first 5 dpf [6].

Table 9.1 Comparison of timepoints when key cardiovascular developmental events happen in zebrafish, mouse, and humans

	Zebrafish	Mouse	Human
Gastrulation	5 hpf	E6.5	17 days
Somitogenesis	10 hpf	E7–E7.5	17–18 days
Cardiac progenitors migrate and fuse at the midline	12–18 hpf	E8	19 days
First contractions of the primitive heart tube	21–22 hpf	E8.5	22 days (3 weeks)
Heart chambers visible	30 hpf	E8.5	22 days (3 weeks)
Looping	30–36 hpf	E8.5	22 days (3 weeks)
Valve formation	45 hpf	E9	30 days
Ventricular septation	–	E15	90 days
Atrial septation	–	After birth (day 21)	After birth (appr. 40 weeks)
Cardiovascular maturity	5 dpf	4 weeks	8–12 years
Vasculogenesis-primitive plexus formation	20 hpf	E8–8.5	1st trimester (up to week 11)

E embryonic day

9.1.2 The Uprising Disease Model

Not only because of these obvious advantages but also with the 1996 mutagenesis screens and the recovery of mutant alleles modeling known human diseases, the zebrafish began its journey toward challenging the till-then dominating mouse model in recapitulating human disease conditions. As a nonmammalian model though, the zebrafish comes with apparent limitation, the most obvious being that zebrafish lack some of the organs found in humans, like lungs and mammary gland. The first zebrafish mutant reported modeling a human pathology and contributed to a better understanding of the underlying pathomechanism was *yquem* (*ype*) [7]. Zebrafish and humans share remarkable conservation in hematopoiesis, and *ype* presented a photosensitive porphyria syndrome as seen in patients. Although porphyria in humans was already associated to a gene encoding uroporphyrinogen decarboxylase (UROD), the study of *ype* now explained the exact enzymatic defect leading to the disease also in humans [7].

Subsequently, publications on zebrafish modeling various human pathologies dramatically increased. While modeling blood disorders in zebrafish dominated initially, zebrafish models for human diseases ranging from ocular disorders and neurodegenerative diseases to muscular dystrophies, cancer, addiction, behavioral disorders, and cardiovascular diseases have been described [8–10]. In particular, concerning congenital heart malformations, like valve deficiencies, to aneurisms, cardiomyopathies, and arrhythmias, zebrafish is now a well-established cardiovascular disease model. Although having a two-chambered heart instead of a four-chambered heart as in mammals, the zebrafish was proved to be particularly valuable for cardiac electropathophysiology (see section below) as compared to mice, since zebrafish action potential shape, heart rate, and the electrocardiogram closely resemble that of humans [11].

9.2   Genetics and Genetic Interrogation in Zebrafish

The ease of performing even large-scale genetic studies in zebrafish, which in the past also contributed significantly to a more detailed understanding of genetics in human disease, makes a high-quality genome sequence and complete annotation essential. The zebrafish is now further frequently used as an independent functional validation tool for findings from human genetics studies [12]. A recent study, initiated in 2001 by the Wellcome Sanger Trust Institute, identified and annotated 26,000 individual genes encoded on the zebrafish's 25 chromosomes compared to the roughly 20,500 genes in the human genome [13]. Sequencing additionally revealed that teleosts underwent a whole-genome duplication, with 25–30% of the duplicated genes still retained in zebrafish [13, 14]. Thus, frequently duplicated genes are found when screening for human disease gene orthologues, adding potentially to a functional validation experimentation since duplicated genes often exhibit subfunctionalization properties. Seventy percent of the human genes have a direct homologue in the zebrafish. More astounding is that 84% of the known human disease genes have a direct homologue in fish [13].

Along with this solid foundation, the zebrafish comes with a highly versatile tool set to interrogate gene functions and identify genes with so far unappreciated functions. As traditionally introduced and envisioned by its pioneers, the zebrafish was and is a powerful forward genetics model.

9.2.1   Forward Genetics

Classical chemical mutagenesis screens, in particular deploying N-ethyl-N-nitrosourea (ENU), are easy and unprecedented in zebrafish compared to any other vertebrate model and enable production of thousands of mutants in a relatively short time (reviewed here [15]). While being easy and producing mutations very efficiently with little genomic bias, chemical mutagenesis screens have the disadvantage that subsequent identification of the gene mutation causative for the observed phenotype is time- and resource-consuming. To overcome this limitation, insertional mutagenesis techniques using replication-deficient pseudotyped retroviruses or transposons were introduced by several groups [16–18]. In addition, and initially adapted from mouse, the *Sleeping Beauty* DNA transposon somatic mutagenesis system was successfully introduced in zebrafish and led to the identification of conserved and novel human cancer genes in zebrafish [19]. Insertional mutagenesis, due to its predesigned insertional cassettes, including reporters, enhancers, silencers, and amplification tags, allows rapid identification of the targeted gene. However, this comes with the price that retroviruses and transposons typically achieve a very much lower mutagenic efficiency than ENU, requiring considerably larger libraries of mutagenized fish.

Using sensitized genetic backgrounds to perform modifier screens is a powerful way to isolate genetic interactions of genes. Since screening of several genomes and large libraries of mutagenized animals is necessary, this was longtime reserved for invertebrate models, including *Drosophila* and *C. elegans*. Kramer et al. performed the first enhancer screen for modifiers affecting dorsoventral patterning by using heterozygous zebrafish mutants for the *chordin* gene and identified initially seven enhancer mutations [20]. Bai et al. screened for genetic suppressors of a mutation in the *moonshine* (*mon*) gene, which encodes for the transcriptional intermediary factor 1 gamma (*tf1γ*) [21]. *mon* mutants exhibit anemia and larval lethality. They utilized a complex haploid screening strategy and identified the *sunrise* mutation that fully normalized globin content in *mon* mutants. Interestingly, sunrise encodes for cdc73, a component of the Pol II transcriptional elongation complex, for the first time linking *tf1γ* function and transcriptional elongation to erythroid development.

9.2.2 Reverse Genetics

The so far numerous forward screens performed in zebrafish to date, despite the significant time, staff, and resource efforts, have led to the identification of plentiful new genes and conserved pathways involved in human disease, including cardio-vascular disease [22–24]. However, not only because zebrafish became a valuable model to functionally validate candidate genes but also because of advances in sequencing and other large-scale gene identification technologies, including high-throughput automated gene expression analysis, i.e., in situ hybridization screenings, reverse genetic tools to interrogate the function of distinct genes have been established [25, 26]. Development of available technologies adapted to zebrafish research accelerated dramatically during recent years (Table 9.2).

First reverse genetics in zebrafish was performed based on technologies that were well established and combined with advanced sequencing leading to targeting induced local lesions in genomes, or TILLING for short [28]. In TILLING, libraries of ENU-mutagenized F1 families are sequenced and sperm cryopreserved. Sperm can be revitalized by in vitro fertilization when a locus of interest is affected by an induced mutation. Hundreds and thousands of cryopreserved sperm from TILLING efforts are stored in huge facilities and catalogued. They are available to the research community through the Zebrafish International Research Center (ZIRC, www.zebrafish.org) at the University of Oregon, USA, and the European Zebrafish Research Center (EZRC, http://www.ezrc.kit.edu) at the Karlsruhe Institute for Technology (KIT), Germany [29].

With the introduction of morpholino-modified antisense oligonucleotides (MO) adapted from *Xenopus,* an easy, fast, efficient, and specific gene silencing tool when well controlled [30] to analyze the function of a distinct gene became available [31]. MOs do not target the genomic locus of a gene but rather bind to the transcript to cause aberrant translation or splicing. They do so either by targeting the

Table 9.2 Overview of available reverse genetics tools

Reverse genetic tools	Mechanism	Pros	Cons	Resources
Morpholino-mediated gene knockdown	Synthesized oligonucleotide analogs disrupt splicing or attenuate translation	Fast, convenient, modified caged MO can be used for conditional gene knockdown, rapid identification of affected gene	Not inheritable, effective within first 3–5 days of development, prone to off-target effect	Gene tools, http://www.gene-tools.com
Tol2-mediated large genomic deletions	Transposon element Tol2 integrates into the genome, causing gene disruption	Integration occurs randomly and efficiently, inheritable, huge acceptance in the community, excellent construct availability	Requires large space and is labor-intensive, tends to have hot spots	Wiki and repository: http://www.tol2kit.genetics.utah.edu, constructs available through Addgene
Retrovirus-mediated insertional mutations	Retroviral element integrates into the genome, causing gene disruption	Integration occurs randomly and efficiently, inheritable, rapid identification of affected gene	Same as above	N/A
ENU-mediated targeting induced local lesions in genomes (TILLING)	ENU mutagenesis followed by sequencing genomic region of interest	Generates a diverse range of mutant alleles for genetic analysis	Same as above	http://www.sanger.ac.uk/Projects/D_rerio/zmp/
ZFN-mediated gene targeting	Zinc finger nucleases-mediated DNA double-strand break	Generates small deletions or insertions at desired genomic region of interest	Constructing ZFNs targeting desired target sites is time- and labor-consuming; generation of stable mutants takes up to 12 months	N/A
TALEN-mediated gene targeting	TALEN nucleases-mediated DNA double-strand break	Same as above	Off-target effects; generation of stable mutants takes up to 12 months	N/A

(continued)

Table 9.2 (continued)

Reverse genetic tools	Mechanism	Pros	Cons	Resources
CRISPR/Cas9-mediated gene targeting	CRISPR/Cas9-mediated DNA double-strand break	Same as above, and the most versatile gene targeting method	Generation of a stable mutant takes up to 12 months; genetic compensation reported [27]	N/A

translational start site to block translation completely or by targeting splice-site junctions in the pre-mRNA causing either exon skipping or intron integration resulting in frameshift and premature stop codon generation [31, 32]. While translation blocking MOs require an antibody specific for the targeted protein to assess efficacy, the efficiency of disruption of splicing by the splice-site MOs can be easily evaluated by RT-PCR. The development of photoactivatable MOs (caged MOs) helped to overcome the limitation that MOs act globally and constitutively [33]. Caged MO activity can be temporarily and even spatially controlled by applying cellular resolution light, i.e., using two-photon microscopes [34]. MOs proved to be a versatile tool not only to inhibit gene function but also to confirm genetic interactions of genes when titrated and co-injected, for gene dosage experiments, to suppress miRNAs by targeting miRNA maturation or for miRNA target validation by interfering with miRNA target $3'$-UTR binding [35–37]. Despite being extremely valuable, MOs come with the disadvantage of having transient effects because they do not cause genetic lesions, allowing gene knockdown only for about 3 days [31]. Further, MO experiments require well-trained and experienced staff to control for potential off-target effects and toxicity [30, 38]. Recently, discrepancies between the observed phenotypes induced by MOs and stable mutants were reported, which can be in part explained by genetic compensation mechanisms occurring in mutants but not when the gene is transiently inactivated as in morphants, possibly thereby obscuring phenotypes otherwise visible after MO use [27, 39].

The latest development of advanced genome-editing technologies complements the existing reverse genetics tools in zebrafish. First, zinc finger nucleases (ZFN), then TALE nucleases, and ultimately CRISPR/Cas9 have been shown to be highly effective in zebrafish, and particularly, CRISPR enables cheap and relatively effortless generation of mutants in virtually every lab [40–45]. Narayanan et al. recently demonstrated the feasibility of simultaneous in vivo mutagenesis of entire gene and miRNA gene families by multiplex CRISPR/Cas9, highlighting the versatility of this system [46]. Multiplexed CRISPR/Cas9 was subsequently employed to demonstrate that miRNAs provide phenotypic robustness and protect the vascular systems from changing environmental conditions [47].

Besides inducing simple lesions, CRISPR/Cas9 enables targeted genome editing allowing introduction and correction of distinct mutations, integration of exogenous sequences like LoxP, and even introduction of reporter genes, including GFP [48]. However, besides the very well and efficiently performing NHEJ (non-homologues end-joining), HDR (homology-directed repair) seems to work rather inefficient in zebrafish [48, 49].

As in other animal models, transgenic strategies are another valuable reverse genetics tool to elucidate gene function. Either wild-type forms or constitutive active or dominant-negative isoforms of the gene of interest can be transiently, by simple RNA injections, or stably, by stably integrated transgenesis, expressed. Generation of a stable transgenic line takes between 6 and 12 months. Here, also a diverse tool set, including GAL4/UAS- and Cre/LoxP-based systems, and a diverse set of promoters, including cell-type-specific promoter elements or tamoxifen and heat shock-responsive elements to drive spatially and temporally controlled expression of the transgene, are available [50].

9.2.3 Genetic Diagnostic Testing Using Zebrafish

Current large-scale genomic sequencing studies of human patients have identified many disease-associated genetic variants at an accelerating pace and at ever-decreasing costs. One of the major challenges that we are facing with all these genomic data is to determine the pathogenicity of newly identified candidate disease genes and genetic variants. This is particularly true for the ones that have not been previously evaluated in an experimental model system. Owing to the abovementioned advantages as a vertebrate model system, zebrafish has emerged as an attractive model to determine the cause-effect relationship between the genetic variants and the human defects and the underlying molecular disease mechanisms.

Giving the versatile genetic tool set in zebrafish, the ability of genetic diagnostic testing ranges from simple gene causality and single variant testing to the identification and characterization of more complex disease genetics. We will focus here on how the zebrafish can be used to test for a variety of genetic testing scenarios; distinct examples can be found in the disease entity sections below.

(I) Using transient MO-mediated knockdown or stable CRISPR/Cas9 knockouts as a tool to identify disease causality of a gene from single genes or from a set of candidates isolated through, i.e., GWAS, enables simple and fast causality screening. Gene candidates mapped for disorders segregating under a recessive mode of inheritance can be easily and rapidly tested. Phenocopy of large parts of the phenotypic spectrum found in the respective patient cohort allows for identification and functional verification of identified null variants with predicted nonsense, frameshift, or splice-site mutations. **(II)** The power of zebrafish is the ability to transiently overexpress genes or gene variants by injection of in vitro generated RNA carrying the desired mutation. This allows very specific in vivo variant testing. In vivo complementation experiments by combining mRNA injection of transcripts carrying the to-be tested mutation with MO-mediated knockdown or stable knockout result in more specific mutation-focused testing. The loss-of-function effect of the variant is proven by the inability to rescue the induced gene loss. This method is very helpful to identify null variants in a scenario of recessively inherited disorder. Injection of variants into wild-type animals enables identification of dominant-negative acting gene mutations. If small amounts of the mutant variant cause a disease phenotype, while injection of equal amounts of the wild-type variant has no effect, this indicates toxicity of the mutant allele and hence dominant-negative mode of action [51]. (**III**) The ease to generate stable mutants in single or multiple genes at once and the availability of a huge variety of mutant alleles through international resource centers as mentioned above enable even the unraveling of more complex genetic traits in vivo. Combining mutant alleles by multiplex targeting through CRISPR or compound mutation generation by crossing of different mutant lines allows demonstration of genetic interactors and modifiers of phenotypic traits [47, 52]. Convenient is the use of MOs, since here the amount of gene reduction can be titrated. Combining carefully titrated MOs targeting different specific genes can thus be used to prove genetic interaction and disease modifiers in an in vivo setting with little time effort [35].

The applicability of zebrafish in the diverse genetic scenarios is summarized in Table 9.3.

9.3 Cardiovascular Disease Models

9.3.1 Congenital Heart Disease Modeling in Zebrafish

Congenital heart diseases (CHDs) are the leading cause of human birth defects, and many of these diseases originate from genetic defects that impact cardiac development and morphogenesis [58, 59]. During development, the heart is the first organ to form and function. After its initial formation at the ventral midline, the vertebrate heart undergoes a series of complex morphogenetic processes that transform the linear heart tube into a functional pumping organ (Fig. 9.1a–g) [5, 63, 64]. Although the zebrafish heart appears to be structural simpler than its counterpart in higher vertebrates, much of the molecular and cellular mechanisms governing cardiac formation and further maturation are conserved between zebrafish and higher vertebrates [5, 23, 60, 64–66].

CHDs in humans feature a wide variety of structural abnormalities that affect nearly every structure of the developing heart including the myocardium, the valves, and the great vessels [59]. If left uncorrected, in severe cases, the structural defects will lead to cardiac dysfunction and thus negatively impact life quality of the affected individual. In zebrafish, myocardial wall abnormalities manifest as either hypoplastic ventricle or hypotrabeculation, both of which cause progressive reduction in cardiac contractility [67]. Cardiac valve malformations often arise from defective endocardial cell migration and/or remodeling, leading to improper endocardial cushion and/or valve leaflet formation [23, 68, 69]. These defects can result in blood regurgitation and altered intracardiac fluid dynamics [68, 69]. Other cardiovascular defects in zebrafish include heterotaxy whereby the relative position of the zebrafish cardiac chambers is altered [70, 71], and malformations of the pharyngeal arch arteries (PAAs), the transient embryonic blood vessels that eventually give rise to the carotid arteries and great vessels of the heart [72, 73].

Table 9.3 Overview of usage scenarios for genetic diagnostic testing using zebrafish and the available methods

Query	Approach options	Findings
Disease causality testing of a single candidate gene Identification of a disease gene from a set of genes, i.e., by GWAS	*MO* (weeks) *or stable KO, CRISPR/TALENS* (months) *MO* (weeks) *or stable KO, CRISPR/TALENS* (months) combined with *transgenic sensor/marker line*	– Identification and validation of a disease gene [51, 53, 54] – Complete phenocopy proves loss-of-function, null variant effect or recessive mode of inheritance [51, 53, 54] – Phenotype in heterozygotes or MO titration proves haploinsufficiency [55] – Isolation of distinct effects and on distinct cell types/tissues/ biological processes [56, 57]
Pathogenicity testing of single identified mutations	WT/variant *RNA injection combined with MO KD or stable KO*	– Inability to rescue with variant mRNA *proves pathogenicity* [57] – Could indicate *recessively inherited* disorder
	WT/variant *RNA injection in WT*	– Development of phenotype with variant injection *proves pathogenicity* [51, 57] – Indicates *dominant-negative* effect of variant [51, 57]
Complex Trait Testing (polygenic evidence) *or* Candidate gene/ signaling network *or* Disease modifier testing	*Multiple mutant/heterozygotes* generation by crossing different mutant lines [52] *Sensitization of a het line* by introduction of a second het mutation [52] *Multiplex CRISPR* targeting [47] *MO combination* experiment [35] *Sensitization* by using *multiple titrated MOs* [35] *Rescue of MO KD* by injecting genetic *interactor or* downstream *affector RNA*	– *Polygenetic cause for a disease* can be proved [52] – Proof for *genetic interaction* [35, 47, 52] – Causation testing of *gene networks* [35, 47, 52] – Identification of *disease modifier* [35, 47, 52] – *Risk factor stratification* [47]

GWAS genome-wide association study, *KD* knockdown, *KO* knockout, *MO* morpholino, *WT* wild type. *Citations provide examples for distinct genetic testing scenarios. The list of cited articles is not exhaustive*

Hypoplastic left heart syndrome (HLHS) is a rare but complex CHD characterized by severe hypoplastic left ventricle and extremely underdeveloped mitral and aortic valves [74, 75]. These severe structural defects render the left side of the heart less efficient in pumping blood to the body, and surgical corrections are needed to improve the survival of HLHS infants [76]. HLHS usually occurs as isolated disease and the associated genetic loci include, among others, those

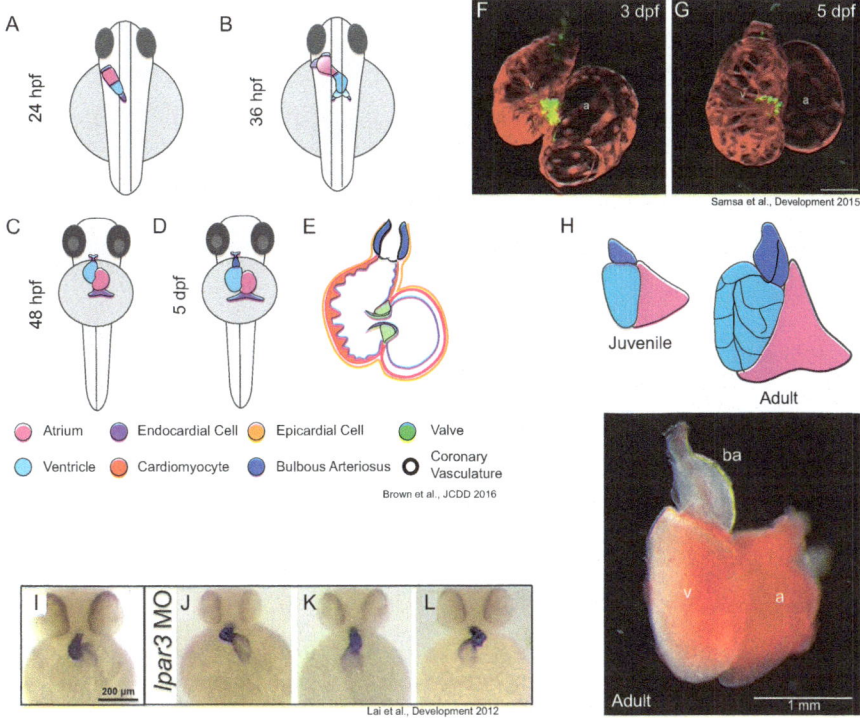

Fig. 9.1 The embryonic and adult zebrafish hearts. (**a**) The zebrafish linear heart tube at 24 hpf. (**b**) The looping zebrafish linear heart tube at 36 hpf. (**c**) The two-chambered zebrafish heart at 48 hpf. (**d**) The zebrafish heart grows in size by 5 dpf. (**e**) Cross-sectional view of 5 dpf zebrafish heart showing ventricular protruding trabeculae, cardiac valves, and the outmost epicardial layer. Modified from Brown et al. JCC 2016 [60]. (**f, g**) Confocal projections of 3 and 5 dpf zebrafish hearts, respectively. From Samsa et al., Development 2015 [61]. (**h**) Schematics of juvenile and adult zebrafish hearts at the top and an isolated adult heart at the bottom. (**i–l**) Control hearts (**i**) and hearts with left-right patterning defects (**j–l**) stained by in situ hybridization labeling the cardiac-specific myosin light chain 2. From Lai et al. Development 2012 [62]

encoding the transcriptional factors NKX2.5 and HAND1 and the gap junction protein GJA1 [77–79]. Interestingly, recent studies in zebrafish revealed an important role for Nkx2.5 and its homologue Nkx2.7 in chamber identity maintenance [80, 81]. Although the numbers of differentiated ventricular and atrial cardiomyocytes (CMs) are not affected by the loss of *nkx2.5* function at the linear heart tube stage, zebrafish *nkx2.5* mutant demonstrates a nearly 50% decrease in the number of ventricular CMs with a corresponding increase in the number of atrial CMs starting from cardiac looping stage. These *nkx2.5* mutant phenotypes are further enhanced by a loss of *nkx2.7* function. These observations suggest that these two Nkx factors play critical role in maintaining ventricular and atrial cell identity during the time of cardiac loop and chamber emergence [80, 81]. Although no studies have reported an enlargement of atria in HLHS human patients, studies

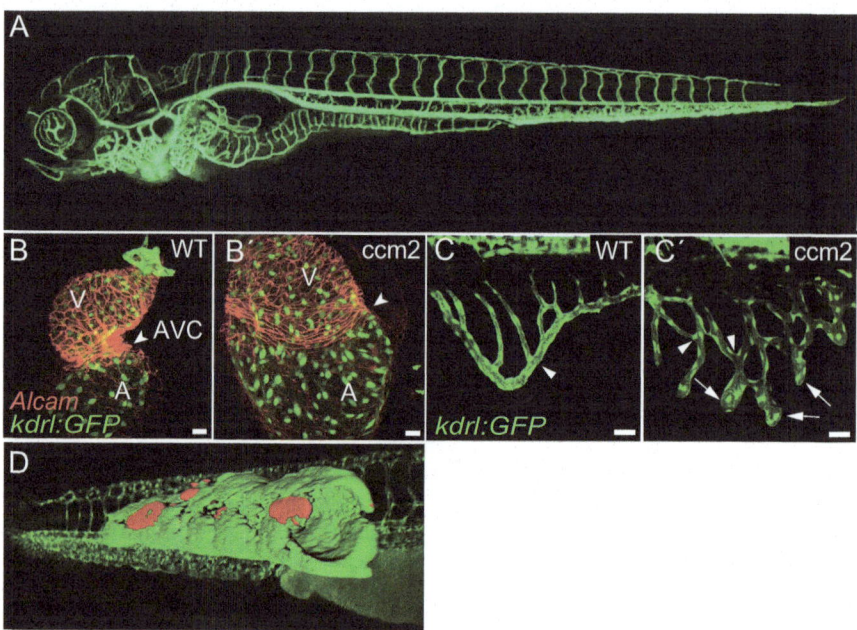

Fig. 9.2 Zebrafish as a vascular model. (**a**) Vasculature of a transgenic Tg(*kdrl:EGFP*) embryo at 5dpf. (**b**, **c**) Cardiovascular defects in zebrafish ccm2 mutants, characterized by endocardial overproliferation (**b′**) and increased sprout- and branch point formation (**c′**) at 48hpf. From Renz et al. [83] (**d**) Zebrafish as a model for atherosclerosis. Lipid deposits labeled by fluorescent lipids in the vascular wall of Tg(*kdrl:EGFP*) transgenic animal [84]

using zebrafish might still provide mechanistic clue as to why HLHS patients develop miniature left ventricles. Combing mouse forward genetics and further validation by CRISPR/Cas9 gene editing in mouse and zebrafish, a recent study showed that Sap130, a histone deacetylase complex subunit, mediates left ventricular hypoplasia, further demonstrating the value of zebrafish as a powerful genetic system to model HLHS [82].

Cardiac valves are critical components in the heart that function to ensure unidirectional blood flow (Fig. 9.2e). Human patients with congenital valve defects would present either obstructed or backward blood flow due to valvular stenosis and valvular insufficiency, respectively [85–87]. In zebrafish, the atrioventricular (AV) valve that forms at the AV junction is the most well-studied valvular structure [23, 68, 69, 88, 89]. At 48 hours postfertilization (hpf), as the AV junction constricts and demarcates the two cardiac chambers, the endocardial cells at the junction begin to differentiate and form a specialized ringlike structure, the AV canal, to help reduce retrograde blood flow [68, 69]. Around 72 hpf, the AV endocardial cells further invaginate and generate primitive valve leaflets allowing for complete prevention of regurgitation [69]. Due to the conserved molecular mechanisms underlying valve formation between zebrafish and higher vertebrates, zebrafish has been utilized to evaluate the potential in vivo effect of genetic variants associated with human congenital valve defects. By sequencing 32 candidate genes implicated in atrioventricular septum (AVS) development, a recent study identified 11 detrimental genetic variants associated with AVS defects [57]. Further functional analysis demonstrated that expression of a patient-specific ALK2 L343P variant in zebrafish embryos led to aberrant expression of endocardial cushion marker genes and disrupted AV canal formation. More recently, Durst et al. identified *DCHS1* missense mutations as an inherited risk factor for mitral valve prolapse (MVP) [53]. *DCHS1* encodes a member of the cadherin superfamily involved in cell polarity. MO-mediated knockdown of the zebrafish homologue *dchs1b* compromised AV canal formation and resulted in retrograde blood flow [53]. Intriguingly, the wild-type human *DCHS1*, but not the two *DCHS1 variants* (P197L, R2513H), *rescued the* AV canal defects caused by *dchs1b* knockdown in zebrafish embryos, further supporting the link between the familiar mutations with MVP [53].

Heterotaxy is a rare birth defect featuring randomization of visceral organ situs [90]. Although situs inversus leads to little or no medical problems, heterotaxy often causes structural abnormalities in the heart and/or other major visceral organs. The heart is the first organ that responds to laterality signals and thus exhibits left-right asymmetry (Fig. 9.1h–k). In addition to the positioning defect, other cardiac defects seen in heterotaxy patients are transposition of the great arteries and septal defect [91–93]. Because of an essential role of the motile cilia in left-right patterning, genetic mutations that affect motile cilia formation and/or function and the downstream nodal signaling that controls left-right patterning result in heterotaxy and cardiac structural abnormalities [94]. To date, targeted candidate exome sequencing has identified many human mutations in ciliary dynein components, such as dynein intermediate chain 1 (DNAI1) [95, 96], dynein heavy chain 5 (DANH5) [95], dynein heavy chain 6 (*DNAH6*) [97], and left-right dynein (DNAH11) [96], in heterotaxy patients. Selection of the candidate genes was based on genetic data obtained from model organisms, in particular *Chlamydomonas* mutants with the defective structure and/or function of flagella [98]. Intriguingly, recent human genetic studies have identified novel heterotaxy-causing genes, some of which have unanticipated roles in cilia structure and function or potentially affect the downstream signaling controlling left-right patterning [99].

9.3.2 Cardiomyopathy Modeling in Zebrafish

In 2002, with the zebrafish mutants *silent heart* (*sih*) and *pickwick* (*pik*) harboring mutations in cardiac troponin and titin, respectively, rendering the hearts a contractile and recapitulating human cardiomyopathy, the usefulness of zebrafish as a cardiomyopathy model became obvious [100, 101]. Both genes are associated with human hypertrophic (HCM) and dilated cardiomyopathy (DCM) [102, 103]. In subsequent years, many more cardiomyopathy-causing genes were evaluated using the zebrafish. Roughly 96% of all known human cardiomyopathy-causing genes have a direct zebrafish orthologue, supporting the usefulness of zebrafish [104]. The zebrafish even helped to identify novel DCM genes, i.e.,

Nexilin, which was first identified as a novel component of the sarcomeric z-disk in zebrafish and then confirmed to be a molecular base of human DCM by a genetic screening of patients with idiopathic DCM [51]. Strikingly, even the ultrastructural characteristics of Nexilin loss described in zebrafish were mimicked in human Nexilin mutation carriers. Subsequently, Nexilin was also linked to human HCM [105]. Vogel et al. systematically evaluated the usefulness of zebrafish as a disease animal model for human DCM [106]. They knocked down a diverse set of known human DCM genes and evaluated the resulting phenotypes. Inactivation of all genes not only reliably caused DCM, but zebrafish morphants even recapitulated gene-specific disease characteristics, highlighting the value of zebrafish as a diagnostic model to evaluate human cardiomyopathy genetics. Genes evaluated in zebrafish quickly expanded from merely sarcomeric proteins to signaling molecules, transcription factors, and cytoskeletal and other regulatory proteins [106–111]. Zebrafish models contributed dramatically to novel mechanistic insights of human cardiomyopathy, as exemplified by a study from Zou et al. The authors used CRISPR technology to generate distinct titin mutants, leading to the identification of a human conserved titin internal promoter that can rescue favorably mutations residing in the N-terminus of the protein. This renders N-terminal mutations less severe than mutations affecting the C-terminus and might serve as an explanation for the observed variability in expressivity and severity in human titin mutation carriers [112].

Besides larval zebrafish, adults recently are used to model cardiomyopathy. Human cardiomyopathy usually affects adult individuals with disease mutations predominantly being recessively inherited. While previously homozygous zebrafish mutants with manifestation of disease during larval developmental stages were predominantly analyzed, adult zebrafish might represent a model closer modeling the human pathology. The adult zebrafish heart consists of all cell types found in the human heart, including cardiomyocytes, fibroblasts and endothelial cells, while the larval heart mainly contains cardiomyocytes. Common pathways involved in disease progression in human heart disease, including the β-adrenergic system, were shown to function more similarly in the adult than in the larval heart [113]. A few adult cardiomyopathy models exist that contributed significantly to a deeper understanding of disease progression and potential therapeutic opportunities [114–117]. A recent study aimed at understanding the basis for the heterogeneity of observed phenotypes in essential myosin light chain (MYL3) mutation carriers, which can be observed even in patients carrying the same mutation [118]. The authors used adult, heterozygous mutant fish, mimicking human dominant mutation carriers. Combining functional analyses with molecular and biochemical assays, the study provides evidence that MYL3 is essential to adapt heart function to physical stress. Heterozygous mutant fish developed cardiomyopathy and increased lethality only after exposure to forced swimming while being unremarkable under normal conditions. Variability in expressivity and severity of MYL3-associated cardiomyopathy in humans hence depends on the degree of the stress the heart experiences during lifetime, and controlling cardiovascular stress might represent a possibility to reduce disease symptoms [118].

While being valuable to more closely resemble human cardiomyopathy, using adult zebrafish comes with disadvantages. The biggest disadvantage is time, since generating stable mutants is significantly more time-consuming than MO-based larval studies. Further, analysis of heart function in larvae is very easy and can be performed by using a simple light microscope. Because adult zebrafish lack transparency, more advanced imaging modalities have to be used, demanding higher costs and highly trained personnel. In accordance to mammalian studies, echocardiography was adapted to analyze heart function in adult zebrafish, now allowing the assessment of conventional echocardiographic parameters to advanced analysis of detailed myocardial mechanics deploying modern speckle tracking [118–120]. And last but not least, the ability of using larval zebrafish to perform high-throughput screens (see section below) is unprecedented and is impossible to match with adult zebrafish. Nevertheless, adult zebrafish cardiomyopathy models are rising and will more and more complement larval studies in the future.

9.3.3 Arrhythmias

While murine hearts beat with up to 900 beats per minute (bpm), zebrafish hearts contract with a frequency of 60–100 bpm, thus much more closely resembling heart rates of humans. This resemblance is furthermore evident in the very similar form and kinetic of a zebrafish ventricular action potential and the echocardiogram [11]. Drugs targeting late repolarization channels induce confound arrhythmias in zebrafish while being only marginally functional in mice. Although being very small, the larval zebrafish heart is able to present a variety of different arrhythmic flavors, including tachycardia, bradycardia, atrioventricular block (AV block), atrial fibrillation, and sinus exit block. First evidence for the applicability of zebrafish as an arrhythmia model came from the *island beat* (*isl*) mutant in 2001, harboring a mutation in the cardiac L-type calcium channel [121]. *isl* larvae display a noncontractile ventricle and an asynchronously contracting atrium, resembling atrial fibrillation. *Breakdance* mutant zebrafish (*bre*), carrying loss-of-function mutations in the zebrafish orthologue of the human HERG channel, model long-QT syndrome 2 (LQT2) [122, 123]. *bre* hearts develop a second-degree AV block, with the ventricle skipping every other beat of the atrium (2,1 rhythm). Using a morpholino targeting *kcnh6* or high dosages of HERG-blocking drugs, such as terfenadine or E-4031, enables to induce a 3:1 and 4:1 and even up to a third-degree AV block. *bre* fish are very useful to screen for modifiers of LQT2 in vivo [124, 125]. Besides LQT, with *reggae* (*reg*) mutant zebrafish, one of the first animal models for the human short-QT syndrome 1 (SQT1) was introduced [126]. *reg* mutant fish carry a gain-of-function mutation in the zebrafish HERG channel and display a whole range of phenotypes, including complete cessation of contractility over up to hours, to atrial fibrillation and sinus exit block. A sodium-calcium exchanger (NCX1) deficiency was reported for the *tremblor* (*tre*) mutant that displays chaotic and dyssynchronized cardiac contractions and atrial fibrillation due to abnormal calcium transients [127]. A second *tre* mutant, besides being important for rhythmicity, indicated a role for NCX1 for normal cardiac development and sarcomere formation, since this mutant displayed malformed hearts and severely disrupted sarcomeres [128].

A direct implication for human genetics came from a zebrafish transgenic line that expressed a mutation in the SCN5A sodium channel frequently associated with conduction disease, sinus node dysfunction, atrial and ventricular arrhythmias, and dilated cardiomyopathy in patients [129]. Mutant SCN5A expressing zebrafish developed bradycardia, conduction system abnormalities, and premature death, suggesting conserved functions in the heart and demonstrating the usefulness of zebrafish as an in vivo screening model to distinguish benign from functional genetic variants found in humans with arrhythmias [129].

9.3.4 Vascular Disease

Zebrafish offers the ability to choose from a diverse range of transgenic reporter lines labeling vascular endothelial cells, lymphatic endothelial cells, or vascular smooth muscle cells and pericytes [68, 130–134]. Combined with its exceptional imaging capabilities, this facilitates unprecedented visualization of organ formation and malformation as well as vascular function and malfunction in vivo (Fig. 9.2a). Studies in zebrafish have contributed greatly to advance our understanding of vascular biology and disease. The prominent role of Notch signaling for arterial-venous specification and its function as a determining factor to select tip and stalk cell fates during angiogenesis was first described in zebrafish [135, 136]. Further, our views of early formation of the first embryonic artery and vein were reinvented with observations in zebrafish that the first embryonic vein forms by selective sprouting of progenitor cells from a common arterial and venous precursor vessel subsequently undergoing fate segregation regulated through the ligand EphrinB2 and its receptor EphB4 [137].

Defects in vascular integrity and resulting hemorrhage were among the first pathologies described in zebrafish, advancing our understanding of factors essential for vessel integrity and of genes involved in human disease conditions involving hemorrhage formation [24, 138–145]. The cerebral cavernous malformation (CCM) protein complex plays a crucial role for normal blood vessel development and vascular integrity [146]. CCMs are vascular lesions characterized by enlarged thin-walled blood vessels and lack of supporting subendothelial cells such as smooth muscle or astrocytic foot processes. In patients, CCMs are primarily found within the neurovasculature of the central nervous system and often cause headaches, seizure, or often lethal cerebral hemorrhages due to a loss of function of at least one of the three genes, KRIT1/CCM1, CCM2/OSM, or CCM3/PDCD10. In zebrafish, *ccm* mutants exhibit proliferation and sprouting defects in endocardial and endothelial cells (Fig. 9.2b, c). Mechanistically, the loss of CCM proteins results in a ß1 integrin-dependent overexpression of the zinc finger transcription factors *Klf2a* and *Klf2b*, which in turn causes an upregulation of endothelial-specific factor *egfl7*, thereby promoting excessive angiogenesis. Pharmacological inhibition of VEGF signaling or *klf2a/b* knockdown by antisense oligonucleotide morpholino injection rescued *ccm* mutant cardiovascular defects [83]. Thus, zebrafish can help to further unravel so far unrecognized pathomechanistic insights of human diseases and enables identification of potential therapeutic targets.

Worldwide, atherosclerosis is the leading cause of death. It is a pathological process of inflammation and progressive deposition of cholesterol, cellular debris, and calcium in the artery walls. Zebrafish, with its unique visualization abilities by combining transgenic reporter lines with biological indicator dyes, can help to identify novel mechanisms and candidate genes and their role in the pathogenesis of the disease to ultimately screen for new druggable targets [147, 148]. Recent studies have shown that zebrafish fed with high-cholesterol diet (HCD) mimic lipid deposition within arterial vessels in humans (Fig. 9.2d) [84]. Combined with transgenic lines marking leukocytes or macrophages, zebrafish becomes a powerful tool to analyze inflammatory modulating drugs on the progression of atherosclerosis in vivo [149].

9.4   Zebrafish in High-Throughput Drug Screens

Over the last 100 years or so, classical drug development has been employing two broad types of small molecule screening strategies, the phenotypic screening and the target-based screening, to identify small molecules that can be used as lead compound for novel pharmaceutical drugs [150]. Historically, new drugs discovery has been mostly relied on phenotype-based screening that utilizes cellular or animal models to search for compounds that induce desirable phenotypic change(s). Since 1980s, owing to the advances in molecular biology and genomics, target-based screening, which aims to identify small molecules against defined molecular targets implicated in human diseases, has immediately gained popularity. Nevertheless, there are a few advantages for phenotypic screening over the target-based screening as a tool of choice for lead discovery [151, 152]. For instance, phenotypic screening doesn't require prior knowledge about the molecular target to identify drugs that produce therapeutic effects. This strategy also has the potential to identify compound that alleviates a diseased phenotype through targeting multiple biological targets. As such, the last two decades has witnessed the emergency of zebrafish as an animal model to the forefront of phenotype-based small molecule screening [152].

In 2000, Peterson et al. published the very first whole-organism-based small molecule screen and identified compounds that affected various aspects of early zebrafish development [153]. The same group also performed the first chemical suppressor screen aiming to reverse the coarctation phenotype observed in zebrafish *gridlock* mutants [154–156]. *gridlock* harbors a hypomorphic mutation in the *hey2* gene that encodes a bHLH transcriptional repressor [155]. After screening 5000 small molecules, two structurally related compounds were identified to suppress the *gridlock* coarctation phenotype likely through upregulating VEGF expression [154]. Given the resemblance of zebrafish cardiac electrophysiology to that of human hearts [11], zebrafish has been served as a particularly valuable model for identifying therapeutics that can rescue arrhythmia defects. By screening 1200 commercially available small molecules, Peal et al. identified two compounds that reproducibly rescued *breakdance* mutant long-QT defect [125]. In recent years, phenotype-based screening using zebrafish has benefited significantly from the latest developed motorized robotics systems. These high-throughput tools and technologies facilitate embryo dispensation, compound delivery, and incubation, as well as image acquisition and analysis of a variety of parameters to grasp the complexity of cardiac function, and will undoubtedly accelerate the identification of cardiovascular disease therapeutics in whole-organism-based systems.

9.5   Conclusions

- A variety of human cardiovascular diseases can be modeled in zebrafish.
- Availability of a highly versatile genetic tool set allows easy and fast interrogation of gene function and evaluation of disease causality of gene candidates and even provides insights into functional genetic networks.
- The unprecedented visualization capabilities of zebrafish enable phenotypic characterization from whole organ level down to cell and subcellular characterization.
- As a future prediction, zebrafish will urge into high-throughput drug discovery and will accelerate the identification of disease, including cardiovascular disease, therapeutics in whole-organism-based systems in the future.

Acknowledgments We would like to thank the members of the Hassel and Liu lab for the vivid discussions and valuable input for the book chapter. We would further like to deeply thank all authors whose work we cited and at the same time apologize to all authors whose papers we could not reference due to space limitations. This work was supported by R56HL133081 to J.L.

References

1. Kimmel CB. Genetics and early development of zebrafish. Trends Genet. 1989;5:283–8.
2. Driever W, Solnica-Krezel L, Schier AF, Neuhauss SC, Malicki J, Stemple DL, Stainier DY, Zwartkruis F, Abdelilah S, Rangini Z, Belak J, Boggs C. A genetic screen for mutations affecting embryogenesis in zebrafish. Development. 1996;123:37–46.
3. Haffter P, Granato M, Brand M, Mullins MC, Hammerschmidt M, Kane DA, Odenthal J, van Eeden FJ, Jiang YJ, Heisenberg CP, Kelsh RN, Furutani-Seiki M, Vogelsang E, Beuchle D, Schach U, Fabian C, Nusslein-Volhard C. The identification of genes with unique and essential functions in the development of the zebrafish, danio rerio. Development. 1996;123:1–36.
4. Kimmel CB, Ballard WW, Kimmel SR, Ullmann B, Schilling TF. Stages of embryonic development of the zebrafish. Dev Dyn. 1995;203:253–310.
5. Stainier DY, Lee RK, Fishman MC. Cardiovascular development in the zebrafish. I. Myocardial fate map and heart tube formation. Development. 1993;119:31–40.
6. Just S, Meder B, Berger IM, Etard C, Trano N, Patzel E, Hassel D, Marquart S, Dahme T, Vogel B, Fishman MC, Katus HA, Strahle U, Rottbauer W. The myosin-interacting protein smyd1 is essential for sarcomere organization. J Cell Sci. 2011;124:3127–36.

7. Wang H, Long Q, Marty SD, Sassa S, Lin S. A zebrafish model for hepatoerythropoietic porphyria. Nat Genet. 1998;20:239–43.
8. Phillips JB, Westerfield M. Zebrafish models in translational research: tipping the scales toward advancements in human health. Dis Model Mech. 2014;7:739–43.
9. Ablain J, Zon LI. Of fish and men: using zebrafish to fight human diseases. Trends Cell Biol. 2013;23:584–6.
10. Santoriello C, Zon LI. Hooked! Modeling human disease in zebrafish. J Clin Invest. 2012;122:2337–43.
11. Verkerk AO, Remme CA. Zebrafish: A novel research tool for cardiac (patho)electrophysiology and ion channel disorders. Front Physiol. 2012;3:255.
12. Davis EE, Frangakis S, Katsanis N. Interpreting human genetic variation with in vivo zebrafish assays. Biochim Biophys Acta. 1842;2014:1960–70.
13. Howe K, Clark MD, Torroja CF, Torrance J, Berthelot C, Muffato M, Collins JE, Humphray S, McLaren K, Matthews L, McLaren S, Sealy I, Caccamo M, Churcher C, Scott C, Barrett JC, Koch R, Rauch GJ, White S, Chow W, Kilian B, Quintais LT, Guerra-Assuncao JA, Zhou Y, Gu Y, Yen J, Vogel JH, Eyre T, Redmond S, Banerjee R, Chi J, Fu B, Langley E, Maguire SF, Laird GK, Lloyd D, Kenyon E, Donaldson S, Sehra H, Almeida-King J, Loveland J, Trevanion S, Jones M, Quail M, Willey D, Hunt A, Burton J, Sims S, McLay K, Plumb B, Davis J, Clee C, Oliver K, Clark R, Riddle C, Elliot D, Threadgold G, Harden G, Ware D, Begum S, Mortimore B, Kerry G, Heath P, Phillimore B, Tracey A, Corby N, Dunn M, Johnson C, Wood J, Clark S, Pelan S, Griffiths G, Smith M, Glithero R, Howden P, Barker N, Lloyd C, Stevens C, Harley J, Holt K, Panagiotidis G, Lovell J, Beasley H, Henderson C, Gordon D, Auger K, Wright D, Collins J, Raisen C, Dyer L, Leung K, Robertson L, Ambridge K, Leongamornlert D, McGuire S, Gilderthorp R, Griffiths C, Manthravadi D, Nichol S, Barker G, Whitehead S, Kay M, Brown J, Murnane C, Gray E, Humphries M, Sycamore N, Barker D, Saunders D, Wallis J, Babbage A, Hammond S, Mashreghi-Mohammadi M, Barr L, Martin S, Wray P, Ellington A, Matthews N, Ellwood M, Woodmansey R, Clark G, Cooper J, Tromans A, Grafham D, Skuce C, Pandian R, Andrews R, Harrison E, Kimberley A, Garnett J, Fosker N, Hall R, Garner P, Kelly D, Bird C, Palmer S, Gehring I, Berger A, Dooley CM, Ersan-Urun Z, Eser C, Geiger H, Geisler M, Karotki L, Kirn A, Konantz J, Konantz M, Oberlander M, Rudolph-Geiger S, Teucke M, Lanz C, Raddatz G, Osoegawa K, Zhu B, Rapp A, Widaa S, Langford C, Yang F, Schuster SC, Carter NP, Harrow J, Ning Z, Herrero J, Searle SM, Enright A, Geisler R, Plasterk RH, Lee C, Westerfield M, de Jong PJ, Zon LI, Postlethwait JH, Nusslein-Volhard C, Hubbard TJ, Roest Crollius H, Rogers J, Stemple DL. The zebrafish reference genome sequence and its relationship to the human genome. Nature. 2013;496:498–503.
14. Postlethwait JH, Woods IG, Ngo-Hazelett P, Yan YL, Kelly PD, Chu F, Huang H, Hill-Force A, Talbot WS. Zebrafish comparative genomics and the origins of vertebrate chromosomes. Genome Res. 2000;10:1890–902.
15. Amsterdam A, Hopkins N. Mutagenesis strategies in zebrafish for identifying genes involved in development and disease. Trends Genet. 2006;22:473–8.
16. Nagayoshi S, Hayashi E, Abe G, Osato N, Asakawa K, Urasaki A, Horikawa K, Ikeo K, Takeda H, Kawakami K. Insertional mutagenesis by the tol2 transposon-mediated enhancer trap approach generated mutations in two developmental genes: Tcf7 and synembryn-like. Development. 2008;135:159–69.
17. Sivasubbu S, Balciunas D, Davidson AE, Pickart MA, Hermanson SB, Wangensteen KJ, Wolbrink DC, Ekker SC. Gene-breaking transposon mutagenesis reveals an essential role for histone h2afza in zebrafish larval development. Mech Dev. 2006;123:513–29.
18. Gaiano N, Amsterdam A, Kawakami K, Allende M, Becker T, Hopkins N. Insertional mutagenesis and rapid cloning of essential genes in zebrafish. Nature. 1996;383:829–32.
19. McGrail M, Hatler JM, Kuang X, Liao HK, Nannapaneni K, Watt KE, Uhl JD, Largaespada DA, Vollbrecht E, Scheetz TE, Dupuy AJ, Hostetter JM, Essner JJ. Somatic mutagenesis with

a sleeping beauty transposon system leads to solid tumor formation in zebrafish. PLoS One. 2011;6:e18826.

20. Kramer C, Mayr T, Nowak M, Schumacher J, Runke G, Bauer H, Wagner DS, Schmid B, Imai Y, Talbot WS, Mullins MC, Hammerschmidt M. Maternally supplied smad5 is required for ventral specification in zebrafish embryos prior to zygotic bmp signaling. Dev Biol. 2002;250:263–79.

21. Bai X, Kim J, Yang Z, Jurynec MJ, Akie TE, Lee J, LeBlanc J, Sessa A, Jiang H, DiBiase A, Zhou Y, Grunwald DJ, Lin S, Cantor AB, Orkin SH, Zon LI. Tif1 gamma controls erythroid cell fate by regulating transcription elongation. Cell. 2010;142:133–43.

22. Asnani A, Peterson RT. The zebrafish as a tool to identify novel therapies for human cardiovascular disease. Dis Model Mech. 2014;7:763–7.

23. Bakkers J. Zebrafish as a model to study cardiac development and human cardiac disease. Cardiovasc Res. 2011;91:279–88.

24. Stainier DY, Fouquet B, Chen JN, Warren KS, Weinstein BM, Meiler SE, Mohideen MA, Neuhauss SC, Solnica-Krezel L, Schier AF, Zwartkruis F, Stemple DL, Malicki J, Driever W, Fishman MC. Mutations affecting the formation and function of the cardiovascular system in the zebrafish embryo. Development. 1996;123:285–92.

25. Thisse C, Thisse B. High-resolution in situ hybridization to whole-mount zebrafish embryos. Nat Protoc. 2008;3:59–69.

26. Vogel G. Genomics. Sanger will sequence zebrafish genome. Science. 2000;290:1671.

27. Rossi A, Kontarakis Z, Gerri C, Nolte H, Holper S, Kruger M, Stainier DY. Genetic compensation induced by deleterious mutations but not gene knockdowns. Nature. 2015;524:230–3.

28. Wienholds E, van Eeden F, Kosters M, Mudde J, Plasterk RH, Cuppen E. Efficient target-selected mutagenesis in zebrafish. Genome Res. 2003;13:2700–7.

29. Sprague J, Doerry E, Douglas S, Westerfield M. The zebrafish information network (zfin): a resource for genetic, genomic and developmental research. Nucleic Acids Res. 2001;29:87–90.

30. Eisen JS, Smith JC. Controlling morpholino experiments: don't stop making antisense. Development. 2008;135:1735–43.

31. Nasevicius A, Ekker SC. Effective targeted gene 'knockdown' in zebrafish. Nat Genet. 2000;26:216–20.

32. Draper BW, Morcos PA, Kimmel CB. Inhibition of zebrafish fgf8 pre-mrna splicing with morpholino oligos: a quantifiable method for gene knockdown. Genesis. 2001;30:154–6.

33. Shestopalov IA, Sinha S, Chen JK. Light-controlled gene silencing in zebrafish embryos. Nat Chem Biol. 2007;3:650–1.

34. Ouyang X, Shestopalov IA, Sinha S, Zheng G, Pitt CL, Li WH, Olson AJ, Chen JK. Versatile synthesis and rational design of caged morpholinos. J Am Chem Soc. 2009;131:13255–69.

35. Montalbano A, Juergensen L, Roeth R, Weiss B, Fukami M, Fricke-Otto S, Binder G, Ogata T, Decker E, Nuernberg G, Hassel D, Rappold GA. Retinoic acid catabolizing enzyme cyp26c1 is a genetic modifier in shox deficiency. EMBO Mol Med. 2016;8:1455–69.

36. Hassel D, Cheng P, White MP, Ivey KN, Kroll J, Augustin HG, Katus HA, Stainier DY, Srivastava D. Microrna-10 regulates the angiogenic behavior of zebrafish and human endothelial cells by promoting vascular endothelial growth factor signaling. Circ Res. 2012;111:1421–33.

37. Choi WY, Giraldez AJ, Schier AF. Target protectors reveal dampening and balancing of nodal agonist and antagonist by mir-430. Science. 2007;318:271–4.

38. Blum M, De Robertis EM, Wallingford JB, Niehrs C. Morpholinos: antisense and sensibility. Dev Cell. 2015;35:145–9.

39. Kok FO, Shin M, Ni CW, Gupta A, Grosse AS, van Impel A, Kirchmaier BC, Peterson-Maduro J, Kourkoulis G, Male I, DeSantis DF, Sheppard-Tindell S, Ebarasi L, Betsholtz C, Schulte-Merker S, Wolfe SA, Lawson ND. Reverse genetic screening reveals poor correlation between morpholino-induced and mutant phenotypes in zebrafish. Dev Cell. 2015;32:97–108.

40. Hwang WY, Fu Y, Reyon D, Maeder ML, Kaini P, Sander JD, Joung JK, Peterson RT, Yeh JR. Heritable and precise zebrafish genome editing using a crispr-cas system. PLoS One. 2013;8:e68708.
41. Chang N, Sun C, Gao L, Zhu D, Xu X, Zhu X, Xiong JW, Xi JJ. Genome editing with rna-guided cas9 nuclease in zebrafish embryos. Cell Res. 2013;23:465–72.
42. Sander JD, Cade L, Khayter C, Reyon D, Peterson RT, Joung JK, Yeh JR. Targeted gene disruption in somatic zebrafish cells using engineered talens. Nat Biotechnol. 2011;29:697–8.
43. Huang P, Xiao A, Zhou M, Zhu Z, Lin S, Zhang B. Heritable gene targeting in zebrafish using customized talens. Nat Biotechnol. 2011;29:699–700.
44. Doyon Y, McCammon JM, Miller JC, Faraji F, Ngo C, Katibah GE, Amora R, Hocking TD, Zhang L, Rebar EJ, Gregory PD, Urnov FD, Amacher SL. Heritable targeted gene disruption in zebrafish using designed zinc-finger nucleases. Nat Biotechnol. 2008;26:702–8.
45. Kim YG, Cha J, Chandrasegaran S. Hybrid restriction enzymes: zinc finger fusions to fok i cleavage domain. Proc Natl Acad Sci U S A. 1996;93:1156–60.
46. Narayanan A, Hill-Teran G, Moro A, Ristori E, Kasper DM, Roden CA, Lu J, Nicoli S. In vivo mutagenesis of mirna gene families using a scalable multiplexed crispr/cas9 nuclease system. Sci Rep. 2016;6:32386.
47. Kasper DM, Moro A, Ristori E, Narayanan A, Hill-Teran G, Fleming E, Moreno-Mateos M, Vejnar CE, Zhang J, Lee D, Gu M, Gerstein M, Giraldez A, Nicoli S. Micrornas establish uniform traits during the architecture of vertebrate embryos. Dev Cell. 2017;40:552–565 e555.
48. Albadri S, Del Bene F, Revenu C. Genome editing using crispr/cas9-based knock-in approaches in zebrafish. Methods. 2017;121–122:77–85.
49. Won M, Dawid IB. Pcr artifact in testing for homologous recombination in genomic editing in zebrafish. PLoS One. 2017;12:e0172802.
50. Kawakami K. Tol2: a versatile gene transfer vector in vertebrates. Genome Biol. 2007;8(Suppl 1):S7.
51. Hassel D, Dahme T, Erdmann J, Meder B, Huge A, Stoll M, Just S, Hess A, Ehlermann P, Weichenhan D, Grimmler M, Liptau H, Hetzer R, Regitz-Zagrosek V, Fischer C, Nurnberg P, Schunkert H, Katus HA, Rottbauer W. Nexilin mutations destabilize cardiac z-disks and lead to dilated cardiomyopathy. Nat Med. 2009;15:1281–8.
52. Wang X, Yu Q, Wu Q, Bu Y, Chang NN, Yan S, Zhou XH, Zhu X, Xiong JW. Genetic interaction between pku300 and fbn2b controls endocardial cell proliferation and valve development in zebrafish. J Cell Sci. 2013;126:1381–91.
53. Durst R, Sauls K, Peal DS, deVlaming A, Toomer K, Leyne M, Salani M, Talkowski ME, Brand H, Perrocheau M, Simpson C, Jett C, Stone MR, Charles F, Chiang C, Lynch SN, Bouatia-Naji N, Delling FN, Freed LA, Tribouilloy C, LeTourneau T, LeMarec H, Fernandez-Friera L, Solis J, Trujillano D, Ossowski S, Estivill X, Dina C, Bruneval P, Chester A, Schott JJ, Irvine KD, Mao Y, Wessels A, Motiwala T, Puceat M, Tsukasaki Y, Menick DR, Kasiganesan H, Nie X, Broome AM, Williams K, Johnson A, Markwald RR, Jeunemaitre X, Hagege A, Levine RA, Milan DJ, Norris RA, Slaugenhaupt SA. Mutations in dchs1 cause mitral valve prolapse. Nature. 2015;525:109–13.
54. Dina C, Bouatia-Naji N, Tucker N, Delling FN, Toomer K, Durst R, Perrocheau M, Fernandez-Friera L, Solis J, Investigators P, Le Tourneau T, Chen MH, Probst V, Bosse Y, Pibarot P, Zelenika D, Lathrop M, Hercberg S, Roussel R, Benjamin EJ, Bonnet F, Lo SH, Dolmatova E, Simonet F, Lecointe S, Kyndt F, Redon R, LeMarec H, Froguel P, Ellinor PT, Vasan RS, Bruneval P, Markwald RR, Norris RA, Milan DJ, Slaugenhaupt SA, Levine RA, Schott JJ, Hagege AA, MVP F, Jeunemaitre X, Leducq Transatlantic MN. Genetic association analyses highlight biological pathways underlying mitral valve prolapse. Nat Genet 2015;47:1206–11.
55. Ding Y, Sun X, Huang W, Hoage T, Redfield M, Kushwaha S, Sivasubbu S, Lin X, Ekker S, Xu X. Haploinsufficiency of target of rapamycin attenuates cardiomyopathies in adult zebrafish. Circ Res. 2011;109:658–69.

56. Group CCHW. Meta-analysis of rare and common exome chip variants identifies s1pr4 and other loci influencing blood cell traits. Nat Genet. 2016;48:867–76.
57. Smith KA, Joziasse IC, Chocron S, van Dinther M, Guryev V, Verhoeven MC, Rehmann H, van der Smagt JJ, Doevendans PA, Cuppen E, Mulder BJ, Ten Dijke P, Bakkers J. Dominant-negative alk2 allele associates with congenital heart defects. Circulation. 2009;119:3062–9.
58. Hoffman JI, Kaplan S. The incidence of congenital heart disease. J Am Coll Cardiol. 2002;39:1890–900.
59. Bruneau BG. The developmental genetics of congenital heart disease. Nature. 2008;451:943–8.
60. Brown DR, Samsa LA, Qian L, Liu J. Advances in the study of heart development and disease using zebrafish. J Cardiovasc Dev Dis. 2016;3:13.
61. Samsa LA, Givens C, Tzima E, Stainier DY, Qian L, Liu J. Cardiac contraction activates endocardial notch signaling to modulate chamber maturation in zebrafish. Development. 2015;142:4080–91.
62. Lai SL, Yao WL, Tsao KC, Houben AJ, Albers HM, Ovaa H, Moolenaar WH, Lee SJ. Autotaxin/lpar3 signaling regulates kupffer's vesicle formation and left-right asymmetry in zebrafish. Development. 2012;139:4439–48.
63. Srivastava D. Making or breaking the heart: from lineage determination to morphogenesis. Cell. 2006;126:1037–48.
64. Stainier DY. Zebrafish genetics and vertebrate heart formation. Nat Rev Genet. 2001;2:39–48.
65. Liu J, Stainier DY. Zebrafish in the study of early cardiac development. Circ Res. 2012;110:870–4.
66. Glickman NS, Yelon D. Cardiac development in zebrafish: coordination of form and function. Semin Cell Dev Biol. 2002;13:507–13.
67. Samsa LA, Yang B, Liu J. Embryonic cardiac chamber maturation: trabeculation, conduction, and cardiomyocyte proliferation. Am J Med Genet C Semin Med Genet. 2013;163C:157–68.
68. Beis D, Bartman T, Jin SW, Scott IC, D'Amico LA, Ober EA, Verkade H, Frantsve J, Field HA, Wehman A, Baier H, Tallafuss A, Bally-Cuif L, Chen JN, Stainier DY, Jungblut B. Genetic and cellular analyses of zebrafish atrioventricular cushion and valve development. Development. 2005;132:4193–204.
69. Scherz PJ, Huisken J, Sahai-Hernandez P, Stainier DY. High-speed imaging of developing heart valves reveals interplay of morphogenesis and function. Development. 2008;135:1179–87.
70. Bisgrove BW, Snarr BS, Emrazian A, Yost HJ. Polaris and polycystin-2 in dorsal forerunner cells and kupffer's vesicle are required for specification of the zebrafish left-right axis. Dev Biol. 2005;287:274–88.
71. Vetrini F, D'Alessandro LC, Akdemir ZC, Braxton A, Azamian MS, Eldomery MK, Miller K, Kois C, Sack V, Shur N, Rijhsinghani A, Chandarana J, Ding Y, Holtzman J, Jhangiani SN, Muzny DM, Gibbs RA, Eng CM, Hanchard NA, Harel T, Rosenfeld JA, Belmont JW, Lupski JR, Yang Y. Bi-allelic mutations in pkd1l1 are associated with laterality defects in humans. Am J Hum Genet. 2016;99:886–93.
72. Paffett-Lugassy N, Singh R, Nevis KR, Guner-Ataman B, O'Loughlin E, Jahangiri L, Harvey RP, Burns CG, Burns CE. Heart field origin of great vessel precursors relies on nkx2.5-mediated vasculogenesis. Nat Cell Biol. 2013;15:1362–9.
73. Nagelberg D, Wang J, Su R, Torres-Vazquez J, Targoff KL, Poss KD, Knaut H. Origin, specification, and plasticity of the great vessels of the heart. Curr Biol. 2015;25:2099–110.
74. Barron DJ, Kilby MD, Davies B, Wright JG, Jones TJ, Brawn WJ. Hypoplastic left heart syndrome. Lancet. 2009;374:551–64.
75. Noonan JA, Nadas AS. The hypoplastic left heart syndrome; an analysis of 101 cases. Pediatr Clin N Am. 1958;5:1029–56.
76. Norwood WI, Lang P, Hansen DD. Physiologic repair of aortic atresia-hypoplastic left heart syndrome. N Engl J Med. 1983;308:23–6.

77. Hinton RB, Martin LJ, Rame-Gowda S, Tabangin ME, Cripe LH, Benson DW. Hypoplastic left heart syndrome links to chromosomes 10q and 6q and is genetically related to bicuspid aortic valve. J Am Coll Cardiol. 2009;53:1065–71.

78. Dasgupta C, Martinez AM, Zuppan CW, Shah MM, Bailey LL, Fletcher WH. Identification of connexin43 (alpha1) gap junction gene mutations in patients with hypoplastic left heart syndrome by denaturing gradient gel electrophoresis (dgge). Mutat Res. 2001;479:173–86.

79. Iascone M, Ciccone R, Galletti L, Marchetti D, Seddio F, Lincesso AR, Pezzoli L, Vetro A, Barachetti D, Boni L, Federici D, Soto AM, Comas JV, Ferrazzi P, Zuffardi O. Identification of de novo mutations and rare variants in hypoplastic left heart syndrome. Clin Genet. 2012;81:542–54.

80. Targoff KL, Colombo S, George V, Schell T, Kim SH, Solnica-Krezel L, Yelon D. Nkx genes are essential for maintenance of ventricular identity. Development. 2013;140:4203–13.

81. Targoff KL, Schell T, Yelon D. Nkx genes regulate heart tube extension and exert differential effects on ventricular and atrial cell number. Dev Biol. 2008;322:314–21.

82. Liu X, Yagi H, Saeed S, Bais AS, Gabriel GC, Chen Z, Peterson KA, Li Y, Schwartz MC, Reynolds WT, Saydmohammed M, Gibbs B, Wu Y, Devine W, Chatterjee B, Klena NT, Kostka D, de Mesy Bentley KL, Ganapathiraju MK, Dexheimer P, Leatherbury L, Khalifa O, Bhagat A, Zahid M, Pu W, Watkins S, Grossfeld P, Murray SA, Porter GA Jr, Tsang M, Martin LJ, Benson DW, Aronow BJ, Lo CW. The complex genetics of hypoplastic left heart syndrome. Nat Genet. 2017;49:1152–9.

83. Renz M, Otten C, Faurobert E, Rudolph F, Zhu Y, Boulday G, Duchene J, Mickoleit M, Dietrich AC, Ramspacher C, Steed E, Manet-Dupe S, Benz A, Hassel D, Vermot J, Huisken J, Tournier-Lasserve E, Felbor U, Sure U, Albiges-Rizo C, Abdelilah-Seyfried S. Regulation of beta1 integrin-klf2-mediated angiogenesis by ccm proteins. Dev Cell. 2015;32:181–90.

84. Fang L, Green SR, Baek JS, Lee SH, Ellett F, Deer E, Lieschke GJ, Witztum JL, Tsimikas S, Miller YI. In vivo visualization and attenuation of oxidized lipid accumulation in hypercholesterolemic zebrafish. J Clin Invest. 2011;121:4861–9.

85. Markwald RR, Norris RA, Moreno-Rodriguez R, Levine RA. Developmental basis of adult cardiovascular diseases: valvular heart diseases. Ann N Y Acad Sci. 2010;1188:177–83.

86. Nishimura RA, Otto CM, Bonow RO, Carabello BA, Erwin JP III, Guyton RA, O'Gara PT, Ruiz CE, Skubas NJ, Sorajja P, Sundt TM III, Thomas JD, Anderson JL, Halperin JL, Albert NM, Bozkurt B, Brindis RG, Creager MA, Curtis LH, DeMets D, Guyton RA, Hochman JS, Kovacs RJ, Ohman EM, Pressler SJ, Sellke FW, Shen WK, Stevenson WG, Yancy CW, American College of C, American College of Cardiology/American Heart A, American Heart A. 2014 AHA/ACC guideline for the management of patients with valvular heart disease: a report of the American College of Cardiology/American Heart Association Task Force on Practice Guidelines. J thorac Cardiovasc Surg. 2014;148:e1–e132.

87. Nishimura RA, Otto CM, Bonow RO, Carabello BA, Erwin JP III, Fleisher LA, Jneid H, Mack MJ, McLeod CJ, O'Gara PT, Rigolin VH, Sundt TM III, Thompson A. 2017 AHA/ACC focused update of the 2014 aha/acc guideline for the management of patients with valvular heart disease: a report of the American College of Cardiology/American Heart Association Task Force on Clinical Practice Guidelines. J Am Coll Cardiol. 2017;70:252–89.

88. Staudt D, Stainier D. Uncovering the molecular and cellular mechanisms of heart development using the zebrafish. Annu Rev Genet. 2012;46:397–418.

89. Pestel J, Ramadass R, Gauvrit S, Helker C, Herzog W, Stainier DY. Real-time 3d visualization of cellular rearrangements during cardiac valve formation. Development. 2016;143:2217–27.

90. Zhu L, Belmont JW, Ware SM. Genetics of human heterotaxias. Eur J Hum Genet. 2006;14:17–25.

91. Amula V, Ellsworth GL, Bratton SL, Arrington CB, Witte MK. Heterotaxy syndrome: impact of ventricular morphology on resource utilization. Pediatr Cardiol. 2014;35:38–46.

92. Unolt M, Putotto C, Silvestri LM, Marino D, Scarabotti A, Valerio M, Caiaro A, Versacci P, Marino B. Transposition of great arteries: new insights into the pathogenesis. Front Pediatr. 2013;1:11.

93. Lin AE, Krikov S, Riehle-Colarusso T, Frias JL, Belmont J, Anderka M, Geva T, Getz KD, Botto LD. National Birth Defects Prevention S. Laterality defects in the national birth defects prevention study (1998–2007): birth prevalence and descriptive epidemiology. Am J Med Genet A. 2014;164A:2581–91.
94. Pennekamp P, Menchen T, Dworniczak B, Hamada H. Situs inversus and ciliary abnormalities: 20 years later, what is the connection? Cilia. 2015;4:1.
95. Kennedy MP, Omran H, Leigh MW, Dell S, Morgan L, Molina PL, Robinson BV, Minnix SL, Olbrich H, Severin T, Ahrens P, Lange L, Morillas HN, Noone PG, Zariwala MA, Knowles MR. Congenital heart disease and other heterotaxic defects in a large cohort of patients with primary ciliary dyskinesia. Circulation. 2007;115:2814–21.
96. Nakhleh N, Francis R, Giese RA, Tian X, Li Y, Zariwala MA, Yagi H, Khalifa O, Kureshi S, Chatterjee B, Sabol SL, Swisher M, Connelly PS, Daniels MP, Srinivasan A, Kuehl K, Kravitz N, Burns K, Sami I, Omran H, Barmada M, Olivier K, Chawla KK, Leigh M, Jonas R, Knowles M, Leatherbury L, Lo CW. High prevalence of respiratory ciliary dysfunction in congenital heart disease patients with heterotaxy. Circulation. 2012;125:2232–42.
97. Li Y, Yagi H, Onuoha EO, Damerla RR, Francis R, Furutani Y, Tariq M, King SM, Hendricks G, Cui C, Saydmohammed M, Lee DM, Zahid M, Sami I, Leatherbury L, Pazour GJ, Ware SM, Nakanishi T, Goldmuntz E, Tsang M, Lo CW. Dnah6 and its interactions with pcd genes in heterotaxy and primary ciliary dyskinesia. PLoS Genet. 2016;12:e1005821.
98. Ostrowski LE, Dutcher SK, Lo CW. Cilia and models for studying structure and function. Proc Am Thorac Soc. 2011;8:423–9.
99. Guimier A, Gabriel GC, Bajolle F, Tsang M, Liu H, Noll A, Schwartz M, El Malti R, Smith LD, Klena NT, Jimenez G, Miller NA, Oufadem M, Moreau de Bellaing A, Yagi H, Saunders CJ, Baker CN, Di Filippo S, Peterson KA, Thiffault I, Bole-Feysot C, Cooley LD, Farrow EG, Masson C, Schoen P, Deleuze JF, Nitschke P, Lyonnet S, de Pontual L, Murray SA, Bonnet D, Kingsmore SF, Amiel J, Bouvagnet P, Lo CW, Gordon CT. Mmp21 is mutated in human heterotaxy and is required for normal left-right asymmetry in vertebrates. Nat Genet. 2015;47:1260–3.
100. Xu X, Meiler SE, Zhong TP, Mohideen M, Crossley DA, Burggren WW, Fishman MC. Cardiomyopathy in zebrafish due to mutation in an alternatively spliced exon of titin. Nat Genet. 2002;30:205–9.
101. Sehnert AJ, Huq A, Weinstein BM, Walker C, Fishman M, Stainier DY. Cardiac troponin t is essential in sarcomere assembly and cardiac contractility. Nat Genet. 2002;31:106–10.
102. Siu BL, Niimura H, Osborne JA, Fatkin D, MacRae C, Solomon S, Benson DW, Seidman JG, Seidman CE. Familial dilated cardiomyopathy locus maps to chromosome 2q31. Circulation. 1999;99:1022–6.
103. Thierfelder L, Watkins H, MacRae C, Lamas R, McKenna W, Vosberg HP, Seidman JG, Seidman CE. Alpha-tropomyosin and cardiac troponin t mutations cause familial hypertrophic cardiomyopathy: a disease of the sarcomere. Cell. 1994;77:701–12.
104. Shih YH, Zhang Y, Ding Y, Ross CA, Li H, Olson TM, Xu X. Cardiac transcriptome and dilated cardiomyopathy genes in zebrafish. Circ Cardiovasc Genet. 2015;8:261–9.
105. Wang H, Li Z, Wang J, Sun K, Cui Q, Song L, Zou Y, Wang X, Liu X, Hui R, Fan Y. Mutations in nexn, a z-disc gene, are associated with hypertrophic cardiomyopathy. Am J Hum Genet. 2010;87:687–93.
106. Vogel B, Meder B, Just S, Laufer C, Berger I, Weber S, Katus HA, Rottbauer W. In-vivo characterization of human dilated cardiomyopathy genes in zebrafish. Biochem Biophys Res Commun. 2009;390:516–22.
107. Ramspacher C, Steed E, Boselli F, Ferreira R, Faggianelli N, Roth S, Spiegelhalter C, Messaddeq N, Trinh L, Liebling M, Chacko N, Tessadori F, Bakkers J, Laporte J, Hnia K, Vermot J. Developmental alterations in heart biomechanics and skeletal muscle function in desmin mutants suggest an early pathological root for desminopathies. Cell Rep. 2015;11:1564–76.

108. Liu J, Bressan M, Hassel D, Huisken J, Staudt D, Kikuchi K, Poss KD, Mikawa T, Stainier DY. A dual role for erbb2 signaling in cardiac trabeculation. Development. 2010;137:3867–75.

109. Meder B, Laufer C, Hassel D, Just S, Marquart S, Vogel B, Hess A, Fishman MC, Katus HA, Rottbauer W. A single serine in the carboxyl terminus of cardiac essential myosin light chain-1 controls cardiomyocyte contractility in vivo. Circ Res. 2009;104:650–9.

110. Bendig G, Grimmler M, Huttner IG, Wessels G, Dahme T, Just S, Trano N, Katus HA, Fishman MC, Rottbauer W. Integrin-linked kinase, a novel component of the cardiac mechanical stretch sensor, controls contractility in the zebrafish heart. Genes Dev. 2006;20:2361–72.

111. Rottbauer W, Just S, Wessels G, Trano N, Most P, Katus HA, Fishman MC. Vegf-plcgamma1 pathway controls cardiac contractility in the embryonic heart. Genes Dev. 2005;19:1624–34.

112. Zou J, Tran D, Baalbaki M, Tang LF, Poon A, Pelonero A, Titus EW, Yuan C, Shi C, Patchava S, Halper E, Garg J, Movsesyan I, Yin C, Wu R, Wilsbacher LD, Liu J, Hager RL, Coughlin SR, Jinek M, Pullinger CR, Kane JP, Hart DO, Kwok PY, Deo RC. An internal promoter underlies the difference in disease severity between n- and c-terminal truncation mutations of titin in zebrafish. elife. 2015;4:e09406.

113. Kossack M, Hein S, Juergensen L, Siragusa M, Benz A, Katus HA, Most P, Hassel D. Induction of cardiac dysfunction in developing and adult zebrafish by chronic isoproterenol stimulation. J Mol Cell Cardiol. 2017;108:95–105.

114. Kalogirou S, Malissovas N, Moro E, Argenton F, Stainier DY, Beis D. Intracardiac flow dynamics regulate atrioventricular valve morphogenesis. Cardiovasc Res. 2014;104:49–60.

115. Asimaki A, Kapoor S, Plovie E, Karin Arndt A, Adams E, Liu Z, James CA, Judge DP, Calkins H, Churko J, Wu JC, MacRae CA, Kleber AG, Saffitz JE. Identification of a new modulator of the intercalated disc in a zebrafish model of arrhythmogenic cardiomyopathy. Sci Transl Med. 2014;6:240ra274.

116. Ding Y, Sun X, Xu X. Tor-autophagy signaling in adult zebrafish models of cardiomyopathy. Autophagy. 2012;8:142–3.

117. Sun X, Hoage T, Bai P, Ding Y, Chen Z, Zhang R, Huang W, Jahangir A, Paw B, Li YG, Xu X. Cardiac hypertrophy involves both myocyte hypertrophy and hyperplasia in anemic zebrafish. PLoS One. 2009;4:e6596.

118. Scheid LM, Mosqueira M, Hein S, Kossack M, Juergensen L, Mueller M, Meder B, Fink RH, Katus HA, Hassel D. Essential light chain s195 phosphorylation is required for cardiac adaptation under physical stress. Cardiovasc Res. 2016;111:44–55.

119. Wang LW, Huttner IG, Santiago CF, Kesteven SH, Yu ZY, Feneley MP, Fatkin D. Standardized echocardiographic assessment of cardiac function in normal adult zebrafish and heart disease models. Dis Model Mech. 2017;10:63–76.

120. Hein SJ, Lehmann LH, Kossack M, Juergensen L, Fuchs D, Katus HA, Hassel D. Advanced echocardiography in adult zebrafish reveals delayed recovery of heart function after myocardial cryoinjury. PLoS One. 2015;10:e0122665.

121. Rottbauer W, Baker K, Wo ZG, Mohideen MA, Cantiello HF, Fishman MC. Growth and function of the embryonic heart depend upon the cardiac-specific l-type calcium channel alpha1 subunit. Dev Cell. 2001;1:265–75.

122. Arnaout R, Ferrer T, Huisken J, Spitzer K, Stainier DY, Tristani-Firouzi M, Chi NC. Zebrafish model for human long qt syndrome. Proc Natl Acad Sci U S A. 2007;104:11316–21.

123. Langheinrich U, Vacun G, Wagner T. Zebrafish embryos express an orthologue of herg and are sensitive toward a range of qt-prolonging drugs inducing severe arrhythmia. Toxicol Appl Pharmacol. 2003;193:370–82.

124. Milan DJ, Kim AM, Winterfield JR, Jones IL, Pfeufer A, Sanna S, Arking DE, Amsterdam AH, Sabeh KM, Mably JD, Rosenbaum DS, Peterson RT, Chakravarti A, Kaab S, Roden DM, MacRae CA. Drug-sensitized zebrafish screen identifies multiple genes, including gins3, as regulators of myocardial repolarization. Circulation. 2009;120:553–9.

125. Peal DS, Mills RW, Lynch SN, Mosley JM, Lim E, Ellinor PT, January CT, Peterson RT, Milan DJ. Novel chemical suppressors of long qt syndrome identified by an in vivo functional screen. Circulation. 2011;123:23–30.

126. Hassel D, Scholz EP, Trano N, Friedrich O, Just S, Meder B, Weiss DL, Zitron E, Marquart S, Vogel B, Karle CA, Seemann G, Fishman MC, Katus HA, Rottbauer W. Deficient zebrafish ether-a-go-go-related gene channel gating causes short-qt syndrome in zebrafish reggae mutants. Circulation. 2008;117:866–75.

127. Langenbacher AD, Dong Y, Shu X, Choi J, Nicoll DA, Goldhaber JI, Philipson KD, Chen JN. Mutation in sodium-calcium exchanger 1 (ncx1) causes cardiac fibrillation in zebrafish. Proc Natl Acad Sci U S A. 2005;102:17699–704.

128. Ebert AM, Hume GL, Warren KS, Cook NP, Burns CG, Mohideen MA, Siegal G, Yelon D, Fishman MC, Garrity DM. Calcium extrusion is critical for cardiac morphogenesis and rhythm in embryonic zebrafish hearts. Proc Natl Acad Sci U S A. 2005;102:17705–10.

129. Huttner IG, Trivedi G, Jacoby A, Mann SA, Vandenberg JI, Fatkin D. A transgenic zebrafish model of a human cardiac sodium channel mutation exhibits bradycardia, conduction-system abnormalities and early death. J Mol Cell Cardiol. 2013;61:123–32.

130. Chen X, Gays D, Milia C, Santoro MM. Cilia control vascular mural cell recruitment in vertebrates. Cell Rep. 2017;18:1033–47.

131. Ando K, Fukuhara S, Izumi N, Nakajima H, Fukui H, Kelsh RN, Mochizuki N. Clarification of mural cell coverage of vascular endothelial cells by live imaging of zebrafish. Development. 2016;143:1328–39.

132. Karpanen T, Schulte-Merker S. Zebrafish provides a novel model for lymphatic vascular research. Methods Cell Biol. 2011;105:223–38.

133. Santoro MM, Pesce G, Stainier DY. Characterization of vascular mural cells during zebrafish development. Mech Dev. 2009;126:638–49.

134. Jin SW, Beis D, Mitchell T, Chen JN, Stainier DY. Cellular and molecular analyses of vascular tube and lumen formation in zebrafish. Development. 2005;132:5199–209.

135. Siekmann AF, Lawson ND. Notch signalling limits angiogenic cell behaviour in developing zebrafish arteries. Nature. 2007;445:781–4.

136. Hellstrom M, Phng LK, Hofmann JJ, Wallgard E, Coultas L, Lindblom P, Alva J, Nilsson AK, Karlsson L, Gaiano N, Yoon K, Rossant J, Iruela-Arispe ML, Kalen M, Gerhardt H, Betsholtz C. Dll4 signalling through notch1 regulates formation of tip cells during angiogenesis. Nature. 2007;445:776–80.

137. Herbert SP, Huisken J, Kim TN, Feldman ME, Houseman BT, Wang RA, Shokat KM, Stainier DY. Arterial-venous segregation by selective cell sprouting: an alternative mode of blood vessel formation. Science. 2009;326:294–8.

138. Xu B, Zhang Y, Du XF, Li J, Zi HX, Bu JW, Yan Y, Han H, Du JL. Neurons secrete mir-132-containing exosomes to regulate brain vascular integrity. Cell Res. 2017;27:882–97.

139. Tobia C, Chiodelli P, Nicoli S, Dell'era P, Buraschi S, Mitola S, Foglia E, van Loenen PB, Alewijnse AE, Presta M. Sphingosine-1-phosphate receptor-1 controls venous endothelial barrier integrity in zebrafish. Arterioscler Thromb Vasc Biol. 2012;32:e104–16.

140. Butler MG, Gore AV, Weinstein BM. Zebrafish as a model for hemorrhagic stroke. Methods Cell Biol. 2011;105:137–61.

141. Fish JE, Santoro MM, Morton SU, Yu S, Yeh RF, Wythe JD, Ivey KN, Bruneau BG, Stainier DY, Srivastava D. Mir-126 regulates angiogenic signaling and vascular integrity. Dev Cell. 2008;15:272–84.

142. Buchner DA, Su F, Yamaoka JS, Kamei M, Shavit JA, Barthel LK, McGee B, Amigo JD, Kim S, Hanosh AW, Jagadeeswaran P, Goldman D, Lawson ND, Raymond PA, Weinstein BM, Ginsburg D, Lyons SE. Pak2a mutations cause cerebral hemorrhage in redhead zebrafish. Proc Natl Acad Sci U S A. 2007;104:13996–4001.

143. Gu Y, Jin P, Zhang L, Zhao X, Gao X, Ning Y, Meng A, Chen YG. Functional analysis of mutations in the kinase domain of the tgf-beta receptor alk1 reveals different mechanisms for induction of hereditary hemorrhagic telangiectasia. Blood. 2006;107:1951–4.

144. Hall CJ, Flores MV, Davidson AJ, Crosier KE, Crosier PS. Radar is required for the establishment of vascular integrity in the zebrafish. Dev Biol. 2002;251:105–17.
145. Liu J, Fraser SD, Faloon PW, Rollins EL, Vom Berg J, Starovic-Subota O, Laliberte AL, Chen JN, Serluca FC, Childs SJ. A betapix pak2a signaling pathway regulates cerebral vascular stability in zebrafish. Proc Natl Acad Sci U S A. 2007;104:13990–5.
146. Kleaveland B, Zheng X, Liu JJ, Blum Y, Tung JJ, Zou Z, Sweeney SM, Chen M, Guo L, Lu MM, Zhou D, Kitajewski J, Affolter M, Ginsberg MH, Kahn ML. Regulation of cardiovascular development and integrity by the heart of glass-cerebral cavernous malformation protein pathway. Nat Med. 2009;15:169–76.
147. Gut P, Baeza-Raja B, Andersson O, Hasenkamp L, Hsiao J, Hesselson D, Akassoglou K, Verdin E, Hirschey MD, Stainier DY. Whole-organism screening for gluconeogenesis identifies activators of fasting metabolism. Nat Chem Biol. 2013;9:97–104.
148. Weger BD, Weger M, Nusser M, Brenner-Weiss G, Dickmeis T. A chemical screening system for glucocorticoid stress hormone signaling in an intact vertebrate. ACS Chem Biol. 2012;7:1178–83.
149. Stoletov K, Fang L, Choi SH, Hartvigsen K, Hansen LF, Hall C, Pattison J, Juliano J, Miller ER, Almazan F, Crosier P, Witztum JL, Klemke RL, Miller YI. Vascular lipid accumulation, lipoprotein oxidation, and macrophage lipid uptake in hypercholesterolemic zebrafish. Circ Res. 2009;104:952–60.
150. Swinney DC, Anthony J. How were new medicines discovered? Nat Rev Drug Discov. 2011;10:507–19.
151. Eggert US. The why and how of phenotypic small-molecule screens. Nat Chem Biol. 2013;9:206–9.
152. MacRae CA, Peterson RT. Zebrafish as tools for drug discovery. Nat Rev Drug Discov. 2015;14:721–31.
153. Peterson RT, Link BA, Dowling JE, Schreiber SL. Small molecule developmental screens reveal the logic and timing of vertebrate development. Proc Natl Acad Sci U S A. 2000;97:12965–9.
154. Peterson RT, Shaw SY, Peterson TA, Milan DJ, Zhong TP, Schreiber SL, MacRae CA, Fishman MC. Chemical suppression of a genetic mutation in a zebrafish model of aortic coarctation. Nat Biotechnol. 2004;22:595–9.
155. Zhong TP, Rosenberg M, Mohideen MA, Weinstein B, Fishman MC. Gridlock, an hlh gene required for assembly of the aorta in zebrafish. Science. 2000;287:1820–4.
156. Weinstein BM, Stemple DL, Driever W, Fishman MC. Gridlock, a localized heritable vascular patterning defect in the zebrafish. Nat Med. 1995;1:1143–7.

Human Induced Pluripotent Stem Cells as Platform for Functional Examination of Cardiovascular Genetics in a Dish

10

Ralf Dirschinger, Tatjana Dorn, and Alessandra Moretti

Contents

10.1   Introduction

The three most ground-breaking developments in biomedical research in the last decades may have been the generation of human induced pluripotent stem cells (hiPSCs) from diseased and healthy individuals, the discovery of the clustered regularly interspaced short palindromic repeats (CRISPR)/CRISPR-associated 9 (Cas9) genome editing system, and the possibility to identify disease-related genetic variants in human populations in genome-wide association studies (GWAS) or by modern sequencing methods. Although these highly complementary

R. Dirschinger (✉) · T. Dorn (✉)
Cardiology Division, First Department of Medicine, Klinikum rechts der Isar – Technical
University of Munich, Munich, Germany
e-mail: r.dirschinger@tum.de; tatjana.dorn@mytum.de

A. Moretti
Cardiology Division, First Department of Medicine, Klinikum rechts der Isar – Technical
University of Munich, Munich, Germany

Munich Heart Alliance, Munich, Germany

© Springer Nature Switzerland AG 2019
J. Erdmann, A. Moretti (eds.), *Genetic Causes of Cardiac Disease*, Cardiac and
Vascular Biology 7, https://doi.org/10.1007/978-3-030-27371-2_10

fields have evolved rapidly in the last decade and already profoundly changed our understanding of cardiovascular genetics, there is tremendous potential to further merge these technologies. This chapter aims to summarize the state of the art with respect to the use of hiPSCs to model cardiovascular genetic diseases, to pinpoint challenges and limitations of this technology, and to outline potential strategies to overcome these by the implementation of evolving genome editing technology and human genetic research tools.

10.2   Studies of Monogenic Cardiovascular Disease in hiPSCs: Insights and Limitations

With the discovery of somatic cell reprogramming by the transient heterologous expression of transcription factors, it suddenly became possible to obtain various cell types from patients and healthy individuals for the in vitro study of genetic disease without the need of genomic modification.

To validate this novel tool, it appeared most reasonable to model genetic diseases with a strong penetrance. Furthermore, the elaborated methods of single-cell electrophysiology were already well established as a potential readout for ion channel disorders. Consequently, the first studies utilizing hiPSC-derived cells for the modeling of cardiovascular disease focused on monogenic, mostly dominant rhythm disorders. The first study to recapitulate a human channelopathy in hiPSC-derived cardiomyocytes (hiPSC-CMs) was published in 2010 by our group [1]. Briefly, hiPSC-CMs from a long-QT syndrome 1 (LQT1) patient, carrying a potassium voltage-gated channel subfamily Q member 1 (KCNQ1) loss-of-function mutation were shown to exhibit significantly prolonged action potential duration, attributable to a reduced I_{Ks} current. A dominant negative trafficking defect of the KCNQ1 channel could be identified as the cause of the severely reduced I_{Ks} current. Subsequently, a wealth of studies have used hiPSC-CMs to model human monogenic arrhythmias, e.g., LQT1 [1–3], LQT2 [4–8], LQT3 [9–12], and LQT8, also known as Timothy syndrome [13], Jervell and Lange-Nielsen syndrome [14], LQTS3/ Brugada overlap syndrome [15], and catecholaminergic polymorphic ventricular tachycardia (CPVT) type 1 [16–21] and 2 [22, 23]—to name a selection.

Another group of cardiovascular monogenic diseases that have been modeled in vitro using hiPSC-derived cells are cardiomyopathies [24–26], e.g., dilated cardiomyopathy (DCM) [27–35], hypertrophic cardiomyopathy (HCM) [36–40], or arrhythmogenic right ventricular cardiomyopathy (ARVC) [41–45]. These studies further emphasized a critical feature of the hiPSC system that may be advantageous or disadvantageous, depending on the situation: hiPSC-derived cells differentiate in vitro along pathways that likely reflect embryonic development. On the upside, this allows the study of developmental defects and enables researchers to distinguish between diseases of the progenitor and diseases of the terminally differentiated cell type. As an example, the altered cell–cell interactions found in ARVC were shown to already affect cardiomyocyte progenitors by priming them toward an adipocytic lineage [45]. On the downside, many hiPSC-derived cells, e.g., cardiomyocytes, resemble their fetal counterparts in humans, not reaching the maturity of adult cardiomyocytes. This limits the validity of hiPSC models for genetic diseases with a late-onset adult phenotype, reflected by the difficulty to provoke a clear-cut mature cellular phenotype in many studies of adult-onset diseases [46].

Many of the studies listed above were conducted in one or two hiPSC lines from patients and few control hiPSC lines from healthy family members or even unrelated donors. This may have been appropriate for monogenic disorders with a strong underlying genetic cause that leads to a disease phenotype regardless of the genetic background. Nevertheless, with almost a decade of successful hiPSC disease modelling, the influence of the genetic background on hiPSC phenotype and the impact of line-to-line variability is becoming increasingly clear. This calls for novel approaches regarding genetic background as well as biological replicates, which will be discussed in the next section.

10.3   Line-to-Line Variability vs. the Influence of Genetic Background: Need for Isogenic Controls

Disease modeling studies using hiPSCs have taught us that individual hiPSC lines exhibit a relatively high variability in their biological potential to differentiate into distinct cell lineages [46, 47]. This line-to-line variability is mostly caused by epigenetic modifications due to reprogramming history and passaging as well as genetic background differences, altogether remaining a considerable challenge of hiPSC technology [26, 46–48]. Hence, mainly robust disease phenotypes of high penetrance can be revealed when comparing patient-derived cells with unrelated controls, whereas mild phenotypic differences may fall into the noise of line-to-line variability and not be detected. Generation of several clonal hiPSC lines from a patient and a control could help to account for cell line variability emerging from genetic and epigenetic alterations during reprogramming and culturing cells over time. To overcome differences in genetic background, generation of isogenic hiPSCs by correcting the disease-associated mutation in patient hiPSCs or by introducing the causative mutation in control hiPSCs has become an important tool to decipher subtle disorder phenotypes and to link a cellular phenotype to a specific mutation [46, 47]. Isogenic lines have identical genetic backgrounds and differ only at one specific locus. The significance of this strategy has been demonstrated exemplarily in a pre-CRISPR/Cas9 study that modeled long QT syndrome using isogenic pairs of patient and (corrected) control hiPSCs as well as control and gene-targeted human embryonic stem cells (hESCs) harboring the same mutation in *KCNH2* gene, generated by conventional recombination targeting strategy, to dissect the influence of genetic background on the phenotypic outcome [49]. Due to the inefficiency of classical homologous recombination approaches in the human system, this approach was rather exceptional and the use of hiPSCs from healthy unrelated individuals or at least unaffected family members was standard in hiPSC-based disease modeling studies for a long time [24, 26, 46–48]. With the recent improvement of genome editing technologies including zinc finger nucleases (ZFNs) [50], transcription

Table 10.1 hiPSC-based cardiovascular disease modeling studies using isogenic controls

Disease	Gene	Method	References
ARVC/D	*SCN5A*	CRISPR/Cas9	[44]
BTHS	*TAZ*	CRISPR/Cas9	[55]
DCM	*PLN*	TALEN	[35]
	TTN	CRISPR/Cas9	[30]
DMD	*DMD*	CRISPR/Cas9	[56]
	DMD	CRISPR/Cas9	[57]
HCM	*ALPK3*	CRISPR/Cas9	[58]
	PRKAG2	CRISPR/Cas9	[59, 60]
	TNNT2	CRISPR/Cas9	[61]
	MYH7	CRISPR/Cas9	[62, 63]
	MYBPC3	CRISPR/Cas9	[63, 64]
HCM/LVNC	*ACTC1*	CRISPR/Cas9 & piggy-bac	[65]
JLNS/LQT1	*KCNQ1*	CRISPR/Cas9	[14]
LQT1	*KCNQ1*	homologous recombination	[66]
LQT2	*KCNH2*	homologous recombination	[49]
LQT15	*CALM2*	CRISPR/Cas9	[67, 68]

ARVC arrhythmogenic right ventricular cardiomyopathy/dysplasia, *BTHS* Barth syndrome, *DCM*, dilated cardiomyopathy, *DMD* Duchenne muscular dystrophy, *HCM* hypertrophic cardiomyopathy, *JLNS* Jervell and Lange-Nielsen syndrome, *LVNC* left ventricular non-compaction cardiomyopathy, *LQT* long QT syndrome

activator-like effector nucleases (TALENs) [51], and CRISPR/Cas9 [52–54] based systems, the genetic manipulation of human stem cells has become easy and efficient allowing and demanding the utilization of isogenic controls for phenotypic and mechanistic characterization of a disease-causing variant. Table 10.1 summarizes studies that have used isogenic PSC lines to model cardiovascular disorders [14, 30, 35, 44, 49, 55–68].

The CRISPR/Cas9 system, with its simplicity, efficacy, and flexibility, stands out among the other gene-editing techniques to modify the genome in human cells and has become the preferred method for either correcting putative disease-causing variants in patient-derived cells or introducing a genetic aberration in unaffected controls. It relies on the guide RNA (gRNA) defining the specificity of the targeting site and offers the possibility by applying different gRNAs to insert multiple mutations at several loci simultaneously [69, 70]. Although CRISPR/Cas9 is a powerful technology, it has technical constraints in the off-target effects resulting from unintended DNA cleavage at sites of unspecific binding of the gRNA-Cas9 complex and requires careful screening for off-target effects after gene editing.

10.4 hiPSC Disease Modeling in the Era of Highly Efficient Genome Editing: Concepts for the Study of Polygenic Diseases, Addressing Variants with Low Penetrance

The usage of isogenic pairs of control and disease hiPSCs that differ solely at a specific genetic variant becomes even more important when modeling polygenic diseases that arise from a complex interaction of risk alleles of relatively small effect sizes with the environmental factors [71, 72]. Genome-wide association studies have identified a plethora of cardiovascular disease-linked genomic loci including deletions/insertions, copy number variations or single nucleotide polymorphisms (SNPs) that may be critical for complex polygenic defects [73, 74]. The hiPSC technology provides a valuable tool for the modeling of complex cardiac disorders, however, the investigation of the disease-causing mechanisms and elucidation of the contribution of individual risk loci to the disease phenotype remain challenging [47]. This could be due to several reasons. Typically, the identified genetic variants reside within non-coding regulatory regions [75] and have rather subtle phenotypic effects, which are difficult to study. Moreover, risk alleles can be found not only in diseased but also in non-affected individuals [46] making it challenging to assess the functionality of a distinct polymorphism. Contrary to monogenic maladies where a mutation of a single gene is disease causative, polygenic disorders develop from a sum of multiple genetic variants acting in a complex network, in which a single variant likely will not be sufficient to trigger the disease, complicating the determination of the causality between the mutation and the phenotype.

Since the main limitation of hiPSC-based models for complex disorders is the lack of robust phenotypes, correlating risk loci with molecular profiles, so-called quantitative traits offers a way of delineating genetic variants by defining the quantitative trait loci (QTL) [47, 76, 77]. Gene expression levels are often used to assess the effect of genetic variation on transcriptional regulation [76]. Other molecular traits, including DNA methylation, epigenetic modification, protein expression, and metabolites can be examined to evaluate the role of risk loci in the disease [47, 76, 77].

As a proof of principle, the group around Kelly Frazer examined associations of the genetic variations with the transcriptional profiles of the hiPSC-CMs from the iPSC Collection for Omic Research (iPSCORE) collection and could demonstrate that gene expression patterns cluster by genetic background and can be used to evaluate SNPs linked with physiological and disease phenotypes [78]. With a similar method, Matsa and colleagues assessed variations in gene expression signatures of hiPSC-CMs from breast cancer patients and could predict patient-specific differences in drug cardiotoxicity based on individual RNA sequencing (RNA-seq) patterns [79].

In another recent study, Gupta et al. discerned the function of the genetic variant associated with five vascular diseases by using a series of isogenic stem cell lines in combination with RNA-seq and genome-wide epigenetic analyses and could identify a non-coding variant that can distally regulate endothelin-1, which is known to play a role in vascular remodeling [80]. This work highlights a strategy how hiPSCs together with gene editing and global profiling of gene expression and epigenetics could be used to decode genetic risk loci in the context of complex diseases [47, 80].

With the aim of addressing the function of GWAS-identified variants within regulatory networks, there is a need to integrate single cell technologies with genetic screens on a genome-wide basis [47, 77]. Perturb-seq [81] and CRISP-seq [82] are novel techniques combining massively parallel single-cell RNA-seq with CRISPR/Cas9-mediated genetic perturbations on a genome scale that could be used to study mechanisms underlying polygenic diseases. These approaches enable the analysis of transcriptional consequences upon CRISPR-based modulation of risk loci in the same cell allowing the modeling of regulatory circuits of these risk variants.

10.5 Upscaling hiPSC-Based Testing: A Potential Paradigm Shift for Drug Development

In several hiPSC disease modeling studies of cardiovascular genetic diseases, drugs acting on the presumed pathophysiologic mechanism were tested [83] and in many cases, features of the phenotype could be modified by the drug, e.g., nifedipine shortened the action potential and reduced the number of early afterdepolarizations (EADs) in LQT2 [4]. Initially, some of the drugs were used to confirm the pathomechanisms proposed in the studies. On the other hand, these experiments demonstrated the potential of hiPSC disease models as drug screening platforms, leading to the detection of new drug candidates by discovering potential novel applications for known compounds, as in the case of roscovitine for the treatment of LQT8 [13] or dantrolene for the treatment of catecholaminergic polymorphic ventricular tachycardia type I (CPVT1) [21].

In principle, the idea of using hiPSC disease models as a platform for pharmacodynamic testing can be extended to multiple candidate drugs or even upscaled to unbiased screens of compound libraries. This however requires the presence of an obvious phenotype that can be modified by the desired drug action and easily quantified, ideally in a high-throughput manner. Again, arrhythmia research may serve as an example of how phenotypic readouts may be optimized for high-throughput analyses: while automated patch clamp recording, multielectrode assays or the use of voltage-sensitive dyes may offer a solution [84–88], a particularly interesting approach for a phenotypic readout has been taken by researchers from our and other groups in the case of channelopathies affecting action potential duration: genetically encoded voltage sensors were successfully utilized to repeatedly visualize and record the action potentials in hiPSC-CMs without harming the cells [66, 89, 90]. Furthermore, the voltage sensors can be expressed under the control of promoters specific for cardiomyocyte subtypes, i.e., ventricular myocytes, atrial myocytes, and nodal myocytes, which helps to overcome the common differentiation heterogeneity and thereby increase the signal-to-noise ratio of the readout [66]. The action potential recordings enable the assessment of single cell action potential duration, which is a meaningful readout for the efficacy of drugs targeting long-QT syndromes. Alternatively, the method can be used to test the cardiotoxicity of drugs that may alter QT duration in some individuals, an important adverse effect of several drugs including certain types of antibiotic and neuroleptic drugs [91].

With growing numbers of cell lines being collected in hiPSC banks, hiPSC-based drug screening could also be realized vice versa by testing single candidate drugs on multiple or even a library of hiPSC lines, i.e., multiple genetic backgrounds. Not only can this help to elucidate whether the efficacy of a drug is present regardless of genetic background, this could also serve as an in vitro platform for pharmacogenomics in the context of precision medicine: in combination with whole genome sequencing of the hiPSC library (which only needs to be conducted once for each hiPSC line) it would be possible to perform in vitro GWAS analyses and identify variants associated with strong or weak susceptibility to the drug action. In perspective, this could serve as a basis to genetically define patient populations who may benefit more or less from a specific pharmacotherapy [48, 92, 93].

Similarly, this strategy could be further extended as a platform for toxicogenomics, i.e., to test candidate drugs for toxicity on a library of hiPSC-derived cells of different cell-types, e.g., cardiomyocytes, hepatocytes [94, 95], or renal proximal tubular cells [96], provided that reproducible differentiation protocols that work for all hiPSC lines can be established and appropriate readout measures can be defined.

Before such multi-dimensional high-throughput/high-content approaches can be realized, several hurdles must be overcome. First, culture conditions must be further standardized to allow automatized cultures of biobanked hiPSC lines. The same is true for differentiation protocols, which need to work equally efficient in the majority of banked hiPSC lines before such projects can be realized. Finally, it has to be pointed out that to date many phenotypic measures used in hiPSC models of cardiovascular diseases like sarcomeric organization or contractile force are difficult to quantify reliably with the same precision and reproducibility as, e.g., action potential duration in single cells. However, as genetic variants affecting drug action or toxicity may have only small individual impact on cellular phenotypes, readout assays with strong discriminatory power will be required to yield reliable results from in vitro pharmacogenomic or toxicogenomic studies. An alternative to phenotypic biologic readouts may be the direct transcriptional analysis by RNA-seq, although the throughput with this cost-intensive approach would be limited and the biologic significance of solely transcriptional readouts is not clear. Thus, further research focusing on the improvement of reliable and upscalable phenotypic assays is warranted.

Taken together, at the current state of technology and the pace of technological development, automatized high-throughput hiPSC-based platforms for drug screening, efficacy, and toxicity can be envisioned to complement or even replace traditional methods of drug development in the future. While there are still important issues that need to be overcome—including cost effectiveness—high-throughput hiPSC-based methods for future drug testing may be an ideal tool to meet aims proposed by the concept of precision medicine.

10.6   3D Culture in Engineered and Biological Systems: Assessing Non-cell-Autonomous Functions

The human heart is a large organ composed of different interacting cell types with a complex architecture. Consequently, not all genetic properties or genetic diseases affecting the heart can be recapitulated by studying individual cardiomyocytes. Yet, due to usual in vitro culture conditions, most hiPSC-based studies so far mainly focus on cell-autonomous functions of cardiomyocytes. However, the interaction of different cells in a three-dimensional system plays an important role for normal cardiac function and in numerous diseases. For instance, arrhythmia mechanisms of macro- or micro-reentry require alterations in action potential propagation from one cell to another, and in many cases the disease mechanism may not be an inherent property of distinct cardiomyocytes, but rather a change in 3D architecture such as scarring or fibrosis or a change in cellular composition. Furthermore, the interaction of contractile cardiomyocytes, conduction system cells, and non-myocyte cells such as fibroblasts, endothelial cells, or pericytes may be underappreciated by current research. Of course it seems extremely challenging to tackle this kind of complexity in a dish in a meaningful manner. However, recent advances in the field of bioengineering and 3D culture are encouraging and raise the hope that 3D systems may become a valid tool for hiPSC disease modeling and drug testing for certain applications.

Cardiac tissue-engineering approaches seeding hiPSC-derived cells in ECM biogels, in many cases following primarily regenerative aims, have led to the development of reproducible three-dimensional tissue patches that appear to have improved physiologic force-development, cell alignment, and maturation, compared to standard 2D cultures [97–101]. Another promising approach is the reseeding of hiPSC-derived cells on decellularized matrix, which gives the cells important environmental cues via the ECM [102, 103]. Finally, the self-organizing capabilities of hiPSC-derived progenitors may be used to form so-called organoids, i.e., structures derived by 3D in vitro culture, that resemble important properties of organs. In the cardiac field, organoid research is still in the early stages of development [104, 105].

In summary, the lack of reliable culture systems that allow the in vitro study of complex physiologic 3D structures and mechanisms has been a limitation of hiPSC disease modeling. State-of-the-art tissue engineering and advances in 3D culture may be able to provide new platforms for hiPSC-based genetic research.

10.7   Conclusion

Due to the limited accessibility of viable cardiac tissue from patients, the in vitro study of cardiac function and disease has been fueled by the discovery of human embryonic stem cells (hESCs) and hiPSCs. Nine years after the first successful cardiac hiPSC disease modeling study by our group, many lessons have been learned, including advantages and disadvantages of the system. The recent discovery and development of highly efficient genome editing tools such as CRISPR/Cas9

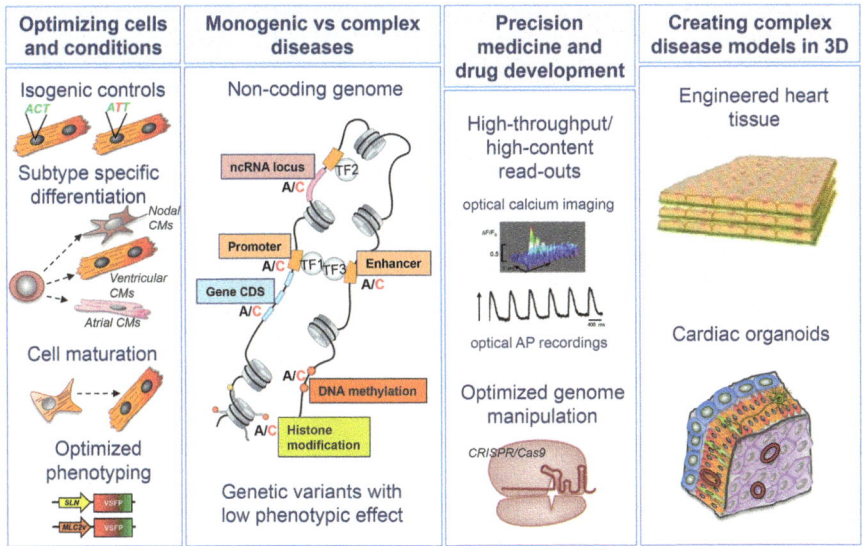

Fig. 10.1 Considerations for the design of hiPSC studies of cardiac genetics

does not only present a solution to some limitations of the technology by allowing the efficient introduction of mutations and generation of isogenic controls, it also may have been the missing link to utilize hiPSC technology for the study of disease-associated common variants identified by GWAS. Furthermore, hiPSC banks themselves may be the source of genetic material for GWAS in pharmacogenomic and toxicogenomic studies, making it a potentially invaluable tool in the context of precision medicine. Improved culturing methods, including 3D systems and more specific phenotyping methods, such as recent optical physiology tools, will be essential to raise this technology to the next level (see Fig. 10.1).

Conflict of Interest   The authors declare no potential conflict of interest relevant to this article.

References

1. Moretti A, Bellin M, Welling A, Jung CB, Lam JT, Bott-Flugel L, Dorn T, Goedel A, Hohnke C, Hofmann F, Seyfarth M, Sinnecker D, Schomig A, Laugwitz KL. Patient-specific induced pluripotent stem-cell models for long-QT syndrome. N Engl J Med. 2010;363:1397–409.
2. Egashira T, Yuasa S, Suzuki T, Aizawa Y, Yamakawa H, Matsuhashi T, Ohno Y, Tohyama S, Okata S, Seki T, Kuroda Y, Yae K, Hashimoto H, Tanaka T, Hattori F, Sato T, Miyoshi S, Takatsuki S, Murata M, Kurokawa J, Furukawa T, Makita N, Aiba T, Shimizu W, Horie M, Kamiya K, Kodama I, Ogawa S, Fukuda K. Disease characterization using LQTS-specific induced pluripotent stem cells. Cardiovasc Res. 2012;95:419–29.
3. Ma D, Wei H, Lu J, Huang D, Liu Z, Loh LJ, Islam O, Liew R, Shim W, Cook SA. Characterization of a novel KCNQ1 mutation for type 1 long QT syndrome and

assessment of the therapeutic potential of a novel IKs activator using patient-specific induced pluripotent stem cell-derived cardiomyocytes. Stem Cell Res Ther. 2015;6:39.

4. Itzhaki I, Maizels L, Huber I, Zwi-Dantsis L, Caspi O, Winterstern A, Feldman O, Gepstein A, Arbel G, Hammerman H, Boulos M, Gepstein L. Modelling the long QT syndrome with induced pluripotent stem cells. Nature. 2011;471:225–9.

5. Jouni M, Si-Tayeb K, Es-Salah-Lamoureux Z, Latypova X, Champon B, Caillaud A, Rungoat A, Charpentier F, Loussouarn G, Baro I, Zibara K, Lemarchand P, Gaborit N. Toward personalized medicine: using cardiomyocytes differentiated from urine-derived pluripotent stem cells to recapitulate electrophysiological characteristics of type 2 long QT syndrome. J Am Heart Assoc. 2015;4:e002159.

6. Lahti AL, Kujala VJ, Chapman H, Koivisto AP, Pekkanen-Mattila M, Kerkela E, Hyttinen J, Kontula K, Swan H, Conklin BR, Yamanaka S, Silvennoinen O, Aalto-Setala K. Model for long QT syndrome type 2 using human iPS cells demonstrates arrhythmogenic characteristics in cell culture. Dis Model Mech. 2012;5:220–30.

7. Matsa E, Rajamohan D, Dick E, Young L, Mellor I, Staniforth A, Denning C. Drug evaluation in cardiomyocytes derived from human induced pluripotent stem cells carrying a long QT syndrome type 2 mutation. Eur Heart J. 2011;32:952–62.

8. Braam SR, Tertoolen L, Casini S, Matsa E, Lu HR, Teisman A, Passier R, Denning C, Gallacher DJ, Towart R, Mummery CL. Repolarization reserve determines drug responses in human pluripotent stem cell derived cardiomyocytes. Stem Cell Res. 2013;10:48–56.

9. Fatima A, Kaifeng S, Dittmann S, Xu G, Gupta MK, Linke M, Zechner U, Nguemo F, Milting H, Farr M, Hescheler J, Saric T. The disease-specific phenotype in cardiomyocytes derived from induced pluripotent stem cells of two long QT syndrome type 3 patients. PLoS One. 2013;8:e83005.

10. Ma D, Wei H, Zhao Y, Lu J, Li G, Sahib NB, Tan TH, Wong KY, Shim W, Wong P, Cook SA, Liew R. Modeling type 3 long QT syndrome with cardiomyocytes derived from patient-specific induced pluripotent stem cells. Int J Cardiol. 2013;168:5277–86.

11. Malan D, Friedrichs S, Fleischmann BK, Sasse P. Cardiomyocytes obtained from induced pluripotent stem cells with long-QT syndrome 3 recapitulate typical disease-specific features in vitro. Circ Res. 2011;109:841–7.

12. Terrenoire C, Wang K, Tung KW, Chung WK, Pass RH, Lu JT, Jean JC, Omari A, Sampson KJ, Kotton DN, Keller G, Kass RS. Induced pluripotent stem cells used to reveal drug actions in a long QT syndrome family with complex genetics. J Gen Physiol. 2013;141:61–72.

13. Yazawa M, Hsueh B, Jia X, Pasca AM, Bernstein JA, Hallmayer J, Dolmetsch RE. Using induced pluripotent stem cells to investigate cardiac phenotypes in Timothy syndrome. Nature. 2011;471:230–4.

14. Zhang M, D'Aniello C, Verkerk AO, Wrobel E, Frank S, Ward-van Oostwaard D, Piccini I, Freund C, Rao J, Seebohm G, Atsma DE, Schulze-Bahr E, Mummery CL, Greber B, Bellin M. Recessive cardiac phenotypes in induced pluripotent stem cell models of Jervell and Lange-Nielsen syndrome: disease mechanisms and pharmacological rescue. Proc Natl Acad Sci U S A. 2014;111:E5383–92.

15. Davis RP, Casini S, van den Berg CW, Hoekstra M, Remme CA, Dambrot C, Salvatori D, Oostwaard DW, Wilde AA, Bezzina CR, Verkerk AO, Freund C, Mummery CL. Cardiomyocytes derived from pluripotent stem cells recapitulate electrophysiological characteristics of an overlap syndrome of cardiac sodium channel disease. Circulation. 2012;125:3079–91.

16. Itzhaki I, Maizels L, Huber I, Gepstein A, Arbel G, Caspi O, Miller L, Belhassen B, Nof E, Glikson M, Gepstein L. Modeling of catecholaminergic polymorphic ventricular tachycardia with patient-specific human-induced pluripotent stem cells. J Am Coll Cardiol. 2012;60:990–1000.

17. Fatima A, Xu G, Shao K, Papadopoulos S, Lehmann M, Arnaiz-Cot JJ, Rosa AO, Nguemo F, Matzkies M, Dittmann S, Stone SL, Linke M, Zechner U, Beyer V, Hennies HC, Rosenkranz S, Klauke B, Parwani AS, Haverkamp W, Pfitzer G, Farr M, Cleemann L,

Morad M, Milting H, Hescheler J, Saric T. In vitro modeling of ryanodine receptor 2 dysfunction using human induced pluripotent stem cells. Cell Physiol Biochem. 2011;28:579–92.

18. Kujala K, Paavola J, Lahti A, Larsson K, Pekkanen-Mattila M, Viitasalo M, Lahtinen AM, Toivonen L, Kontula K, Swan H, Laine M, Silvennoinen O, Aalto-Setala K. Cell model of catecholaminergic polymorphic ventricular tachycardia reveals early and delayed afterdepolarizations. PLoS One. 2012;7:e44660.

19. Di Pasquale E, Lodola F, Miragoli M, Denegri M, Avelino-Cruz JE, Buonocore M, Nakahama H, Portararo P, Bloise R, Napolitano C, Condorelli G, Priori SG. CaMKII inhibition rectifies arrhythmic phenotype in a patient-specific model of catecholaminergic polymorphic ventricular tachycardia. Cell Death Dis. 2013;4:e843.

20. Zhang XH, Haviland S, Wei H, Saric T, Fatima A, Hescheler J, Cleemann L, Morad M. Ca2+ signaling in human induced pluripotent stem cell-derived cardiomyocytes (iPS-CM) from normal and catecholaminergic polymorphic ventricular tachycardia (CPVT)-afflicted subjects. Cell Calcium. 2013;54:57–70.

21. Jung CB, Moretti A, Mederos Y, Schnitzler M, Iop L, Storch U, Bellin M, Dorn T, Ruppenthal S, Pfeiffer S, Goedel A, Dirschinger RJ, Seyfarth M, Lam JT, Sinnecker D, Gudermann T, Lipp P, Laugwitz KL. Dantrolene rescues arrhythmogenic RYR2 defect in a patient-specific stem cell model of catecholaminergic polymorphic ventricular tachycardia. EMBO Mol Med. 2012;4:180–91.

22. Novak A, Barad L, Lorber A, Gherghiceanu M, Reiter I, Eisen B, Eldor L, Itskovitz-Eldor J, Eldar M, Arad M, Binah O. Functional abnormalities in iPSC-derived cardiomyocytes generated from CPVT1 and CPVT2 patients carrying ryanodine or calsequestrin mutations. J Cell Mol Med. 2015;19:2006–18.

23. Novak A, Barad L, Zeevi-Levin N, Shick R, Shtrichman R, Lorber A, Itskovitz-Eldor J, Binah O. Cardiomyocytes generated from CPVTD307H patients are arrhythmogenic in response to beta-adrenergic stimulation. J Cell Mol Med. 2012;16:468–82.

24. Eschenhagen T, Mummery C, Knollmann BC. Modelling sarcomeric cardiomyopathies in the dish: from human heart samples to iPSC cardiomyocytes. Cardiovasc Res. 2015;105:424–38.

25. Kamdar F, Klaassen Kamdar A, Koyano-Nakagawa N, Garry MG, Garry DJ. Cardiomyopathy in a dish: using human inducible pluripotent stem cells to model inherited cardiomyopathies. J Card Fail. 2015;21:761–70.

26. Hoes MF, Bomer N, van der Meer P. Concise review: the current state of human in vitro cardiac disease modeling: a focus on gene editing and tissue engineering. Stem Cells Transl Med. 2019;8:66–74.

27. Sun N, Yazawa M, Liu J, Han L, Sanchez-Freire V, Abilez OJ, Navarrete EG, Hu S, Wang L, Lee A, Pavlovic A, Lin S, Chen R, Hajjar RJ, Snyder MP, Dolmetsch RE, Butte MJ, Ashley EA, Longaker MT, Robbins RC, Wu JC. Patient-specific induced pluripotent stem cells as a model for familial dilated cardiomyopathy. Sci Transl Med. 2012;4:130ra47.

28. Siu CW, Lee YK, Ho JC, Lai WH, Chan YC, Ng KM, Wong LY, Au KW, Lau YM, Zhang J, Lay KW, Colman A, Tse HF. Modeling of lamin A/C mutation premature cardiac aging using patient-specific induced pluripotent stem cells. Aging. 2012;4:803–22.

29. Tse HF, Ho JC, Choi SW, Lee YK, Butler AW, Ng KM, Siu CW, Simpson MA, Lai WH, Chan YC, Au KW, Zhang J, Lay KW, Esteban MA, Nicholls JM, Colman A, Sham PC. Patient-specific induced-pluripotent stem cells-derived cardiomyocytes recapitulate the pathogenic phenotypes of dilated cardiomyopathy due to a novel DES mutation identified by whole exome sequencing. Hum Mol Genet. 2013;22:1395–403.

30. Hinson JT, Chopra A, Nafissi N, Polacheck WJ, Benson CC, Swist S, Gorham J, Yang L, Schafer S, Sheng CC, Haghighi A, Homsy J, Hubner N, Church G, Cook SA, Linke WA, Chen CS, Seidman JG, Seidman CE. HEART DISEASE. Titin mutations in iPS cells define sarcomere insufficiency as a cause of dilated cardiomyopathy. Science. 2015;349:982–6.

31. Lee YK, Lau YM, Cai ZJ, Lai WH, Wong LY, Tse HF, Ng KM, Siu CW. Modeling treatment response for lamin A/C related dilated cardiomyopathy in human induced pluripotent stem cells. J Am Heart Assoc. 2017;6

32. Broughton KM, Li J, Sarmah E, Warren CM, Lin YH, Henze MP, Sanchez-Freire V, Solaro RJ, Russell B. A myosin activator improves actin assembly and sarcomere function of human-induced pluripotent stem cell-derived cardiomyocytes with a troponin T point mutation. Am J Physiol Heart Circ Physiol. 2016;311:H107–17.
33. Wyles SP, Li X, Hrstka SC, Reyes S, Oommen S, Beraldi R, Edwards J, Terzic A, Olson TM, Nelson TJ. Modeling structural and functional deficiencies of RBM20 familial dilated cardiomyopathy using human induced pluripotent stem cells. Hum Mol Genet. 2016;25:254–65.
34. Streckfuss-Bomeke K, Tiburcy M, Fomin A, Luo X, Li W, Fischer C, Ozcelik C, Perrot A, Sossalla S, Haas J, Vidal RO, Rebs S, Khadjeh S, Meder B, Bonn S, Linke WA, Zimmermann WH, Hasenfuss G, Guan K. Severe DCM phenotype of patient harboring RBM20 mutation S635A can be modeled by patient-specific induced pluripotent stem cell-derived cardiomyocytes. J Mol Cell Cardiol. 2017;113:9–21.
35. Karakikes I, Stillitano F, Nonnenmacher M, Tzimas C, Sanoudou D, Termglinchan V, Kong CW, Rushing S, Hansen J, Ceholski D, Kolokathis F, Kremastinos D, Katoulis A, Ren L, Cohen N, Gho JM, Tsiapras D, Vink A, Wu JC, Asselbergs FW, Li RA, Hulot JS, Kranias EG, Hajjar RJ. Correction of human phospholamban R14del mutation associated with cardiomyopathy using targeted nucleases and combination therapy. Nat Commun. 2015;6:6955.
36. Lan F, Lee AS, Liang P, Sanchez-Freire V, Nguyen PK, Wang L, Han L, Yen M, Wang Y, Sun N, Abilez OJ, Hu S, Ebert AD, Navarrete EG, Simmons CS, Wheeler M, Pruitt B, Lewis R, Yamaguchi Y, Ashley EA, Bers DM, Robbins RC, Longaker MT, Wu JC. Abnormal calcium handling properties underlie familial hypertrophic cardiomyopathy pathology in patient-specific induced pluripotent stem cells. Cell Stem Cell. 2013;12:101–13.
37. Han L, Li Y, Tchao J, Kaplan AD, Lin B, Li Y, Mich-Basso J, Lis A, Hassan N, London B, Bett GC, Tobita K, Rasmusson RL, Yang L. Study familial hypertrophic cardiomyopathy using patient-specific induced pluripotent stem cells. Cardiovasc Res. 2014;104:258–69.
38. Tanaka A, Yuasa S, Mearini G, Egashira T, Seki T, Kodaira M, Kusumoto D, Kuroda Y, Okata S, Suzuki T, Inohara T, Arimura T, Makino S, Kimura K, Kimura A, Furukawa T, Carrier L, Node K, Fukuda K. Endothelin-1 induces myofibrillar disarray and contractile vector variability in hypertrophic cardiomyopathy-induced pluripotent stem cell-derived cardiomyocytes. J Am Heart Assoc. 2014;3:e001263.
39. Prondzynski M, Kramer E, Laufer SD, Shibamiya A, Pless O, Flenner F, Muller OJ, Munch J, Redwood C, Hansen A, Patten M, Eschenhagen T, Mearini G, Carrier L. Evaluation of MYBPC3 trans-Splicing and Gene Replacement as Therapeutic Options in Human iPSC-Derived Cardiomyocytes. Mol Ther Nucleic Acids. 2017;7:475–86.
40. Ojala M, Prajapati C, Polonen RP, Rajala K, Pekkanen-Mattila M, Rasku J, Larsson K, Aalto-Setala K. Mutation-specific phenotypes in hiPSC-derived cardiomyocytes carrying either myosin-binding protein C or alpha-tropomyosin mutation for hypertrophic cardiomyopathy. Stem Cells Int. 2016;2016:1684792.
41. Kim C, Wong J, Wen J, Wang S, Wang C, Spiering S, Kan NG, Forcales S, Puri PL, Leone TC, Marine JE, Calkins H, Kelly DP, Judge DP, Chen HS. Studying arrhythmogenic right ventricular dysplasia with patient-specific iPSCs. Nature. 2013;494:105–10.
42. Ma D, Wei H, Lu J, Ho S, Zhang G, Sun X, Oh Y, Tan SH, Ng ML, Shim W, Wong P, Liew R. Generation of patient-specific induced pluripotent stem cell-derived cardiomyocytes as a cellular model of arrhythmogenic right ventricular cardiomyopathy. Eur Heart J. 2013;34:1122–33.
43. Caspi O, Huber I, Gepstein A, Arbel G, Maizels L, Boulos M, Gepstein L. Modeling of arrhythmogenic right ventricular cardiomyopathy with human induced pluripotent stem cells. Circ Cardiovasc Genet. 2013;6:557–68.
44. Te Riele AS, Agullo-Pascual E, James CA, Leo-Macias A, Cerrone M, Zhang M, Lin X, Lin B, Sobreira NL, Amat-Alarcon N, Marsman RF, Murray B, Tichnell C, van der Heijden JF, Dooijes D, van Veen TA, Tandri H, Fowler SJ, Hauer RN, Tomaselli G, van den Berg MP, Taylor MR, Brun F, Sinagra G, Wilde AA, Mestroni L, Bezzina CR, Calkins H, Peter van Tintelen J, Bu L, Delmar M, Judge DP. Multilevel analyses of SCN5A mutations in

arrhythmogenic right ventricular dysplasia/cardiomyopathy suggest non-canonical mechanisms for disease pathogenesis. Cardiovasc Res. 2017;113:102–11.

45. Dorn T, Kornherr J, Parrotta EI, Zawada D, Ayetey H, Santamaria G, Iop L, Mastantuono E, Sinnecker D, Goedel A, Dirschinger RJ, My I, Laue S, Bozoglu T, Baarlink C, Ziegler T, Graf E, Hinkel R, Cuda G, Kaab S, Grace AA, Grosse R, Kupatt C, Meitinger T, Smith AG, Laugwitz KL, Moretti A. Interplay of cell-cell contacts and RhoA/MRTF - A signaling regulates cardiomyocyte identity. EMBO J. 2018;37

46. Hockemeyer D, Jaenisch R. Induced pluripotent stem cells meet genome editing. Cell Stem Cell. 2016;18:573–86.

47. Soldner F, Jaenisch R. Stem cells, genome editing, and the path to translational medicine. Cell. 2018;175:615–32.

48. Musunuru K, Sheikh F, Gupta RM, Houser SR, Maher KO, Milan DJ, Terzic A, Wu JC. American Heart Association Council on Functional G, Translational B, Council on Cardiovascular Disease in the Y, Council on C and stroke N Induced pluripotent stem cells for cardiovascular disease modeling and precision medicine: a scientific statement from the American Heart Association. Circul Genom Precision Med. 2018;11:e000043.

49. Bellin M, Casini S, Davis RP, D'Aniello C, Haas J, Ward-van Oostwaard D, Tertoolen LG, Jung CB, Elliott DA, Welling A, Laugwitz KL, Moretti A, Mummery CL. Isogenic human pluripotent stem cell pairs reveal the role of a KCNH2 mutation in long-QT syndrome. EMBO J. 2013;32:3161–75.

50. Urnov FD, Miller JC, Lee YL, Beausejour CM, Rock JM, Augustus S, Jamieson AC, Porteus MH, Gregory PD, Holmes MC. Highly efficient endogenous human gene correction using designed zinc-finger nucleases. Nature. 2005;435:646–51.

51. Boch J, Scholze H, Schornack S, Landgraf A, Hahn S, Kay S, Lahaye T, Nickstadt A, Bonas U. Breaking the code of DNA binding specificity of TAL-type III effectors. Science. 2009;326:1509–12.

52. Jinek M, Chylinski K, Fonfara I, Hauer M, Doudna JA, Charpentier E. A programmable dual-RNA-guided DNA endonuclease in adaptive bacterial immunity. Science. 2012;337:816–21.

53. Jinek M, East A, Cheng A, Lin S, Ma E, Doudna J. RNA-programmed genome editing in human cells. elife. 2013;2:e00471.

54. Komor AC, Kim YB, Packer MS, Zuris JA, Liu DR. Programmable editing of a target base in genomic DNA without double-stranded DNA cleavage. Nature. 2016;533:420–4.

55. Wang G, McCain ML, Yang L, He A, Pasquali FS, Agarwal A, Yuan H, Jiang D, Zhang D, Zangi L, Geva J, Roberts AE, Ma Q, Ding J, Chen J, Wang DZ, Li K, Wang J, Wanders RJ, Kulik W, Vaz FM, Laflamme MA, Murry CE, Chien KR, Kelley RI, Church GM, Parker KK, Pu WT. Modeling the mitochondrial cardiomyopathy of Barth syndrome with induced pluripotent stem cell and heart-on-chip technologies. Nat Med. 2014;20:616–23.

56. Macadangdang J, Guan X, Smith AS, Lucero R, Czerniecki S, Childers MK, Mack DL, Kim DH. Nanopatterned human iPSC-based model of a dystrophin-null cardiomyopathic phenotype. Cell Mol Bioeng. 2015;8:320–32.

57. Kyrychenko V, Kyrychenko S, Tiburcy M, Shelton JM, Long C, Schneider JW, Zimmermann WH, Bassel-Duby R, Olson EN. Functional correction of dystrophin actin binding domain mutations by genome editing. JCI Insight. 2017;2

58. Phelan DG, Anderson DJ, Howden SE, Wong RC, Hickey PF, Pope K, Wilson GR, Pebay A, Davis AM, Petrou S, Elefanty AG, Stanley EG, James PA, Macciocca I, Bahlo M, Cheung MM, Amor DJ, Elliott DA, Lockhart PJ. ALPK3-deficient cardiomyocytes generated from patient-derived induced pluripotent stem cells and mutant human embryonic stem cells display abnormal calcium handling and establish that ALPK3 deficiency underlies familial cardiomyopathy. Eur Heart J. 2016;37:2586–90.

59. Zhan Y, Sun X, Li B, Cai H, Xu C, Liang Q, Lu C, Qian R, Chen S, Yin L, Sheng W, Huang G, Sun A, Ge J, Sun N. Establishment of a PRKAG2 cardiac syndrome disease model and mechanism study using human induced pluripotent stem cells. J Mol Cell Cardiol. 2018;117:49–61.

60. Ben Jehuda R, Eisen B, Shemer Y, Mekies LN, Szantai A, Reiter I, Cui H, Guan K, Haron-Khun S, Freimark D, Sperling SR, Gherghiceanu M, Arad M, Binah O. CRISPR correction of the PRKAG2 gene mutation in the patient's induced pluripotent stem cell-derived cardiomyocytes eliminates electrophysiological and structural abnormalities. Heart Rhythm. 2018;15:267–76.

61. Wang L, Kim K, Parikh S, Cadar AG, Bersell KR, He H, Pinto JR, Kryshtal DO, Knollmann BC. Hypertrophic cardiomyopathy-linked mutation in troponin T causes myofibrillar disarray and pro-arrhythmic action potential changes in human iPSC cardiomyocytes. J Mol Cell Cardiol. 2018;114:320–7.

62. Cohn R, Thakar K, Lowe A, Ladha FA, Pettinato AM, Romano R, Meredith E, Chen YS, Atamanuk K, Huey BD, Hinson JT. A contraction stress model of hypertrophic cardiomyopathy due to sarcomere mutations. Stem Cell Rep. 2019;12:71–83.

63. Mosqueira D, Mannhardt I, Bhagwan JR, Lis-Slimak K, Katili P, Scott E, Hassan M, Prondzynski M, Harmer SC, Tinker A, Smith JGW, Carrier L, Williams PM, Gaffney D, Eschenhagen T, Hansen A, Denning C. CRISPR/Cas9 editing in human pluripotent stem cell-cardiomyocytes highlights arrhythmias, hypocontractility, and energy depletion as potential therapeutic targets for hypertrophic cardiomyopathy. Eur Heart J. 2018;39:3879–92.

64. Seeger T, Shrestha R, Lam CK, Chen C, McKeithan WL, Lau E, Wnorowski A, McMullen G, Greenhaw M, Lee J, Oikonomopoulos A, Lee S, Yang H, Mercola M, Wheeler M, Ashley EA, Yang F, Karakikes I, Wu JC. A premature termination codon mutation in MYBPC3 causes hypertrophic cardiomyopathy via chronic activation of nonsense-mediated decay. Circulation. 2019;139:799–811.

65. Smith JGW, Owen T, Bhagwan JR, Mosqueira D, Scott E, Mannhardt I, Patel A, Barriales-Villa R, Monserrat L, Hansen A, Eschenhagen T, Harding SE, Marston S, Denning C. Isogenic pairs of hiPSC-CMs with hypertrophic cardiomyopathy/LVNC-associated ACTC1 E99K mutation unveil differential functional deficits. Stem Cell Rep. 2018;11:1226–43.

66. Chen Z, Xian W, Bellin M, Dorn T, Tian Q, Goedel A, Dreizehnter L, Schneider CM, Ward-van Oostwaard D, Ng JK, Hinkel R, Pane LS, Mummery CL, Lipp P, Moretti A, Laugwitz KL, Sinnecker D. Subtype-specific promoter-driven action potential imaging for precise disease modelling and drug testing in hiPSC-derived cardiomyocytes. Eur Heart J. 2017;38:292–301.

67. Limpitikul WB, Dick IE, Tester DJ, Boczek NJ, Limphong P, Yang W, Choi MH, Babich J, DiSilvestre D, Kanter RJ, Tomaselli GF, Ackerman MJ, Yue DT. A precision medicine approach to the rescue of function on malignant calmodulinopathic long-QT syndrome. Circ Res. 2017;120:39–48.

68. Yamamoto Y, Makiyama T, Harita T, Sasaki K, Wuriyanghai Y, Hayano M, Nishiuchi S, Kohjitani H, Hirose S, Chen J, Yokoi F, Ishikawa T, Ohno S, Chonabayashi K, Motomura H, Yoshida Y, Horie M, Makita N, Kimura T. Allele-specific ablation rescues electrophysiological abnormalities in a human iPS cell model of long-QT syndrome with a CALM2 mutation. Hum Mol Genet. 2017;26:1670–7.

69. Cong L, Ran FA, Cox D, Lin S, Barretto R, Habib N, Hsu PD, Wu X, Jiang W, Marraffini LA, Zhang F. Multiplex genome engineering using CRISPR/Cas systems. Science. 2013;339:819–23.

70. Gonzalez F, Zhu Z, Shi ZD, Lelli K, Verma N, Li QV, Huangfu D. An iCRISPR platform for rapid, multiplexable, and inducible genome editing in human pluripotent stem cells. Cell Stem Cell. 2014;15:215–26.

71. McClellan J, King MC. Genetic heterogeneity in human disease. Cell. 2010;141:210–7.

72. Mitchell KJ. What is complex about complex disorders? Genome Biol. 2012;13:237.

73. Buniello A, MacArthur JAL, Cerezo M, Harris LW, Hayhurst J, Malangone C, McMahon A, Morales J, Mountjoy E, Sollis E, Suveges D, Vrousgou O, Whetzel PL, Amode R, Guillen JA, Riat HS, Trevanion SJ, Hall P, Junkins H, Flicek P, Burdett T, Hindorff LA, Cunningham F, Parkinson H. The NHGRI-EBI GWAS Catalog of published genome-wide association studies, targeted arrays and summary statistics 2019. Nucleic Acids Res. 2019;47:D1005–12.

74. Rau CD, Lusis AJ, Wang Y. Genetics of common forms of heart failure: challenges and potential solutions. Curr Opin Cardiol. 2015;30:222–7.
75. Maurano MT, Humbert R, Rynes E, Thurman RE, Haugen E, Wang H, Reynolds AP, Sandstrom R, Qu H, Brody J, Shafer A, Neri F, Lee K, Kutyavin T, Stehling-Sun S, Johnson AK, Canfield TK, Giste E, Diegel M, Bates D, Hansen RS, Neph S, Sabo PJ, Heimfeld S, Raubitschek A, Ziegler S, Cotsapas C, Sotoodehnia N, Glass I, Sunyaev SR, Kaul R, Stamatoyannopoulos JA. Systematic localization of common disease-associated variation in regulatory DNA. Science. 2012;337:1190–5.
76. Heinig M. Using gene expression to annotate cardiovascular GWAS loci. Front Cardiovasc Med. 2018;5:59.
77. Lau E, Paik DT, Wu JC. Systems-wide approaches in induced pluripotent stem cell models. Annu Rev Pathol. 2019;14:395–419.
78. Panopoulos AD, D'Antonio M, Benaglio P, Williams R, Hashem SI, Schuldt BM, DeBoever C, Arias AD, Garcia M, Nelson BC, Harismendy O, Jakubosky DA, Donovan MKR, Greenwald WW, Farnam K, Cook M, Borja V, Miller CA, Grinstein JD, Drees F, Okubo J, Diffenderfer KE, Hishida Y, Modesto V, Dargitz CT, Feiring R, Zhao C, Aguirre A, McGarry TJ, Matsui H, Li H, Reyna J, Rao F, O'Connor DT, Yeo GW, Evans SM, Chi NC, Jepsen K, Nariai N, Muller FJ, Goldstein LSB, Izpisua Belmonte JC, Adler E, Loring JF, Berggren WT, D'Antonio-Chronowska A, Smith EN, Frazer KA. iPSCORE: a resource of 222 iPSC lines enabling functional characterization of genetic variation across a variety of cell types. Stem Cell Rep. 2017;8:1086–100.
79. Matsa E, Burridge PW, Yu KH, Ahrens JH, Termglinchan V, Wu H, Liu C, Shukla P, Sayed N, Churko JM, Shao N, Woo NA, Chao AS, Gold JD, Karakikes I, Snyder MP, Wu JC. Transcriptome profiling of patient-specific human iPSC-cardiomyocytes predicts individual drug safety and efficacy responses in vitro. Cell Stem Cell. 2016;19:311–25.
80. Gupta RM, Hadaya J, Trehan A, Zekavat SM, Roselli C, Klarin D, Emdin CA, Hilvering CRE, Bianchi V, Mueller C, Khera AV, Ryan RJH, Engreitz JM, Issner R, Shoresh N, Epstein CB, de Laat W, Brown JD, Schnabel RB, Bernstein BE, Kathiresan S. A genetic variant associated with five vascular diseases is a distal regulator of endothelin-1 gene expression. Cell. 2017;170:522–533 e15.
81. Dixit A, Parnas O, Li B, Chen J, Fulco CP, Jerby-Arnon L, Marjanovic ND, Dionne D, Burks T, Raychowdhury R, Adamson B, Norman TM, Lander ES, Weissman JS, Friedman N, Regev A. Perturb-seq: dissecting molecular circuits with scalable single-cell RNA profiling of pooled genetic screens. Cell. 2016;167:1853–1866 e17.
82. Jaitin DA, Weiner A, Yofe I, Lara-Astiaso D, Keren-Shaul H, David E, Salame TM, Tanay A, van Oudenaarden A, Amit I. Dissecting immune circuits by linking CRISPR-pooled screens with single-cell RNA-seq. Cell. 2016;167:1883–1896 e15.
83. Goedel A, My I, Sinnecker D, Moretti A. Perspectives and challenges of pluripotent stem cells in cardiac arrhythmia research. Curr Cardiol Rep. 2017;19:23.
84. Scheel O, Frech S, Amuzescu B, Eisfeld J, Lin KH, Knott T. Action potential characterization of human induced pluripotent stem cell-derived cardiomyocytes using automated patch-clamp technology. Assay Drug Dev Technol. 2014;12:457–69.
85. Navarrete EG, Liang P, Lan F, Sanchez-Freire V, Simmons C, Gong T, Sharma A, Burridge PW, Patlolla B, Lee AS, Wu H, Beygui RE, Wu SM, Robbins RC, Bers DM, Wu JC. Screening drug-induced arrhythmia [corrected] using human induced pluripotent stem cell-derived cardiomyocytes and low-impedance microelectrode arrays. Circulation. 2013;128:S3–13.
86. Asakura K, Hayashi S, Ojima A, Taniguchi T, Miyamoto N, Nakamori C, Nagasawa C, Kitamura T, Osada T, Honda Y, Kasai C, Ando H, Kanda Y, Sekino Y, Sawada K. Improvement of acquisition and analysis methods in multi-electrode array experiments with iPS cell-derived cardiomyocytes. J Pharmacol Toxicol Methods. 2015;75:17–26.
87. Lee P, Klos M, Bollensdorff C, Hou L, Ewart P, Kamp TJ, Zhang J, Bizy A, Guerrero-Serna G, Kohl P, Jalife J, Herron TJ. Simultaneous voltage and calcium mapping of genetically purified

human induced pluripotent stem cell-derived cardiac myocyte monolayers. Circ Res. 2012;110:1556–63.

88. Lopez-Izquierdo A, Warren M, Riedel M, Cho S, Lai S, Lux RL, Spitzer KW, Benjamin IJ, Tristani-Firouzi M, Jou CJ. A near-infrared fluorescent voltage-sensitive dye allows for moderate-throughput electrophysiological analyses of human induced pluripotent stem cell-derived cardiomyocytes. Am J Physiol Heart Circ Physiol. 2014;307:H1370–7.

89. Shinnawi R, Huber I, Maizels L, Shaheen N, Gepstein A, Arbel G, Tijsen AJ, Gepstein L. Monitoring human-induced pluripotent stem cell-derived cardiomyocytes with genetically encoded calcium and voltage fluorescent reporters. Stem Cell Rep. 2015;5:582–96.

90. Song L, Awari DW, Han EY, Uche-Anya E, Park SH, Yabe YA, Chung WK, Yazawa M. Dual optical recordings for action potentials and calcium handling in induced pluripotent stem cell models of cardiac arrhythmias using genetically encoded fluorescent indicators. Stem Cells Transl Med. 2015;4:468–75.

91. Arunachalam K, Lakshmanan S, Maan A, Kumar N, Dominic P. Impact of drug induced long QT syndrome: a systematic review. J Clin Med Res. 2018;10:384–90.

92. Zhang J, Li H, Trounson A, Wu JC, Nioi P. Combining hiPSCs and human genetics: major applications in drug development. Cell Stem Cell. 2017;21:161–5.

93. Warren CR, Cowan CA. Humanity in a dish: population genetics with iPSCs. Trends Cell Biol. 2018;28:46–57.

94. Cai J, DeLaForest A, Fisher J, Urick A, Wagner T, Twaroski K, Cayo M, Nagaoka M, Duncan SA. Protocol for directed differentiation of human pluripotent stem cells toward a hepatocyte fate. Cambridge, MA: StemBook; 2008.

95. Cayo MA, Mallanna SK, Di Furio F, Jing R, Tolliver LB, Bures M, Urick A, Noto FK, Pashos EE, Greseth MD, Czarnecki M, Traktman P, Yang W, Morrisey EE, Grompe M, Rader DJ, Duncan SA. A drug screen using human iPSC-derived hepatocyte-like cells reveals cardiac glycosides as a potential treatment for hypercholesterolemia. Cell Stem Cell. 2017;20:478–489 e5.

96. Kandasamy K, Chuah JK, Su R, Huang P, Eng KG, Xiong S, Li Y, Chia CS, Loo LH, Zink D. Prediction of drug-induced nephrotoxicity and injury mechanisms with human induced pluripotent stem cell-derived cells and machine learning methods. Sci Rep. 2015;5:12337.

97. Tiburcy M, Hudson JE, Balfanz P, Schlick S, Meyer T, Chang Liao ML, Levent E, Raad F, Zeidler S, Wingender E, Riegler J, Wang M, Gold JD, Kehat I, Wettwer E, Ravens U, Dierickx P, van Laake LW, Goumans MJ, Khadjeh S, Toischer K, Hasenfuss G, Couture LA, Unger A, Linke WA, Araki T, Neel B, Keller G, Gepstein L, Wu JC, Zimmermann WH. Defined engineered human myocardium with advanced maturation for applications in heart failure modeling and repair. Circulation. 2017;135:1832–47.

98. Fujita B, Zimmermann WH. Myocardial tissue engineering strategies for heart repair: current state of the art. Interact Cardiovasc Thorac Surg. 2018;27:916–20.

99. Lemoine MD, Mannhardt I, Breckwoldt K, Prondzynski M, Flenner F, Ulmer B, Hirt MN, Neuber C, Horvath A, Kloth B, Reichenspurner H, Willems S, Hansen A, Eschenhagen T, Christ T. Human iPSC-derived cardiomyocytes cultured in 3D engineered heart tissue show physiological upstroke velocity and sodium current density. Sci Rep. 2017;7:5464.

100. Breckwoldt K, Letuffe-Breniere D, Mannhardt I, Schulze T, Ulmer B, Werner T, Benzin A, Klampe B, Reinsch MC, Laufer S, Shibamiya A, Prondzynski M, Mearini G, Schade D, Fuchs S, Neuber C, Kramer E, Saleem U, Schulze ML, Rodriguez ML, Eschenhagen T, Hansen A. Differentiation of cardiomyocytes and generation of human engineered heart tissue. Nat Protoc. 2017;12:1177–97.

101. Weinberger F, Mannhardt I, Eschenhagen T. Engineering cardiac muscle tissue: a maturating field of research. Circ Res. 2017;120:1487–500.

102. Ott HC, Matthiesen TS, Brechtken J, Grindle S, Goh SK, Nelson W, Taylor DA. The adult human heart as a source for stem cells: repair strategies with embryonic-like progenitor cells. Nat Clin Pract Cardiovasc Med. 2007;4(Suppl 1):S27–39.

103. Guyette JP, Charest JM, Mills RW, Jank BJ, Moser PT, Gilpin SE, Gershlak JR, Okamoto T, Gonzalez G, Milan DJ, Gaudette GR, Ott HC. Bioengineering human myocardium on native extracellular matrix. Circ Res. 2016;118:56–72.
104. Nugraha B, Buono MF, Emmert MY. Modelling human cardiac diseases with 3D organoid. Eur Heart J. 2018;39:4234–7.
105. Nugraha B, Buono MF, von Boehmer L, Hoerstrup SP, Emmert MY. Human cardiac organoids for disease modeling. Clin Pharmacol Ther. 2019;105:79–85.

Systems Medicine as a Transforming Tool for Cardiovascular Genetics

11

Melanie Boerries and Tanja Zeller

Contents

M. Boerries
Faculty of Medicine, Institute of Medical Bioinformatics and Systems Medicine, University Medical Center, University of Freiburg, Freiburg, Germany

German Cancer Consortium (DKTK), Partner Site Freiburg, Freiburg, Germany

German Cancer Research Center (DKFZ), Heidelberg, Germany
e-mail: m.boerries@dkfz-heidelberg.de

T. Zeller (✉)
Clinic for General and Interventional Cardiology, University Heart Center Hamburg, Hamburg, Germany

German Center for Cardiovascular Research (DZHK), Partner site Hamburg/Lübeck/Kiel, Hamburg, Germany
e-mail: t.zeller@uke.de

© Springer Nature Switzerland AG 2019
J. Erdmann, A. Moretti (eds.), *Genetic Causes of Cardiac Disease*, Cardiac and Vascular Biology 7, https://doi.org/10.1007/978-3-030-27371-2_11

Abstract

Cardiovascular diseases represent one of the most important causes of morbidity and mortality worldwide. The pathogenesis of cardiovascular disease is complex and remains elusive. Remarkable progress in deciphering the molecular mechanisms of cardiovascular disease has been achieved in different fields of research, ranging from basic-experimental research to molecular epidemiology to clinical research. Each of these isolated fields have successfully improved the pathophysiological understanding in cardiovascular disease. Within the last years, systems medicine has emerged to study the complex genetic, molecular, and physiological interactions leading to cardiovascular diseases by integrating and combining multilevel data sets from different research fields.

This chapter provides an overview of the current understanding of systems medicine and the computational and epidemiological tools applied. First, applications of systems medicine in cardiovascular research are described and challenges and opportunities that arise with systems medicine as a promising tool for cardiovascular genetics are discussed.

11.1   Introduction

Cardiovascular disease (CVD) comprises numerous complications of the heart, many of which are related to impaired blood flow that might lead to congestive disorders, arrhythmia, or heart valve problems [1]. Causes for organ-wide failure or disorders often have their origin at the subcellular, molecular level, whose effects spread to the level of the whole heart, eventually leading to its spontaneous failure. Thus, to understand the origin and progression of CVD, multiple levels of cellular function—from genomics to transcriptomics and proteomics—to cellular organization—from cell-cell communication, adhesion, and movement—to organ organization including heart physiology and the phenotypic and clinical characteristics of an individual, have to be considered—and must be studied in a dynamic context. Knowledge from each individual level has already been accumulated—although somewhat fragmented—and has been made increasingly accessible with digitalization [2]. To provide a detailed and comprehensive picture, now, we have to learn how to "put the individual pieces together" by learning how to best link and systematically integrate multidimensional datasets.

The concept of systems biology tackles exactly these questions [3, 4]. Systems biology aims at understanding inter- and intracellular dynamic processes through the integration of various high- and low-throughput quantitative data by using mathematical approaches. The concept of a mathematical description of molecular and cellular processes was already applied in simple biological systems more than 50 years ago [5–7]. Driven forward by the development of molecular biology and genetics as research disciplines, the subsequent decades showed large advancements to link molecular events to diseases. In recent years, the generation and evolution of high-dimensional and global molecular data, which are summarized as omics data (i.e., genomics, transcriptomics, proteomics, metabolomics, lipidomics, gylcomics, as well as phenomics or radiomics), have revolutionized approaches to complex biological phenomena. Owing to this development, the last years have witnessed spectacular successes in the identification of new loci involved in the susceptibility to CVD [8–12]. The underlying pathophysiological mechanisms of CVD remain, however, still to be determined. Recently, the idea of a systematic integration of multidimensional datasets in biomedical research and clinical management has percolated through virtually all medical disciplines and has been expanded as the field of systems medicine.

Systems medicine can be described as the implementation of systems biology approaches into medical research (https://www.casym.eu [13]). According to this description, a more comprehensive and detailed picture and thus an improvement of disease understanding is taken by the combination of multidimensional omics data from omics-based science, systems biology, network theory, and clinical experience and medical informatics tools for translating the knowledge into improved patient care [14].

Consequently, systems medicine as a new research field does not only rely on bioinformatics and mathematics, but also on fruitful collaborations across disciplinary boundaries additionally involving computational experts, clinicians, engineers, data manager(s), as well as epidemiologists and researchers in life science (s) (Fig. 11.1).

This chapter provides an overview of the status of systems medicine in the field of cardiovascular disease, with a particular view on the computational methods and tools currently applied, and discusses the opportunities and challenges that come along with the implementation of systems medicine into cardiology.

11.2 Methods in Systems Medicine

Despite technological advances, the main conceptual challenge remains the integration of complex data across entities, space, and time. The difficulty of data integration is deeply rooted in the emergence of novel behavior stemming from molecular and cellular self-organization, which does not allow extrapolation of knowledge across biological processes. For example, current knowledge of protein dynamics cannot explain the emergence of the transcriptional landscape of a cell. To circumvent this problem, the modeling of biological systems generally pursues two major approaches: a top-down and a bottom-up approach. The top-down approach starts from statistical analysis of large-scale omics data with the goal to discern regulatory patterns between different experimental conditions or patient samples. While it allows an unbiased, holistic view of the biology, it usually provides only a limited understanding of molecular mechanisms and thus remains mostly correlative instead of causative. In the case of CVD, a top-down approach can stratify patients with respect to their disease state and pathway activity at a given time point, without, however, considering their pathophysiology in detail or understanding the individual cell, e.g., a cardiomyocyte, all of which might be causing the disease (Fig. 11.2). A

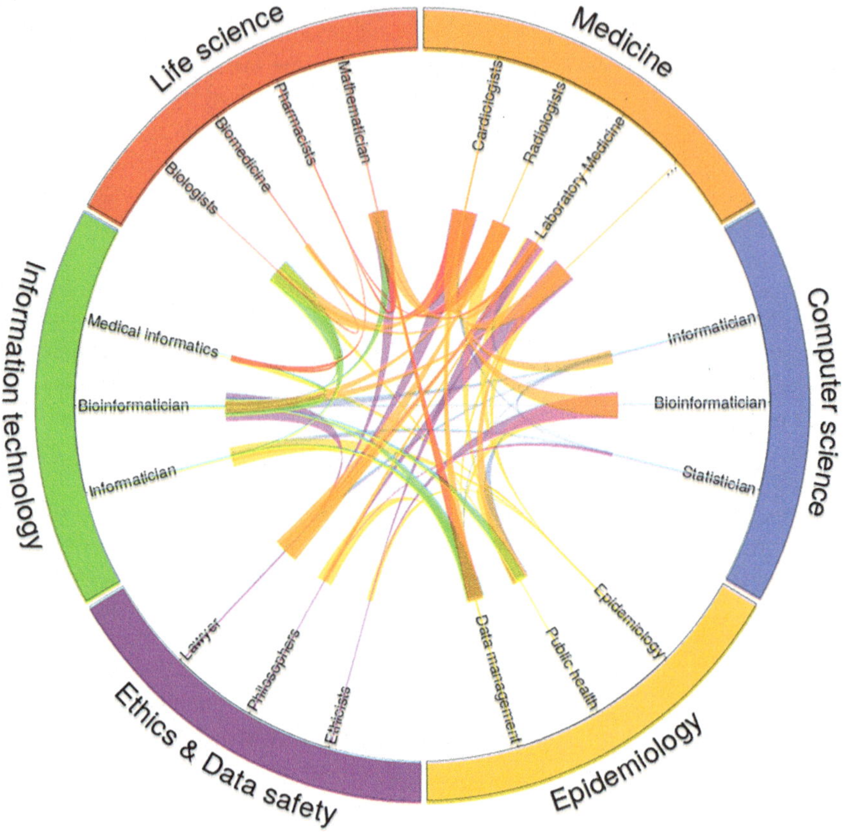

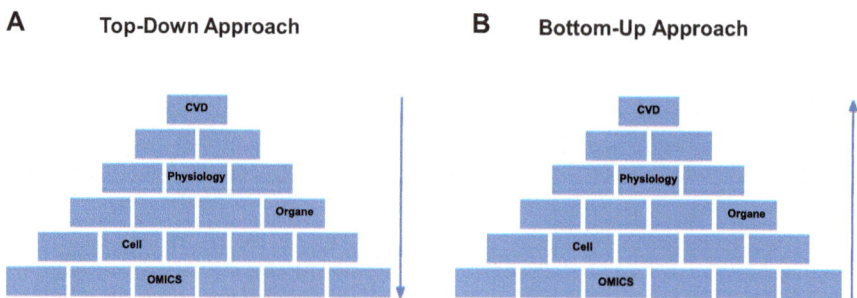

Fig. 11.1 Interdisciplinarity in systems medicine. Different disciplines jointly work together in a highly connected environment

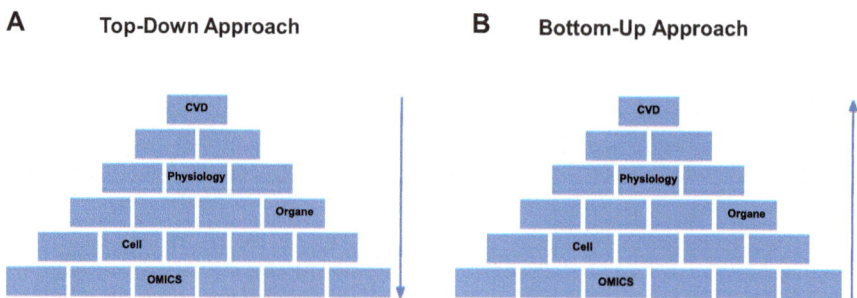

Fig. 11.2 Principles of systems biology. (**a**) The top-down approach allows an unbiased, holistic view of a biological system, without understanding the molecular mechanisms in detail. In contrast, the bottom-up approach attempts to understand and capture molecular mechanisms and interaction with the aim to transfer this knowledge on larger biological network (**b**). Figure modified from Michael Liebmann, Guest Commentary, Bio IT World, March, 2004

bottom-up approach tries to understand disease starting from molecular interactions and integrating these into larger biological networks. Such an approach requires quantitative dynamic data, which is often not available in sufficient detail, often limiting model up-scaling, and in a limited medical impact.

11.2.1 Networks and Pathways

A different approach toward understanding global properties of biological systems is the use of network theory. Cellular behavior is controlled through protein–protein interactions. This so-called interactome can be abstracted into a network, whereby proteins corresponding to nodes and edges represent possible interactions between them. The virtue of this approach is the use of network theory tools to study the importance of individual proteins, to allow dynamic simulation, and to study the effect of network rewiring on the network state, namely, cellular phenotype [15, 16]. The resulting network topology provides insight into organizational principles of the cell. Biological networks are hierarchically organized with few central hubs being connected to many nodes and with many nodes having only few interactors [17]. This makes biological networks robust to random deletions of nodes, i.e., failure of a protein/gene mutation, but fragile to targeted interventions with hub genes or proteins, which often then leads to disease. Likewise, these hub genes/proteins are potential drug targets for a given phenotype [18]. The topology of biological networks has often been exploited for functional enrichment and annotation of genes/proteins of interest [19]. Random [20] and network diffusion methods [21, 22] allow to detect modules and subnetworks among differentially regulated genes or sets of gene mutations, which can subsequently be studied in details. In particular, network diffusion is an interesting approach to find genetic predisposition toward disease and has been successfully applied to cancer [23] or autism [24].

Using this network approach, it is possible to broaden the view from a single protein to larger interaction networks and to explore complex interactions and disease progression in the necessary unbiased way.

It needs to be kept in mind that network analyses crucially depend on the underlying knowledge on network nodes and edges. Various public protein–protein interaction databases exist, as the Biological General Repository for Interaction Datasets [25] (http://www.thebiogrid.org), the Biomolecular Interaction Network Database [26] (https://omictools.com/bind-tool), the Database of Interacting Proteins [27] (http://dip.doe-mbi.ucla.edu), the Molecular INTeraction database [28] (https://omictools.com/mint-3-tool), the IntAct molecular interaction database [29] (http://www.ebi.ac.uk/intact), or the Human Protein Reference Database [30] (http://www.hprd.org). A common protein–protein interaction (PPI) tool is the STRING database (https://string-db.org/), which includes indirect (functional) as well as direct interactions (physical) and also predicted protein–protein associations based on co-expression data [31]. From these PPI networks, molecular function and disease association can be predicted based on network topology [32]. However, the identification of relevant network regions (modules), which are linked to diseases and biological functions, is still a major challenge and goal in the field of systems biology/medicine [33]. The following aspects make it even more difficult: (a) search for a subnetwork (module) depends on the size of the biological network and is in general time consumable, (b) the usage of different omics platforms (sequencing, mass spectrometry, or microarrays) comprise the problems of normalization, batch correction, annotation, and also completeness. Moreover, our biological understanding is incomplete, e.g., only approximately 15% of all protein–protein interactions are known. To gain insight into the interaction properties, network diffusion-based approaches have been developed to calculate a global measure of network proximity by simulation the diffusion of quantity throughout a network. This idea is implemented in the tools HotNet [22] or Response Net [34]. Network-based stratification (NBS) integrates mutations into PPIs to stratify cancer into informative subtypes. This elucidates whether patients share similar network regions based on their mutations. These networks are also predictive on patient survival or tumor response therapy and can be used for RNA expression patterns with the aim to obtain similar information in the absence with DNA sequencing [35]. In this context, weighted gene co-expression network analysis (WGCNA) [36] uses network construction to identify biological functional relevant modules and its respective key drivers. Using this approach, Horvath et al. could identify a molecular target in glioblastoma as a key gene within a gene co-expression module that has a major impact on cell proliferation. The emphasis on pathway modules reduces the internal complexity, e.g., comparison of 20 modules as compared to 20,000 genes. Moreover, intramodular connectivity can be used to identify key drivers, so-called hub nodes. In this context, WGCNA were used to identify the two genes, G6PD and S100A7, that are related to coronary artery disease [37].

Another network approach is transcriptional regulatory associations in pathways (TRAP). In contrast to other gene–gene network approaches, this network inference algorithm is able to identify gene-pathway transcriptional regulatory relationship, in which a single gene is determined as a transcriptional regulator [38]. In this algorithm, regulator-pathway associations are derived from transcriptional co-regulation in more than 3000 individual microarray experiments with transcription factor-coding genes, taken from the Riken Transcription Factor Database and Regulator Pathways.

In order to obtain a functional insight into gene regulation, Subramanian et al. developed a powerful analytical method called gene set enrichment analysis (GSEA) that is based on focusing on gene sets. The latter is a group of genes that share common biological properties/function. GSEA determines whether the genes are randomly distributed or specifically assigned to a gene set [39]. Such an analysis benefits greatly from the annotation of biological databases, e.g., ConsensusPath DB or Kyoto Encyclopedia of Genes and Genomes (KEGG). ConsensusPathDB is a meta pathway database that provides a comprehensive collection of human (as well as mouse and yeast) molecular interaction data together with a web interface (http://consensuspathdb.org) that offers several computational methods and visualization tools to analyze these data. This database comprises network modules interaction, biochemical pathways, and functional information to support data interpretation of high-throughput data with the aim to prioritize specific biological function and signaling pathways [40, 41]. KEGG is a database resource of large molecular datasets based on genome sequencing and other high-throughput experimental technologies to obtain a high-level functional understanding of biological systems, such as a cell or an organism [42] (http://www.genome.jp/kegg). Another database is Reactome that focuses on human pathways, biological processes, and reactions information [43]. It also offers a web interface including tools for pathway browsing and data analysis (https://reactome.org). The Gene Ontology Consortium (http://www.geneontology.org) has the aim to generate an extensible, ordered, and controlled vocabulary that comprises many eukaryotic processes and is subgrouped into three main ontologies: biological process, molecular function, and cellular component [44].

Further databases for specific diseases and gene expression exist, such as the genotype-tissue expression (GTEx, https://www.gtexportal.org) [45] that examines association between genetic variation and gene expression in human tissues, or the Human Protein Atlas (https://www.proteinatlas.org) [46] and The Cancer Genome Atlas (TCGA, https://cancergenome.nih.gov). The latter links genetic alteration to specific cancer types, whereas the protein Human Protein Atlas maps proteins in cells, tissues, and organs based on both protein and transcriptome data sets. Due to such databases and tools [47], the analysis and understanding of high throughput data, also with regard to CVD, can greatly be improved.

11.2.2 Cohort Studies and Biobanks

In order to identify novel risk factors for CVD, large epidemiological cohort studies are particularly suitable. In a cohort study, cardiovascular phenotypes and subsequent fatal and non-fatal events can be studied over time and provide powerful results in identifying relationships between the characteristics of a population and that population's behavior. Two types of cohort studies are prominently used in cardiovascular disease. Prospective studies are carried out from the present time into the future, whereas retrospective studies are carried out at the present time but look to the past to examine disease events and outcomes [48]. The best-known epidemiological cohort study in the cardiovascular field is the Framingham Heart Study (FHS), initiated in 1948, which is currently investigating the third generation of the original participants. The FHS is the longest running prospective cohort study and has contributed enormously to the understanding of various cardiovascular risk factors and to how these factors relate to the overall and cardiovascular-related mortality [49, 50]. The FHS could only be generated based on large populations, generation of a large amount of big data, and by the support of several health institutes. On the basis of the knowledge generated by the FHS, other epidemiological cohort studies were implemented in the last years and have provided additional information on cardiovascular disease risk [51–55].

In the last decades, bio specimens became an important additional resource in epidemiological studies and are collected and stored into various so-called biobanks.

The main bio specimens collected include blood samples, urine, feces, tonsil swaps, saliva, tear fluid, tissue samples and biopsies, as well as genetic material. These materials made modern molecular analyses such as omics analyses feasible on the large scale as these specimens are easily collected and various related phenotypic and linked clinical data are available.

Today, large biobanks are an essential part of a cohort's infrastructure [56, 57] and build the basis for a large part of the biomedical research and consequently are critical for translating advances in molecular biology and new possibilities in the context of systems medicine [56].

11.3   Systems Medicine in Cardiovascular Disease

In the cardiovascular biomarker field, recent studies that relied on the concept of systems medicine have been proposed and/or applied. A few studies have already successfully identified novel mechanisms of CVD and/or were translated into clinical application.

11.3.1 Systems Medicine Approaches to Reveal Novel Molecular Pathways

11.3.1.1 Myocardial Infarction

As a major cause of death, myocardial infarction (MI) is best predicted in middle-aged adults by a positive family history, strongly supporting a genetic basis of MI [58]. Investigating the genetic background of MI with tools applied in systems medicine, a novel pathway and potential drug targets for treatment of cardiovascular disease were identified in elegant studies combining omics and clinical data with bioinformatical and experimental models. In an extended family with several members diagnosed of coronary artery disease (CAD), exome sequencing and linkage analysis identified rare loss-of-function and missense mutations in the genes GUCY1A3 and the chaperonin containing TCP1 subunit 7 (CCT7), respectively [12].

The GUCY1A3 gene encodes the alpha1 subunit of the soluble guanylyl cyclase (sGC). The sGC complex, a heterodimer of the alpha1 and a beta1 subunit, acts as the receptor for nitric oxide (NO) and catalyzes the formation of the second messenger cGMP [59]. cGMP has several cellular functions, including the inhibition of platelet aggregation, thereby representing an important feature of thrombus formation in MI, and smooth muscle cell relaxation [59]. Further evidence point to a critical involvement of the NO-sGC-cGMP pathway in mediating CAD and MI risk [59].

Transfection of the GUCY1A3 mutation into human embryonic kidney cells as well as downregulation of CCT7 by siRNA were preformed to elucidate the functional implications of these variants. These experimental approaches led to a strong reduction of soluble guanylyl cyclase (α1-sGC) α1 levels [12]. Using platelets extracted from family members carrying either single GUCY1A3 or CCT7 mutations, a double mutation, or none of the rare alleles, α1-sGC levels and NO-dependent cGMP generation were investigated. Carriers of the digenic mutation exhibited a significant reduction of α1-sGC levels and cGMP formation. Subsequently, Gucy1A3-deficient mice showed an increased thrombus formation, indicating an increased risk of MI via dysfunctional nitric oxide signaling in rare allele carrying family members.

Further functional evidence that the variants in the risk locus affect GUCY1A3 gene expression was provided by [59]. Using human samples and cell lines, the SNP rs7692387, located in a DNaseI hypersensitive site, was identified as being involved in the regulation of GUCY1A3 gene expression. The GUCY1A3 risk variant rs7692387 seems to act by a modulation of the binding site for a transcription factor, ZEB1, thereby directly affecting the expression of the target gene.

As a consequence of impaired GUCY1A3 expression, homozygous risk allele carriers exhibited decreased inhibitory effects of sGC stimulation on cell migration, the production of cGMP was inhibited, and platelet aggregation after exposure to an NO donor was impaired. In agreement with these human data, lower Gucy1a3 expression correlated with more aortic atherosclerosis in mice. By the integration of a series of data derived from experimental settings, human clinical data, and bio specimens, a novel link between impaired soluble-guanylyl-cyclase-dependent nitric oxide signaling and MI risk was identified, possibly providing a new therapeutic target for reducing the risk of MI [12].

11.3.2 Systems Medicine Approaches to Reveal Novel Cardiovascular Biomarkers

11.3.2.1 Blood Pressure

Huan et al. [60, 61] conducted a multilevel integration analysis on one of the major cardiovascular risk factors, blood pressure (BP), to identify novel candidate genes involved in BP regulation. Computationally, the authors combined multilevel datasets including genetic, transcriptomic, and phenotype data. Several genes were identified in relation to BP, which jointly explain 5–9% of BP variation. To further seek for molecular key drivers of BP regulation, co-expression sub-networks that were jointly connected by the SH2B adaptor protein 3 (*SH2B3*) were identified [58, 61]. In order to elucidate the molecular role of *SH2B3* and the relation to hypertension, Sh2b3−/− mice were investigated in response to low-dose angiotensin II treatment [62]. In untreated Sh2b3−/− mice, kidneys and aortas showed greater levels of inflammation, oxidative stress, and glomerular injury. These effects were accelerated after angiotensin II infusion. A strong indication that the predominant effect of SH2B3 on BP is mediated by hematopoietic cells was shown by experiments of bone marrow transplantations of SH2b3−/− into wild-type which reproduced the hypertensive phenotype. Subsequent studies identified the genes of the BP co-expression networks [58, 63] including CRIP1 (cysteine-rich protein 1). These studies further elucidated that CRIP1 transcript expression additionally correlated to measures of cardiac hypertrophy and identified circulating CRIP1 protein levels as a potential biomarker for increased risk for incident stroke, a sequel of high BP [63].

11.3.2.2 Microbiome

During the last years, the gut microbiome has attracted increasing attention in many medical fields and emerged as a central factor affecting human health and disease [64] and has been linked to cardiovascular disease. In a metagenome-wide association study, a cross-disease cohort integrative analysis was performed, including data from subjects affected by atherosclerotic disease, cardiometabolic diseases, obesity, type 2 diabetes, liver cirrhosis, or the autoimmune disease rheumatoid arthritis [65].

Compared to the healthy controls, the gut microbiome of atherosclerotic disease patients was significantly different in multivariate analyses and showed separation in principle component and distance-based redundancy analyses, which was supported by the relative reduction in the genera Bacteroides and Prevotella, and an enrichment in Streptococcus and Escherichia in atherosclerotic disease patients. Co-abundance network of metagenomically (genetically) linked groups confirmed the major compositional differences in individuals with and without atherosclerotic disease.

To explore the diagnostic value of the gut microbiome composition in relation to atherosclerotic disease, random forest classifiers including several different metagenomics linked groups and KEGG (The Kyoto Encyclopedia of Genes and Genomes) functional modules were selected to construct a mathematical model that predicts clinical indices. A discriminatory ability to distinguish between individuals with and without atherosclerotic disease with an area under the receiver operating curve of 0.86 was observed, demonstrating the presence of atherosclerotic disease-associated features in the gut microbiome that may be further developed into non-invasive and inexpensive biomarkers [65]. Insights into the potential possible changes within the gut microbiome in the disease state were provided by KEGG pathway analyses, showing—among others—a higher potential for sugar and amino acid transport, a lower potential for vitamin synthesis altered potential for homocysteine and tetrahydrofolate metabolism, and an enrichment in virulence factors.

As medication can influence the gut microbiome, subsequent analyses were performed and random-forest classifiers for distinguishing between atherosclerotic disease patients treated with and without medication. Overall, the results propose that although drug treatment may affect the composition of the gut microbiota and thus constitute a confounding factor, medication weakened the disease signal, meaning an even more significant difference would be expected in a cohort free of medication [65].

These results were extended to other related diseases such as obesity, type 2 diabetes, liver cirrhosis, and rheumatoid arthritis, and multi-disease microbiome-based classifiers were extracted from these additional microbiome datasets. The results showed that the gut microbiome of each disease exhibited unique features in functional capacity, in species, and gene compositions.

By using human bio specimens, clinical data and bioinformatical approaches, this study implicates a role of the gut microbiome in heart disease. The data represent a comprehensive resource for further investigations on the role of the gut microbiome in promoting or preventing cardiovascular disease as well as other related diseases.

There is a considerable power of using the microbiome to separate diseased subjects from healthy subjects as well as to predict responses to treatment or the development of diseases in the absence of treatment [66]. To further guide the utility of the microbiome for robust clinical use, data must be validated in larger and more diverse populations, and methodologies must be standardized across different laboratories [66].

11.4 Challenges in Systems Medicine

Systems medicine aims to explore medicine beyond linear relationships and single parameters and includes multiple parameters and spatial conditions to achieve a holistic perspective [67].

Over the last years, emerging and interdisciplinary approaches with extensive tools, novel analytical methods and strategies have rapidly been developed.

Promises are made that systems medicine will lead to an innovation in diagnosis and prognosis of cardiovascular diseases and a better understanding of disease mechanisms and pathology and thereby improved therapeutic options.

Nevertheless, at the current state, the adaptation of systems medicine into medical research and, in the long run, into "real-world" clinical practice is a complex endeavor and researchers are facing multiple challenges (Fig. 11.3).

Here, we will discuss the main challenges that arise from working with large amounts of data and from working in an interdisciplinary team.

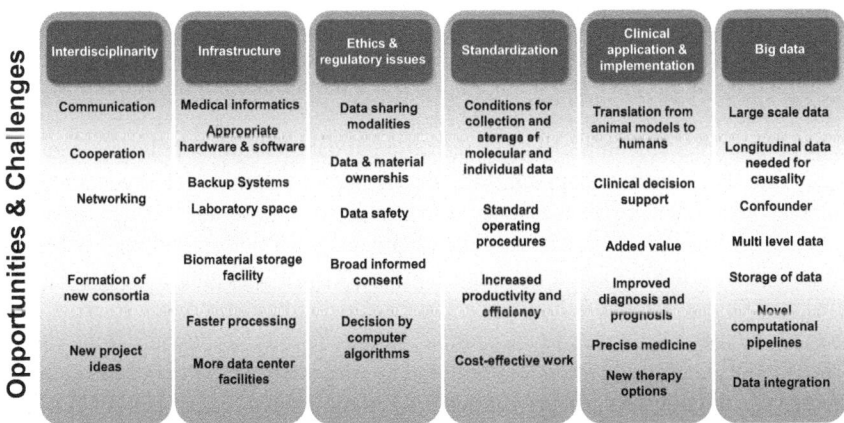

Fig. 11.3 Challenges and opportunities in systems medicine

11.4.1 Big Data: Harmonization of Data/Methods/Formats

With the advent of high-throughput technologies, massive data sets are generated that need to be handled, processed, harmonized, and shared [68].

For instance, with the falling sequencing costs, soaring accuracy, and a steadily expanding base of scientific knowledge a vast amount of data, perhaps in the range of petabytes, is generated each year [69]. Thus, some researchers worry that the flood of data could overwhelm the computational pipelines needed for analysis and generate unprecedented demand for storage [69].

Therefore, new infrastructures and information technology systems need to be developed and implemented into the research facilities and clinics. In particular, clinical routine will require different levels of granularity of the information— ranging from a short overview of the patient record in the clinical information system up to an in-depth discussion of potential therapies [70] or to an inclusion for electronic decision support systems [71, 72]. Subsequently, standardization of conditions for collection and storage of molecular and individual data as well as the respective metadata need to be established.

The benefits of harmonizing and pooling research databases will be enormous. Although over the past half-decade both governments and researchers emphasized the importance of harmonizing data as well as collaborative usage of data and samples, the integration and harmonization of different data sets, especially high-throughput data, will continue to be a major challenge [73, 74]. This situation is exacerbated by ethical, legal, and civil rights with regard to the sharing or aggregation of data at an individual level. New database management systems are at the forefront of providing solutions to some of these issues [75]. The idea is that these tools can link different distributed databases with the aim to ensure a secure and effective data analysis. This endeavor can be supported through an interactive and strong collaboration of the different project partners and also through a clear and defined annotation of the diverse data. In this context, BioSHarre (biobank standardization and harmonization for research excellence in the European Union) is a project of the Seventh Framework Program (FP7) and aims to develop tools for data harmonization and standardized IT systems for existing biobanks and cohorts across Europe and apply them to conduct pan-European epidemiological research [76] (https://www.bioshare.eu/).

In spite of this, it is important that the different data sets are stored right from the start with the correct annotation and format in the database. This can only be implemented if the consensus on data formats and data annotation and also metadata information is clearly defined. Such data standards have first started with MIAME (minimum information about a microarray experiment) [77] and have found their way into Omics databases like Gene Expression omnibus or Array Express and recently with the Sequence Read Archive. Sequence data on humans from whole exome/whole genome data add an additional layer of controlled access due to ethical reasons. Such databases are the European Genome-Phenome Archive (EGA) [78] the NCBI (NCBI, National Center for Biotechnology Information; https://www.ncbi.nlm.nih.gov/) where the information is usually stored.

11.4.2 Data Exchange, Data Safety, and Ethical Aspects

Worldwide, large discussions are currently ongoing regarding the privacy and ownership of data and sharing of data and bio specimens. Traditional models of consent became impractical as more and more large numbers of individuals are included in cohort studies and a much larger amount of data is generated from one individual (e.g., by genome-wide microarray or sequencing approaches). Furthermore, due to the interdisciplinary setting, these large data sets and bio specimens might be shared with internal as well as external partners. Therefore, newer consent strategies include the use of broad or open consent [79] to cover the multilevel assessment of personal, clinical, and biological data as well as the storage and use of data and bio specimens in biobanks [80].

Further issues need to be discussed when medical informatics tools, e.g., electronic decision support and medical informatics systems, might directly be integrated and translated into clinical practice to support patient care [14, 81–83]. For example, is it feasible that clinical decisions will be derived (only) from computer algorithms? What about concerns related to privacy, data protection, and ownership of data?

11.4.3 Interdisciplinary Communication

Besides all the technical difficulties, an important challenge is the multidisciplinarity that inevitably evolves with systems medicine. Different scientific and clinical disciplines are involved in systems medicine, ranging from epidemiologists and data managers to informaticians, bioinformaticians, statisticians, and life scientists, and most importantly to clinicians. Even within one of these specific groups, several experts (e.g., cardiologist, radiologist, experts in intensive care) are needed for proper interpretation of the data and clinical situation.

A first step in working in an interdisciplinary team is to overcome problems of communication. Each research area has its own language leading to difficulty of understanding and working with each other [13]. However, the closer these disciplines work together, understand, and learn from each other, the more successful will be the progress in system medicine.

11.5 Conclusion and Perspectives

11.5.1 The Future of Systems Medicine in Cardiovascular Research

Systems medicine emerged as a powerful tool to study complex diseases by the integration of multidimensional datasets [47]. In the near future, multidimensional, large-scale data are generated, jointly analyzed with phenotypical and clinical characteristics, hypotheses are functionally examined in basic research models, and findings are translated into epidemiological cohort studies to gain knowledge about the cardiovascular clinical meaning. The expectations are high that the research results will be translated into routine clinical practice.

For an efficient implementation of systems medicine in the cardiovascular field, important cornerstones are: (1) a sustained interplay and communication between experts of multiple disciplines, (2) an optimized usage of resources and infrastructures to efficiently share data, biomaterial, and knowledge, and (3) the extension of the research beyond traditional domains of discovery and disease etiology to accelerate translation into the clinics [84].

In the cancer field, some of these cornerstones have already been implemented such as the interdisciplinary tumor boards. Therein, the individual patient diagnoses are discussed in order to identify therapeutic opportunities based on extended molecular diagnostics in combination with genome, transcriptome, epigenome sequencing followed by bioinformatics analysis. Such boards are characterized not only by great interdisciplinarity, integrating various clinical research areas, such as oncology, pathology, system biology/system medicine, but also by standardized workflows for patient selection, sample preparation, analysis methods, and reporting. For example, the Molecular Tumor Board of the German Cancer Consortium (DKTK) MASTER program has set itself the task to clarify whether matching treatments with individual molecular profiles lead to improved disease outcome based on the clinical application of whole exome/whole genome and RNA sequencing. This program emphasizes that the main advantage of precision medicine is the collection of characterized molecular data from large patient cohorts and its integration of genetic and clinical information of individual patients. Even if these data are often incomplete, they can be reused for many important future medical questions and research [85].

These kind of "boards" have not been established in the field of cardiovascular disease so far. Certainly, the collection of cardiovascular samples for sequence analysis is much more difficult compared to tumor samples, however, an interdisciplinary board of, e.g., cardiologists, pathologists, nephrologists, systems biologists, bioinformaticians, or further clinical disciplines will tremendously increase the molecular understanding of cardiovascular diseases and therefore will open up new avenues in diagnosis, therapy, and prevention. An example of a proposed cardiovascular board is provided in Fig. 11.4.

Given all the promises, opportunities, but also challenges of systems medicine, the upcoming years will see that researchers of different disciplines are working together to translate results from cardiovascular research into better health strategies and the use of systems medicine as a transforming tool for cardiovascular research (Fig. 11.5).

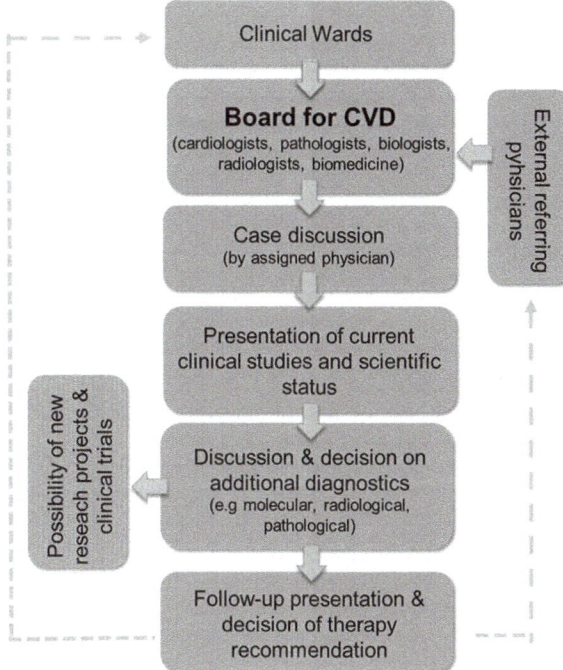

Fig. 11.4 Proposed molecular-cardiovascular board. The interdisciplinary board discusses individual patient diagnoses to identify therapeutic opportunities based on extended molecular diagnostics to open new avenues for both research projects and clinical trials

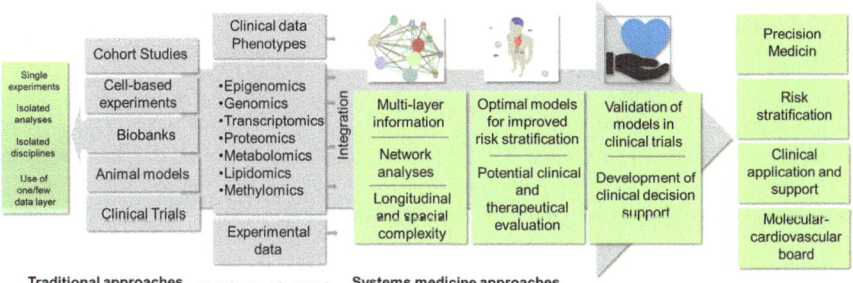

Fig. 11.5 Systems medicine as a transforming tool in cardiovascular disease. From traditional approaches of single experiments and isolated analyses to large-scale, integrated systems-based approaches to facilitate the translation of discoveries to clinical utility

References

1. Benjamin EJ, Blaha MJ, Chiuve SE, Cushman M, Das SR, Deo R, De Ferranti SD, Floyd J, Fornage M, Gillespie C, Isasi CR, Jimenez MC, Jordan LC, Judd SE, Lackland D, Lichtman JH, Lisabeth L, Liu S, Longenecker CT, Mackey RH, Matsushita K, Mozaffarian D, Mussolino ME, Nasir K, Neumar RW, Palaniappan L, Pandey DK, Thiagarajan RR, Reeves MJ, Ritchey M, Rodriguez CJ, Roth GA, Rosamond WD, Sasson C, Towfighi A, Tsao CW, Turner MB, Virani SS, Voeks JH, Willey JZ, Wilkins JT, Wu JH, Alger HM, Wong SS, Muntner P. Heart disease and stroke statistics-2017 update: A report from the American Heart Association. Circulation. 2017;135:e146–603.
2. Rebhan M. Towards a systems approach for chronic diseases, based on health state modeling. F1000Res. 2017;6:309.
3. Barabasi AL, Oltvai ZN. Network biology: understanding the cell's functional organization. Nat Rev Genet. 2004;5:101–13.
4. Kitano H. Computational systems biology. Nature. 2002;420:206–10.
5. Adam G, Delbruck M. Reduction of dimensionality in biological diffusion processes. San Francisco: W.H. Freeman & Company; 1968.
6. Hopfield JJ. Kinetic proofreading: a new mechanism for reducing errors in biosynthetic processes requiring high specificity. Proc Natl Acad Sci U S A. 1974;71:4135–9.
7. Luria SE, Delbruck M. Mutations of bacteria from virus sensitivity to virus resistance. Genetics. 1943;28:491–511.
8. Auer PL, Stitziel NO. Genetic association studies in cardiovascular diseases: Do we have enough power? Trends Cardiovasc Med. 2017;27:397–404.
9. Howson JMM, Zhao W, Barnes DR, Ho WK, Young R, Paul DS, Waite LL, Freitag DF, Fauman EB, Salfati EL, Sun BB, Eicher JD, Johnson AD, Sheu WHH, Nielsen SF, Lin WY, Surendran P, Malarstig A, Wilk JB, Tybjaerg-Hansen A, Rasmussen KL, Kamstrup PR, Deloukas P, Erdmann J, Kathiresan S, Samani NJ, Schunkert H, Watkins H, Do R, Rader DJ, Johnson JA, Hazen SL, Quyyumi AA, Spertus JA, Pepine CJ, Franceschini N, Justice A, Reiner AP, Buyske S, Hindorff LA, Carty CL, North KE, Kooperberg C, Boerwinkle E, Young K, Graff M, Peters U, Absher D, Hsiung CA, Lee WJ, Taylor KD, Chen YH, Lee IT, Guo X, Chung RH, Hung YJ, Rotter JI, Juang JJ, Quertermous T, Wang TD, Rasheed A, Frossard P, Alam DS, Majumder AAS, Di Angelantonio E, Chowdhury R, Chen YI, Nordestgaard BG, Assimes TL, Danesh J, Butterworth AS, Saleheen D. Fifteen new risk loci for coronary artery disease highlight arterial-wall-specific mechanisms. Nat Genet. 2017;49:1113–9.
10. Nelson CP, Goel A, Butterworth AS, Kanoni S, Webb TR, Marouli E, Zeng L, Ntalla I, Lai FY, Hopewell JC, Giannakopoulou O, Jiang T, Hamby SE, Di Angelantonio E, Assimes TL, Bottinger EP, Chambers JC, Clarke R, Palmer CNA, Cubbon RM, Ellinor P, Ermel R, Evangelou E, Franks PW, Grace C, Gu D, Hingorani AD, Howson JMM, Ingelsson E, Kastrati A, Kessler T, Kyriakou T, Lehtimaki T, Lu X, Lu Y, Marz W, McPherson R, Metspalu A, Pujades-Rodriguez M, Ruusalepp A, Schadt EE, Schmidt AF, Sweeting MJ, Zalloua PA, Alghalayini K, Keavney BD, Kooner JS, Loos RJF, Patel RS, Rutter MK, Tomaszewski M, Tzoulaki I, Zeggini E, Erdmann J, Dedoussis G, Bjorkegren JLM, Schunkert H, Farrall M, Danesh J, Samani NJ, Watkins H, Deloukas P. Association analyses based on false discovery rate implicate new loci for coronary artery disease. Nat Genet. 2017;49:1385–91.
11. Nikpay M, Goel A, Won HH, Hall LM, Willenborg C, Kanoni S, Saleheen D, Kyriakou T, Nelson CP, Hopewell JC, Webb TR, Zeng L, Dehghan A, Alver M, Armasu SM, Auro K, Bjonnes A, Chasman DI, Chen S, Ford I, Franceschini N, Gieger C, Grace C, Gustafsson S, Huang J, Hwang SJ, Kim YK, Kleber ME, Lau KW, Lu X, Lu Y, Lyytikainen LP, Mihailov E, Morrison AC, Pervjakova N, Qu L, Rose LM, Salfati E, Saxena R, Scholz M, Smith AV, Tikkanen E, Uitterlinden A, Yang X, Zhang W, Zhao W, De Andrade M, De Vries PS, Van Zuydam NR, Anand SS, Bertram L, Beutner F, Dedoussis G, Frossard P, Gauguier D, Goodall AH, Gottesman O, Haber M, Han BG, Jalilzadeh S, Kessler T, Konig IR, Lannfelt L, Lieb W,

Lind L, Lindgren CM, Lokki ML, Magnusson PK, Mallick NH, Mehra N, Meitinger T, Memon FU, Morris AP, Nieminen MS, Pedersen NL, Peters A, Rallidis LS, Rasheed A, Samuel M, Shah SH, Sinisalo J, Stirrups KE, Trompet S, Wang L, Zaman KS, Ardissino D, Boerwinkle E, Borecki IB, Bottinger EP, Buring JE, Chambers JC, Collins R, Cupples LA, Danesh J, Demuth I, Elosua R, Epstein SE, Esko T, Feitosa MF, Franco OH, et al. A comprehensive 1,000 Genomes-based genome-wide association meta-analysis of coronary artery disease. Nat Genet. 2015;47:1121–30.

12. Erdmann J, Stark K, Esslinger UB, Rumpf PM, Koesling D, De Wit C, Kaiser FJ, Braunholz D, Medack A, Fischer M, Zimmermann ME, Tennstedt S, Graf E, Eck S, Aherrahrou Z, Nahrstaedt J, Willenborg C, Bruse P, Braenne I, Nothen MM, Hofmann P, Braund PS, Mergia E, Reinhard W, Burgdorf C, Schreiber S, Balmforth AJ, Hall AS, Bertram L, Steinhagen-Thiessen E, Li SC, Marz W, Reilly M, Kathiresan S, McPherson R, Walter U, Ott J, Samani NJ, Strom TM, Meitinger T, Hengstenberg C, Schunkert H. Dysfunctional nitric oxide signalling increases risk of myocardial infarction. Nature. 2013;504:432–6.

13. Wolkenhauer O, Auffray C, Jaster R, Steinhoff G, Dammann O. The road from systems biology to systems medicine. Pediatr Res. 2013;73:502–7.

14. Fischer T, Brothers KB, Erdmann P, Langanke M. Clinical decision-making and secondary findings in systems medicine. BMC Med Ethics. 2016;17:32.

15. Barabasi AL, Albert R. Emergence of scaling in random networks. Science. 1999;286:509–12.

16. Watts DJ, Strogatz SH. Collective dynamics of 'small-world' networks. Nature. 1998;393:440–2.

17. Strogatz SH. Exploring complex networks. Nature. 2001;410:268–76.

18. Rajalingam K, Schreck R, Rapp UR, Albert S. Ras oncogenes and their downstream targets. Biochim Biophys Acta. 2007;1773:1177–95.

19. Hu JX, Thomas CE, Brunak S. Network biology concepts in complex disease comorbidities. Nat Rev Genet. 2016;17:615–29.

20. Chipman KC, Singh AK. Predicting genetic interactions with random walks on biological networks. BMC Bioinform. 2009;10:17.

21. Bersanelli M, Mosca E, Remondini D, Castellani G, Milanesi L. Network diffusion-based analysis of high-throughput data for the detection of differentially enriched modules. Sci Rep. 2016;6:34841.

22. Vandin F, Upfal E, Raphael BJ. Algorithms for detecting significantly mutated pathways in cancer. J Comput Biol. 2011;18:507–22.

23. Network CGAR. Comprehensive molecular characterization of clear cell renal cell carcinoma. Nature. 2013;499:43–9.

24. Mosca E, Bersanelli M, Gnocchi M, Moscatelli M, Castellani G, Milanesi L, Mezzelani A. Network diffusion-based prioritization of autism risk genes identifies significantly connected gene modules. Front Genet. 2017;8:129.

25. Stark C, Breitkreutz BJ, Reguly T, Boucher L, Breitkreutz A, Tyers M. BioGRID: a general repository for interaction datasets. Nucleic Acids Res. 2006;34:D535–9.

26. Bader GD, Donaldson I, Wolting C, Ouellette BF, Pawson T, Hogue CW. Bind – The biomolecular interaction network database. Nucleic Acids Res. 2001;29:242–5.

27. Xenarios I, Rice DW, Salwinski L, Baron MK, Marcotte EM, Eisenberg D. DIP: the database of interacting proteins. Nucleic Acids Res. 2000;28:289–91.

28. Zanzoni A, Montecchi-Palazzi L, Quondam M, Ausiello G, Helmer-Citterich M, Cesareni G. MINT: a Molecular INTeraction database. FEBS Lett. 2002;513:135–40.

29. Hermjakob H, Montecchi-Palazzi L, Lewington C, Mudali S, Kerrien S, Orchard S, Vingron M, Roechert B, Roepstorff P, Valencia A, Margalit H, Armstrong J, Bairoch A, Cesareni G, Sherman D, Apweiler R. IntAct: an open source molecular interaction database. Nucleic Acids Res. 2004;32:D452–5.

30. Peri S, Navarro JD, Amanchy R, Kristiansen TZ, Jonnalagadda CK, Surendranath V, Niranjan V, Muthusamy B, Gandhi TK, Gronborg M, Ibarrola N, Deshpande N, Shanker K, Shivashankar HN, Rashmi BP, Ramya MA, Zhao Z, Chandrika KN, Padma N, Harsha HC,

Yatish AJ, Kavitha MP, Menezes M, Choudhury DR, Suresh S, Ghosh N, Saravana R, Chandran S, Krishna S, Joy M, Anand SK, Madavan V, Joseph A, Wong GW, Schiemann WP, Constantinescu SN, Huang L, Khosravi-Far R, Steen H, Tewari M, Ghaffari S, Blobe GC, Dang CV, Garcia JG, Pevsner J, Jensen ON, Roepstorff P, Deshpande KS, Chinnaiyan AM, Hamosh A, Chakravarti A, Pandey A. Development of human protein reference database as an initial platform for approaching systems biology in humans. Genome Res. 2003;13:2363–71.

31. Szklarczyk D, Franceschini A, Wyder S, Forslund K, Heller D, Huerta-Cepas J, Simonovic M, Roth A, Santos A, Tsafou KP, Kuhn M, Bork P, Jensen LJ, Von Mering C. STRING v10: protein-protein interaction networks, integrated over the tree of life. Nucleic Acids Res. 2015;43:D447–52.

32. Wang X, Gulbahce N, Yu H. Network-based methods for human disease gene prediction. Brief Funct Genomics. 2011;10:280–93.

33. Barabasi AL, Gulbahce N, Loscalzo J. Network medicine: a network-based approach to human disease. Nat Rev Genet. 2011;12:56–68.

34. Lan A, Smoly IY, Rapaport G, Lindquist S, Fraenkel E, Yeger-Lotem E. ResponseNet: revealing signaling and regulatory networks linking genetic and transcriptomic screening data. Nucleic Acids Res. 2011;39:W424–9.

35. Hofree M, Shen JP, Carter H, Gross A, Ideker T. Network-based stratification of tumor mutations. Nat Methods. 2013;10:1108–15.

36. Zhang B, Horvath S. A general framework for weighted gene co-expression network analysis. Stat Appl Genet Mol Biol. 2005;4:Article 17.

37. Liu J, Jing L, Tu X. Weighted gene co-expression network analysis identifies specific modules and hub genes related to coronary artery disease. BMC Cardiovasc Disord. 2016;16:54.

38. Kwong LN, Costello JC, Liu H, Jiang S, Helms TL, Langsdorf AE, Jakubosky D, Genovese G, Muller FL, Jeong JH, Bender RP, Chu GC, Flaherty KT, Wargo JA, Collins JJ, Chin L. Oncogenic NRAS signaling differentially regulates survival and proliferation in melanoma. Nat Med. 2012;18:1503–10.

39. Subramanian A, Tamayo P, Mootha VK, Mukherjee S, Ebert BL, Gillette MA, Paulovich A, Pomeroy SL, Golub TR, Lander ES, Mesirov JP. Gene set enrichment analysis: a knowledge-based approach for interpreting genome-wide expression profiles. Proc Natl Acad Sci U S A. 2005;102:15545–50.

40. Herwig R, Hardt C, Lienhard M, Kamburov A. Analyzing and interpreting genome data at the network level with ConsensusPathDB. Nat Protoc. 2016;11:1889–907.

41. Kamburov A, Stelzl U, Lehrach H, Herwig R. The ConsensusPathDB interaction database: 2013 update. Nucleic Acids Res. 2013;41:D793–800.

42. Kanehisa M, Goto S. KEGG: kyoto encyclopedia of genes and genomes. Nucleic Acids Res. 2000;28:27–30.

43. Croft D, O'kelly G, Wu G, Haw R, Gillespie M, Matthews L, Caudy M, Garapati P, Gopinath G, Jassal B, Jupe S, Kalatskaya I, Mahajan S, May B, Ndegwa N, Schmidt E, Shamovsky V, Yung C, Birney E, Hermjakob H, D'eustachio P, Stein L. Reactome: a database of reactions, pathways and biological processes. Nucleic Acids Res. 2011;39:D691–7.

44. Ashburner M, Ball CA, Blake JA, BOTSTEIN D, Butler H, Cherry JM, Davis AP, Dolinski K, Dwight SS, Eppig JT, Harris MA, Hill DP, Issel-Tarver L, Kasarskis A, Lewis S, Matese JC, Richardson JE, Ringwald M, Rubin GM, Sherlock G. Gene ontology: tool for the unification of biology. The Gene Ontology Consortium. Nat Genet. 2000;25:25–9.

45. Carithers LJ, Moore HM. The Genotype-Tissue Expression (GTEx) Project. Biopreserv Biobank. 2015;13:307–8.

46. Thul PJ, Lindskog C. The human protein atlas: A spatial map of the human proteome. Protein Sci. 2018;27(1):233–44.

47. Haase T, Bornigen D, Muller C, Zeller T. Systems medicine as an emerging tool for cardiovascular genetics. Front Cardiovasc Med. 2016;3:27.

48. Song JW, Chung KC. Observational studies: cohort and case-control studies. Plast Reconstr Surg. 2010;126:2234–42.

49. Chen G, Levy D. Contributions of the Framingham Heart Study to the Epidemiology of Coronary Heart Disease. JAMA Cardiol. 2016;1:825–30.
50. Long MT, Fox CS. The Framingham Heart Study--67 years of discovery in metabolic disease. Nat Rev Endocrinol. 2016;12:177–83.
51. Investigators TA. The Atherosclerosis Risk in Communities (ARIC) Study: design and objectives. The ARIC investigators. Am J Epidemiol. 1989;129:687–702.
52. Koch B, Schaper C, Ittermann T, Volzke H, Felix SB, Ewert R, Glaser S. Reference values for lung function testing in adults--results from the study of health in Pomerania (SHIP). Dtsch Med Wochenschr. 2009;134:2327–32.
53. Lowel H, Lewis M, Hormann A, Keil U. Case finding, data quality aspects and comparability of myocardial infarction registers: results of a south German register study. J Clin Epidemiol. 1991;44:249–60.
54. Wild PS, Zeller T, Beutel M, Blettner M, Dugi KA, Lackner KJ, Pfeiffer N, Munzel T, Blankenberg S. The Gutenberg Health Study. Bundesgesundheitsblatt Gesundheitsforschung Gesundheitsschutz. 2012;55:824–9.
55. Zeller T, Hughes M, Tuovinen T, Schillert A, Conrads-Frank A, Ruijter H, Schnabel RB, Kee F, Salomaa V, Siebert U, Thorand B, Ziegler A, Breek H, Pasterkamp G, Kuulasmaa K, Koenig W, Blankenberg S. BiomarCaRE: rationale and design of the European BiomarCaRE project including 300,000 participants from 13 European countries. Eur J Epidemiol. 2014;29:777–90.
56. Mayrhofer MT, Holub P, Wutte A, Litton JE. BBMRI-ERIC: the novel gateway to biobanks. From humans to humans. Bundesgesundheitsblatt Gesundheitsforschung Gesundheitsschutz. 2016;59:379–84.
57. Simeon-Dubach D, Zeisberger SM, Hoerstrup SP. Quality assurance in biobanking for pre-clinical research. Transfus Med Hemother. 2016;43:353–7.
58. Rotival M, Zeller T, Wild PS, Maouche S, Szymczak S, Schillert A, Castagne R, Deiseroth A, Proust C, Brocheton J, Godefroy T, Perret C, Germain M, Eleftheriadis M, Sinning CR, Schnabel RB, Lubos E, Lackner KJ, Rossmann H, Munzel T, Rendon A, Erdmann J, Deloukas P, Hengstenberg C, Diemert P, Montalescot G, Ouwehand WH, Samani NJ, Schunkert H, Tregouet DA, Ziegler A, Goodall AH, Cambien F, Tiret L, Blankenberg S. Integrating genome-wide genetic variations and monocyte expression data reveals trans-regulated gene modules in humans. PLoS Genet. 2011;7:e1002367.
59. Kessler T, Wobst J, Wolf B, Eckhold J, Vilne B, Hollstein R, Von Ameln S, Dang TA, Sager HB, Moritz Rumpf P, Aherrahrou R, Kastrati A, Bjorkegren JLM, Erdmann J, Lusis AJ, Civelek M, Kaiser FJ, Schunkert H. Functional characterization of the GUCY1A3 coronary artery disease risk locus. Circulation. 2017;136:476–89.
60. Huan T, Esko T, Peters MJ, Pilling LC, Schramm K, Schurmann C, Chen BH, Liu C, Joehanes R, Johnson AD, Yao C, Ying SX, Courchesne P, Milani L, Raghavachari N, Wang R, Liu P, Reinmaa E, Dehghan A, Hofman A, Uitterlinden AG, Hernandez DG, Bandinelli S, Singleton A, Melzer D, Metspalu A, Carstensen M, Grallert H, Herder C, Meitinger T, Peters A, Roden M, Waldenberger M, Dorr M, Felix SB, Zeller T, Vasan R, O'donnell CJ, Munson PJ, Yang X, Prokisch H, Volker U, Van Meurs JB, Ferrucci L, Levy D. A meta-analysis of gene expression signatures of blood pressure and hypertension. PLoS Genet. 2015a;11:e1005035.
61. Huan T, Meng Q, Saleh MA, Norlander AE, Joehanes R, Zhu J, Chen BH, Zhang B, Johnson AD, Ying S, Courchesne P, Raghavachari N, Wang R, Liu P, O'donnell CJ, Vasan R, Munson PJ, Madhur MS, Harrison DG, Yang X, Levy D. Integrative network analysis reveals molecular mechanisms of blood pressure regulation. Mol Syst Biol. 2015b;11:799.
62. Saleh MA, McMaster WG, Wu J, Norlander AE, Funt SA, Thabet SR, Kirabo A, Xiao L, Chen W, Itani HA, Michell D, Huan T, Zhang Y, Takaki S, Titze J, Levy D, Harrison DG, Madhur MS. Lymphocyte adaptor protein Lnk deficiency exacerbates hypertension and end-organ inflammation. J Clin Invest. 2015;125:1189–202.

63. Zeller T, Schurmann C, Schramm K, Muller C, Kwon S, Wild PS, Teumer A, Herrington D, Schillert A, Iacoviello L, Kratzer A, Jagodzinski A, Karakas M, Ding J, Neumann JT, Kuulasmaa K, Gieger C, Kacprowski T, Schnabel RB, Roden M, Wahl S, Rotter JI, Ojeda F, Carstensen-Kirberg M, Tregouet DA, Dorr M, Meitinger T, Lackner KJ, Wolf P, Felix SB, Landmesser U, Costanzo S, Ziegler A, Liu Y, Volker U, Palmas W, Prokisch H, Guo X, Herder C, Blankenberg S, Homuth G. Transcriptome-wide analysis identifies novel associations with blood pressure. Hypertension. 2017;70:743–50.

64. Clemente JC, Ursell LK, Parfrey LW, Knight R. The impact of the gut microbiota on human health: an integrative view. Cell. 2012;148:1258–70.

65. Jie Z, Xia H, Zhong SL, Feng Q, Li S, Liang S, Zhong H, Liu Z, Gao Y, Zhao H, Zhang D, Su Z, Fang Z, Lan Z, Li J, Xiao L, Li R, Li X, Li F, Ren H, Huang Y, Peng Y, Li G, Wen B, Dong B, Chen JY, Geng QS, Zhang ZW, Yang H, Wang J, Zhang X, Madsen L, Brix S, Ning G, Xu X, Liu X, Hou Y, Jia H, He K, Kristiansen K. The gut microbiome in atherosclerotic cardiovascular disease. Nat Commun. 2017;8:845.

66. Gilbert JA, Quinn RA, Debelius J, Xu ZZ, Morton J, Garg N, Jansson JK, Dorrestein PC, Knight R. Microbiome-wide association studies link dynamic microbial consortia to disease. Nature. 2016;535:94–103.

67. Ahn AC, Tewari M, Poon CS, Phillips RS. The limits of reductionism in medicine: could systems biology offer an alternative? PLoS Med. 2006;3:e208.

68. Gijzen H. Development: Big data for a sustainable future. Nature. 2013;502:38.

69. Eisenstein M. Big data: The power of petabytes. Nature. 2015;527:S2–4.

70. Lowes M, Kleiss M, Lueck R, Detken S, Koenig A, Nietert M, Beissbarth T, Stanek K, Langer C, Ghadimi M, Conradi LC, Homayounfar K. The utilization of multidisciplinary tumor boards (MDT) in clinical routine: results of a health care research study focusing on patients with metastasized colorectal cancer. Int J Color Dis. 2017;32(10):1463–9.

71. Ginsburg GS, Willard HF. Genomic and personalized medicine: foundations and applications. Transl Res. 2009;154:277–87.

72. Mirnezami R, Nicholson J, Darzi A. Preparing for precision medicine. N Engl J Med. 2012;366:489–91.

73. Harris JR, Burton P, Knoppers BM, Lindpaintner K, Bledsoe M, Brookes AJ, Budin-Ljosne I, Chisholm R, Cox D, Deschenes M, Fortier I, Hainaut P, Hewitt R, Kaye J, Litton JE, Metspalu A, Ollier B, Palmer LJ, Palotie A, Pasterk M, Perola M, Riegman PH, Van Ommen GJ, Yuille M, Zatloukal K. Toward a roadmap in global biobanking for health. Eur J Hum Genet. 2012;20:1105–11.

74. Walport M, Brest P. Sharing research data to improve public health. Lancet. 2011;377:537–9.

75. Muilu J, Peltonen L, Litton JE. The federated database--a basis for biobank-based post-genome studies, integrating phenome and genome data from 600,000 twin pairs in Europe. Eur J Hum Genet. 2007;15:718–23.

76. Doiron D, Burton P, Marcon Y, Gaye A, Wolffenbuttel BHR, Perola M, Stolk RP, Foco L, Minelli C, Waldenberger M, Holle R, Kvaloy K, Hillege HL, Tasse AM, Ferretti V, Fortier I. Data harmonization and federated analysis of population-based studies: the BioSHaRE project. Emerg Themes Epidemiol. 2013;10:12.

77. Brazma A, Hingamp P, Quackenbush J, Sherlock G, Spellman P, Stoeckert C, Aach J, Ansorge W, Ball CA, Causton HC, Gaasterland T, Glenisson P, Holstege FC, Kim IF, Markowitz V, Matese JC, Parkinson H, Robinson A, Sarkans U, Schulze-Kremer S, Stewart J, Taylor R, Vilo J, Vingron M. Minimum information about a microarray experiment (MIAME)-toward standards for microarray data. Nat Genet. 2001;29:365–71.

78. Lappalainen I, Almeida-King J, Kumanduri V, Senf A, Spalding JD, Ur-Rehman S, Saunders G, Kandasamy J, Caccamo M, Leinonen R, Vaughan B, Laurent T, Rowland F, Marin-Garcia P, Barker J, Jokinen P, Torres AC, De Argila JR, Llobet OM, Medina I, Puy MS, Alberich M, De La Torre S, Navarro A, Paschall J, Flicek P. The European Genome-phenome Archive of human data consented for biomedical research. Nat Genet. 2015;47:692–5.

79. Caulfield T, Murdoch B. Genes, cells, and biobanks: Yes, there's still a consent problem. PLoS Biol. 2017;15:e2002654.
80. Grabe HJ, Assel H, Bahls T, Dorr M, Endlich K, Endlich N, Erdmann P, Ewert R, Felix SB, Fiene B, Fischer T, Flessa S, Friedrich N, Gadebusch-Bondio M, Salazar MG, Hammer E, Haring R, Havemann C, Hecker M, Hoffmann W, Holtfreter B, Kacprowski T, Klein K, Kocher T, Kock H, Krafczyk J, Kuhn J, Langanke M, Lendeckel U, Lerch MM, Lieb W, Lorbeer R, Mayerle J, Meissner K, Zu Schwabedissen HM, Nauck M, Ott K, Rathmann W, Rettig R, Richardt C, Salje K, Schminke U, Schulz A, Schwab M, Siegmund W, Stracke S, Suhre K, Ueffing M, Ungerer S, Volker U, Volzke H, Wallaschofski H, Werner V, Zygmunt MT, Kroemer HK. Cohort profile: Greifswald approach to individualized medicine (GANI_MED). J Transl Med. 2014;12:144.
81. Auffray C, Balling R, Barroso I, Bencze L, Benson M, Bergeron J, Bernal-Delgado E, Blomberg N, Bock C, Conesa A, Del Signore S, Delogne C, Devilee P, Di Meglio A, Eijkemans M, Flicek P, Graf N, Grimm V, Guchelaar HJ, Guo YK, Gut IG, Hanbury A, Hanif S, Hilgers RD, Honrado A, Hose DR, Houwing-Duistermaat J, Hubbard T, Janacek SH, Karanikas H, Kievits T, Kohler M, Kremer A, Lanfear J, Lengauer T, Maes E, Meert T, Muller W, Nickel D, Oledzki P, Pedersen B, Petkovic M, Pliakos K, Rattray M, I Màs JR, Schneider R, Sengstag T, Serra-Picamal X, Spek W, Vaas LA, Van Batenburg O, Vandelaer M, Varnai P, Villoslada P, Vizcaino JA, Wubbe JP, Zanetti G. Making sense of big data in health research: Towards an EU action plan. Genome Med. 2016;8:71.
82. Hazin R, Brothers KB, Malin BA, Koenig BA, Sanderson SC, Rothstein MA, Williams MS, Clayton EW, Kullo IJ. Ethical, legal, and social implications of incorporating genomic information into electronic health records. Genet Med. 2013;15:810–6.
83. Lamas E, Barh A, Brown D, Jaulent MC. Ethical, Legal and Social Issues related to the health data-warehouses: re-using health data in the research and public health research. Stud Health Technol Inform. 2015;210:719–23.
84. Vasan RS, Benjamin EJ. The future of cardiovascular epidemiology. Circulation. 2016;133:2626–33.
85. Horak P, Klink B, Heining C, Groschel S, Hutter B, Frohlich M, Uhrig S, Hubschmann D, Schlesner M, Eils R, Richter D, Pfutze K, Georg C, Meissburger B, Wolf S, Schulz A, Penzel R, Herpel E, Kirchner M, Lier A, Endris V, Singer S, Schirmacher P, Weichert W, Stenzinger A, Schlenk RF, Schrock E, Brors B, Von Kalle C, Glimm H, Frohling S. Precision oncology based on omics data: The NCT Heidelberg experience. Int J Cancer. 2017;141:877–86.

Sex Differences in Prevalent Cardiovascular Disease in the General Population

12

Daniel Engler, Natascha Makarova, and Renate B. Schnabel

Contents

Abstract

In this chapter, we provide latest insights about sex differences in prevalent cardiovascular diseases (CVD) in the general population. CVD is considered as one of the most prevalent diseases and the leading cause of death in both men and women. Sex-related factors have an important impact of the differences in the

D. Engler (✉) · N. Makarova · R. B. Schnabel
Department of General and Interventional Cardiology, University Heart Center Hamburg-Eppendorf, Hamburg, Germany

DZHK (German Center for Cardiovascular Research), Partner Site Hamburg/Kiel/Lübeck, Hamburg, Germany
e-mail: d.engler@uke.de

© Springer Nature Switzerland AG 2019
J. Erdmann, A. Moretti (eds.), *Genetic Causes of Cardiac Disease*, Cardiac and Vascular Biology 7, https://doi.org/10.1007/978-3-030-27371-2_12

development, the presentation of symptoms, the awareness for the disease, and the progression and management of CVD. Common CVDs in the general population such as ischemic heart disease, atrial fibrillation, and heart failure demonstrate the importance of sex-specific approaches that have an effect on clinical outcomes. The relationship between sex-specific attributable risk factors and the development of CVD often pass unnoticed. However, differences in classical risk factor distribution can only partly explain observed sex differences. Genetics may contribute to the understanding of sex differences in CVD. Existing and emerging technologies take genetic examinations at increasingly high resolution at the population level.

12.1 Introduction

When discussing the differences in the development of cardiovascular disease (CVD) in women and men, two terms need to be defined: "sex" and "gender." The term "sex" summarizes the biological differences between men and women. This sex-differentiation is characterized by the structure and function of the cardiovascular system. This includes chromosomes, sex organs, and hormonal contributions among others [1]. Whereas the term "gender" summarizes psychosocial and behavioral factors for women and men in the context of their cultural and societal role [2].

It is very likely that the combination of "sex" and "gender" have an impact on the development, progression, and management of CVD and are both of importance (Fig. 12.1) [3].

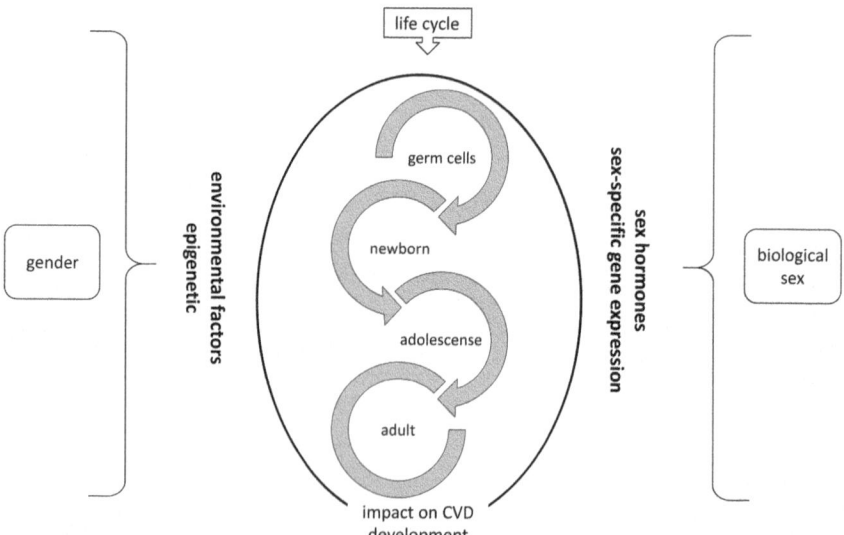

Fig. 12.1 Interaction between sex and gender over a lifetime and their impact on CVD development in the general population

For both men and women, CVD is the leading cause of death worldwide. A further increase of mortality due to CVD is observed at a global level [4]. According to the recent report of the European Heart Network, CVD in women is responsible for 2.1 million deaths (49% of all deaths), while CVD in men accounts for 1.8 million deaths (40% of all deaths) in the European population [5]. This divergence in death rates can be partly explained by a higher risk of competing events, i.e., the risk of men dying from other causes [6]. Men have earlier onset of CVD compared to women [7]. Women lose fewer years of life due to CVD as they develop the disease approximately 7 to 10 years later [8].

Despite these facts, CVD has often been stereotyped as a men's disease. Differences in risk factor profile, clinical presentation, and disease perception may play a role here. For instance, women are considerably more likely than men to present with atypical symptoms (e.g., fatigue, sleep disturbance, nausea, and abdominal, neck, jaw, or shoulder pain) and absence of typical chest pain [9]. These differences of clinical presentation therefore turn into behavioral differences between men and women if they experience a CVD event. In the past, research showed that the delay in seeking CVD treatment in women is significantly longer than men. This time loss may worsen the outcome of the disease [10, 11].

The knowledge about sex differences between the female and male heart such as the difference of physiological and pathophysiological conditions increasing constantly. These differences can be found in all domains of cardiovascular health and disease, including coronary disease, heart rhythm, and heart failure. With regard to the epidemiology, pathophysiology, presentation, and outcomes of CVD, one can determine different clinical characteristics between the sexes [12]. Sex differences in the cardiovascular system can be summarized by general differences with regard to the following factors:

Cardiovascular system: sex-related characteristics in the prevalent general population	
Physiology	• Sympathetic activity in women is reduced. • Parasympathetic activity in women is enhanced. • Plasma concentrations of norepinephrine in women is lower.

Anatomy	• Women's vs men's anatomic dimensions on average (age and race adjusted):
	  – Left ventricular mass: men > women,
	  – Ventricular wall thickness: men > women,
	  – Atrial and ventricular dimensions: men > women,
	  – Vessel size: men >women.

Cardiovascular function	• Stroke volume in women is 10% less on average and thus women have lower cardiac output. • Higher pulse rate in women (3–5 beats/minute). • Higher ejection fraction in women.
Electrocardiographic and electrophysiologic indices Electrocardiographic and electrophysiologic indices	• Women on average have a longer corrected QT interval and a shorter sinus node recovery time. • Drug-induced torsades de pointes is more common in women.

(continued)

	• Sudden cardiac death and atrial fibrillation are more common in men.
Hormonal status	• Estrogen and progesterone represents the most important sex hormones in women; testosterone have a decisive importance in men. • The menstrual cycle can affect hematologic and electrocardiographic indices.

Cardiovascular adaptive mechanisms	• In response to stress:
	  – By increasing the pulse rate, women more strongly regulate the cardiac output.
	  – Due to an increased vascular resistance, men develop an increased blood pressure.
	• Orthostatic hypotension and syncope is more related to women.

Hematologic indices	• Women have a lower hematocrit, due to a lower number of circulating red blood cells per unit volume of plasma. • Women have a reduced capacity for carrying oxygen due to a lower of hemoglobin. • The consumption of oxygen in women is lower.

Note: Modified with permission from Fink, S. W. Cardiovascular Disease in Women. In: Richardson M, Chessman KH, Chant C, et al. (Eds.). Pharmacotherapy Self-Assessment Program, seventh Edition. Book 1: Cardiology. Lenexa, Kansas City: American College of Clinical Pharmacy, 2010; 182

Multiple variables such as the societal roles and behaviors, biological and physiological characteristics, genes, epigenetics, gene expression, and hormonal status may have sex-specific interactions that determine the pathogenesis and course of the disease.

Over the last decades, research and clinical practice have improved in their focus on sex differences in CVD. Although many of the basic mechanisms underlying sex differences in CVD remain unknown, epidemiological data have clearly shown that there are sex differences in the community. This overview provides insights into the sex differences in risk factors and CVD distribution in the general population.

12.2   Sex and Cardiovascular Risk Factor

12.2.1 Risk Factors in General

Numerous large-scale epidemiological studies (e.g., Framingham Heart Study, INTERHEART, etc.) showed that more than 90% of all cardiovascular events could be attributed to a relatively small number of often modifiable risk factors. These risk factors include: smoking, hypertension, dyslipidemia, diabetes, abdominal obesity, high-risk diet, psychosocial factors, lack of physical activity, and alcohol abstinence compared to moderate consumption [13–15].

In general, the association of these major risk factors with CVD do not differ between the sexes in their direction but may have different magnitude and significance. Also, there are significant sex interactions [6]. For instance, women with diabetes have more than 40% higher risk of incident coronary heart disease (CHD) than men with diabetes [12, 15]. Pregnancy-related hazards such as gestational hypertension [16], as well as frequently occurring endocrine disorders, e.g., polycystic ovary syndrome (PCOS) in women of reproductive age, are accompanied by an increased risk of developing CVD in later life [17]. A recent Norwegian cohort study that included over 15,000 pregnant women confirms the previous findings. Women who experience hypertension in pregnancy have a twofold to threefold higher risk of increased systolic blood pressure in the future. Furthermore, they experience significantly higher body mass index and wider waist circumference compared to women with normal blood pressure during pregnancy [16]. Over the last decades, there have been mixed trends in CVD risk factors for both women and men. On the one hand, there is a decrease in the prevalence of very high levels of cholesterol; on the other hand, a dramatic increase in the prevalence of obesity, particularly among women, has been observed. The proportion of women with high blood pressure increased, whereas it decreased among men [18].

12.2.2 Blood Pressure and Hypertension

One in three adults worldwide have hypertension (hypertension as systolic pressure greater than 140 mm Hg or diastolic blood pressure greater than 90 mm Hg) [19]. In general, the differences between women and men the prevalence of the disease seems to be equally distributed. A recent systematic analysis of population-based studies from 90 countries showed an age-standardized prevalence of hypertension in adults aged $\geq$20 years in 2010 was 31.9% in men and 30.1% in women [20]. In younger age groups, the incidence rate of hypertension in men is higher than in women (18–29 years: women = 1.3%; men = 8.3%) [21]. Men experience hypertension more often and also have higher low density lipopotein (LDL) cholesterol levels than similar aged women in premenopausal status [22]. In older age groups (> 65 years) the disease is more prevalent in women [23, 24].

In women, white coat hypertension appears to be more frequent. Men show a higher proportion of masked hypertension [25]. The most common type of hypertension in old women is isolated systolic hypertension. Women with hypertension present with stiffer myocardium and vessels compared to men at old age. These conditions may be a possible link to a higher prevalence of heart failure with preserved ejection fraction (HFpEF) in women [26]. Potential sex differences in the pathophysiology of hypertension have been highlighted. The regulation of arterial pressure and renal function by the renin-angiotensin system (RAS) is noticeably different between the sexes [27]. The RAS regulates extracellular fluid homeostasis through renal function and arterial pressure. Several studies show that premenopausal women as compared to aged-matched men are more protected from hypertension due to a differential balance in the RAS [28, 29]. Despite such differences in the pathophysiology of hypertension, there is no sex-specific approach in the treatment of hypertension yet [26].

Based on the attributable risk of hypertension for myocardial infarction in the INTERHEART study, normalization of blood pressure values can decrease the risk of disease in both sexes. In men, the risk can be reduced by 15.7% compared to women in whom a reduction by 25.4% could be calculated [30]. At the population level, the awareness and knowledge of high blood pressure is limited. An ongoing US national health and nutrition examination study, shows a low rate of awareness with regard to hypertension treatment and control among both sexes [31]. It induces the need to improve health education in both men and women. It remains to be shown whether sex-specific strategies are needed for this purpose.

12.2.3 Diabetes Mellitus

The pathophysiological consequences of diabetes mellitus extend to all components of the cardiovascular system: the microvasculature, the larger arteries, the heart, as well as the kidneys [32]. Diabetes mellitus, type 1 and type 2, is an established risk factor for CVD that harbors a significant increase in CVD mortality [32]. About 70% of all deaths in patients with diabetes are related to CVD [33]. New estimates indicate a total number of individuals with diabetes will rise from 6.4%, 285 million adults, in 2010 to 7.7%, 439 million adults, by 2030 [34]. On a global scale, the overall prevalence of diabetes is somewhat higher in men than in women. In particular, age is an important factor of sex-specific difference in the prevalence of diabetes. Whereas men have a higher prevalence of diabetes in the age group of <60 years, women have a significantly higher number of the disease in the older age group [35]. Consistent evidence exists that women with this condition experience a greater relative risk for CVD events and CVD death compared to their male counterparts [36]. Framingham Heart Study data showed a twofold increased risk of CVD in men with diabetes and a more than 3.5-fold higher risk in women [37]. An excess risk in women with diabetes has also been observed for heart failure, and peripheral artery occlusive disease. Data of the INTERHEART study confirm that diabetes was more strongly associated with a risk of myocardial infarction in women with a relative risk of 4.26 versus 2.67 in men [38]. A recent meta-analysis of prospective population-based studies also reported a 44% higher relative risk in women compared to men [14]. More than 15% excess risk of fatal coronary heart disease is also observed among women with type 1 diabetes compared to men [39].

12.2.4 Body Mass Index and Obesity

The development of CVD is associated with obesity and overweight. However, the development of obesity and overweight differ between men and women due to the manifestation of obesity-related conditions such as hyperlipidemia, insulin resistance, and type 2 diabetes mellitus. For instance, weight gain occurs differently in men and women. Men generally have higher body weight than women, but the proportion of fat is greater in women.

According to the data of the global burden of disease (GBD), the prevalence of obesity (BMI ≥ 30 kg/m^2) has more than doubled worldwide since 1980 and is now 5% in children and 12% in adults [40]. Two billion adults (>18 years) worldwide are overweight whereof 39% men and 40% of women (>18 years) are overweight (BMI ≥ 25 kg/m^2) and 11% of men and 15% of women were obese [40]. In the western societies, the rates of overweight and obesity have dramatically increased in the recent decades for both men and women [41]. Associations can be seen between a high value of BMI in individuals and the gross domestic product (GDP) level of a respective country. The prevalence of obesity in upper middle income counties (24%) in both sexes is more than threefold higher compared to lower middle income countries (7%). Several overlapping physiological systems and processes predict the development of CVD related to overweight and obesity which includes sex differences. In the Framingham Heart Study, obesity increased the relative risk of coronary artery disease (CAD) by 64% in women as opposed to 46% in men [42].

12.2.5 Cholesterol

In general, there is no sex difference in the prevalence of elevated total cholesterol worldwide [43]. Among adults, 37% men and 40% women present with elevated cholesterol levels. On a global scale, the mean total cholesterol changed little between 1980 and 2008, falling by less than 0.1 mmol/L per decade in men and women [44]. The lipid profiles of young women tend to be more beneficial compared to men. High-density lipoprotein cholesterol concentrations are higher on average whereas low-density lipoprotein cholesterol is lower in women compared to men. However, with age this advantage diminishes, in part due to hormonal changes after the menopausal transition of women [45]. In a meta-analysis of 55,000 participants a decrease in total cholesterol by 1 mmol/L was associated with about lower ischemic heart disease mortality in both sexes in all age categories [46]. In contrast, a recent meta-analysis of 97 cohort studies with over one million participants showed that there is evidence of a small but significantly stronger effect in men (CI = 95%; RR = 1.24) compared to women (CI = 95% CI; RR = 1.20) with raised cholesterol to develop coronary heart disease [47]. A systematic work-up of potential sex-specific effects of major lipids on cardiovascular risk is missing [6].

12.2.6 Smoking and Alcohol Intake

Smoking and alcohol intake are common in many populations. However, the consumption characteristics vary considerably between sexes. On a global level, the prevalence of smoking is much higher in men (48%) than it is in women (10%) [48]. In general, women smoked fewer cigarettes than men and tended to start smoking later in life. However, women and men who start smoking cigarettes nowadays smoke nearly the same number of cigarettes and take up smoking at the same age. Multiple studies have shown a stronger association in women between long-term smoking and the risk of developing CVD than in men [49–51]. A meta-analysis by Huxley and Woodward which included about four million individuals and at least 67,000 coronary heart disease events from 26 prospective trials found that women had a 25% increased risk related to cigarette smoking compared with men [39]. Both the change of smoking patterns in women over the last century and the higher risk of CVD when smoking may therefore result in increasing numbers of smoking-related CVD compared to men. Considering the population attributable risk to develop ischemic heart disease, it seems to be possible that the prevalence of ischemic heart disease will particularly increase in the future for this target group [52].

The reasons for the differential associations of smoking may include sex hormones. Combined use of contraceptives and smoking pose a significant risk of ischemic heart disease [39, 51]. In a small cohort study of 346 women with PCOS, including 98 smokers and 248 non-smokers, an increased level of fasting insulin, free testosterone, and free androgen index was observed in women with PCOS who smoke [53]. These increased levels may enhance insulin resistance that potentially results in a cardiovascular health risk. Another possible association is shown due to differential gender effect of smoking and arginine vasopressin (AVP) [54]. AVP has a direct impact of the cardiovascular system [55]. A small study observed a higher level of AVP in smoking in women that could contribute to negative effect on cardiovascular health in women [55]. These hormones possibly lead to a higher risk of CVD [54, 55]. However, it should be considered that these findings are only based on limited data and need validation. Therefore, the current understanding of the negative impact of smoking on CVD in females needs further investigation.

Besides smoking, alcohol intake is also a relevant risk factor for the development of CVD for both men and women. Alcohol consumption in the population is attributed to 3.3 million deaths per year and men have a higher mortality rate attributed to alcohol (7.6%) compared to their female counterparts (4.0%). In general, the intake of alcohol is described with a U-shape association. Both very low or high alcohol consumption are negative for CVD health whereas moderate alcohol consumption appears to be beneficial, except for atrial fibrillation for which a linear relation has been shown consistently. Several studies demonstrated that consuming moderate amounts of alcohol has been associated with reduced risk of coronary heart disease, stroke, and congestive heart failure. There are speculations on sex differences in alcohol intake and CVD risk. However, due to diverse study results about general alcohol consumption there is no scientific certainty whether women or men have a different risk of CVD development related to alcohol consumption [56, 57].

12.2.7 Psychosocial Factors

Psychosocial factors are well established for CVD risk. Depression is known for the adverse effects on CVD risk factors including hypertension, obesity, and smoking. According to the World Health Organisation (WHO) at a global scale, more women are affected by depression than men and have an earlier age of onset [58, 59].

In fact, all of them predict the development of CVD. Recent studies report that more than 300 million people of all ages experience depression worldwide. Women show twice the rates of stress-related psychiatric disorders, including posttraumatic stress disorders and depression, compared to men [60]. Several studies have identified psychosocial factors (e.g., depression, perceived stress at home or work, low amount of control, and major life events) as predictors of future cardiovascular events in initially healthy populations [61–63]. A large case control study showed that women with psychosocial or mental health problems also have a 40% higher population attributable risk to develop myocardial infarction compared to women without these conditions. In contrast, men with psychosocial problems have a population attributable risk of 25% to develop myocardial infarction [30]. Furthermore, half of the women aged <55 years who overcome a myocardial infarction experience or already had a history of depression [64]. The cardiovascular risk increases almost twofold in women with depression [65]. In a follow-up of a community-based survey, the risk of women aged >40 years with depression to develop CVD is reported significantly higher than in men in the same age group [60].

Physical and sexual abuse and child neglect are determined as significant risk factors for the development of CVD. Affected individuals in the general population are mostly women. Independent of other risk factors, individuals with a history of child abuse have a 50% increased risk of CVD [66]. It seems that young women who experienced sexual abuse have a greater risk of developing CVD than men [67].

Furthermore, sex differences in CVD have been, at least in part, related to environmental and social differences between men and women (e.g., occupational hazards, habits, social stress) [68, 69]. The combination of work stress and private difficulties has also been associated with an increased risk in CVD events in women [70]. A recent review of Varianaco and Bremers describes that young women are disproportionally vulnerable to the adverse effects of marital stress and the risk of CVD [59].

In addition to long-term mental health predictors of CVD, acute stressors can also contribute to CVD onset. An acute stress situation reduces the perfusion of the epicardial coronary arteries. This mental stress-induced myocardial ischemia has been associated with adverse prognosis in patients with coronary artery disease [71]. Few studies of patients with coronary artery disease or a history of myocardial infarction report an approximately twofold higher incidence rate of mental stress-induced myocardial ischemia in women compared to men [72, 73]. However, for the general population psychosocial factors have rarely been evaluated for their ability to explain sex differences in CVD.

12.2.8 Risk Factors Unique to Sex

Today, increased attention is given to sex-specific risk factors of CVD such as hypertensive disorders of pregnancy (HDP) and gestational diabetes. The American Heart Association declares "HDP as a major risk factor for the development of the disease" [74]. The two types of HDP manifestation are gestational hypertension and preeclampsia. The prevalence of hypertensive disorders in pregnancy is between 2.8% and 5.2% in Europe [75]. A prospective cohort study ($n = 15,000$) from Magnusson et al. confirms that gestational hypertension in women that recurs during several pregnancies is significantly associated with a higher blood pressure later in life [76]. Therefore, one can assume that women with a history of hypertensive disorders in pregnancy, and particularly women with recurrent pregnancy disorders, should be followed closely and treated as necessary to prevent premature CVD [76]. Several studies had already shown that the risk of developing CVD later in life is two times greater in pregnant women with preeclampsia [77–79]. To date, it remains largely unknown whether HDP is an independent risk factor for CVD or whether the frequent co-existence of classical cardiovascular risk factors are responsible for the associations. Further epidemiological research will be needed to elucidate the mechanisms of both pathways of HDP and possible causal relations including genetic predispositions.

The other well-known pregnancy-associated risk factor is gestational diabetes that significantly increases the risk of developing future CVD. The prevalence of gestational diabetes is associated ethnic differences. For instance, in Europe, gestational diabetes is more common in women with an Asian background compared to women with a European background [80]. In a population-based matched case–control study with 2600 women with a cardiovascular event, the odds ratio of CVD in women with gestational diabetes was 1.51-times higher [81]. Recent population-based studies that examined the association between gestational diabetic women also report a more than twofold risk to develop CVD compared to women without diabetes in the pregnancy [82, 83]. But the mechanisms contributing to the vascular dysfunction seen in gestational women remain uncertain.

Menarche, menopause, and the length of the reproductive period appear to play a role for CVD risk. Early menarche (10 years or younger) has been identified as a risk factor for developing CVD in women. A recent cohort study ($n = 1.2$ million) that included women with no known CVD at follow-up demonstrated that women aged 10 years compared to their counterparts aged 13 years had an approximately 30% higher risk of developing CVD [84]. In a meta-analysis, every 1 year increase in age at menarche was associated with a 3% reduction in CVD-related mortality [85]. Women aged <44 years who experience early natural menopause have an increased CVD risk. In a recent meta-analysis with over 300,000 women, the findings indicate a two times higher risk of coronary heart disease, a 25% higher risk of CVD mortality, and overall mortality in women who experience premature menopause (before age 45) as compared to women in whom menopause occurs later [86]. Menopause per se may not be directly related to an increase in CVD risk after menopause, but the prevalence of CVD risk factors and co-morbidities rises in the postmenopausal phase. The risk factor distribution in women and men becomes similar. A prospective cohort of 34,000 individuals in Norway observed a decline in the gender gap of CVD with age in part due to an increasingly similar risk factor profile in post-menopausal women and middle-aged men, e.g., for hypertension and dyslipidemia [87].

12.2.9 Sex-Specific Biological Factors

The difference between men and women in the epidemiology, pathophysiology, clinical manifestations, effects of therapy, and outcomes of CVD can partly be explained by differences related to sex chromosomes. According to the current state of the scientific knowledge, XX and XY zygotes differ only in their sex chromosomes. This genetic disposition predisposes to sex differences between men and women [88]. The prediction of mechanisms like the gonadal effects of Sry, non-gonadal effects of Y genes, X gene dosage, and imprinting of X genes are associated with sex differences in the development of CVD [89]. One of the key players seems to be the sex-determining region on the Y chromosome (Sry). This region acts within gonadal primordia and leads to organ differentiation into testes. Afterward, the cardiovascular system is exposed to a lifelong difference in the levels of gonadal hormones, e.g., testosterone, estradiol, and progesterone [88]. Male-specific nongonadal effects of the Y chromosomes are, for instance, the expression of Sry in catecholaminergic cells which results in a predisposition to the development of hypertension in men [90]. Further explanation that causes sex differences in the expression of specific X genes could possibly be the mechanism that XY cells receive a maternal imprint on X genes and XX cells receive imprints from both parents. This imbalance in parental imprinting seems to be a possible cause of the difference in the expression of specific X genes that enhance differences in the development of the disease in men and women [89]. Recent research also demonstrated potential sex differences in the phenotype caused by X chromosomal genes due to sex-specific dose and X escape [91].

12.3   Differences in Disease and Outcomes

This chapter focuses on three major sex-specific differences in CVD area—ischemic heart disease, atrial fibrillation, and heart failure that are common in the general population (Fig. 12.2).

12.3.1 Ischemic Heart Disease

Ischemic heart disease is a complex, multifactorial disease and is defined as "an impairment due to ischaemia in the myocardium, which can arise from stenosis of the major coronary arteries, impaired microcirculation, or a demand/supply

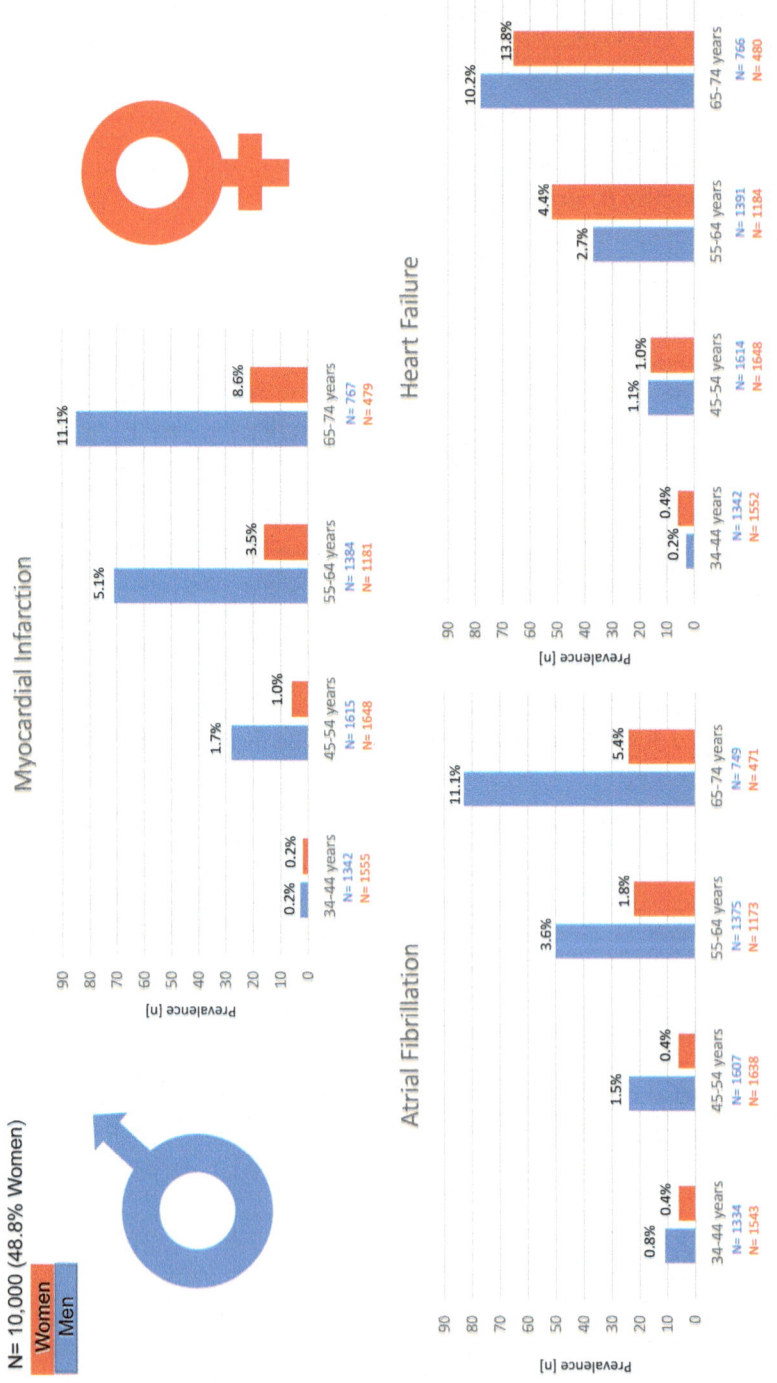

Fig. 12.2 Sex-differences in the prevalence of cardiovascular disease in the community-based Gutenberg Health Study by age decade for myocardial infraction, atrial fibrillation, and heart failure (combined HFpEF and HFrEF) (http://www.gutenberghealthstudy.org/ghs/overview.html?L=1)

imbalance." Ischemic heart disease is the leading cause of cardiovascular death in both men and women in the western societies [12].

Coronary artery disease (CAD) is the most common type of ischemic heart disease. On a global scale, it is the leading cause of death for both men and women. The worldwide mortality rate is about 7.2 million deaths per year as a result of an atherosclerotic process of the epicardial coronary arteries. Recent report present almost the same coronary artery disease prevalence for men (6.3%) and women (5.3%) aged 40 to 59. However, men experience coronary artery disease more often if they are aged between 60 to 79 years compared to their female counterparts (19.9% vs 9.7%).

Ischemic heart disease results in myocardial ischemia and injury, most prominently myocardial infarction [92]. Men experience myocardial infarction (ST-elevation and non-ST-elevation) and acute coronary syndromes three to four times more frequently than women below the age of 60. However, if an ischemic heart disease event occurs after the age of 75 the majority of the patients are women [93]. Ischemic heart disease develops on average 7–10 years later in women compared to their male counterparts. In 20–30% of typical infarction emergencies, women presented a pathophysiological obstruction or a lack of microvascular vessel extension of the heart. Compared to men (5–10%), the prevalence of these causes seems to be doubled in women [94]. Usually, these syndromes are diagnosed by the measurements of perfusion and coronary flow. But these measurements are not part of the CVD standard diagnostic [95]. However, in the recent decade the diagnostic possibilities to clarify the coronary microcirculation have improved. Women more frequently present non-obstructive coronary artery disease that is not related to atherosclerotic plaques. Acute coronary syndrome pooled trials already show that women have a 7% higher risk of non-obstructive coronary artery disease [62]. Besides plaque erosion and calcific nodule, plaque rupture is the most common cause of fatal myocardial infarction for both women (55%) and men (76%). The recent PROSPECT study can confirm that women present 10% less plaque rupture than men if they have an acute coronary syndrome [96]. Clinical manifestations such as angina pectoris, myocardial infarction, and sudden cardiac death reveal sex differences in the progression of the disease [97]. A recent study reports significant sex-differences in coronary atherosclerosis progression due to a higher score in the progression rate of severe proximal plaque, segment stenosis, and segment involvement in men compared to women with suspected coronary artery disease [98]. A possible explanation for the differences in ischemic heart disease progression lies in the higher incidents of comorbidities such as obesity and inflammation as well as more adverse changes in the coagulation in women [99]. Furthermore, the endothelial function may lead to a greater cardio metabolic risk in women with diabetes. In a meta-analysis by Hemingway et al. which included more than 14,000 angina patients from 31 countries, the prevalence of angina was similar or slightly higher in women and up to 20% higher in pre- and postmenopausal women [100]. Life-threatening conditions of acute coronary artery disease (myocardial infarction; sudden cardiac death) increase with age for both men and women. However, a recent meta-analysis revealed that young women (30–54 years) with acute myocardial infarction have more comorbidities, a longer hospital stay, and higher in-hospital mortality compared to men of the same age [101].

Chest pain is the predominate presentation of the acute coronary syndrome for both men and women. However, there are possible gender differences in symptom presentation of the acute coronary syndrome. It seems that men with acute coronary syndrome are more likely to present chest pain and discomfort, whereas women with acute coronary syndrome present more atypical symptoms such as dyspnea, weakness, fatigue, and indigestion compared to men [9]. However, there is inconsistency in the findings reported in the literature on whether women present with chest pain as a symptom of acute coronary syndrome less frequently. There is scientific evidence that the symptom of chest pain in acute coronary syndrome declines considerably with age which may account for lower chest pain frequency in women because they usually present at older age [21]. However, in current chest pain unit cohorts, the distribution of typical chest pain is fairly comparable between both sexes. Prior reports showed significant sex differences in outcomes after acute coronary syndrome, with adverse prognosis in women which had been explained by older age and more co-morbidities. In the meantime, the short-term outcomes have become similar in men and women in contemporary chest pain unit cohorts although therapeutic strategies remain different. For example, invasive therapy is more commonly used in men. Future changes in awareness and treatment of ischemic heart disease can be expected.

12.3.2 Atrial Fibrillation

The worldwide prevalence of atrial fibrillation in the general population is on average higher in men (approximately 596 per 100,000) compared to women (373 per 100,000) [102]. In the last decades, one can observe a decrease of atrial fibrillation–associated mortality rate in men compared to women. The higher mortality in women in the developing countries is responsible for the overall higher atrial fibrillation–related mortality among women compared with men. In both sexes, an increase in atrial fibrillation prevalence and also in the incidence has been documented due to a better awareness of the disease, the use of routine electrocardiography, and extended electrocardiographic monitoring in the community among other factors. On a global scale, many individuals remain undiagnosed due to the intermittent and often asymptomatic nature of the disease. Thus, it can be assumed that the true prevalence is likely to be substantially higher [103].

In general, atrial fibrillation is associated with a significantly increased risk of morbidity and mortality. Causes of morbidity and death are due to a fivefold increased risk of stroke and a doubled risk for myocardial infarction compared to the population without atrial fibrillation. A recent meta-analysis of 30 population-based cohort studies with approximately four million participants report that atrial fibrillation is a stronger risk factor for all-cause mortality, stroke, cardiovascular mortality, cardiac events, and heart failure in women compared to men [104].

Cohort data of the Biomarker for Cardiovascular Risk Assessment in Europe (BiomarCaRE) consortium reported significant sex differences in the risk of developing atrial fibrillation. BiomarCaRE is a population-based and longitudinal study design with long-term follow-up information that has good power to examine gender interactions. Fewer atrial fibrillation cases were observed in women (4.4%) than in men (6.4%) over a median follow-up of more than 12 years. Cumulative incidence increased markedly after the age of 50 years in men and about a decade later, after 60 years, in women. The lifetime risk of atrial fibrillation accumulates to more than 25–30% in both sexes. Obesity and hypertension range among the most important modifiable risk factors. BMI appears to be more strongly related to atrial fibrillation in men compared to women [105]. The occurrence of atrial fibrillation is accompanied by a poor prognosis for cardiovascular outcomes and mortality. The BiomarCaRE data provide evidence that differences in atrial fibrillation incidence observed by gender may be explained by gender-specific distribution of risk factors and by differential associations of traditional risk factors. A substantial proportion of the atrial fibrillation burden can be explained by classical cardiovascular disease risk factors in both genders. While blood pressure, smoking, alcohol consumption, and prevalent cardiovascular disease are largely similar predictors of incident atrial fibrillation in both genders, total cholesterol concentrations may show gender differences. A higher BMI and obesity are significantly more hazardous for the development of atrial fibrillation in men and need better awareness and targeted intervention. Observed gender differences in the association of BMI and total cholesterol with atrial fibrillation need to be evaluated for their pathophysiology and relevance in gender-specific prevention strategies [105].

12.3.3 Heart Failure

Heart failure results from different pathological processes. According to the European Society of Cardiology (ESC), heart failure can be defined as "a clinical syndrome characterized by typical symptoms that may be accompanied by signs caused by a structural and/or functional cardiac abnormality, resulting in a reduced cardiac output and/or elevated intra cardiac pressures" [106]. Left-sided heart failure can be differentiated as heart failure with preserved ejection fraction (HFpEF) or heart failure with reduced ejection fraction (HFrEF) due to ischemic or non-ischemic genesis. The frequency of HFpEF conversion into HFrEF in the general population is unknown and a topic of ongoing studies. Most population-based studies report an increasing incidence of heart failure [4] and a new epidemic has been predicted while the incidence of HFrEF appears to be deceasing [107]. A recent meta-analysis of 25 different western populations (>60 years) reports a prevalence of 4.9% of HFpEF compared to 3.3% that have HFrEF. The prevalence of HFrEF in men is higher compared to women and the prevalence of HFpEF in the population increases with age and is less common in men. Due to the onset of heart failure in later life, one can expect that women have a higher probability to experience the syndrome [62]. In younger age groups HFpEF is rare (almost 0.1% in men and about 1% in women).

For older individuals (>80 years), the prevalence of HFpEF in men is 4–6% and for women it is 8–10% [108]. In the population-based Gutenberg Health Study, sex differences in the prevalence of the two heart failure subtypes are clearly visible (Fig. 12.3).

Heart failure with HFpEF accounts for 50% of all heart failure cases and is characterized by diastolic dysfunction and high prevalence of left ventricular hypertrophy. The manifestation of HFpEF in older men and women (>60 years) differs in clinical outcomes. In a sample of approximately 4000 patients of the I-PRESERVE study, men with HFpEF had significantly less chronic kidney disease and hypertension than women. However, men experience more ischemic events and have an increased risk of atrial fibrillation [109]. Under stress conditions, the remodeling of the ventricles is different. Whereas women with the combination of the risk factors obesity and hypertension experience central (concentric) hypertrophy, men with the same risk factors tend to develop eccentric hypertrophy [110].

Women are more likely to develop HFpEF due to the specific risk profile of HFpEF. In addition, in female patients, HFrEF caused by non-obstructive coronary disease is more common whereas in men atherosclerotic coronary artery disease prevails in the causal chain [107]. The most common risk factors for the development of HFpEF are hypertension, diabetes, and obesity [111]. In the general population, hypertension has the highest attributable risk for heart failure among women. Pregnancy and cancer treatment for breast cancer are sex-specific risk factors for the development of HFrEF triggered by non-ischemic events. Hypertension may also be associated with peripartum cardiomyopathy. HFrEF occurs in the terminal phase of pregnancy or shortly after birth and its cause remains unclear. A combination of environmental and genetic factors has been suggested. Recent evidence suggests that inflammation may play a relevant role in the development of peripartum cardiomyopathy [112]. Furthermore, anthracyclines (type 1) and potentially reversible toxicity (type 2) are determined as possible risk factors for the development of cardiomyopathy due to toxic chemotherapy treatment against breast cancer [113]. Chemotherapy treatment such as anthracyclines (type 1), for breast cancer, and newer targeted agents such as trastuzumab (type 2) have established risks of cardiotoxicity [114]. In women, 8% of all acute heart failure cases are attributed to the takotsubo syndrome. It is characterized by acute and profound, but transient, regional left ventricular dysfunction in the absence of obstructive coronary disease and is commonly associated with emotional or physical stress [115]. About 90% of the patients are women and most of them experience psychological stress. The mortality rate is around 5% and the recurrence of the disease is about 10% [116].

A major issue of the sex difference of heart failure patients is that women who experience heart failure have a higher rate of adverse drug events compared to men, in particular for diuretics, anticoagulants, digoxin, and angiotensin-converting enzyme inhibitors which show that sex-specific treatment and management of the disease may be needed [117].

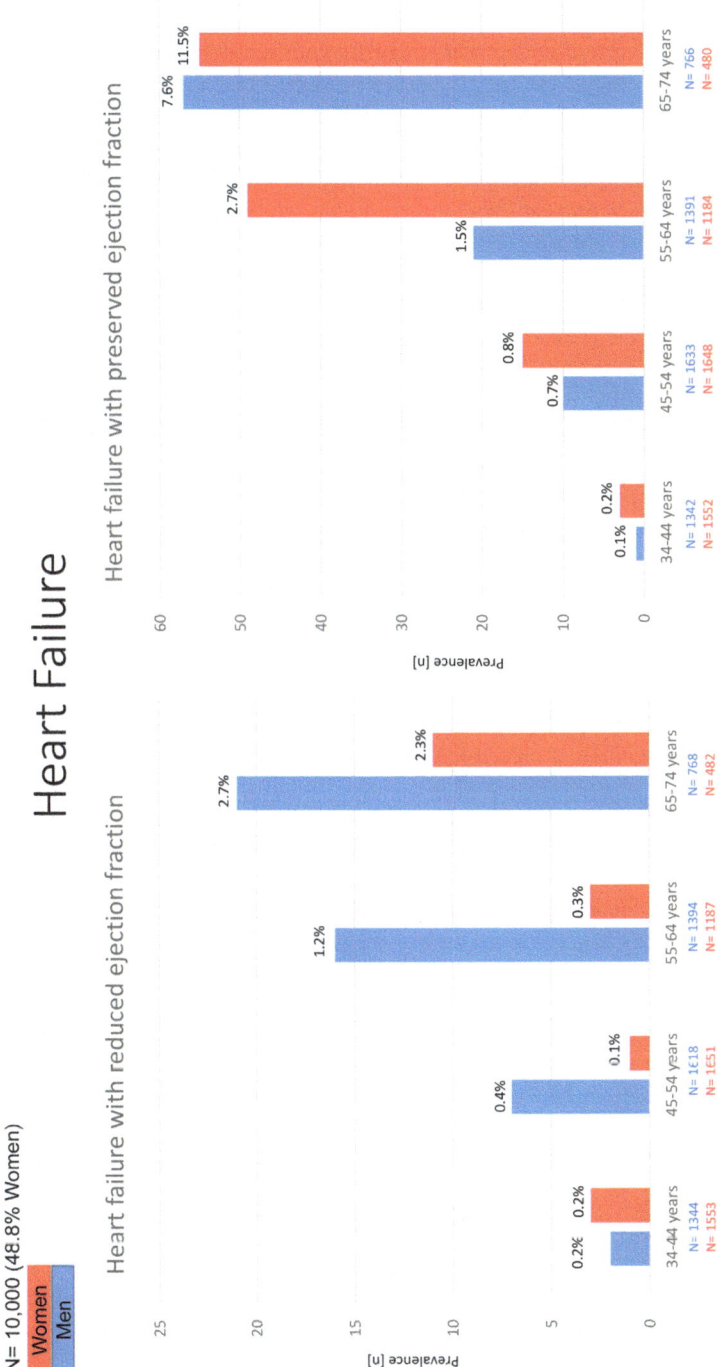

Fig. 12.3 Prevalence of heart failure in the community-based Gutenberg Health Study by age, decade, and sex for heart failure with reduced ejection fraction (HFrEF) and heart failure with preserved ejection fraction (HFpEF) (http://www.gutenberghealthstudy.org/ghs/overview.html?L=1)

12.4 Sex Differences in CVD Prevention

In the past, the importance of prevention in women with cardiovascular disease was underestimated and women were largely underrepresented in CVD clinical research [118]. Only in the last decades, the practice of CVD prevention in women has been transformed. As an example, the data release of the Women Health Initiative (WHI) in particular addressing hormone replacement therapy reported that menopausal hormone therapy did not prevent incident coronary heart disease and increased the risk of stroke in healthy women [119].

There are also general behavioral factors that may have impact on the prevention of CVD in men and women. Women tend to be more aware of a healthy lifestyle and they are more focused on healthy behaviors. At the same time, the risk of CVD is still underestimated by women and their treating physicians, which is a serious misperception that hampers CVD prevention [94]. In particular, for very young and very old women this circumstance results in a poor emergency management in women [94]. There is a strong need to investigate lifestyle methods to prevent CVD effective, in particular, approaches with long-term sustainability among diverse groups of women. An increase of sex-specific data will inform future guidelines and enhance the translation of research findings into practice. Gender disparities in preventive care need to be overcome. For instance, the number of individuals that participate in secondary prevention programs, such as heart rehabilitation, is on average lower in women compared to men. These gender disparities can partly be explained due to personal barriers such as lack of time and family obligations [120]. Many studies have investigated gender disparities in the acute management of cardiovascular disease with less emphasis on disparities in cardiovascular prevention [121]. For instance, women are less likely to receive electrocardiography when presenting to the emergency departments [122]. Screening for smoking status and education on cessation are important for lowering cardiovascular risk for all patients, but particularly for women.

In conclusion, understanding the gender differences CVD risk and risk factors is essential for developing long-term preventive measures to reduce mortality, public health burden, and healthcare costs related to CVD in both women and men [123].

12.5 Challenges in Research on Sex Differences

Crucial in future sex-specific research in CVD at the population and patient level is to collect sex-specific information and evaluate it in adequately powered studies. In the past, women have been underrepresented in diagnostic, prognostic, and therapeutic CVD studies [26]. That means that more women need to be enrolled in clinical trials or need to be over-sampled and results of those investigations need to be reported by sex. A separate monitoring of both sexes in studies does not appear to be sufficient. According to the recommendations of the EU-funded project EUGenMed (www.eugenmed.eu/), there is a broad range of possible strategies and methods to target sex and gender-specific issues in CVD research [124]:

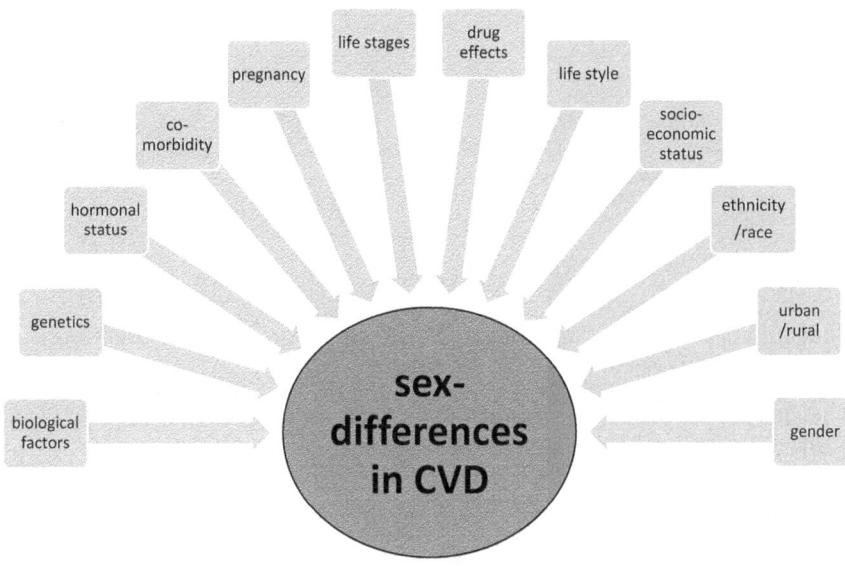

Fig. 12.4 Holistic view on sources of the sex differences in CVD in the general population

- Oversampling one sex
- Stratification
- Prospective planning for meta-analyses based on stratified patient data
- Studying interactions of risk factors with multilevel Cox regression models
- Planning of robust study designs that take into account relevant covariates from the beginning to avoid inefficient data collection

More extensive information needs to be gathered to better understand sex differences in CVD in the general population using a holistic approach (Fig. 12.4). Current efforts of publishing women-specific guidelines have been successful in increasing awareness and improving treatment and prevention of CVD risk in women. Differences in classical risk factor distribution can only partly explain observed sex differences. Genetics may contribute to the understanding of sex differences in CVD. Existing and emerging technologies take genetic examinations at increasingly high resolution at the population level. They are promising tools and may provide novel insights into the biology of sex differences of CVD in the community beyond classic epidemiology.

References

1. Oertelt-Prigione S, Gohlke B-O, Dunkel M, Preissner R, Regitz-Zagrosek V. GenderMedDB: an interactive database of sex and gender-specific medical literature. Biol Sex Differ. 2014;5 (1):7.

2. Bird CE, Rieker PP. Gender matters: an integrated model for understanding men's and women's health. Soc Sci Med. 1999;48:745–55.
3. Arain FA, Kuniyoshi FH, Abdalrhim AD, Miller VM. Sex/gender medicine: the biological basis for personalized care in cardiovascular medicine. Circ J. 2009;73:1774–82.
4. Mozaffarian D, Benjamin EJ, Go AS, Arnett DK, Blaha MJ, Cushman M, et al. Heart disease and stroke statistics-2016 update: a report from the American Heart Association. Circulation. 2016;133(4):e38–360.
5. Wilkins E, Wilson L, Wickramasinghe K, Bhatnagar P, Leal J, Luengo-Fernandez R, Burns R, Rayner M, Townsend N. European cardiovascular disease statistics 2017. Brussels: European Heart Network; 2017.
6. Appelman Y, van Rijn BB, ten Haaf ME, Boersma E, Peters SAE. Sex differences in cardiovascular risk factors and disease prevention. Atherosclerosis. 2015;241(1):211–8.
7. Mikkola TS, Gissler M, Merikukka M, Tuomikoski P, Ylikorkala O. Sex differences in age-related cardiovascular mortality. PLoS One. 2013;8(5):e63347.
8. Clark AM, DesMeules M, Luo W, Duncan AS, Wielgosz A. Socioeconomic status and cardiovascular disease: risks and implications for care. Nat Rev Cardiol. 2009;6(11):712–22. https://doi.org/10.1038/nrcardio.2009.163.
9. Canto JG, Rogers WJ, Goldberg RJ, Peterson ED, Wenger NK, Vaccarino V, et al. Association of age and sex with myocardial infarction symptom presentation and in-hospital mortality. JAMA. 2012;307(8):813–22.
10. Collins SD. Acute myocardial infarction in women: is there a sex disparity between door-to-balloon time and clinical outcomes? Cardiovasc Revasc Med. 2012;13(2):125–7.
11. Davis LL, Mishel M, Moser DK, Esposito N, Lynn MR, Schwartz TA. Thoughts and behaviors of women with symptoms of acute coronary syndrome. Heart Lung. 2013;42 (6):428–35.
12. Groban L, Lindsey SH, Wang H, Alencar AK. Chapter 5 – Sex and gender differences in cardiovascular disease A2 – Neigh, Gretchen N. In: Mitzelfelt MM, editor. Sex differences in physiology. Boston: Academic; 2016. p. 61–87.
13. Garcia M, Mulvagh SL, Merz CNB, Buring JE, Manson JE. Cardiovascular disease in women: clinical perspectives. Circ Res. 2016;118(8):1273–93.
14. Peters SAE, Huxley RR, Woodward M. Diabetes as risk factor for incident coronary heart disease in women compared with men: a systematic review and meta-analysis of 64 cohorts including 858,507 individuals and 28,203 coronary events. Diabetologia. 2014;57 (8):1542–51.
15. Wei YC, George NI, Chang CW, Hicks KA. Assessing sex differences in the risk of cardiovascular disease and mortality per increment in systolic blood pressure: a systematic review and meta-analysis of follow-up studies in the United States. PLoS One. 2017;12(1): e0170218.
16. Alsnes IV, Vatten LJ, Fraser A, Bjørngaard JH, Rich-Edwards J, Romundstad PR, et al. hypertension in pregnancy and offspring cardiovascular risk in young adulthood. Prospective and sibling studies in the HUNT study (Nord-Trøndelag Health Study) in Norway. Hypertension. 2017;69(4):591–8.
17. Gunning MN, Fauser BCJM. Are women with polycystic ovary syndrome at increased cardiovascular disease risk later in life? Climacteric. 2017;20(3):222–7.
18. Sanchis-Gomar F, Perez-Quilis C, Leischik R, Lucia A. Epidemiology of coronary heart disease and acute coronary syndrome. Ann Transl Med. 2016;4(13):256.
19. Forouzanfar MH, Liu P, Roth GA, et al. Global burden of hypertension and systolic blood pressure of at least 110 to 115 mm hg, 1990-2015. JAMA. 2017;317(2):165–82.
20. Mills KT, Bundy JD, Kelly TN, Reed JE, Kearney PM, Reynolds K, et al. Global disparities of hypertension prevalence and control: a systematic analysis of population-based studies from 90 countries. Circulation. 2016;134(6):441–50.

21. Mozaffarian D, Benjamin EJ, Go AS, Arnett DK, Blaha MJ, Cushman M, et al. Heart disease and stroke statistics—2015 update: a report from the American Heart Association. Circulation. 2015;131(4):e29–e322.
22. Everett B, Zajacova A. Gender differences in hypertension and hypertension awareness among young adults. Biodemography Soc Biol. 2015;61(1):1–17.
23. Neuhauser H, Thamm M, Ellert U. Blutdruck in Deutschland 2008–2011. Bundesgesundheitsbl Gesundheitsforsch Gesundheitsschutz. 2013;56(5):795–801.
24. Nwankwo T, Yoon SS, Burt V, Gu Q. Hypertension among adults in the United States: National Health and Nutrition Examination Survey, 2011-2012. NCHS Data Brief. 2013;133:1–8.
25. Mancia G, Fagard R, Narkiewicz K, Redon J, Zanchetti A, Böhm M, et al. 2013 ESH/ESC Guidelines for the management of arterial hypertension The Task Force for the management of arterial hypertension of the European Society of Hypertension (ESH) and of the European Society of Cardiology (ESC). Eur Heart J. 2013;34(28):2159–219.
26. Regitz-Zagrosek V, Kararigas G. Mechanistic pathways of sex differences in cardiovascular disease. Physiol Rev. 2017;97(1):1–37.
27. Hilliard LM, Sampson AK, Brown RD, Denton KM. The "his and hers" of the renin-angiotensin system. Curr Hypertens Rep. 2013;15(1):71–9.
28. Komukai K, Mochizuki S, Yoshimura M. Gender and the renin-angiotensin-aldosterone system. Fundam Clin Pharmacol. 2010;24(6):687–98.
29. Sullivan JC. Sex and the renin-angiotensin system: inequality between the sexes in response to RAS stimulation and inhibition. Am J Physiol Regul Integr Comp Physiol. 2008;294(4):R1220–6.
30. Yusuf S, Hawken S, Ôunpuu S, Dans T, Avezum A, Lanas F, et al. Effect of potentially modifiable risk factors associated with myocardial infarction in 52 countries (the INTERHEART study): case-control study. Lancet. 2004;364(9438):937–52.
31. Stewart JC, France CR, Sheffield D. Hypertension awareness and pain reports: data from the NHANES III. Ann Behav Med. 2003;26(1):8–14.
32. Grundy SM, Benjamin IJ, Burke GL, Chait A, Eckel RH, Howard BV, et al. Diabetes and cardiovascular disease. A statement for healthcare professionals from the American Heart Association. Circulation. 1999;100(10):1134–46.
33. Szuszkiewicz-Garcia MM, Davidson JA. Cardiovascular disease in diabetes mellitus. Endocrinol Metab Clin. 2014;43(1):25–40.
34. Shaw JE, Sicree RA, Zimmet PZ. Global estimates of the prevalence of diabetes for 2010 and 2030. Diabetes Res Clin Pract. 2010;87(1):4–14.
35. Wild S, Roglic G, Green A, Sicree R, King H. Global Prevalence of Diabetes. Estimates for the year 2000 and projections for 2030. Diabetes Care. 2004;27(5):1047–53.
36. Anagnostis P, Majeed A, Johnston DG, Godsland IF. Cardiovascular risk in women with type 2 diabetes mellitus and prediabetes: is it indeed higher than men? Eur J Endocrinol. 2014;171 (6):R245–55.
37. Kannel WB. Framingham study insights on diabetes and cardiovascular disease. Clin Chem. 2011;57(2):338–9.
38. Anand SS, Islam S, Rosengren A, Franzosi MG, Steyn K, Yusufali AH, et al. Risk factors for myocardial infarction in women and men: insights from the INTERHEART study. Eur Heart J. 2008;29(7):932–40.
39. Huxley RR, Hirakawa Y, Hussain MA, Aekplakorn W, Wang X, Peters SAE, et al. Age- and sex-specific burden of cardiovascular disease attributable to 5 major and modifiable risk factors in 10 asian countries of the western pacific region. Circ J. 2015;79(8):1662–U1262.
40. Gregg EW, Shaw JE. Global health effects of overweight and obesity. N Engl J Med. 2017;377 (1):80–1.
41. Resnick EM, Simon VR, Iskikian SO, Marts SA. Future research in sex differences in obesity and cardiovascular disease. Report by the Society for Women's Health Research. J Inverstig Med. 2007;55(2):75–85.

42. Wilson PW, D'Agostino RB, Sullivan L, Parise H, Kannel WB. Overweight and obesity as determinants of cardiovascular risk: the Framingham experience. Arch Intern Med. 2002;162 (16):1867–72.
43. World Health Organization. (2009). Global health risks: mortality and burden of disease attributable to selected major risks.
44. WHO. Global Status Report on noncommunicable diseases 2014. Attaining the nine global noncommunicable diseases targets; a shared responsibility; 2014.
45. Trémollières FA, Pouilles J-M, Cauneille C, Ribot C. Coronary heart disease risk factors and menopause: a study in 1684 French women. Atherosclerosis. 1999;142(2):415–23.
46. Lewington S, Whitlock G, Clarke R, Sherliker P, Emberson J, Halsey J, et al. Blood cholesterol and vascular mortality by age, sex, and blood pressure: a meta-analysis of individual data from 61 prospective studies with 55,000 vascular deaths. Lancet. 2007;370 (9602):1829–39.
47. Peters SAE, Singhateh Y, Mackay D, Huxley RR, Woodward M. Total cholesterol as a risk factor for coronary heart disease and stroke in women compared with men: a systematic review and meta-analysis. Atherosclerosis. 2016;248:123–31.
48. Hitchman SC, Fong GT. Gender empowerment and female-to-male smoking prevalence ratios. Bull World Health Organ. 2011;89(3):195–202.
49. Huxley RR, Woodward M. Cigarette smoking as a risk factor for coronary heart disease in women compared with men: a systematic review and meta-analysis of prospective cohort studies. The Lancet. 2011;378(9799):1297–305.
50. Mucha L, Stephenson J, Morandi N, Dirani R. Meta-analysis of disease risk associated with smoking, by gender and intensity of smoking. Gend Med. 2006;3(4):279–91.
51. Peters SA, Huxley RR, Woodward M. Smoking as a risk factor for stroke in women compared with men: a systematic review and meta-analysis of 81 cohorts, including 3,980,359 individuals and 42,401 strokes. Stroke. 2013;44(10):2821–8.
52. Ford ES, Capewell S. Coronary heart disease mortality among young adults in the U.S. from 1980 through 2002. J Am Coll Cardiol. 2007;50(22):2128–32.
53. Cupisti S, Häberle L, Dittrich R, Oppelt PG, Reissmann C, Kronawitter D, et al. Smoking is associated with increased free testosterone and fasting insulin levels in women with polycystic ovary syndrome, resulting in aggravated insulin resistance. Fertil Steril. 2010;94(2):673–7.
54. Guaderrama MM, Corwin EJ, Kapelewski CH, Klein LC. Sex differences in effects of cigarette smoking and 24-hr abstinence on plasma arginine vasopressin. Addict Behav. 2011;36(11):1106–9.
55. Allen AM, Oncken C, Hatsukami D. Women and smoking: the effect of gender on the epidemiology, health effects, and cessation of smoking. Curr Addict Rep. 2014;1(1):53–60.
56. Skov-Ettrup LS, Eliasen M, Ekholm O, Grønbæk M, Tolstrup JS. Binge drinking, drinking frequency, and risk of ischaemic heart disease: a population-based cohort study. Scand J Soc Med. 2011;39(8):880–7.
57. Zheng Y-L, Lian F, Shi Q, Zhang C, Chen Y-W, Zhou Y-H, et al. Alcohol intake and associated risk of major cardiovascular outcomes in women compared with men: a systematic review and meta-analysis of prospective observational studies. BMC Public Health. 2015;15 (1):773.
58. World Health Organization. Global burden of disease: 2004 update 46, 51; 2008.
59. Vaccarino V, Bremner JD. Behavioral, emotional and neurobiological determinants of coronary heart disease risk in women. Neurosci Biobehav Rev. 2017;74(Part B):297–309.
60. Wyman L, Crum RM, Celentano D. Depressed mood and cause-specific mortality: a 40-year general community assessment. Ann Epidemiol. 2012;22(9):638–43.
61. Hackett RA, Steptoe A. Psychosocial factors in diabetes and cardiovascular risk. Curr Cardiol Rep. 2016;18(10):95.
62. Humphries KH, Izadnegahdar M, Sedlak T, Saw J, Johnston N, Schenck-Gustafsson K, et al. Sex differences in cardiovascular disease – impact on care and outcomes. Front Neuroendocrinol. 2017;46:46–70.

63. Steptoe A, Kivimaki M. Stress and cardiovascular disease. Nat Rev Cardiol. 2012;9 (6):360–70. https://doi.org/10.1038/nrcardio.2012.45.
64. Smolderen KG, Strait KM, Dreyer RP, D'Onofrio G, Zhou S, Lichtman JH, et al. Depressive symptoms in younger women and men with acute myocardial infarction: insights from the VIRGO study. J Am Heart Assoc. 2015;4(4)
65. Whang W, Kubzansky LD, Kawachi I, Rexrode KM, Kroenke CH, Glynn RJ, et al. Depression and risk of sudden cardiac death and coronary heart disease in women. J Am Coll Cardiol. 2009;53(11):950–8.
66. Rich-Edwards JW, Mason S, Rexrode K, Spiegelman D, Hibert E, Kawachi I, et al. Physical and sexual abuse in childhood as predictors of early-onset cardiovascular events in women. Circulation. 2012;126(8):920–7.
67. Korkeila J, Vahtera J, Korkeila K, Kivimäki M, Sumanen M, Koskenvuo K, et al. Childhood adversities as predictors of incident coronary heart disease and cerebrovascular disease. Heart. 2010;96(4):298–303.
68. Puckrein GA, Egan BM, Howard G. Social and medical determinants of cardiometabolic health: the big picture. Ethn Dis. 2015;25(4):521–4.
69. Shi A, Tao Z, Wei P, Zhao J. Epidemiological aspects of heart diseases. Exp Ther Med. 2016;12(3):1645–50.
70. Maas AH, van der Schouw YT, Regitz-Zagrosek V, Swahn E, Appelman YE, Pasterkamp G, et al. Red alert for women's heart: the urgent need for more research and knowledge on cardiovascular disease in women: proceedings of the workshop held in Brussels on gender differences in cardiovascular disease, 29 September 2010. Eur Heart J. 2011;32(11):1362–8.
71. Wei J, Rooks C, Ramadan R, Shah AJ, Bremner JD, Quyyumi AA, et al. Meta-analysis of mental stress–induced myocardial ischemia and subsequent cardiac events in patients with coronary artery disease. Am J Cardiol. 2014;114(2):187–92.
72. Samad Z, Boyle S, Ersboll M, Vora AN, Zhang Y, Becker RC, et al. Sex differences in platelet reactivity and cardiovascular and psychological response to mental stress in patients with stable ischemic heart disease. J Am Coll Cardiol. 2014;64(16):1669–78.
73. Vaccarino V, Wilmot K, Mheid IA, Ramadan R, Pimple P, Shah AJ, et al. Sex differences in mental stress-induced myocardial ischemia in patients with coronary heart disease. J Am Heart Assoc. 2016;5(9)
74. Mosca L, Benjamin EJ, Berra K, Bezanson JL, Dolor RJ, Lloyd-Jones DM, et al. Effectiveness-based guidelines for the prevention of cardiovascular disease in women— 2011 update. J Am Coll Cardiol. 2011b;57(12):1404–23.
75. Umesawa M, Kobashi G. Epidemiology of hypertensive disorders in pregnancy: prevalence, risk factors, predictors and prognosis [Review Series]. Hypertens Res. 2017;40(3):213–20.
76. Magnussen EB, Vatten LJ, Smith GD, Romundstad PR. Hypertensive disorders in pregnancy and subsequently measured cardiovascular risk factors. Obstet Gynecol. 2009;114(5):961–70.
77. Ahmed R, Dunford J, Mehran R, Robson S, Kunadian V. Pre-eclampsia and future cardiovascular risk among women: a review. J Am Coll Cardiol. 2014;63(18):1815–22.
78. Heida KY, Franx A, van Rijn BB, Eijkemans MJC, Boer JMA, Verschuren MWM, et al. Earlier age of onset of chronic hypertension and type 2 diabetes mellitus after a hypertensive disorder of pregnancy or gestational diabetes mellitus. Hypertension. 2015;66(6):1116–22.
79. Hermes W, Franx A, van Pampus MG, Bloemenkamp KWM, Bots ML, van der Post JA, et al. Cardiovascular risk factors in women who had hypertensive disorders late in pregnancy: a cohort study. Am J Obstet Gynecol. 2013;208(6):474.e471–8.
80. Dornhorst A, Paterson CM, Nicholls JS, Wadsworth J, Chiu DC, Elkeles RS, et al. High prevalence of gestational diabetes in women from ethnic minority groups. Diabet Med. 1992;9 (9):820–5.
81. Fadl H, Magnuson A, Östlund I, Montgomery S, Hanson U, Schwarcz E. Gestational diabetes mellitus and later cardiovascular disease: a Swedish population based case–control study. BJOG Int J Obstet Gynaecol. 2014;121(12):1530–6.

82. Canoy D, Cairns BJ, Balkwill A, Wright FL, Khalil A, Beral V, et al. Hypertension in pregnancy and risk of coronary heart disease and stroke: a prospective study in a large UK cohort. Int J Cardiol. 2016;222:1012–8.
83. Shostrom DCV, Sun Y, Oleson JJ, Snetselaar LG, Bao W. History of gestational diabetes mellitus in relation to cardiovascular disease and cardiovascular risk factors in US women. Front Endocrinol. 2017;8:144.
84. Canoy D, Beral V, Balkwill A, Wright FL, Kroll ME, Reeves GK, et al. Age at menarche and risks of coronary heart and other vascular diseases in a large UK cohort. Circulation. 2015;131 (3):237–44.
85. Charalampopoulos D, McLoughlin A, Elks CE, Ong KK. Age at menarche and risks of all-cause and cardiovascular death: a systematic review and meta-analysis. Am J Epidemiol. 2014;180(1):29–40.
86. Muka T, Oliver-Williams C, Kunutsor S, et al. Association of age at onset of menopause and time since onset of menopause with cardiovascular outcomes, intermediate vascular traits, and all-cause mortality: a systematic review and meta-analysis. JAMA Cardiol. 2016;1(7):767–76.
87. Albrektsen G, Heuch I, Løchen M, et al. Lifelong gender gap in risk of incident myocardial infarction: the tromsø study. JAMA Intern Med. 2016;176(11):1673–9.
88. Arnold AP. The end of gonad-centric sex determination in mammals. Trends Genet. 2012;28 (2):55–61.
89. Arnold AP, Cassis LA, Eghbali M, Reue K, Sandberg K. Sex hormones and sex chromosomes cause sex differences in the development of cardiovascular diseases. Arterioscler Thromb Vasc Biol. 2017;37(5):746–56.
90. Ely D, Underwood A, Dunphy G, Boehme S, Turner M, Milsted A. Review of the Y chromosome, Sry and hypertension. Steroids. 2010;75(11):747–53.
91. Balaton BP, Brown CJ. Escape artists of the X chromosome. Trends Genet. 2016;32 (6):348–59.
92. Vaccarino V. Ischemic heart disease in women: many questions, few facts. Circ Cardiovasc Qual Outcomes. 2010;3(2):111–5.
93. Puymirat E, Simon T, Steg P, et al. Association of changes in clinical characteristics and management with improvement in survival among patients with st-elevation myocardial infarction. JAMA. 2012;308(10):998–1006.
94. Regitz-Zagrosek V. Geschlecht und Herz-Kreislauf-Erkrankungen. Internist. 2017;58 (4):336–43.
95. Mygind ND, Michelsen MM, Pena A, Frestad D, Dose N, Aziz A, et al. Coronary microvascular function and cardiovascular risk factors in women with angina pectoris and no obstructive coronary artery disease: The iPOWER study. J Am Heart Assoc. 2016;5(3):e003064.
96. Lansky AJ, Ng VG, Maehara A, Weisz G, Lerman A, Mintz GS, et al. Gender and the extent of coronary atherosclerosis, plaque composition, and clinical outcomes in acute coronary syndromes. JACC Cardiovasc Imaging. 2012;5(3):S62–72.
97. Vaccarino V, Badimon L, Corti R, de Wit C, Dorobantu M, Hall A, et al. Ischaemic heart disease in women: are there sex differences in pathophysiology and risk factors? Position Paper from the Working Group on Coronary Pathophysiology and Microcirculation of the European Society of Cardiology. Cardiovasc Res. 2011;90(1):9–17.
98. Gu H, Gao Y, Wang H, Hou Z, Han L, Wang X, et al. Sex differences in coronary atherosclerosis progression and major adverse cardiac events in patients with suspected coronary artery disease. J Cardiovasc Comput Tomogr. 2017;11(5):367–72.
99. Kaul P, Savu A, Nerenberg KA, Donovan LE, Chik CL, Ryan EA, et al. Impact of gestational diabetes mellitus and high maternal weight on the development of diabetes, hypertension and cardiovascular disease: a population-level analysis. Diabet Med. 2015;32(2):164–73.
100. Hemingway H, Langenberg C, Damant J, Frost C, Pyörälä K, Barrett-Connor E. Prevalence of angina in women versus men. A systematic review and meta-analysis of international variations across 31 countries. Circulation. 2008;117(12):1526–36.

101. Gupta A, Wang Y, Spertus JA, Geda M, Lorenze N, Nkonde-Price C, et al. Trends in acute myocardial infarction in young patients and differences by sex and race, 2001 to 2010. J Am Coll Cardiol. 2014;64(4):337–45.
102. Chugh SS, Havmoeller R, Narayanan K, Singh D, Rienstra M, Benjamin EJ, et al. Worldwide epidemiology of atrial fibrillation. A Global Burden of Disease 2010 Study. Circulation. 2014;129(8):837–47.
103. Friberg L, Engdahl J, Frykman V, Svennberg E, Levin L-Å, Rosenqvist M. Population screening of 75- and 76-year-old men and women for silent atrial fibrillation (STROKESTOP). Europace. 2013;15(1):135–40.
104. Emdin CA, Wong CX, Hsiao AJ, Altman DG, Peters SA, Woodward M, et al. Atrial fibrillation as risk factor for cardiovascular disease and death in women compared with men: systematic review and meta-analysis of cohort studies. BMJ. 2016;352:h7013.
105. Magnussen C, Niiranen TJ, Ojeda FM, Gianfagna F, Blankenberg S, Njolstad I, et al. Sex differences and similarities in atrial fibrillation epidemiology, risk factors, and mortality in community cohorts: results from the biomarcare consortium (biomarker for cardiovascular risk assessment in Europe). Circulation. 2017;136(17):1588–97.
106. Ponikowski P, Voors AA, Anker SD, Bueno H, Cleland JGF, Coats AJS, et al. 2016 ESC Guidelines for the Diagnosis and Treatment of Acute and Chronic Heart Failure. Revista Española de Cardiología (English Edition). Eur Heart J. 2016;69(12):1167.
107. Gerber Y, Weston SA, Redfield MM, et al. A contemporary appraisal of the heart failure epidemic in olmsted county, minnesota, 2000 to 2010. JAMA Intern Med. 2015;175 (6):996–1004.
108. Lam CSP, Donal E, Kraigher-Krainer E, Vasan RS. Epidemiology and clinical course of heart failure with preserved ejection fraction. Eur J Heart Fail. 2011;13(1):18–28.
109. Lam CSP, Carson PE, Anand IS, Rector TS, Kuskowski M, Komajda M, et al. Sex differences in clinical characteristics and outcomes in elderly patients with heart failure and preserved ejection fraction: the I-PRESERVE trial. Circ Heart Fail. 2012;5(5):571–8.
110. Rider OJ, Lewandowski A, Nethononda R, Petersen SE, Francis JM, Pitcher A, et al. Gender-specific differences in left ventricular remodelling in obesity: insights from cardiovascular magnetic resonance imaging. Eur Heart J. 2013;34(4):292–9.
111. Eaton CB, Pettinger M, Rossouw J, Martin LW, Foraker R, Quddus A, et al. Risk factors for incident hospitalized heart failure with preserved versus reduced ejection fraction in a multiracial cohort of postmenopausal women. Circ Heart Fail. 2016;9(10)
112. Arany Z, Elkayam U. Peripartum cardiomyopathy. Circulation. 2016;133(14):1397–409.
113. Suter TM, Ewer MS. Cancer drugs and the heart: importance and management. Eur Heart J. 2013;34(15):1102–11.
114. Zagar TM, Cardinale DM, Marks LB. Breast cancer therapy-associated cardiovascular disease [Review]. Nat Rev Clin Oncol. 2016;13(3):172–84.
115. Potu KC, Raizada A, Gedela M, Stys A. Takotsubo cardiomyopathy (broken-heart syndrome): a short review. S D Med. 2016;69(4):169–71.
116. Templin C, Ghadri JR, Diekmann J, Napp LC, Bataiosu DR, Jaguszewski M, et al. Clinical features and outcomes of takotsubo (stress) cardiomyopathy. N Engl J Med. 2015;373 (10):929–38.
117. Catananti C, Liperoti R, Settanni S, Lattanzio F, Bernabei R, Fialova D, et al. Heart failure and adverse drug reactions among hospitalized older adults. Clin Pharmacol Ther. 2009;86 (3):307–10.
118. Westerman S, Wenger NK. Women and heart disease, the underrecognized burden: sex differences, biases, and unmet clinical and research challenges. Clin Sci. 2016;130(8):551–63.
119. Rossouw JE, Anderson GL, Prentice RL, LaCroix AZ, Kooperberg C, Stefanick ML, et al. Risks and benefits of estrogen plus progestin in healthy postmenopausal women: principal results From the Women's Health Initiative randomized controlled trial. JAMA. 2002;288 (3):321–33.

120. Huxley VH. Sex and the cardiovascular system: the intriguing tale of how women and men regulate cardiovascular function differently. Adv Physiol Educ. 2007;31(1):17–22.
121. Mosca L, Barrett-Connor E, Kass Wenger N. Sex/gender differences in cardiovascular disease prevention. What a difference a decade makes. Circulation. 2011a;124(19):2145–54.
122. Mosca L, Appel LJ, Benjamin EJ, Berra K, Chandra-Strobos N, Fabunmi RP, et al. Evidence-based guidelines for cardiovascular disease prevention in women. Circulation. 2004;109 (5):672–93.
123. Magnussen C, Niiranen TJ, Ojeda F, Gianfagna F, Blankenberg S, Njølstad I, Vartianinen E, Sans S, Pasterkamp G, Hughes M, Costanzo S, Donati MB, Jousilathi P, Linneberg A, Palosaari T, de Gaetano G, Bobak M, den Ruijter HM, Mathiesen E, Jørgensen T, Söderberg S, Kuulasmaa K, Zeller T, Iacoviello L, Salmaa V and Schnabel BR. Sex differences and similarities in atrial fibrillation epidemiology, risk factors, and mortality in community cohorts. Circulation. 2017;136:1588–97. https://doi.org/10.1161/circulationaha.117028981.
124. Regitz-Zagrosek V, Oertelt-Prigione S, Prescott E, Franconi F, Gerdts E, Foryst-Ludwig A, et al. Gender in cardiovascular diseases: impact on clinical manifestations, management, and outcomes. Eur Heart J. 2016;37(1):24–34.

Printed by Printforce, the Netherlands